SYSTEMATICS OF THE GREEN ALGAE

Proceedings of an International Symposium held at
The Polytechnic of North London 29–31 March 1983.

THE SYSTEMATICS ASSOCIATION
SPECIAL VOLUME No. 27

SYSTEMATICS OF THE GREEN ALGAE

Edited by

D. E. G. IRVINE

*Department of Food and Biological Sciences,
The Polytechnic of North London,
London, England*

and

D. M. JOHN

*Department of Botany,
British Museum (Natural History),
London, England*

1984

Published for the
SYSTEMATICS ASSOCIATION
by
ACADEMIC PRESS
(Harcourt Brace Jovanovich, Publishers)

LONDON ORLANDO SAN DIEGO NEW YORK
TORONTO MONTREAL SYDNEY TOKYO

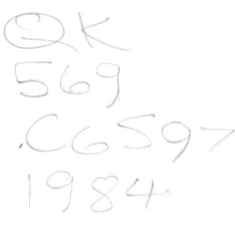

COPYRIGHT © 1984, BY THE SYSTEMATICS ASSOCIATION.
ALL RIGHTS RESERVED.
NO PART OF THIS PUBLICATION MAY BE REPRODUCED OR
TRANSMITTED IN ANY FORM OR BY ANY MEANS, ELECTRONIC
OR MECHANICAL, INCLUDING PHOTOCOPY, RECORDING, OR ANY
INFORMATION STORAGE AND RETRIEVAL SYSTEM, WITHOUT
PERMISSION IN WRITING FROM THE PUBLISHER.

ACADEMIC PRESS INC. (LONDON) LTD.
24-28 Oval Road,
London NW1 7DX

United States Edition published by
ACADEMIC PRESS, INC.
Orlando, Florida 32887

British Library Cataloguing in Publication Data

Systematics of the green algae.–(The Systematics
　　Association special volume; no. 27)
　　1. Chlorophyceae–Classification
　　I. Irvine, D. E. G.　II. John, D. M.　III. Series
　　598.4'7'012　　QK569.C6

Library of Congress Cataloging in Publication Data
Main entry under title:

Systematics of the green algae.

　　(The Systematics Association special volume, ISSN
0309–2593 ; no. 27)
　　Based on a Systematics Association symposium held at
the Polytechnic of North London, Mar. 29-31, 1983.
　　Includes bibliographies and index.
　　1. Chlorophyceae--Classification--Congresses.
2. Charophyta--Classification--Congresses.　I. Irvine,
David E. G.　II. John, D. M.　III. Systematics Association.
IV. Series.
QK569.C6S97　1984　　589.4'7'012　　83-83435
ISBN 0–12–374040–1 (alk. paper)

PRINTED IN THE UNITED STATES OF AMERICA

84 85 86 87　　9 8 7 6 5 4 3 2 1

Contributors

Brook, A. J., *School of Life Sciences, University of Buckingham, Buckingham MK18 1EG, England*

Chapman, R. L., *Department of Botany, Louisiana State University, Baton Rouge, Louisiana 70803, USA*

Collins, J. C., *Department of Botany, University of Liverpool, Liverpool L69 3BX, England*

Elliott, G. F., *Department of Palaeontology, British Museum (Natural History), London SW7 5BD, England*

Floyd, G. L., *Department of Botany, Ohio State University, Columbus, Ohio 43210, USA*

Francke, J. A., *Biological Laboratory, Free University, 1007 MC Amsterdam, The Netherlands*

Hillis-Colinvaux, L., *Department of Zoology, Ohio State University, Columbus, Ohio 43210, USA*

Hoek, C. van den, *Marine Botany Research Group, Biological Centre, University of Gröningen, 9750 AA Haren, The Netherlands*

John, D. M., *Department of Botany, British Museum (Natural History), London SW7 5BD, England*

Kessler, E., *Institut für Botanik und Pharmazeutische Biologie, Universität Erlangen–Nürnberg, D-8520 Erlangen, Federal Republic of Germany*

Khan, M., *Department of Botany, Kamla Nehru Institute of Science and Technology, Sultanpur 228001 (U.P.), India*

Kouwets, F. A. C., *Vries-Laboratorium, University of Amsterdam, 1018 DD Amsterdam, The Netherlands*

Lokhorst, G. M., *Rijksherbarium, 2313 ZT Leiden, The Netherlands*

Mattox, K. R., *Department of Botany, Miami University, Oxford, Ohio 45056, USA*

Melkonian, M., *Botanisches Institut, Universität Münster, D-4400 Münster, Federal Republic of Germany*

O'Kelly, C. J.,* *Department of Botany, Ohio State University, Columbus, Ohio 43210, USA*

---

*Present address: Department of Botany, La Trobe University, Bundoora, Victoria 3083, Australia.

Phillips, J. A., *Department of Botany, Monash University, Clayton, Victoria 3168, Australia*

Reymond, O. L.,† *Département de Biologie Végétale, Université de Genève, Laboratoire de Microbiologie Générale, CH-1211 Genève 4, Switzerland*

Roberts, K. R., *Department of Biology, The University of Southwestern Louisiana, Lafayette, Louisiana 70504, USA*

Round, F. E., *Department of Botany, University of Bristol, Bristol BS8 1UG, England*

Russell, G., *Department of Botany, University of Liverpool, Liverpool L69 3BX, England*

Sarma, Y. S. R. K., *Centre of Advanced Study in Botany, Banaras Hindu University, Varanasi 221005, India*

Schlösser, U. G., *Sammlung von Algenkulturen, Pflanzenphysiologisches Institut, Universität Göttingen, D-3400 Göttingen, Federal Republic of Germany*

Silva, P. C., *Department of Botany, University of California, Berkeley, California 94720, USA*

Simons, J., *Biological Laboratory, Free University, 1007 MC Amsterdam, The Netherlands*

Stewart, K. D., *Department of Botany, Miami University, Oxford, Ohio 45056, USA*

Young, A. J., *Department of Botany, University of Liverpool, Liverpool L69 3BX, England*

---

†Present address: Institut d'Histologie et d'Embryologie, Université de Lausanne, CH-1011 Lausanne, Switzerland.

Preface

Interest in the systematics of the well-defined assemblage of chlorophyll a– and b–containing algae commonly known as the "Green Algae" has been rekindled in recent years by the rapid accumulation of new descriptive information on details of their ultrastructure. This has all come from a repeat of the superb light microscopical studies of the nineteenth and earlier twentieth centuries by a new generation of researchers now using the electron microscope and interference light microscopy in conjunction with a vast array of equipment recently available to the cell biologist. The synthesis and evaluation of this newly acquired cytological data are leading to attempts to reorganize fundamentally the traditional classification of the green algae and to the creation of what might be termed a "new taxonomy". Furthermore, there has been a revival of interest in the phylogeny of this group, which has long been appreciated as standing nearer than any other to the main line of evolution of the higher plants.

Given the new wave of interest in green algal systematics it was decided to approach several eminent specialists in the field to see whether the time was right to hold a symposium on this topic. We received an unequivocally enthusiastic response to our suggestion and have been gratified by the support we have been given in formulating and arranging this Symposium. Naturally opinions differ as to the algae to be designated as "Green Algae", and for the purposes of this Symposium we decided to exclude the Euglenophyta (-phyceae) but to include the Charophyta (-phyceae *sensu stricto*) even though in some conservative treatments the latter are classified separately. In this Symposial Volume little attempt has been made to standardize the taxonomic level used by the authors for the major divisions of the green algae as this is unlikely to cause any confusion or misunderstanding.

The Symposium provided a forum in which specialists reviewed the current status of the systematics of green algal taxa, both extant and fossil, at all levels, and showed how the systematics of the higher groupings (families, orders, classes) and their suggested phylogenetic relationships with others are affected by newly acquired ultrastructural and, to a lesser extent, cultural and biochemical information. We believe that the Symposium achieved all its objectives and that this resulting volume makes a useful contribution to green algal systematics by bringing together in a single work much of our current thinking concerning this group.

We wish to thank Academic Press for their willing cooperation with us as editors at all times, the chairmen of the sessions, and those contributors who kept to the space allocated to them. It was unfortunately necessary to edit valuable material from some manuscripts, and to those authors we apologize, and to limit the index drastically because of the strict restraints imposed upon us concerning the length of the published volume. Despite our strenuous efforts in this direction it still exceeded the page limit generally applied to symposial volumes in the series, but the Systematics Association decided to make an exception in this case and to allow us to exceed this limit as the volume covered the whole group and dealt with important new classificatory proposals. Much of the success of this international Symposium was due in no small part to the assistance given us by many helpers, but those deserving of special mention are Mrs O. E. J. Etherington of the Department of Geography and Geology, Polytechnic of North London, Mrs J. A. Moore and Dr L. R. Johnson of the Department of Botany, British Museum (Natural History), and Dr P. F. Brandon, Head of the Department of Geography and Geology, Polytechnic of North London, who kindly allowed this Symposium to be held in his Department. We give special thanks to Mr D. R. Croome, Deputy Director of the Polytechnic of North London, for his opening address to the Symposium. Finally, we are most grateful for the financial support given us by our sponsors, the Systematics Association, and for support and guarantees provided by the Royal Society, the British Council and the British Phycological Society, as well as from various sources to many of our contributors.

September 1984

D. E. G. Irvine
D. M. John

Contents

List of Contributors	v
Preface	vii

INTRODUCTION

1 The Systematics of the Chlorophyta: An Historical Review Leading to some Modern Concepts [Taxonomy of the Chlorophyta III]
F. E. ROUND — 1

CYTOSYSTEMATICS OF THE GREEN ALGAE

2 Classification of the Green Algae: A Concept Based on Comparative Cytology
K. R. MATTOX and K. D. STEWART — 29

3 Flagellar Apparatus Ultrastructure in Relation to Green Algal Classification
M. MELKONIAN — 73

4 Correlations among Patterns of Sporangial Structure and Development, Life Histories, and Ultrastructural Features in the Ulvophyceae
C. J. O'KELLY and G. L. FLOYD — 121

REVIEWS OF THE SYSTEMATICS OF SELECTED HIGHER GROUPINGS

5 The Systematics of the Cladophorales
C. VAN DEN HOEK — 157

6 Current Ideas on Classification of the Ulotrichales Borzi
G. M. LOKHORST — 179

7 On the Systematics of the Chaetophorales
D. M. JOHN — 207

8 An Assessment of the Current State of Our Knowledge of the Trentepohliaceae
R. L. CHAPMAN — 233

9   Comparative Studies in a Polyphyletic Group, The Desmidiaceae—
    30 Years On
       A. J. BROOK   251
10 Systematics of the Siphonales
       L. HILLIS-COLINVAUX   271
11 Modern Developments in the Classification of Some Fossil Green
    Algae
       G. F. ELLIOTT   297
12 Cytogeography and Cytosystematics of Charophyta
       M. KHAN and Y. S. R. K. SARMA   303

SYSTEMATICS AND CYTOLOGY OF SELECTED GENERA

13 The Flagellar Apparatus in *Batophora* and *Trentepohlia* and Its Phylogenetic Significance
       K. R. ROBERTS   331
14 Ultrastructural Characterization of Taxa in the Genus *Enteromorpha*
       A. J. YOUNG, J. C. COLLINS and G. RUSSELL   343
15 The Validity of Morphological and Anatomical Characters in Distinguishing Species of *Ulva* in Southern Australia
       J. A. PHILLIPS   353
16 Morphology and Systematics of *Stigeoclonium* Kütz. (Chaetophorales)
       J. A. FRANCKE and J. SIMONS   363
17 Taxonomical and Ultrastructural Survey of the Genus *Desmatractum* West & West (Chlorococcales)
       O. L. REYMOND and F. A. C. KOUWETS   379

CHEMOTAXONOMY OF THE GREEN ALGAE

18 A General Review on the Contribution of Chemotaxonomy to the Systematics of Green Algae
       E. KESSLER   391
19 Species-Specific Sporangium Autolysins (Cell-Wall-Dissolving Enzymes) in the Genus *Chlamydomonas*
       U. G. SCHLÖSSER   409

EXTRINSIC FACTORS AND GREEN ALGAL SYSTEMATICS

20 The Role of Extrinsic Factors in the Past and Future of Green Algal Systematics
       P. C. SILVA   419
    Taxonomic Index   435
    List of Systematics Association Publications   450

SYSTEMATICS OF THE GREEN ALGAE

1 | The Systematics of the Chlorophyta: An Historical Review Leading to some Modern Concepts [Taxonomy of the Chlorophyta III]

F. E. ROUND

Department of Botany, University of Bristol, Bristol, England

Abstract: A brief historical outline traces the development of systematic studies on the green algae from the earliest valid descriptions of genera to the latest organization into classes. The problem of dealing with one of the most complex and diverse groups of algae is discussed and the criteria for subdivision involving both gross and ultrastructural detail are outlined. Classification in an overall system of green plants (Viridiplantae of Cavalier-Smith) is proposed. The major orders are placed into classes and a number of divisions suggested. The importance of attempting to reconstruct phyletic lineages back to unicellular flagellate ancestors, placing the discussion on an earlier basis, is stressed.

> *Evolution is the essence of systematics*
> (WHITTAKER AND MARGULIS, 1978)

INTRODUCTION

An historical perspective is desirable when any subject is discussed, and for the green algae it is helpful to understand the conflicts which have punctuated the writings and coloured the scientific approaches of the last two centuries and are raging even today. There has been and still is much confusion; some of this may be resolved if we invoke a second historical perspec-

---

Systematics Association Special Volume No. 27, "Systematics of the Green Algae", edited by D. E. G. Irvine and D. M. John, 1984, pp. 1–27. Academic Press, London and Orlando.
ISBN 0 12 374040 1
Copyright © by the Systematics Association
All rights of reproduction in any form reserved

tive and consider the possible lines of early evolutionary history of these green algae. This duality is well within the definition of systematics; many definitions have been proposed, and two suitable ones are "the scientific study of the kinds and diversity of organisms and of any and all relationships among them" (Simpson, 1961) and "systematics, which is concerned primarily with form (in the broad sense) and secondarily with time" (Nelson and Platnick, 1981). It may be easier to discern relationships if we take an evolutionary stance which involves time, form and diversity. A systematic treatment of the green algae must indicate which clusters have sufficient features in common to form discrete groups and also indicate any genera or clusters which, although allied to others, nevertheless have peculiar features and thus form problematic entities. Often these entities have been squeezed into larger taxa and concealed, whereas in my view the "splitting" approach highlights these anomalous groups and points the way to new lines of research. One problem I have encountered, especially in reading the modern literature, is the situation in which some specific attribute is discussed as though it were a feature of all green algae when clearly it is only a feature of a subset. It is a difficulty which may cause problems of communication in this symposium. In the botanical vernacular "green algae" refers to organisms possessing chlorophyll a and b—often grouped as Chlorophyta(phyceae), Charophyta(phyceae) and Prasinophyta(phyceae) in a conservative sense and including the Euglenophyta(phyceae) and Prochlorophyta(phyceae) in the broadest sense, though of course the latter are prokaryotic algae and the euglenoids belong in a quite separate position. But there are yet other taxa (see below), and confusion results if it is not clear exactly what a taxon includes or excludes.

The main thrust of work on the green algae has been along morphological/cytological/life history paths, but these approaches when pursued in isolation tell us little of the inter-relationships. Their comparative study does allow us to distinguish similar clusters, but the relative evolutionary distances cannot be discerned. In fact, no detailed studies along biochemical lines have been undertaken on green algae, such as those used by Schwartz and Dayhoff (1978) working on genera from different groups of organisms and by numerous workers considering, for example, the inter-relationships of primates. However, it must never be forgotten that these studies are based on samples from the tips of the evolutionary series; the information may be presented as a phenogram, but this must not be interpreted as a phylogeny (Cavalier-Smith, 1980). It will be more difficult, if not impossible, to get comparable biochemical information from fossils, although it is surprising how many data have been obtained from some fossils (King, 1974). If the existing morphological data can be used to define an evolutionary series, then subsequent biochemical studies will not be quite as random or based on laboratory "weeds" as are Schwartz and Dayhoff's data. This is not to decry the value of that work in any way but, as Mattox and Stewart (1977) commented, it is desirable that biochemical studies be extended "from a few 'standard' organisms to a carefully selected variety", even if the results, as for the primates, tend to confirm the predictions from gross anatomical detail (Cherfas, 1981). I write "gross" intentionally, since there is a recent tendency in green algal studies to neglect the overall morphological features—ultrastructural studies must take their place with all other attributes and should not over-balance the systematics of the green algae.

Although the systematics of the green algae has been confused, this has not prevented their use in studying many biological problems; in fact they have been used more than any algae from any group with the following five green algae providing a vast range of biological data: *Chlamydomonas* (genetics), *Chlorella* (biochemistry), *Acetabularia* (morphogenetics), *Nitella* (biophysics/salt uptake) and *Euglena* (general biochemistry and rhythms). It is time we sorted out the systematics.

EARLY HISTORY

Before Linnaeus's time algae had figured in various publications (early Chinese, Greek and Roman); the earliest mention in a systematic sense is probably that in Bauhin (1620).

In the *Species Plantarum* of Linnaeus (1753), green algae were recognised in the class Cryptogamia only as *Ulva, Conferva, Chara* and *Volvox* (the last excusably as an animal, but can we excuse the modern biologists who still place "chlamydomonads" in the animal kingdom?). In 1758 Linnaeus also mentioned *Acetabularia* and *Codium,* but not as algae. *Chara* was allocated to several plant groups before coming to rest as an alga where, in spite of its unique features, it must surely remain. *Volvox* was probably the earliest recorded green flagellate since it is mentioned in a letter from Leeuwenhoek to the Royal Society in 1700. Lamouroux (1813) established an "ordre" Ulvacées and he was perhaps the first to conceive a green group of algae (of macroscopic form—*Ulva, Bryopsis, Caulerpa*). Agardh in 1817 added to the Lamouroux system a new subset, the Confervoideae, which included filamentous green algae. Gray (1821) established the Characeae and Dumortier (1822) the Conjugataceae. The 1750–1825 period was one of pioneer description of algae and the crude beginning of systematic clustering. Some workers placed algae in clusters which are totally outside modern concepts and need not be considered here.

Green algae as a natural assemblage, separate from other algae, were first recognised by Harvey (1836) and named Chlorospermeae. Blue-green algae were included in Harvey's Chlorospermeae, but not the desmids which were placed within the Diatomaceae. By 1838 Kützing had separated the desmids into a group termed Chamaephyceae (dwarf algae). Already in this period Kützing (1843) could record a very impressive range of green algae which he split into a whole range of families, many of which are recognised entities even today: Desmideae, Palmelleae, Hormideae, Ulotricheae, Conferveae, Zygnemeae, *Hydrodictyon* (as a family), Protonemeae (including *Gongrosira*), Draparnaldieae, Ulvaceae, Enteromorpheae, Vauchericae (not now a green set), Caulerpeae, Codieae, Anadyomeneae, Dasycladeae and Chareae; the last six are clustered in a sub-order Coeloblaste. The term "zönoblastische" was used by Ettl and Komárek (1982) for algae with numerous nuclei and is presumably a modern form of the old Coeloblaste. In 1845 Kützing changed the name Chlorospermeae to Chlorophyceae, though as a discrete group it was marred by the inclusion of genera from bacteria, fungi, and red and brown algae, a not entirely surprising feature considering the optics of the time; even today students confuse groups when observing only pigmentation. He also termed them Gymnospermeae and commented on the fact that the "seeds" were on the surface and not surrounded by a fruit body—a feature of modern definitions (cf. the definition in Bold and Wynne, 1978). Macroscopic forms such as *Fucus* he had in a section called the Angiospermeae. At this time the groups Palmellaceae and Protococcaceae were established for colonial and unicellular green algae and we find considerable discussion of their status in this early literature (cf. the present discussion, e.g. Ettl and Komárek, 1982).

In these early years the conjugate green algae often appeared as a distinct entity, although other groups were often linked with them. Running through the whole of the nineteenth century is the concept of five clusters of green algae: a protococcoid, a palmelloid, a confervoid (Chlorophycean), a conjugate (Zygnematophycean) and a macroscopic-multicellular (Charophycean) cluster. The "chlamydomonads" were given further status (Volvocaceae) by Cohn (1856), and then Rabenhorst (1863) grouped the Volvocaceae, Palmellaceae and Protococcaceae into the Coccophyceae [cf. the modern Volvocophyceae of Schoenischen (1925) and Volvocineae of Pascher (1931)]. Even the siphonaceous group was recognised around this time by Greville (1830) as "orders" Siphoneae (*Codium, Bryopsis, Vaucheria, Botrydrium*), and Kützing (1849) distinguished the groups Sphaeropleaceae, Cladophoraceae, Valoniaceae, Caulerpaceae and Dasycladaceae; Stitzenberger (1860) placed the three last clusters into the Siphonophyceae. Thus these early workers established a precedent for the multiplicity of classes which have now been described*.

---

*These are listed by Silva (1980), but it might be advantageous to mention here the number of classes which have been established by various authors: some merely descriptive, some valid

About this time De Bary (1858) united the filamentous Zygnemataceae and Desmidiaceae into Conjugatae and thus by 1868 Rabenhorst could distinguish the Coccophyceae, Zygophyceae (=Conjugatae), Siphophyceae (Botrydiaceae and Vaucheriaceae) and the filamentous Nematophyceae (Ulvaceae, Sphaeropleaceae, Confervaceae, Oedogoniaceae, Ulothrichaceae, Chroolepidaceae (=Trentepohliaceae), and Chaetophoraceae)—almost a twentieth-century grouping. Wille (1890–1891) turned the clock back in that he had only two major sections, one containing the coccoid/colonial forms (Protococcoideae) and one the filamentous (Confervoideae) and siphonaceous (Siphoneae) forms. An unfortunate piece of terminology (the Thallophyta) became common in the latter half of the nineteenth century to include both algae and fungi and regrettably still appears in some student texts. In 1895 Borzi elevated the simple filamentous series to the Ulotrichales. This 1825–1900 period was one of intense activity, describing algae and clustering them into families, classes, etc. Many other interesting problems were discussed in the literature (e.g. fertilisation, alternation of generations, animal/plant distinctions, pigment complement) and cultural/physiological work commenced, often using green unicells.

At the turn of the century, Blackman (1900), building on earlier, scattered work, crystallised the concept of motile, coccoid ("a heterogeneous remainder of primitive forms"), tetrasporine and stephanokont lines of development. His writing enabled workers to grasp something of the phylogenetic relations of what once seemed a chaos of forms and he came to the conclusion that they are all derived from a *Chlamydomonas*-like organism. His concept of "volvocine" forms reproducing by means of a set number of divisions (coenobial), and of tetrasporine forms in which vegetative cell division can be continuous (though not always), is an interesting one. Blackman did, however, consider that some of the genera were out of place and that separation into tetrasporine and confervoid lines was not defensible. He also conceived a line of development from coccoid to siphonaceous forms. In 1901, Willie separated the branching filamentous genera into the Chaetophorales. A key paper by Bohlin (1901) appeared between Blackman's (1900) discussion of evolutionary lines and the systematic treatment of Blackman and Tansley (1902). Bohlin's contribution was to suggest that

---

and some invalid, viz. Akontae, Bryopsidophyceae, Chaetophorophyceae, Chlorococcophyceae, Codiolophyceae, Coleochaetophyceae, Confervophyceae, Conjugatophyceae, Euchlorophyceae, Glaucophyceae [these possess many exceptional features, some of which may yet ally them with the chlorophyll *b* series (Moestrup, 1982), though Cavalier-Smith (1981) places them alongside the red algae]; Isokontae, Oedogoniophyceae, Placodermae, Protococcophyceae, Saccodermae, Siphoneae, Siphonocladophyceae, Stephanokontae, Tetrasporophyceae, Ulotrichophyceae, Ulvophyceae, Volvocophyceae, Zygophyceae and Zygnematophyceae.

the Confervoid group be split, with the Confervales in the newly described Heterokontae (Luther, 1899) and the Ulotrichales, Microsporales and Stephanokontae (=Oedogoniales; he suggested this term be used) left in the Chlorophyceae as three equal groups. He also discussed many of the propositions which today are being re-investigated, e.g. he stated quite clearly that (I paraphrase), "zoospores must be considered as embryonic forms of great importance in systematics. . . . within algal systematics the zoospores provide the first and foremost basis for subdivision ("indelningsgrunden"). . . . when there are no zoospores, then the characters of the spermatozooids are of importance for systematics". Bohlin also pointed to the value of pigments and considered that the pigmented forms were developed from colourless forms and could not change from one into another. He was not sure that all "protococcoid" algae (certainly not the conjugate algae) were derived from the same flagellate, i.e. he saw the possibility of multiple origins.

Blackman and Tansley (1902) porposed the term "isokont" to apply to algae with equally developed flagella, but it can apply to other groups (see also discussion in Moestrup, 1982). They also incorporated Bohlin's (1901) concepts into the first overall scheme, which incidentally appears to have been devised largely because "it fell to their lot" to deliver an advanced course of lectures to students in Cambridge and London. They were not primarily phycologists, but they proposed the Ulvales as a separate entity and maintained the Ulotrichales for the unbranched and branched filamentous forms. Although their work is much quoted it is mainly a catalogue of genera. The major lines of algal morphological advance were elaborated by Pascher (1914, 1931), who also elevated the classes to divisions and in his later paper presented what now appear to be rather modern subdivisions. He used the ending -*ineae* but termed them classes: Volvocineae, Tetrasporineae, Protococcineae, Ulotrichineae, Siphonineae, and Siphonocladineae. He also commented perceptively that it was not possible to find connections between Chlorophyceae and other algal groups (thus pre-empting the Chlorophyte/Chromophyte series in the multi-divisional system of Cavalier-Smith, 1982) and also that the charophytes could not be drawn into the Chlorophyta—a view echoed by many workers since. After Pascher, green algal systematics tended to rest on Fritsch's (1935) monumental and apparently logical and simplified nine-order system, which most modern phycologists have had instilled into them. But Fritsch was careful to distinguish those genera which were anomalous and to discuss these somewhere in his treatment of each order. During the last 25 years many of these "out-of-place genera" have been re-studied and re-allocated, though some are still awaiting systematic evaluation. This period from 1900 to 1950 was a stable one with few major advances—a period when systematic studies were rather out of favour.

From 1950 onwards, three different approaches have revitalised systematic studies of the green algae; one is the re-birth of "cytology" using electron microscopy, a second is the stimulus arising from Margulis' writing on endosymbiosis, and the third is the intense study of biochemical/molecular genetic systems. All these approaches require integration and are to a great degree dependent on one another. Out of them is emerging a new systematics and a new phylogeny. This present period may be termed the "Age of Ultrastructure" in green algal systematics, just as Corliss (1979) termed it in the Protozoa, where also it has led to a proliferation of groupings above the family level.

SYSTEMATIC CONCEPTS

The repeated attempts at sub-division and re-clustering in the early studies gave a sure indication that something was amiss in the conceptual framework of the classification. Almost no other algal group has suffered such vicissitude, e.g. the limits of the Phaeophyta and Rhodophyta (see Boney, 1978) have never been really questioned and indeed they seem to form very discrete entities in spite of numerous internal conflicts which do not disturb the overall consistency. Compare this with the green group, where there is anything but consistency*, although attempts have been made to impose consistency by lumping groups together. To quote the extremely perceptive G. S. West (1916), "in no other group of plants are there such wide differences in form and cytological structure: or such varied life histories as can be found in one section of the Green Algae"; he continued, "the most important result of this recent work is the recognition by the more experienced of algological investigators that the various groups of the green algae have originated by a progressive evolution, either directly or indirectly from flagellated ancestors and that the cytological structure of the motile zoogonidia funishes a reliable key to phylogenetic relationships". Carried through to the present day, Moestrup (1982), commenting on the ultrastructure of the transition region of the flagella, writes "the impression is that of variation in the primitive groups [of green algae] more so, in fact, than in any of the 'chromophyte' algal classes that have been examined carefully". Of course there are common features of biochemistry (e.g. pigmentation, photosynthetic products, starch (α-1,4-glucose) contained with-

---

*". . . diversity of habit and habitat is very striking and in this respect the Chlorophyceae surpass any other algal class" (Fritsch, 1935).

in the plastid*, but also major differences of morphology (e.g. the charophyte organisation, the methods of cross wall formation, the ultrastructure of wall components), of reproductive systems (e.g. charophyte and conjugate) and of biochemistry (e.g. special accessory pigments) coupled with a vast range of life-history and gross morphological variation. Both wall components (see Nisizawa and Sasaki, 1975) and intermediary metabolites (see Kremer, 1980) can vary greatly between groups, though in general the clusters which diverge from the bulk of Chlorophyceae are the siphonaceous forms and some of the aerophilous genera which have evolved to a stage of producing sugar alcohols (cf. some of the most exposed seaweeds).

There is no other algal group which exhibits such phenotypic variation imposed upon a few common (primitive!) features—the common features being also those of higher green organisms (Bryophyta, Pteridophyta *sensu lato* and even the seed plants). With this in mind it is surely of little advantage to use these common features for unification unless one adopts the system combining all green plants possessing chlorophyll *a* and *b* into a single division. Cavalier-Smith (1981, 1982) does, in fact, do just this with his kingdom Viridiplantae; earlier, Casper (1974) made all green plants a phylum (Volvocicae–Chlorophyta) and before that Whittaker (1969) had a somewhat confused kingdom, the Plantae. A similar concept is inherent in another recent paper by Bremer and Wantorp (1981), who have a subkingdom Chlorobionta with two divisions: (1) Chlorophyta (algal) and (2) Streptophyta (higher plants—subdivisions Embryophytina, but also a subdivision Charophytina). It is difficult to see the justification for placing the "charophytes" along with the higher plants which have either apical cells cutting off daughter cells in several planes or multicellular meristems (neither of these occur in charophytes or other green algae), and those with vascular tissue are distinguished as "tracheophytes". Cavalier-Smith also establishes two other algal kingdoms, Biliphyta (algae containing biliprotein pigments) and Chromophyta (containing chlorophyll *c*). His classification presenting nine monophyletic kingdoms is, I believe, the most satisfactory and well-argued system for organisms as a whole so far published. My criticism concerns the terminology—phyta is a divisional ending and I suggest Biliphyta should be named Biliplantae, Chromophyta should be Chromoplantae, and Cryptophyta should be Cryptoplantae. Other systems, es-

---

*If Cavalier-Smith (1982) is correct, the euglenoids originated from the same endosymbiotic event but the symbiont lost its ability to form the α-1,4-linked starch since it already had a cytoplasmic β-1,3 system. The discovery of a multilayered structure in euglenoids also points to a common ancestry at least as far as the ancestral host group of some of the green algae is concerned (see later).

pecially those separating algae into a kingdom Protoctista divorced from a kingdom Plantae (e.g. Margulis, 1981), are extremely artificial.

The systematic investigation of the green algae has so far been largely a descriptive process; this applies not only to the nineteenth century when only the light microscope was in use but also to the twentieth century with the electron microscope. The complexity of the information derivable from the latter should not blind us to the fact that it is still simply that of descriptive morphology/cytology in the great tradition of Sachs, Hofmeister, Goebel, etc. and can only yield an alpha-taxonomy. This is not to belittle this work—it is the alphabet on which everything is founded. It does, however, tend in the words of Stewart and Mattox (1980) to provide "both too much and too little" knowledge and "the protistologist-phylogenist must temper his natural enthusiasm with the realisation that his concepts are, if not actually false, almost certainly simplistic". The delineation of the species in all their complexity as revealed by light microscopy, electron microscopy and biochemistry is basically only a simple descriptive science and to some workers it is the only taxonomic reality—"only species exist, everything above this level is artificially constructed by man . . .". Groups cannot evolve—only species—and to quote Gould (1981) "species arise (in geological time) with their differences established at the start and do not change substantially thereafter". The word "substantially" is critical here, for there clearly are changes, e.g. to produce non-interfertile populations (with different chromosome numbers) of *Pandorina* (Coleman, 1977). In a sense, the observational detail necessary to describe species is the simple side of biology (though it is surprising how variable the attempts are) and the real art starts with the intellectual effort necessary to cluster species into genera, genera into families, and so on. This art is nevertheless susceptible to scientific substantiation (perhaps even verification?) from further detailed comparative biochemical features. When the bulk of confirmatory evidence becomes overwhelming the 'clusters" of species become segregated into higher taxa. The final "true" picture would emerge if it were possible to uncover all the fossil intermediates and in some way trace the lineages back to the original symbiotic events. Unfortunately, the taxonomist seeks the most stable, conservative feature of taxa, so that the descriptions disguise a mass of variation at once morphological, physiological, biochemical and ecological. Algal systematics lags behind bacterial where, of necessity, physiological/biochemical criteria are commonplace, and I can think of only one instance in which a green alga is defined on non-morphological features and that is Loeblich's (1982) statement, based on the original description, that *Dunaliella salina* is "any *Dunaliella* organism that has the capability of turning red (with a carotenoid to chlorophyll ratio greater than 6:1)". Whether we like it or not, these more extensive phenetic data have to be obtained for all genera. The basic task is to provide the descriptive framework; it is vitally important and should not be belittled. It is an end in itself and if it provides a totally reliable, usable store of information and at the same time enables the identification of species and the ability to exchange information on these, it has achieved the basic aim. Once the clustering of species into higher units commences the subject changes its nature and the elements of relationship, taxonomic distance, origin and evolution begin to impinge. If one believes in the concept of evolution then these aspects are no less real than are the species. It is this side of taxonomy with its predictive value which to me is most fascinating. As Whittaker and Margulis (1978) noted, "classification was based on a view from above; concepts based on the higher plants and animals were extended downward and imposed upon the protists". A much clearer picture emerges if one works upwards from the origin, even if it has to be from somewhat hypothetical ancestors; there could be several independent lines, each from an original symbiotic event, or several arising from rapid (lateral) branching after one symbiosis, or several arising either by gradual branching (gradualistic view) or sudden branching (punctuational view). Presumably all algal divisions commenced as unicells—we tend to assume that these were flagellate but should amoeboid/coccoid forms be ignored? The original symbiotic event involving chlorophyll-containing units is always assumed to have been with a flagellate ancestor, and the overall similarity of the flagellar apparatus in green algae (see Moestrup, 1982) indicates a basic cluster of flagellates into which the green symbionts entered. The flagellar structure of modern forms varies only slightly within certain defined evolutionary lines, suggesting merely evolutionary "noise" and a long period of stasis in the history of the genera, such as would conform with the general punctuational viewpoint of Eldredge and Gould. In fact, I would strongly recommend that the green algal series be regarded as long-stable lines exhibiting only minor variation in many basic features and having the same kind of evolutionary history as that postulated for the diatoms by Round and Crawford (1981). This also means that the basic features (e.g. of flagellar structure, flagellar roots, cross wall formation, mitotic processes) are primitive characters and their value in phylogenetic/systematic studies is limited to overall clustering and defining at the level below the kingdom in Cavalier-Smith's scheme, i.e. at the divisional level (compare the even less valuable feature of possession of chlorophyll *a* and *b*). It is the derived characters which are important in the subdivision of the classes. The groups sharing primitive characters are paraphyletic and those sharing only derived characters are monophyletic in the sense of Hennig (see, for example, the discussion in Patterson, 1982); to use shared primitive characters can result in misleading conclusions. The only cluster relationships between species are uniquely shared characters (homologies). One needs definitions of "primitive" and "derived", yet it is difficult to make these. One way is to regard as primitive those basic features which extend over several clearly diverse groups, e.g. flagella, basal bodies, plastids, chlorophyll *a*. Derived features are then those which are confined to subsets within the groups. The danger of circular argument is, however, always present. Monophyletic groups are defined as those containing all, and only, the descendents of a common ancestor (Cavalier-Smith, 1981; Patterson, 1982), but if the symbiotic hypothesis is accepted there are two common ancestors—one a colourless and one a pigmented organism. This concept was made very plain by Mereschkowsky (1905, 1910), who wrote of the "Amoeboplasm" (the animal plasma with powers of movement) and the "Mykoplasma" (the cyanophyte which enters the amoeboplasm and has no innate motility system) and he quite categorically stated that plants and animals (and fungi) exist as three kingdoms—"Jeder Organismus ist darum entweder ein Tier, eine Pflanze oder, ein Mykoid" (1910); "Es sind fremde Korper, fremde Organismen die in farblose Plasma der Zelle eingedrungen, und mit derselben in symbiotisches Zusammenleben getreten sind"; and "Diese Theorie nach ist eine Pflanzenzelle nicht anders, als eine Tierzelle mit in sie eingedrungenen Cyanophycean infolge dessen *ist die Pflanzenwelt von des Tierwelt abzuleiten*. Die Urpflanze waren nichts anders als Amoebien oder Flagellaten, in welche Cyanophyceae eingwandert waren . . . *so ist der Ursprung* der Pflanzenwelt ein in hohem Grad polyphyletischer".

Few would argue that the plant world is polyphyletic and I am not sure that we can yet dismiss the idea that "green algae" are not polyphyletic. If we take Cavalier-Smith's Viridiplantae as monophyletic, it is necessary to postulate a branching into several subgroups to produce the variety of form seen today. This may have occurred, but I suggest that, rather than a single colourless organism accepting a pigmented unit, it is more likely that several slightly different colourless organisms accepted the *same* kind of pigmented organism. If this did happen then it could explain the existence of chlorophyte, zygnemaphyte (conjugate green algae), prasinophyte and siphonophyte series without the necessity to derive one from another. I am not suggesting that derivation (branching) did not occur, but merely that multiple symbioses could provide an alternative explanation. If the occurrence of one symbiosis is generally accepted then I see no reason to dismiss the possibility of others—indeed, it would be an unusual situation if in the "post-primeval soup" only one such event occurred.

If we make the simple assumption that evolution of all groups has proceeded along the line: flagellate→coccoid→unbranched filamentous→branched filamentous→thalloid, or with siphonaceous inserted as a blind end presumably from a coccoid origin (but there are alternate possibilities), then we should search for the ancestors and build up the lines along a Pascherian scheme. Of course many unicellular forms may have died out (as is probably the case in other divisions, e.g. the Phaeophyta), but conditions seem to have been suitable for the preservation of many lineages of these primarily aquatic organisms and for the retention of the primeval forms relatively unchanged over most of geological time; this is now also accepted by Cavalier-Smith (1981, 1982). The most likely evidence for comparing the lineages is now coming from ultrastructural studies and this will probably result in the breaking up of the coccoid and flagellate green algal series. These have always seemed a heterogeneous collection of algae. Comparative discussion of these groups has been difficult, and when we achieve a more definitive clustering of genera [see, for example, Ettl and Komárek (1982), where they relocated some coccoid genera in the flagellate Chlamydophyceae], this will become easier—a genus out of place in a cluster simply confuses the whole issue. More splitting of the flagellate/coccoid series, and perhaps further recombination, is most likely until finally all the extant lines will be defined. I am sure that the lineages will prove to be either freshwater or marine [a view earlier expressed in Round (1981) and in Hoek (1981)] with little extension from one to the other, and no more than one might expect when the diversity and potentialities from the genetic background are taken into account (cf. a similar situation in the Bacillariophyta; Round and Sims, 1982). A way of discovering the possibilities is to work backwards from the tips of the "clades" of green algae to determine where the gaps are which need to be considered, and then to reverse the working order and start from the possible origins even if we have to postulate hypothetical ancestors—"the most logical starting point for an evolutionary tree is at its base" (Taylor, 1978).

The first major problem then is how (if at all) one should split up the algal groups containing chlorophyll a and b into "clades" or lineages. If one assumes that such splitting is feasible, then have they all been derived from a single ancestor, i.e. are they monophyletic, or derived from several ancestors (polyphyletic)? Or, put another way, can we discern anything of the early history and evolution of the green algae? This is the second and very difficult problem we have to confront. It is probably a fact that most major groups differentiated in the Precambrian, though most fossils of this period are extremely difficult to allocate to specific divisions. But by the Devonian many groups had become so well defined that they can be clearly associated with modern genera (see examples in Tappan, 1980). The evidence seems to indicate a fairly rapid evolution in the Proterozoic, which probably saw the appearance of all the groups, though the fossil record of some seems to end in the Devonian. This was succeeded by relative stasis and only minor evolutionary development. What controls this speciation or lack of speciation (e.g. consider the confinement of the monotypic *Platydorina caudata* to a few States in the Ohio/Mississippi basin) and what are the barriers to distribution and speciation?

Assuming a Proterozoic origin for green algae and a process of serial endosymbiosis or modification as suggested by Cavalier-Smith (1981, 1982), then presumably this process occurred in unicells. If this is so, then *a priori* there must have been either a single unicellular ancestor for the green algae (it is difficult to conceive symbiotic events once multicellularity has been achieved) which then in some way diverged into a series of radiating groups, or there were multiple unicellular ancestors, one for each of the green clusters. Although the second hypothesis is more convenient it is considered unlikely in the system elaborated by Cavalier-Smith (1982). However, as mentioned above, I am not convinced that Cavalier-Smith is right on this point.

Asking the question I first posed in 1963—How many major divisions can be defined amongst the algae possessing chlorophyll *a* and *b*? (to which I answered four)—it now seems this still may be the minimum necessary since the euglenoids do appear to have very distinct features warranting separate status based on their origin by incorporation of a green symbiont, as postulated by Cavalier-Smith (who elevates them to a Kingdom). The other three green algal clusters (Chlorophyta, Prasinophyta and Charophyta) are still valid, especially if we accept an overall classification of these into the group Viridiplantae. If this is not done then Chlorophyta could legitimately be the division for all "green" plants, though it is unlikely to be accepted by the vascular plant systematists. If we accept that the "green" plants form a natural series, i.e. Viridiplantae, bound together by a set of conservative (primitive) features, then divisions must be distinguished on features of major morphological/reproductive/life history bases. These divisions should also be of comparable organisational status with those of the "red" and "brown" series of plants. Recently, Margulis (1974) and Whittaker and Margulis (1978) have suggested elevating the Zygnemaphyceae to the phylum Zygophyta and the siphonaceous algae to the Siphonophyta. Casper (1974) had a rather similar system though with a slightly different status for the groups, thus sub-phylum for Volvocicae (flagellate, coccoid and filamentous), Oedogoniicae, Bryopsidicae, Zygnematicae, Tetraselmidicae, Charicae and Rhyniicae (i.e. "echte Pflanzen"). These approaches are all based on gross features which tend to be neglected by some workers. The other approach based on very detailed ultrastructural studies has resulted first in the fusion of groups; thus Stewart and Mattox (1975) only recognised the chlorophyte and charophyte series (drawing the Zygnemaphyceae, Prasinophyta, Klebsormidiales and Coleochaetales together into this latter division), but later (Mattox and Stewart, this volume,

Chapter 2) they enlarge this to five (Micromonadophyceae, Pleurastrophyceae, Ulvophyceae, Chlorophyceae and Charophyceae). I also incline to this wider view, taking into account gross morphological features and, I now believe, ecological factors. The latter are generally dismissed by systematists who use only phenetic information. Corliss (1974) pleaded that this is too narrow a view for the Protozoa. This usage of ecological factors certainly applies to clusters of green algal genera; it operates at the divisional level, e.g. freshwater Charophyta and marine Prasinophyta (a small number of species are freshwater, but a small and variable leakage of species into the opposing environment does feature even in a group such as the Rhodophyta). As in the diatoms, there is also a tendency for a very few freshwater green groups to invade brackish waters (e.g. the charophytes) and of marine groups to invade fresh water, though this is fairly rare (e.g. *Enteromorpha* and *Cladophora*). These ecologically defined clusters I suggest form ancient lines which developed special features at an early stage of evolution and have retained them ever since. Here I find it hard to conceive of the mainly marine "prasinophytes" in the same cluster as the freshwater "zygnemaphytes."

SYSTEMATIC CHARACTERS

During the history of green algal systematics the stress has changed. Initially it was totally on morphology; at first only macroscopic form was available as a criterion, and then gross microscopic form. This phase lasted 150 years. All the systematic treatments at the ordinal level from Bohlin at the turn of the century through Pascher, Fritsch and Smith to the modern treatments of Bold and Wynne (1978), Hoek (1978), etc. use basic morphology as the key criterion. When more than a simple morphological step was involved, e.g. in the case of the "charophytes", then the Fritschian order was raised to the class or phylum level. Gradually life history was also drawn into the scheme and in fact this has continued until quite recently with the erection of the class Codiolophyceae (Kornmann, 1973). There is a tendency in modern work to ignore these two key features—at least this is the impression gained from reading the literature. In an overall evolutionary scheme, separation of clusters on morphological/life history grounds may divide groups which are really parts of one lineage (this is due to "the view from above"). I tended (1963, 1971) to separate clusters on these bases since I regarded the characters as being "derived", whereas workers using ultrastructural detail tend to fuse the groups on what I suspect are "primitive" characters. Much was made of the flagellate stage (zoospore/gamete) in systematics between 1850 and 1950, yet it is only since 1950 that this feature has really come of age along with some other features which I shall now deal with briefly. I do not imply that consideration of these started recently, only that they have recently achieved greater importance. Further details and discussion of the following points can be found in Hoek (1981) and will not be repeated here.

1. Flagella

It now appears that there are at least four basic types of flagellate cell in the green algae. The two commonest have cruciate roots (x-2-x-2) or a unilateral multilayered structure (MLS) associated with the basal bodies [see discussions in Melkonian (1980) and Stewart and Mattox (1980)]. The former group tends to be associated with bilateral symmetry and the cells usually have an eyespot, (e.g. *Chlamydomonas* and some prasinophytes), whilst the other group has the flagella inserted on one side and never has plastidic eyespots. A third type occurs in the Trentepohliales where the cell is bilaterally symmetrical/cruciate and has two MLS's associated with the two major roots. A fourth type involving cruciate roots and MLS's occurs in some prasinophytes, e.g. *Nephroselmis* and *Mesostigma*. Statements are often made that one type arose from another without any evidence from intermediates or from other confirmatory data. Is it not more likely that the present-day state is simply the result of symbiotic events involving chlorophyll-*a/b* units with several slightly different colourless flagellate types? After all, many symbiotic events seem to have occurred to produce the range of flagellate form in the algae as a whole. However, against this "multiple host hypothesis" Cavalier-Smith (1982) argued for a single symbiotic event for the euglenoids and chlorophytes and an early divergence of the two series. Incidentally, a problem with both series is how the eyespot arose in one instance in the cytoplasm and in the other in the plastid. Perhaps it is reasonable to assume assimilation of an eyespot-containing unit in the chlorophyceae and a second (later) ingestion of an eyespot-containing plastid in the euglenoids which then lost its chlorophyll apparatus since it would have been redundant. On the other hand, Cavalier-Smith argues for subsequent independent origins of eyespots, but how does this occur in a symbiotic organelle and from whence the evolutionary pressure to evolve this?* A further point to consider is the presence or absence of scales on the flagella; they are now known to be fairly common, but does possession in two series also imply common ancestry, which is generally assumed to be from a biflagellate ancestor?

On the whole the two-stranded roots appear to be common in almost every chlorophyte possessing the x-2-x-2 system, whereas the x- can vary, presumably as a result of mutation. The more striking variations listed by

---

*For a detailed account of the green algal eyespot see Melkonian and Robenek (1984).

Moestrup (1978, 1982) tend to occur in clusters which on other grounds have been given some special taxonomic rank, e.g. in the Oedogoniales there is a three-stranded root and one cross-banded root per flagellum. It is this kind of anomaly which in my view reinforces the separation of the Oedogoniales into a class, especially when it is coupled with the pecularities of the cell division, life cycle, etc.; it may be significant that keeled flagella occur on trentepohlialean and characean motile cells. Another external feature of significance is the presence or absence of hairs, with fine tubular hairs occurring on prasinophycean flagella.

The importance of flagellar form is now coupled with extremely fine detail of basal body/root structure. It may be worth remembering that all these flagella features are "host" features and should be compared, not only amongst algal cells, but with extant types in the heterotrophic lines (i.e. the Protozoa). Comparisons at this level may be of "primitive" characters which have not changed greatly. If this is so then the accent on these may be misplaced when they are used to imply close relationship, but if of different lineages then the accent may be correct.

2. Wall Structure and Scales

Cellulose occupies the key position in the green algae and it is the major wall component in filamentous genera, conjugate greens, charophytes and the Cladophorales. Groups which contain genera utilising other polysaccharides are the Volvocales and Chlorococcales, and these require further study. It is a little worrying that the archetypal "flagellate" of the chlorophytes is *Chlamydomonas,* which does not contain the usual microfibrillar cellulose but a complex hydroxyproline/arabinose/galactose glycoprotein (Roberts et al., 1972). However, rather than completely upsetting the ancestral concept it may merely indicate that this structure is the result of considerable evolution within the chlamydomonad series, or provides further evidence that a whole series of flagellates with walls of varied type received the "autotrophic symbiont" and from these the range throughout the chlorococcal/tetrasporal/filamentous/siphonous clusters evolved. An interesting problem is to what degree the "wall type" of the algal flagellates is simply a product of the "host" or has been partially or wholly modified by the "symbiont". Indeed, such variation would be in line with the play on the x-2-x-2 pattern of flagellar roots (Lembi, 1980). It is worth noting here that cellulose seems to be absent from the prasinophyte genera and instead pectin/galactose/arabinose components are common. Without doubt the most intriguing group in relation to wall structure is the siphonacean series where, apart from the Cladophorales and Siphonocladales, which have cellulose (Nisizawa and Sasaki, 1975), the wall is composed of mannan or xylan. The siphonaceous genera with heteromorphic life histories (the *Derbesia/Halicystis/Bryopsis* complex) are interesting in that the gametophyte wall is mainly xylan and the sporophyte mannan; McCandless (1978) and Parker (1970) used this feature as an important one in the recognition of the Derbesiales. In *Acetabularia* the vegetative wall is mannan but the gametangial cysts have cellulose walls. Prasiolales also have xylo-mannan. Additionally in the Bryopsidophyceae, there is considerable deposition of calcium carbonate, which is a feature rare in other green algae but recurrent in the charophytes.

The other algal divisions have a greater consistency of wall structure and the variety seen in the green series could be used to support the view that several divisions are involved. The consistency should extend from the flagellate to the higher levels of morphological organisation. Clearly, it is the overall patterns of biochemical structure which are important, and these data should always be correlated with other phenetic features.

3. Cytokinesis

This fundamental feature of cell biology was hardly considered as a systematic tool until the classic work of Pickett-Heaps (1969, 1972).

Is it a coincidence that there are four flagellar root systems and four mitosis/cytokinesis mechanisms (Hoek, 1981), and that also there are similarities to link the mitosis/cytokinesis mechanisms in pairs (I and IV, II and III; see Fig. 3.4 in Hoek)? Types I and IV lose the nuclear membrane and have interzonal spindles with a phragmaplast in IV, whilst II and III retain the nuclear membrane and have a phycoplast.

- Type I encompasses Volvocales, Klebsormidiales and Ulotrichales [Modifications of this type are found in some siphonaceous algae].
- Type II encompasses *Dunaliella*, some Volvocales, Chlorococcales, Ulotrichales (sarcinoid and filamentous) and Acrosiphoniales (modified type).
- Type III encompasses some freshwater Ulotrichales and Oedogoniales (modified type).
- Type IV encompasses *Coleochaete* and *Chara*.

Spirogyra has modified Type IV but *Mougeotia* has Type I which suggests that the above linking into pairs might be valid. Type I associates the genera often classed as Ulvophyceae (*sensu* Stewart and Mattox) but as Hoek (1981) commented, whilst the features of mitosis/cytokinesis appear to be important at the ordinal level there are still many problems, and in the prasinophytes Types I and II occur (but see the new "splits" of Mattox and Stewart, this volume, Chapter 2).

4. Plastid Structure

Variation in this feature is almost confined to the Bryopsidophyceae, where it does correlate fairly closely with the generally-accepted subdivision into orders, e.g. Cladophorales and Siphonocladales are decidedly similar (see also Hoek, this volume, Chapter 5). *Urospora* is distinctive and supports the erection of the Acrosiphoniales. The siphonaceous group can be subdivided into two major groups: (1) with chloroplasts only (*Codium, Derbesia, Bryopsis*), and (2) with chloroplasts and amyloplasts (*Dichotomosiphon, Chlorodesmis, Halimeda, Udotea, Caulerpa, Avrainvillea*). The Dasycladales form a further discrete cluster with many distinctive features, e.g. no pyrenoids, starch inside and outside plastids, and variable arrangement of thylakoids (Hori and Ueda, 1975, 1976).

Few other groups have been investigated, but the single genus *Ulothrix* has eight types of pyrenoid (Lokhorst and Star, 1980), and thus variation can exist within a single genus yet be consistent within whole groups within the Bryopsidophyceae.

5. Pigment Composition and Plastid Origin

The chlorophylls give little help in separating the subgroups. The xanthophylls of the Chlorophyceae *sensu lato* are similar to those of the higher plants (i.e. lutein, neoxanthin, violoxanthin and zeaxanthin). Charophyceae have no xanthophylls according to the table in Hirose (1975) and in addition the carotenes vary somewhat.

The endosymbiotic origin of plastids in algae is now overwhelmingly accepted by phycologists, but there remains a problem of single or multiple origins, with Calvalier-Smith (1982) favouring the former and Stewart and Mattox (1980) the latter. Stewart and Mattox stated that "it is now obvious that the different lines of evolution of higher green algae are different from each other because they had independent origins from different kinds of green flagellates". I find it difficult to conceive of a single symbiotic event, and consider that on probability grounds alone many symbioses must have taken place (cf. the very extensive range of present-day animals containing symbionts in various degrees of reduction), and that "genetic fit" between host and symbiont is a perfectly good concept for the restriction of chlorophyll *b*–containing symbionts coupling only with hosts possessing flattened mitochondrial cristae as postulated by Stewart and Mattox. I suspect, however, that these authors' views and that of Cavalier-Smith are perfectly compatible in that a single type of endosymbiont ("only one prokaryote was ever converted into a plastid"; Cavalier-Smith, 1982) was utilised but that it entered several different hosts; this reconciles both views and allows a branching system but also allows all the known similarities and variability. As Cavalier-Smith said, it will be interesting "to compare the detailed organisation of the plastid DNA in the six kingdoms" and, I would add, equally important to compare it in the classes of green algae.

6. Assimilates

Kremer (1980) makes the point that the Chlorophyceae seem to be the only heterogeneous group in relation to assimilates. Most have glucose, fructose and sucrose. Charophyceae have glucose and sucrose. Aerophilous genera tend to have sugar alcohols, and the most striking of all is *Trentepohlia* with glycerol, erythritol, ribitol, arabinitol, mannitol and volemitol (cf. the occurrence of sugar alcohols in genera of other divisions living under stress from exposure). The prasinophytes have mannitol and glucose—mannitol is not a common compound in higher plants.

DISCUSSION

The questions I first posed in 1963 are still with us: how many divisions and how many classes? The question of how many divisions involves the rarely discussed concept of a division, but I still believe that groups such as the Rhodophyta, Phaeophyta, and Bryophyta hold the clues, i.e. groups with overall life history/reproductive/physiological and even ecological uniformity or, as Cavalier-Smith (1982) put it, "based on conservative aspects of fundamental cell structure". To this can now probably be added an origin from a geologically-early symbiotic event involving only a single photosynthetic symbiont entering slightly different hosts followed by a long period of relative stasis in either freshwater or marine environments. The green groups still pose problems when we try to impose the criteria used to establish other divisions, yet surely similar criteria must be applied. This leads back to the number of basic symbioses which are involved. But let us return to the problem of divisional status. If we retain Bryophyta and divisions showing greater morphological complexity then I believe we need a number of green algal divisions, since it is generally agreed that the higher cryptogams arose from *one* of the lines of green algae. Should we wish to maintain a systematic viewpoint which encompasses all the green plants with certain basic features in common from the seed plants downwards, then a single division could be maintained for these. I do, however, greatly prefer Cavalier-Smith's view that these comprise a kingdom—the Viridiplantae. I agree also with Cavalier-Smith (1981) that the "green" euglenoids do not belong in the Viridiplantae and form a separate kingdom. The Viridiplantae of Cavalier-Smith contain Chlorophyta (green algae), Bryophyta (mosses and liverworts) and Tracheophyta (vascular plants), and on very gross features of organisation I can accept this. However, I do consider that it is more satisfactory to subdivide at this level on more than just gross morphology in order to produce the predictive system which Cavalier-Smith quite rightly believes is a fundamental feature of systematics. Thus Tracheophyta merely indicates the existence of vascular tissue, whereas the use of Pteridophyta, Cycadophyta, Pinophyta, etc., or whatever nomenclature is accepted, does confer a tighter set of predictive concepts. Turning to the green algae, if we have a single division, the Chlorophyta, it should be monophyletic, but present evidence suggests that it is not. Some workers avoid the issue by referring to classes (e.g. Stewart and Mattox, 1975; Mattox and Stewart, this volume, Chapter 2), but classes surely have to be included in a higher taxon. The "charophytes" seem to me a side-line but one of considerable importance, and on gross morphological features I really wonder whether they are on the main line to the vascular plants, the MLS notwithstanding. A cytologically early split to give charophytes and bryophytes seems more likely. All the Viridiplantae have plate-like mitochondrial cristae, but is it not possible that a *group* of colourless organisms with this feature took in the symbiont necessary to produce Viridiplantae? After all, many animal groups today have either brown or green symbionts. If it is proven that Cavalier-Smith's Chlorophyta and Silva's (1982) Chlorophycota (both including Chlorophyceae, Prasinophyceae and Charophyceae) are monophyletic, then I am happy to accept it. But at this moment in time I do not regard the evidence as conclusive and therefore prefer to retain a number of divisions (Chlorophyta, Prasinophyta, Charophyta) with a slight inclination to accept the fusion of Prasinophyta and Charophyta. A fourth division, the Gamophyta (=Zygnematophyta or Conjugatophyta), is proposed for the conjugate green algae by Margulis (1981), so this group has ranged in status from order through class to division and further work is necessary to make sure of its status. A further division, Ulvophyta, may prove necessary (see Fig. 1 below).

The problem of classes remains as fraught as ever. Here again it is difficult, if not impossible, to discuss the "class concept" without a definition. Pascher (1931), and more recently Fott (1974), clearly regarded a class as encompassing lines of organisms ranging from the simplest (i.e. flagellate) through to the more complex and this would seem to be a sensible approach if one believes in the symbiotic theory. However, it does not mean that a class cannot exist either without the simplest forms (which may not have survived to the present day, e.g. in the Phaeophyceae) or without the complex. I would suggest, therefore, that an algal class be loosely defined as a subset of organisms encompassing a "range of organisation from simple to complex (but not necessarily inclusive of these) in which the basic biochemistry and morphology of the motile cells is of one type". If one insists on seeing the full range of morphology in a class this will lead to many problems, amongst which is the combining of subsets into larger sets. Silva's (1980) survey (see pp. 4–5) lists the basic entities (classes) variously discussed during the last 150 years, and crystallised as I prepared this account by Ettl and Komárek (1982) into a flagellate series (Chlamydophyceae), a filamentous series (Chlorophyceae or Ulotrichophyceae or Ulvophyceae), a heteromorphic series (Codiolophyceae), a conjugate series [Zygnema(to)phyceae], and a siphonous series (Bryopsidophyceae). On an overall vegetative morphology/reproductive basis I would also strongly maintain a stephanokont group (Oedogoniophyceae). Where the Zygnema(to)phyceae fit in is problematical, but their individuality is hardly in doubt. The Codiolophyceae give me more problems and I await further studies, though there are undoubtedly important features combining genera in this class. The Charophyceae (in the old sense) pose fewer problems for they are an elegantly defined group with no anomalous genera and a very long geological history. The old Prasinophyceae (first proposed as a class by Christensen, 1962) appear well defined at their lower flagellate/coccoid end, but only further detailed work will show the exact relationships now that the Micromonadophyceae and Pleurastrophyceae have been distinguished. The combining of the conjugate green algae and "charophytes" within a single series is, I feel, not yet conclusively established, though loosely I can accept an MLS-containing sector out of which the land plants arose. I am convinced that all these lines must be conceived in an historical/geological sense.

After having written this section I received Mattox and Stewart's paper (this volume, Chapter 2) and I note their change of heart from Chlorophyceae and Charophyceae in 1975 to Charophyceae, Chlorophyceae, Ulvophyceae, Micromonadophyceae and Pleurastrophyceae (thus approaching Ettl and Komárek's seven classes). Their classification is clearly the most satisfactory to date, based as it is on features of motile cell ultrastructure and very tellingly correlating with ecology and thus substantiating the concept of long-term separation of gene pools in marine or freshwater habitats. My only query concerns the basis of the hierarchy of groups. Mattox and Stewart combine sets into classes on ultrastructural features, which results in a "downgrading" of general gross morphological features. To give an example, I would not argue with placing the Oedogoniales in the same set as the Volvocales, etc. but would question the merit of equal standing; likewise with the Charales and Zygnematales. The point of argument here is really the phylogenetic/phenetic distance between the sets.

And finally, how many orders? In 1963 I suggested 27(31) and in 1971 this was increased to 38, but these included the splitting of the Prasinophyceae and Zygnemaphyceae into orders. Nowadays more knowledgeable specialists than I recognise 28 (Daily, 1982; Norris, 1982; Silva, 1982) or 30 (Ettl and Komárek, 1982), excluding the Charales. In Mattox and Stewart (this volume, Chapter 2) there are 20, excluding those of the Micromonadophyceae. The most recent new ones are the Ctenocladales of Silva (1982) and the Tetraselmidales/Pleurastrales of Mattox and Stewart (this volume, Chapter 2). Can we discern which groups are diverging (e.g. Cladophorales/Siphonocladales) or should we maintain only a single group for these? This involves the common problem of distinguishing between continuous variation or disjunct series—the art of systematics? A possible scheme of classification is suggested in Fig. 1 together with a tentative indication of the geological history. The classes are those of Mattox and Stewart (this volume, Chapter 2) with my alternative additions. The Trentepohliales are left in doubt—they have been attributed to both the Charophyceae and Ulvophyceae; on ecological grounds the former would be preferred. Only a minimum of orders are indicated—for other possibilities see works previously referred to. All the classes can be placed in an overall division Chlorophyta if higher plant systematists accept this; if not, then I suggest the series at the top of the diagram with the Micromonadophyceae/Pleurastrophyceae and a relationship to the division Prasinophyta as still outstanding problems, especially since Mattox and Stewart regard the Micromonadophyceae as a group of genera which will probably have to be distributed into other classes. The other most recent scheme (Ettl and Komárek, 1982) retains the Codiolophyceae and maintains the flagellate/coccoid forms with walled motile cells in the class Chlamydophyceae, and the remainder of the coccoid (the Chlorellales of Bold and Wynne, 1978) and filamentous series in the Chlorophyceae. All these possibilities should be kept in mind and the argument kept open as the systematic study develops into an even more natural scheme.

My final comment is to reiterate that early workers, relying on gross features, produced the best schemes possible, whereas latterly electron microscopy has become prominent and has yielded much sounder systems, marred only by lack of consideration of the gross features used by the pioneers. Corliss (1974) made a plea to protozoologists not to ignore the classical literature and the same plea can be made to workers in chlorophytan phylogeny. If "evolution is the essence of systematics" then equally evolution implies change, and change is certainly also a feature of the classification of the green algae—there are a few periods when it has been relatively static but the present is not one.

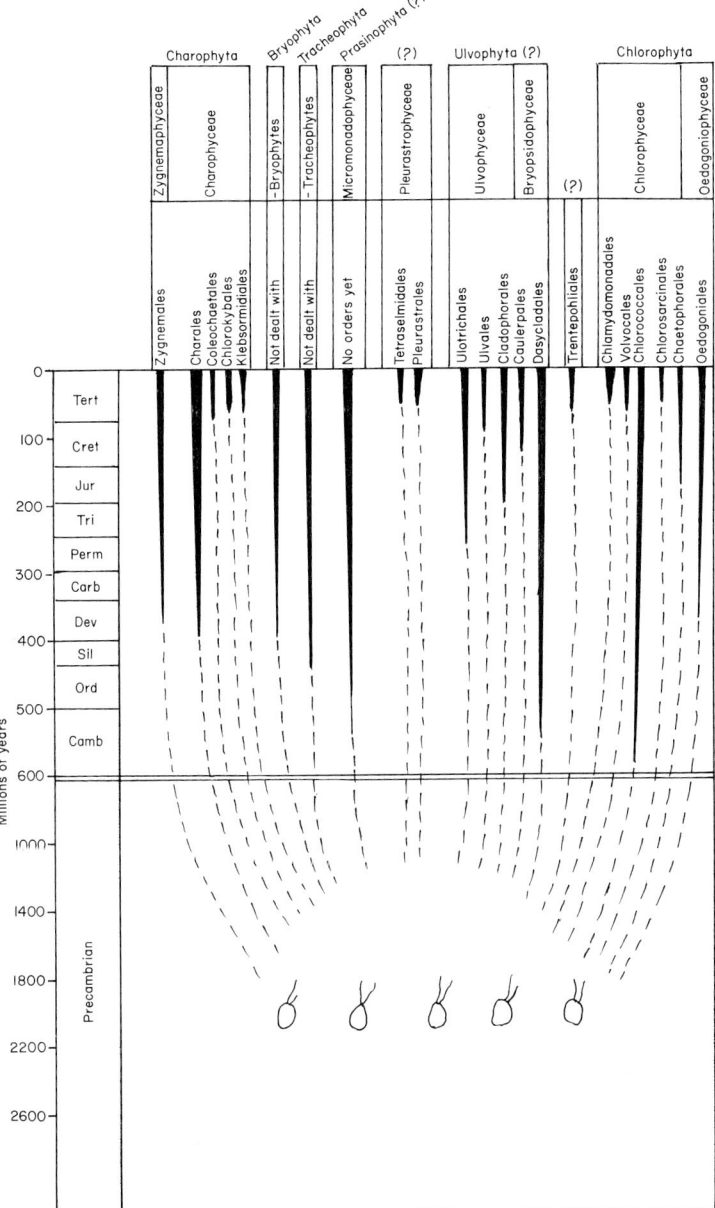

Fig. 1. The lines of evolution of green plants. Note that not all probably valid orders of the green algae are listed within the classes. Branching may have occurred especially in the earlier stages of evolution, but there are very few data to support this. The number of flagellate ancestors is also open to discussion (see text).

REFERENCES

Agardh, C. A. (1817). "Synopsis algarum Scandinaviae." Officina Berlingiana, Lund.
Bauhin, C. (1620). "Prodromus theatri botanici . . ." Joannis Treudelii, Frankfurt.
Blackman, F. E. (1900). The primitive algae and the flagellate. An account of modern work bearing on the evolution of the algae. *Ann. Bot. (London)* **14**, 647–688.
Blackman, F. E. and Tansley, A. G. (1902). A revision of the classification of the green algae. *New Phytol.* **1**, 17–24, 67–77, 89–96, 114–120, 133–144, 163–168, 189–192, 213–220, 258–264.
Bohlin, K. (1901). "Utkast till de Gröna Algernas och Arkegoniaternas Fylogeni." Almqvist & Wiksell, Stockholm.
Bold, H. C. and Wynne, M. J. (1978). "Introduction to the Algae." Prentice-Hall, New Jersey.
Boney, A. D. (1978). Taxonomy of red and brown algae. *In* "Modern Approaches to the Taxonomy of Red and Brown Algae" (D.E.G. Irvine and J. H. Price, eds), Syst. Assoc. Spec. Vol. No. 10, pp. 1–19. Academic Press, London.
Borzi, A. (1895). "Studi Algologici," Fasc. II. Reber, Palermo.
Bremer, K. and Wantorp, H. E. (1981). A cladistic classification of green plants. *Nord. J. Bot.* **1**, 1–3.
Casper, S. J. (1974). "Grundzüge eines natürlichen Systems der Mikroorganismen." Fischer, Jena.
Cavalier-Smith, T. (1980). Cell compartmentation and the origin of eukaryote membranous organelles. *In* "Endocytobiology: Endosymbiosis and Cell Biology, a Synthesis of Recent Research" (W. Schwemmler and H. E. A. Schenk, eds), pp. 893–916. de Gruyter, Berlin.
Cavalier-Smith, T. (1981). Eukaryote kingdoms: Seven or nine. *BioSystems* **14**, 461–481.
Cavalier-Smith, T. (1982). The origin of plastids. *Biol. J. Linn. Soc.* **17**, 289–306.
Cherfas, J. (1981). Proving the pattern of life. *In* "Darwin Up-to-Date" (J. Cherfas, ed), pp. 39–42. I.P.C. Magazines, London.
Christensen, T. (1962). Alger. *In* "Botanik" (T. W. Böcher, M. Lange, and T. Sørensen, eds) Vol. 2, No. 2, pp. 1–178. Munksgaard, Copenhagen.
Cohn, F. (1856). Observations sur les Volvocinées et spécialement sur l'organisation et la propagation du *Volvox globator*. *Annls Sci. Nat.* (Bot.) Sèr. **4,5**, 323–332.
Coleman, A. W. (1977). Sexual and genetic isolation in the cosmopolitan algal species *Pandorina morum*. *Am. J. Bot.* **64**, 361–368.
Corliss, J. O. (1974). Time for evolutionary biologists to take more interest in protozoan phylogenetics. *Taxon* **23**, 497–522.
Corliss, J. O. (1979). The impact of electron microscopy on Ciliate systematics. *Am. Zool.* **19**, 573–587.
Daily, F. K. (1982). Charophyceae. *In* "Synopsis and Classification of Living Organisms" (S. P. Parker, ed.), Vol. I, pp. 161–162. McGraw-Hill, New York.
De Bary, A. (1858). "Untersuchungen über die Familie der Conjugaten, . . ." Förstnersche, Leipzig.
Dumortier, B. C. J. (1822). "Commentationes Botanicae. Observations botaniques, dédiées à la Société d'Horticulture de Tournay." Tournay.
Ettl, H. and Komárek, J. (1982). Was besteht man unter dem Begriff "coccale Grünalgen"? *Arch. Hydrobiol., Supp.* **60**, 345–374.
Fott, B. (1974). The phylogeny of eukaryotic algae. *Taxon* **23**, 446–461.
Fritsch, F. E. (1935). "The Structure and Reproduction of the Algae", Vol. I. Cambridge Univ. Press, London and New York.

Gould, S. J. (1981). Punctuated equilibrium—a different way of seeing. In "Darwin Up-to-Date" (J. Cherfas, ed), pp. 26–30. I.P.C. Magazines, London.
Gray, S. F. (1821). "A Natural Arrangement of British Plants . . .," Vols. 1. and 2. Baldwin, Cradock & Joy, London.
Greville, R. K. (1830). "Algae britannicae . . ." MacClachlan & Stewart, Edinburgh.
Harvey, W. H. (1836). Algae. In "Flora Hibernica" (J. T. Mackay, ed), Vol. 2, Part 3, pp. 157–254. William Curry, Dublin.
Hirose, H. (1975). Photoreactive pigments of algae and algal phylogeny. In "Advance of Phycology in Japan" (J. Tokida and H. Hirose, eds), pp. 52–65. Fischer, Jena.
Hoek, C. van den (1978). "Algen. Einführung in die Phykologie" Thieme, Stuttgart.
Hoek, C. van den (1981). Chlorophyta: Morphology and classification. In "The Biology of Seaweeds" (C. S. Lobban and M. J. Wynne, eds), pp. 86–132. Blackwell, Oxford.
Hori, T. and Ueda, R. (1975). The fine structure of algal chloroplasts and algal phylogeny. In "Advance of Phycology in Japan" (J. Tokida and H. Hirose, eds), pp. 11–42. Fischer, Jena.
Hori, T. and Ueda, R. (1976). Electron microscope studies on the fine structure of plastids in siphonous green algae with special reference to their phylogenetic relationships. Sci. Rep. Tokyo Kyoiku Daigaku **12**, Sect. B. 225–244.
King, K. (1974). Comparative amino acid composition of some siliceous microfossils. Year Book—Carnegie Inst. Washington **73**, 595–596.
Kornmann, P. (1973). Codiolophyceae, a new class of Chlorophyta. Helgol. wiss. Meeresunters. **25**, 1–13.
Kremer, B. P. (1980). Taxonomic implications of algal photoassimilate patterns. Br. Phycol. J. **15**, 399–409.
Kützing, F. T. (1843). "Phycologica generalis . . ." Brockhaus, Leipzig.
Kützing, F. T. (1845). "Phycologia germanica." W. Köhne, Nordhausen.
Kützing, F. T. (1849). "Species algarum." Brockhaus, Leipzig.
Lamouroux, J. V. F. (1813). Essai sur les genres de la famille des Thalassiophytes non articulées. Ann. Mus. Hist. Nat. Paris **20**, 21–47, 115–139, 267–293.
Lembi, P. A. (1980). Unicellular chlorophytes. In "Phytoflagellates" (E. R. Cox, ed), pp. 5–59. Elsevier/North-Holland, New York.
Linnaeus, C. (1753) "Species plantarum, . . .," Vols. 1 and 2. Impensis Laurentii Salvii, Holmii.
Linnaeus, C. (1758). "Systema naturae," 10th ed., Vol. I. Impensis Laurentii Salvii, Holmii.
Loeblich, L. A. (1982). Photosynthesis and pigments influenced by light intensity and salinity in the halophile *Dunaliella salina* (Chlorophyta). J. mar. Biol. Assoc. U.K. **62**, 493–508.
Lokhorst, G. M. and Star, W. (1980). Pyrenoid ultrastructure in *Ulothrix (Chlorophyceae)*. Acta Bot. Neerl. **29**, 1–15.
Luther, A. (1899). Uber *Chlorosaccus* einer neue Gattung der Susswasseralgen. Bih. K. Svenska Vetensk. Akad. Handl. **24**, Part III, 1–22.
McCandless, E. L. (1978). The importance of cell wall constituents in algal taxonomy. In "Modern Approaches to the Taxonomy of Red and Brown Algae" (D.E.G. Irvine and J. H. Price, eds), Syst. Assoc. Spec. Vol. No. 10, pp. 63–85. Academic Press, London.
Margulis, L. (1974). Five-Kingdom classification and the origin and evolution of cells. Evol. Biol. **7**, 45–78.
Margulis, L. (1981). "Symbiosis in Cell Evolution. Life and its Environment on the Early Earth." Freeman, San Francisco, California.
Mattox, K. R. and Stewart, K. D. (1977). Cell division in the scaly green flagellate *Heteromastix angulata* and its bearing on the origin of the Chlorophyceae. Am. J. Bot. **64**, 931–945.

Melkonian, M. (1980). Ultrastructural aspects of basal body associated fibrous structures in green algae: A critical review. *BioSystems* **12**, 85–104.
Melkonian, M., and Robenek, H. (1984). The eyespot apparatus of flagellated green algae: A critical review. *Prog. Phyc. Res.* **3**, (in press).
Mereschkowsky, C. (1905). Über Natur und Ursprung der Chromatophoren in Pflanzenreich. *Biol. Zentralbl.* **25**, 593–604.
Mereschkowsky, C. (1910). Theorie der zwei Plasmaarten als Grundlage der Symbiogenesis, einer neuen Lehre von der Entstehung der Organismen. *Biol. Zentralbl.* **30**, 278–303, 321–346, 353–367.
Moestrup, Ø. (1978). On the phylogenetic validity of the flagellar apparatus in green algae and other chlorophyll *a* and *b* containing plants. *BioSystems* **10**, 117–144.
Moestrup, Ø. (1982). Flagella structure in algae: A review, with new observations particularly on the Chrysophyceae, Phaeophyceae (Fucophyceae), Euglenophyceae, and *Reckertia*. *Phycologia* **21**, 427–528.
Nelson, G. and Platnick, N. (1981). "Systematics and Biogeography. Cladistics and Vicariance." Columbia Univ. Press, New York.
Nisizawa, K. and Sasaki, S. F. (1975). Cell wall composition of algae from a phylogenetic point of view. *In* "Advance of Phycology in Japan" (J. Tokida and H. Hirose, eds), pp. 42–47, 52. Fischer, Jena.
Norris, R. F. (1982). Prasinophyceae. *In* "Synopsis and Classification of Living Organisms" (S. P. Parker, ed), Vol. I, p. 1166. McGraw-Hill, New York.
Parker, B. C. (1970). Significance of cell wall chemistry to phylogeny in the algae. *Ann. N.Y. Acad. Sci.* **175**, 417–428.
Pascher, A. (1914). Über Flagellaten und Algen. *Ber. Dtsch. bot. Ges.* **32**, 136–160.
Pascher, A. (1931). Systematische Übersicht über die mit Flagellaten in Zusammenhang stehenden Algenreihen und Versuch einer Einreihung dieser Algenstämme in die Stämme des Pflanzenreichs. *Beih. bot. Zentralbl., Abt. 2* **48**, 317–332.
Patterson, C. (1982). Cladistics and classification. *In* "Darwin Up-to-Date" (J. Cherfas, ed), pp. 35–39. I.P.C. Magazines, London.
Pickett-Heaps, J. D. (1969). The evolution of the mitotic apparatus: An attempt at comparative ultrastructural cytology in dividing plant cells. *Cytobios* **1**, 259–280.
Pickett-Heaps, J. D. 1972. Variation in mitosis and cytokinesis in plant cells: Its significance in the phylogeny and evolution of ultrastructural systems. *Cytobios* **5**, 59–77.
Rabenhorst, L. (1863). "Kryptogamen-Flora von Sachsen, der Ober-Lausitz, Thüringen und Nordböhmen mit Berüchsichtigung der benachbarten Länder," Vol. I. Kummer, Leipzig.
Rabenhorst, L. (1868). "Flora europaea algarum aquae dulcis et submarinae," Vol. III. Kummer, Leipzig.
Roberts, K. M., Gurney-Smith, M. and Hills, G. J. (1972) Structure, composition and morphogenesis of the cell wall of *Chlamydomonas reinhardtii*. I. Ultrastructure and preliminary chemical analysis. *J. Ultrastruct. Res.* **40**, 599–613.
Round, F. E. (1963). The taxonomy of the Chlorophyta. *Br. phycol. Bull.* **2**, 224–235.
Round, F. E. (1971). The taxonomy of the Chlorophyta. II. *Br. phycol. J.* **6**, 235–264.
Round, F. E. (1980). The evolution of pigmented and unpigmented unicells—a reconsideration of the protista. *BioSystems* **12**, 61–69.
Round, F. E. (1981). "The Ecology of Algae." Cambridge Univ. Press, London and New York.
Round, F. E. and Crawford, R. M. (1981). The lines of evolution of the Bacillariophyta. I. Origin. *Proc. R. Soc. London, B* **211**, 237–260.
Round, F. E. and Sims, P. A. (1982). The distribution of diatom genera in marine and freshwater environments and some evolutionary considerations. *In* "Proceedings of the

Sixth Symposium on Recent and Fossil Diatoms" (R. Ross, ed), pp. 301–320. Koeltz, Konigstein.
Schoenischen, W. (1925). "Einfachste Lebensformen des Tier-und Pflanzenreiches," 5th ed., Vol. 1. Lichterfelde, Berlin.
Schwartz, R. M. and Dayhoff, M. O. (1978). Origins of prokaryotes, eukaryotes, mitochondria and chloroplasts. *Science,* **199,** 395–403.
Silva, P. S. (1980). Names of classes and families of living algae. *Regnum Veg.* **103,** 1–156.
Silva, P. S. (1982). Chlorophycota. *In* "Synopsis and Classification of Living Organisms" (S. P. Parker, ed), Vol. I, pp. 133–161. McGraw-Hill, New York.
Simpson, G. G. (1961). "Principles of Animal Taxonomy." Columbia Univ. Press, New York.
Stewart, K. D. and Mattox, K. R. (1975). Comparative cytology, evolution and classification of the green algae with some consideration of the origin of other organisms with chlorophylls *a* and *b*. *Bot. Rev.* **41,** 105–135.
Stewart, K. D. and Mattox, K. R. (1980). Phylogeny of phytoflagellates. *In* "Phytoflagellates" (E. R. Cox, ed), pp. 433–462. Elsevier/North-Holland, New York.
Stitzenberger, E. (1860). "Dr. Ludwig Rabenhorst's Algen Sachsens resp. Mitteleuropa's, Decade I-C. Systematisch geordnet (mit Zugrundelegung eines neuen Systems)." Dresden.
Tappan, H. (1980). "The Paleobiology of Plant Protists." Freeman, San Francisco, California.
Taylor, F. J. R. (1978). Problems in the development of an explicit hypothetical phylogeny of the lower eukaryotes. *BioSystems* **10,** 67–89.
West, G. S. (1916) "Algae," Vol. 1. Cambridge Univ. Press, London and New York.
Whittaker, R. H. (1969) New concepts of kingdoms of organisms. *Science* **163,** 150–159.
Whittaker, R. H. and Margulis, L. (1978). Protist classification and the kingdoms of organisms. *BioSystems* **10,** 3–18.
Wille, N. (1890–1891). Conjugatae [und] Chlorophyceae. *In* "Die Natürlichen Pflanzenfamilien . . .," Part I, Sect. 2, pp. 1–175. Engelmann, Leipzig.
Wille, N. (1901). Algologische notizen. VII and VIII. *Nytt Mag. Naturvid.* **39,** 1–22.

2 | Classification of the Green Algae: A Concept Based on Comparative Cytology

K. R. MATTOX and K. D. STEWART

*Department of Botany,
Miami University,
Oxford, Ohio, USA*

Abstract: A class-level system for the Chlorophyta is proposed based on comparative cytological investigations of the last two decades. In some cases the classification is extended to order and family, where fine structural characteristics are combined with certain classical criteria. The classification is further based on the idea that class-level distinction in algal divisions with flagellated cells should be based on the characteristics of the flagellated cell, as it already is in most cases. Advanced green algae appear to be derived from four different types of green flagellates. Some green flagellates are assigned to classes that also include advanced genera. Green flagellates that cannot presently be assigned in that manner are placed in a fifth class containing flagellates only. The structure of the flagellar apparatus is considered to be of major importance, but other features now possessed by green flagellates and apparently also possessed by flagellates ancestral to advanced genera (modes of cell division and type of cell covering) are also taken into account. Relevant aspects of evolution among green flagellates are briefly discussed.

INTRODUCTION

The green algae have long been considered the most diverse of algal groups, and now electron microscopy has not only added to the known diversity but has, in a sense, multiplied it. The various levels of organization beyond that of the flagellate—coccoid, sarcinoid, filamentous, and parenchyma-

---

Systematics Association Special Volume No. 27, "Systematics of the Green Algae", edited by D. E. G. Irvine and D. M. John, 1984, pp. 29–72. Academic Press, London and Orlando.
ISBN 0 12 374040 1   *Copyright © by the Systematics Association*
All rights of reproduction in any form reserved

tous—are now seen to be independently derived in three or four lines of evolution within the green algae. Plasmodesmata and other intercellular connections, the manifestations of true multicellularity (Stewart et al., 1973; Stewart and Mattox, 1975), are known to have arisen in at least four different lines. Continued use of levels of organization for the basis of orders or families, as in older classifications, would now lead to the multiplication of taxa, with orders or families of that kind in each of several higher groups.

In earlier years it was widely assumed that green flagellates exhibited only minor variations on a basically *Chlamydomonas*-like cell. It is now obvious that the different lines of evolution of higher green algae are different from each other because they had independent origins from different kinds of green flagellates, only one of which was close to *Chlamydomonas* or *Carteria*. The structure of the flagellar apparatus and the nature of cell coverings in the swarmers of higher green algae, along with the structural and developmental features of mitosis and cytokinesis in their vegative cells, provide clear evidence concerning the nature of their ancestral flagellates. The apparent reason for this is that the most fundamental evolutionary changes in general cellular structure occurred among green flagellates rather than after the origin of higher groups. Most of the differences in cell division, the flagellar apparatus, and cell coverings that have recently been used to distinguish between classes of higher green algae also occur in green flagellates. Green flagellates possess further variations not known among higher green algae, so that if one considers only basic cellular structure and development, the flagellates are even more diverse than their descendents.

Our primary purpose here is the proposal of a class-level system for the Chlorophyta based on comparative cytological investigations of the last two decades. When warranted by evidence, the system is extended to order and family, where a number of classically employed criteria are used along with the fine structural data. Since evolutionary diversification among green flagellates appears to be the basis for differences between groups of higher green algae, it seems necessary to provide a brief discussion of phylogenetic trends in green flagellates before the classification is outlined.

PHYLOGENY OF GREEN FLAGELLATES

One of the safest assumptions that can be made about the characteristics of the most primitive or ancestral green flagellate is that it possessed those characteristics that define the Chlorophyta—chlorophylls *a* and *b,* the "star" and "H-piece" of the flagellar transition region, starch storage in the chloroplast, a chloroplast envelope of two membranes, and the lack of chloroplast ER. That combination of characteristics is distinctive enough that there can be little doubt that the Chlorophyta are monophyletic, although they are sometimes casually or uncritically suggested to be otherwise (as in Heath, 1981, p. 267).

The ancestral green flagellate (AGF) probably also possessed a covering of scales, because a wide variety of green flagellates has scales on the body, flagella, or both. Green flagellates that possess a "wall" do not have a wall with the type of fibrillar substructure characteristic of true walls, but rather have a theca composed of tiny, nearly isodiametric subunits. Thecae apparently evolved by the modification of a scaly covering. In *Tetraselmis* (=*Platymonas*) the theca has been shown to be fabricated by the extracellular assembly of subunits that are similar to early stages of scale development in other scaly green flagellates and to those of the flagellar scales of *Tetraselmis* itself (Domozych *et al.*, 1981a). The scales or scale precursors of *Tetraselmis* "interweave" along their edges so that a coherent cell covering is formed. In effect, then, both the evolution and development of green algal thecae are either from the fusion of the scales themselves or at least from "interwoven" areas between them. That evolutionary history is apparently shared by the cell coverings of *Chlamydomonas* and all other "walled" volvocalean flagellates (Domozych *et al.*, 1981b,c). Since the structure, development, and evolutionary history of the covering in *Tetraselmis* and volvocalean flagellates are different from those of the cell wall in non-motile stages of higher green algae, it is recommended that it be called a theca rather than a wall.

The AGF doubtless had an interzonal mitotic spindle that persisted during cytokinesis, because that is the condition in nearly all eukaryotic microorganisms, including many green flagellates. The only green flagellates known to have an early collapsing interzonal mitotic spindle and a phycoplast are those that also possess a theca (D. S. Domozych, unpublished) and naked flagellates whose internal structure is so clearly like thecate green flagellates that they can be assumed to lack the theca secondarily (e.g. *Asteromonas*, Floyd, 1978; *Dunaliella*, Marano, 1976). All scaly green flagellates apparently have a persistent interzonal spindle, and this striking correlation indicates that the collapsing spindle and phycoplast evolved along with the theca as the result of the theca's restriction of cell elongation during cell division (Mattox and Stewart, 1977).

Green flagellates exhibit an amazing diversity in cell shape, cell symmetry, and flagellar insertion (Fig. 1). In external form the cell can be radially symmetrical (e.g. *Chlamydomonas,* Fig. 1m), asymmetrical, or bilaterally symmetrical (*Pedinomonas,* Fig. 1e; *Micromonas,* Fig. 1d; and *Trichloris,* Fig. 1j). In many the flagella are apically or anteriorly inserted, but in others they are subapical or lateral [*Nephroselmis* (=*Heteromastix*), Fig. 1f; *Trichloris*]. The microtubular flagellar root system can be asymmetrical, with only one

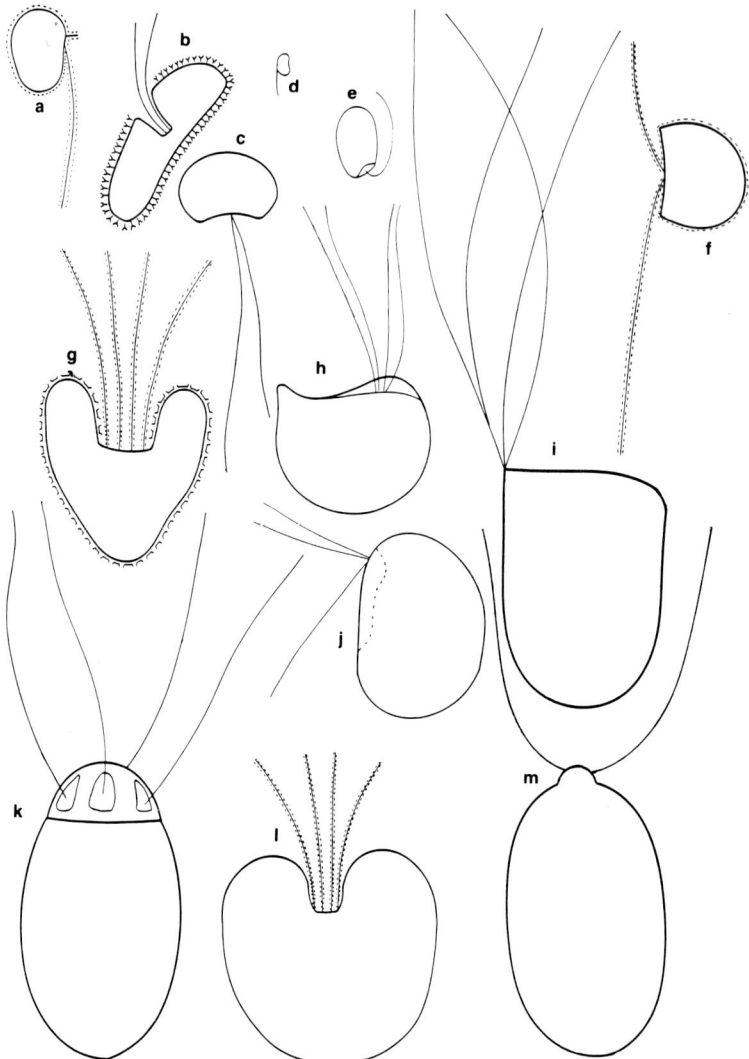

Fig. 1. Assorted green flagellates selected to show diversity in flagellar position, relative length of members of a flagellar pair, number of flagella, direction of motion, external symmetry, cell covering, and size. When known, the flagellates are drawn in a way so that the direction of swimming would be toward the top of the page (not known for *Cymbomonas* and *Trichloris*). (a) *Mantoniella*, scale-covered; (b) *Mesostigma*, scale-covered; (c) *Scourfieldia*, naked; (d) *Micromonas pusilla*, naked; (e) *Pedinomonas minor*, naked; (f) *Nephroselmis* (=*Heteromastix*), scale-covered; (g) *Pyramimonas*, scale-covered; (h) *Cymbomonas tetramitiformis*, unknown whether naked or scaly; (i) *Cymbomonas adriatica*, unknown whether naked or scaly; (j) *Trichloris*, unknown whether naked or scaly; (k) *Carteria*, thecate; (l) *Tetraselmis* (=*Platymonas*), thecate; (m)

of each kind of root (*Mantoniella*, Barlow and Cattolico, 1980; *Nephroselmis*, Moestrup and Ettl, 1979), or there can be two "copies" of each of two different roots in an X-shaped pattern called a cruciate root system (as in *Chlamydomonas* and the flagellar root system of the swarmers in Fig. 2b–d and Fig. 3). The cruciate pattern most commonly occurs in those flagellates that exhibit a radial or near-radial cell symmetry externally. The cruciate system is not truly radially symmetrical, however, nor even bilaterally symmetrical; it can be symmetrically divided in only one plane, and this yields identical, not mirror-image, halves that will superimpose if one of the halves is rotated 180° (see Fig. 3). Floyd *et al.* (1980) have previously discussed this arrangement and have called it "180 degree rotational symmetry." The problem here is to determine which of these arrangements is the more primitive. Did the AGF have only one of each kind of flagellar root, or did it have two of each kind in a cruciate system? Did it have apically or laterally inserted flagella? A comparison of symmetry and flagellar root systems in green flagellates with those of other flagellate groups, both algal and protozoan, favors the conclusion that the AGF was asymmetrical, had only one of each kind of flagellar root, and subapically inserted flagella. The flagellar root system in all other groups of phyto- and zooflagellates, as far as is known, has only one of each kind of microtubular flagellar root. Some green flagellates and some prymnesiophyte (haptophyte) flagellates are exceptional in having apically inserted, isokont flagella and an external form that approaches radial symmetry. Since the structure of other green flagellates is like that more widespread among flagellate groups, the favored conclusion is that an external form that is radially or nearly radially symmetrical and root system that appears to be doubled (so that two of each kind of root occurs) is the more derived type and originated within the Chlorophyta. The facts presently known suggest that one of the flagella of the AGF had two different microtubular roots associated with it. This flagellum and its associated roots became duplicated or "doubled" to give rise to the cruciate root system, while other roots and flagella of the AGF were lost. Perhaps when some of the more obscure genera of asymmetrical or bilateral green flagellates with laterally inserted flagella (e.g. *Trichloris* and *Cymbomonas*, Fig. 1h–j) are carefully studied, they will prove to have a

---

Chlamydomonas, thecate. Sources: (a) Drawn from description of Barlow and Catollico (1980); (b) redrawn from Manton and Ettl (1965); (c) drawn from description of Manton (1975); (d) drawn from descriptions by Manton and Parke (1960); (e) drawn from micrographs published by Pickett-Heaps and Ott (1974); (f) drawn from descriptions and micrographs of Mattox and Stewart (1977) and Moestrup and Ettl (1979); (g) redrawn from various sources; (h,i) redrawn from Schiller (1925); (j) redrawn from Fritsch (1935); (k) redrawn from Bourrelly (1973); (l) drawn from description of Stewart *et al.* (1974); (m) redrawn from various sources.

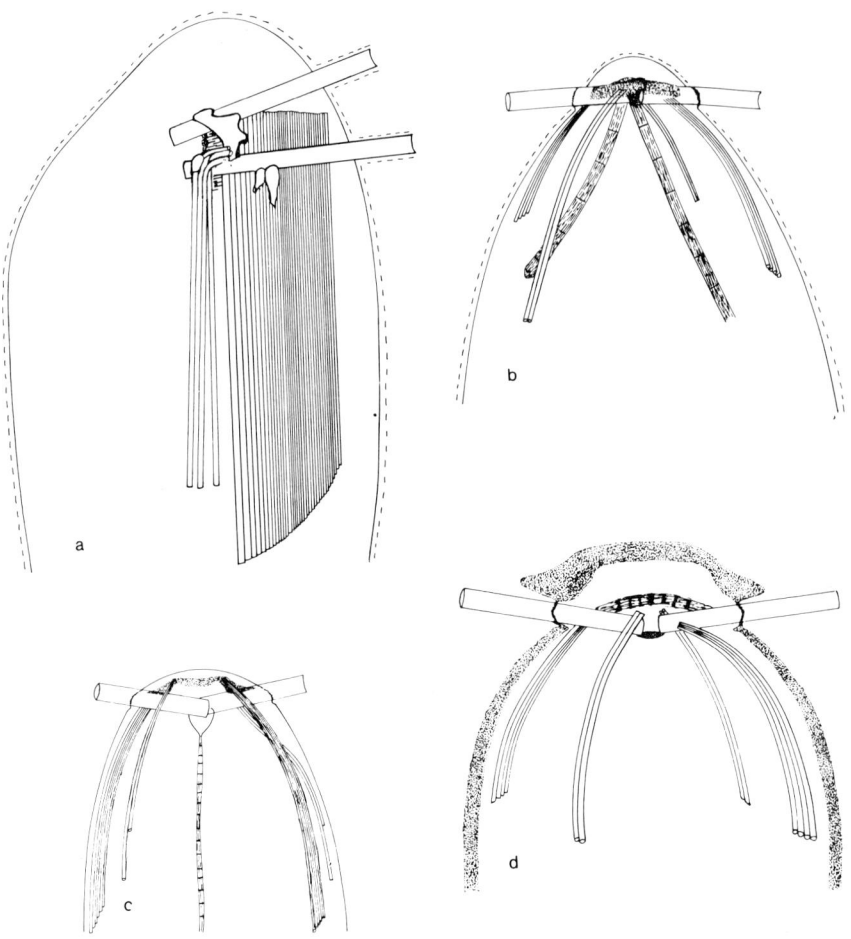

Fig. 2. Schematic diagrams of the side view of swarmers produced by the four classes of green algae that include advanced genera. (a) Charophyceae: scaly cell with one large root (the MLS), one smaller root, and flagella extending from the point of insertion to the right of the cell. (b) Ulvophyceae: four microtubular roots (two of each kind) in a cruciate arrangement and a scaly covering. (c) Pleurastrophyceae: cruciate roots and no cell covering. (d) Chlorophyceae: cruciate roots and covered by a theca.

non-cruciate root system more like that of other flagellate groups than like the majority of green algae.

The argument given for green algae in the last paragraph can be supported by analogy with the situation in the prymnesiophytes, the other "exceptional" group. Many genera of prymnesiophytes approach an external radial symmetry and have anteriorly inserted, isokont flagella, reminis-

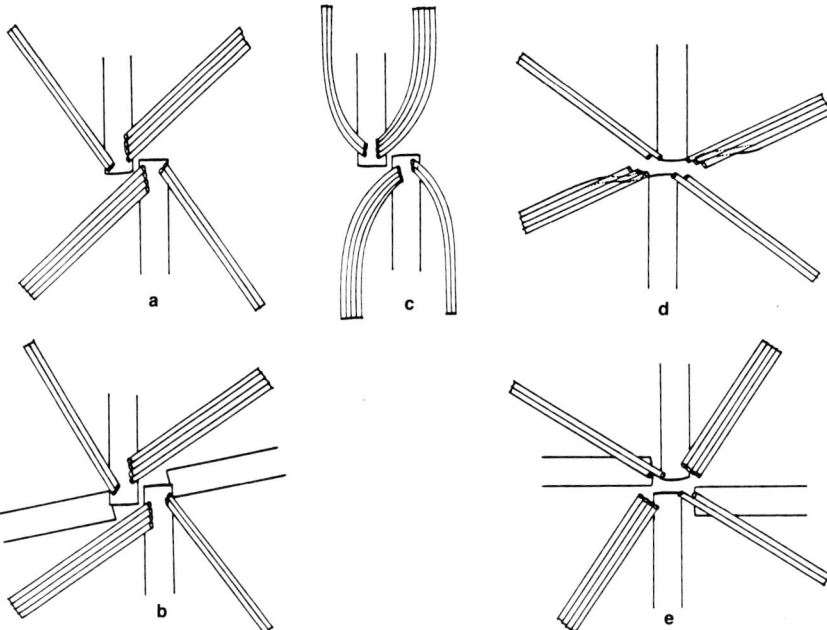

Fig. 3. Schematic diagrams of the arrangement of flagellar basal bodies and microtubular roots as viewed from the anterior in swarmers produced by the three classes of green algae that have cruciate microtubular root systems and advanced genera. (a,b) Ulvophyceae: bi- and quadriflagellate swarmers; the flagellar basal bodies overlap in the biflagellate cell (a); when turned so that one of the larger roots is in the upper right, the distal ends of the basal bodies in the biflagellate cell lie in 11 and 5 o'clock positions (counterclockwise orientation); in the quadriflagellate cell (b) the two basal bodies to which the roots are attached overlap and show the same counterclockwise orientation. (c) Pleurastrophyceae: biflagellate swarmers only; basal bodies show same counterclockwise orientation as the Ulvophyceae; roots become nearly parallel after emerging from basal bodies and are restricted to two sides of the swarmer. (d,e) Chlorophyceae: bi- and quadriflagellate swarmers; when turned so that one of the larger roots is in the upper right, the flagella are at least sightly offset so that the distal ends of the basal bodies in the biflagellate cell (d) lie in 1 and 7 o'clock positions (clockwise orientation); in the quadriflagellate cell (e) the two basal bodies to which the roots are attached show the same clockwise orientation.

cent of many green flagellates. Prymnesiophytes of the order Pavlovales, however, are asymmetric and have anisokont flagella laterally inserted in a depression. Thus, the prymnesiophytes and green flagellates are analogous in having flagellates of very different appearance in the same group. Since the Pavlovales have a cell form widespread in other groups, it is likely that they are the most primitive prymnesiophytes and that the near symmetry of some isokont prymnesiophytes is derived. Whether the isokont prymnesiophytes attained their apparent symmetry by doubling of part of the flagellar apparatus, as proposed for the green algal cruciate root system, or by some other modification of the cytoskeleton, cannot be evaluated at the moment, because a complete reconstruction of the flagellar apparatus of an isokont prymnesiophyte has not been published. It seems likely that all flagellate algal groups evolved from zooflagellates (see Stewart and Mattox, 1980), and if so, the primitive status of the Pavlovales within the Prymnesiophyceae is further attested to by the fact that a protozoan exists that is related to the Prymnesiophyceae, and it has symmetry and flagellar insertion like that of the Pavlovales. The ancestor of the Prymnesiophyceae was very likely similar to the unusual zooflagellate *Colponema*. Mignot and Brugerolle (1975) have shown that *Colponema* has peripheral alveoli, hairs of distinctive type (i.e. not similar to tubular mastigonemes) on only one of the two flagella, and a "star" (Fig. 7 in Mignot and Brugerolle, 1975) in the flagellar transition. All of those features can be found in the Pavlovales except the "star", which has nevertheless been observed in another prymnesiophyte order (Manton, 1964). The type of flagellar hair found in *Colponema* and its occurrence on only one of the flagella are known only in that organism and in some Pavlovales (e.g. see Gayral and Fresnel, 1979). The "star" in *Colponema* and that found in a prymnesiophyte are similar to each other, with both slightly different from the green algal "star". In addition, the external form of *Colponema,* with the flagella attached in a subapical channel, matches rather perfectly with that of *Pavlova*. Finally, the structure of the flagellar transition region of *Colponema,* as seen in longitudinal section, has some similarities to that of the Prymnesiophyceae (see Hibberd, 1980). There can be little doubt that *Colponema* is closely related to the prymnesiophytes, and if the zoological and botanical systems were not separate, they would probably be classified in the same group. Hibberd (1976) has emphasized that the Prymnesiophyceae are very different from the Chrysophyceae. We think that the difference is due to the fact that the prymnesiophytes originated from a *Colponema*-like zooflagellate while Chrysophyceae probably arose from a very different kind of protozoan, which might have been similar to *Bicosoeca,* whose fine structure has been described by Mignot (1974) and Moestrup and Thomsen (1976). The origin of the Chlorophyta seems likely to have had some analogous similarities to that of the Prymnesiophyceae, but unfortunately there is no known zooflagellate that can be as closely linked to the Chlorophyta as *Colponema* to the Prymnesiophyceae.

One of the microtubular flagellar roots of the AGF was apparently of the type that is called a "multilayered structure" (MLS). The MLS has distinctive lamellar structures associated with the proximal end of a band of microtubules. This root has survived in the flagellated male gametes of the

Embryophyta and is one of the characteristics that link charophycean green algae to the origin of land plants. Among green flagellates the MLS has been reported only in *Mesostigma* (Rogers et al., 1981). *Mesostigma* has two MLSs, but swarmers of the Charophyceae have only one. The reason that an MLS is suspected to have been present in the AGF is that an MLS occurs in some zooflagellates (see Rogers et al., 1981). In each case the zooflagellate has only one MLS. This is further evidence that many green flagellates and all the swarmers of higher green algae except those of the presently recognized Charophyceae (i.e., all of those with cruciate root systems) are "doubled cells." It might also indicate that the swarmers of the Charophyceae retain more of the features of the AGF than any green flagellate yet studied in detail. Melkonian has found some very interesting and important structural evidence (personal communication) that the smaller root of the widespread cruciate root system is a homologue of the MLS. He provides further discussion of the MLS here (this volume, Chapter 3).

The AGF probably had a pit or at least a groove or depression at the point of flagellar insertion. Pits or grooves occur in scaly green flagellates that can be considered primitive for other reasons and are common among both zooflagellate and other phytoflagellate groups. The presence of deep anterior flagellar pits in some of the better known green flagellates with scaly flagella [e.g. *Pyramimonas* and *Tetraselmis* (=*Platymonas*), see Fig. 1b, g, l] has been used as one of the characteristics to separate scaly green flagellates as a separate class, Prasinophyceae, from volvocalean flagellates, which tend to have flagella that project from an anterior protrusion called the papilla. Although a papilla might at first seem to be very different from a pit, there is some reason to believe that a papilla, at least in the Chlorophyceae, is a rather direct modification of a pit (Fig. 4). In green flagellates with scaly flagella and deep pits, the flagella lie parallel to each other and project in the anterior direction (Fig. 4a). The naked green flagellate *Hafniomonas* (Fig. 4b) has a pit that appears to be intermediate in structure between pits like those of *Tetraselmis* (Fig. 1l) and large papillas of the kind that occur in some species of *Carteria*, (Figs. 1k, 4c). *Hafniomonas* was included in the genus *Pyramimonas* until recently, when Ettl and Moestrup (1980) described its fine structure and showed that it possesses a flagellar apparatus very much like that of some species of *Carteria*, as described by Lembi (1975). Thus, *Hafniomonas*, which is described as having a pit, must be fairly closely related to *Carteria*, whose species sometimes have very large papillas. Proposed stages in the evolution of the papilla in the Chlorophyceae, as illustrated in Fig. 4, involve the "folding down" of flagella so that they lie in clefts along the pit wall, as they do in *Hafniomonas* (Fig. 4b). The evolutionary fusion of the pit lobes above the flagella would produce a papilla of the type shown for a *Carteria* species in Fig. 4c. Another line of evidence

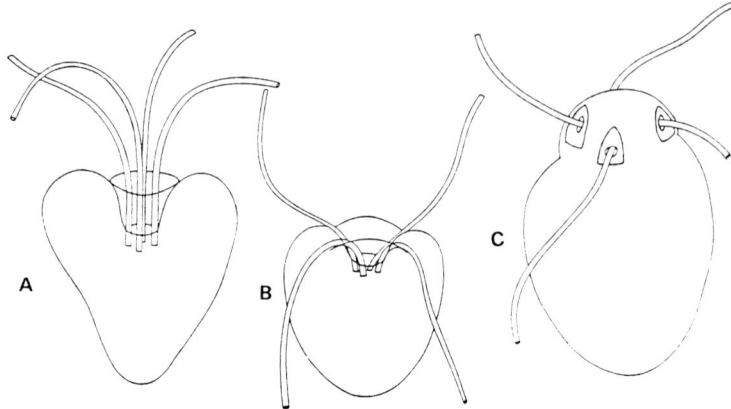

Fig. 4. Schematic drawings of three genera of green flagellates that show a possible evolutionary origin of the papilla in the Chlorophyceae. (A) A flagellate of the *Pyramimonas* or *Tetraselmis* type that possesses the apparently primitive features of a flagellar pit and parallel basal bodies so that flagella extend directly toward the anterior. (B) *Hafniomonas*, in which the flagellar pit is still present, but the basal bodies lie at an angle to each other so that the flagella lie along the clefts in the sides of the flagellar pit. (C) *Carteria* sp. in which the flagella emerge from the sides of an anterior papilla. Sources: (A) Drawn from various sources; (B) descriptions and illustrations of Ettl and Moestrup (1980); (C) redrawn from Bourrelly (1973).

indicating the homology of pits and papillas is that when a papilla is formed in *Carteria crucifera* (Domozych et al., 1981d), a system of microtubules reminiscent of those that line the pit in *Pyramimonas* supports the cytoplasmic bulge that molds the developing theca into a papilla. If the evolution of the papilla proceeded in the manner described here, it follows that some volvocalean flagellates with very small papillas might be evolutionarily farther removed from pitted forms than those that have large papillas. It seems that a structure similar to a papilla would result if the sides of a pit simply became reduced, and so the evolution of the chlorophycean papilla, as proposed here, might be somewhat different from the origin of the papilla in the Pleurastrophyceae and Ulvophyceae of this classification. The patterns of variation between scaly green flagellates and volvocalean flagellates do not provide sharp breaks, and the separation of the Prasinophyceae along the traditional lines is not possible or desirable.

Although there are many other aspects of evolution of green flagellates that could be discussed, the only other one that seems necessary here is the orientation of flagellar basal bodies in biflagellate cells. In those advanced green algae that have swarmers with a cruciate root system, biflagellate swarmers with offset or overlapping basal bodies have the basal bodies positioned in a manner so that the distal ends of the basal bodies lie in "11 o'clock" and "5 o'clock" positions as viewed from the anterior of the cell (the ulvophycean type; see Fig. 3a, b) or else they lie in the "1 o'clock" and "7 o'clock" positions (the chlorophycean type: see Fig. 3d, e). These points have only recently become clear from the application of techniques to distinguish actual from mirror-image arrangements as described by Floyd et al. (1980). Literature describing the two different arrangements and their correspondence to taxonomic categories is very recent (Hoops et al., 1982; Roberts et al., 1982; Melkonian and Berns, 1983; O'Kelly and Floyd, 1983; M. Melkonian and H. Preisig, personal communication). The 1/7 type occurs in flagellates and swarmers of the Chlorophyceae, whereas the 11/5 arrangement is characteristic of the Charophyceae and Ulvophyceae of this classification. O'Kelly and Floyd (1983, 1984) have recommended that the terms "counterclockwise absolute orientation" and "clockwise absolute orientation" replace the 11/5 and 1/7 o'clock descriptions, respectively, and have applied those terms to quadriflagellate as well as biflagellate cells. In a well-documented discussion (1984), they provided evidence that counterclockwise absolute orientation is the more primitive and that clockwise absolute orientation is associated with the origin of the Chlorophyceae. They have also provided convincing explanations of how the clockwise rotation of basal body arrangement could have occurred through evolutionary alteration of developmental stages to give rise to the clockwise orientation of the Chlorophyceae. They showed that some chlorophycean algae are evolutionary intermediates in that the basal bodies are almost directly opposite each other. Even clockwise rotation to that degree, however, indicates that an alga is chlorophycean, because all ulvophycean algae have a distinct counterclockwise orientation. The work of O'Kelly and Floyd seems likely to have a lasting influence on concepts of green algal phylogeny and classification. We therefore support their recommendations on terminology, because they express an important evolutionary concept as well as provide a widely applicable notation. However, the terms are shortened to "counterclockwise orientation" (CCW) and "clockwise orientation" (CW) for the remainder of this paper. These matters and other aspects of the flagellar apparatus are more fully discussed by Melkonian (this volume, Chapter 3).

Since the phycoplast occurs both in green algae with CCW orientation (the Pleurastrophyceae of this classification) and in those with CW orientation (the Chlorophyceae), it seems clear that the phycoplast evolved before the evolutionary rotation of basal body arrangement and is to be associated more specifically with the evolution of the theca than with the origin of the Chlorophyceae (see Mattox and Stewart, 1977). There appears to be no evidence for a multiple origin of the phycoplast and theca, nor is a multiple origin necessary to explain the characteristics of any group of green algae.

A CLASSIFICATION OF THE GREEN ALGAE

The classification outlined in Table I assigns advanced green algae to four classes. The term "advanced" is used here to designate green algae whose vegetative state is non-motile and non-flagellated. Green flagellates that are obviously related to advanced green algae are placed in the same class with them. Those green flagellates that do not appear to be closely related to the four classes containing advanced green algae are placed in a fifth class composed of flagellates only. Sedentary or sessile genera that still retain flagellate characteristics (e.g. an eyespot, a theca, or pseudocilia) are treated as flagellates.

In designing the classification we were guided by the conviction that a subgroup of the Chlorophyta should be regarded as a class if its characteristics clearly indicate that it evolved from a green flagellate that was different in important respects from the green flagellate ancestors of other subgroups. It is an important feature of the classification that the main characteristics used to define the classes that include advanced green algae are characteristics possessed as well by green flagellates. The structure of the flagellar apparatus is considered to provide features of fundamental importance, but other features that reflect basic differences in flagellate organization and development, mainly the nature of cell division (phycoplast or persistent interzonal mitotic spindle) and the nature of the cell covering of flagellated stages (scales or theca), are also major differences between the flagellate ancestors of higher green algae and between the groups of higher green algae themselves. As discussed above, the evolution of the phycoplast and theca apparently preceded evolutionary changes in the flagellar apparatus that led to the CW orientation of basal bodies in the Chlorophyceae. In an evolutionary classification, earlier divergences will be more important in defining the classes of a division than later ones. For that reason, it is important to use all the relevant characteristics of green flagellates and avoid basing the classification solely on the structure of the flagellar apparatus. Using the structure of the basic motile cell type for class-level distinctions is not only in accord with modern practice for other algal divisions, but seems to be appropriate in that evolutionary diversification among the flagellates of a division that led to diversity in advanced genera must represent some of the earliest and most fundamental divergences detectable in the division. Figure 5 represents the presumed evolutionary events among green flagellates that led to the origin of the four classes that include advanced genera. Primitive green flagellates with persistent interzonal mitotic spindles (the Micromonadophyceae of this classification) are not included in Fig. 5 since some evolutionary events among these flagellates were probably not along lines leading to advanced green algae, and the information available for

Table I. Classification of the green algae.

I. MICROMONADOPHYCEAE

Scaly or naked flagellates with interzonal mitotic spindles that are persistent during cytokinesis.

II. CHAROPHYCEAE

Motile cells, if produced, scaly or naked and asymmetric with laterally or subapically inserted flagella that extend to the right of the cell from the point of insertion; motile cells always biflagellate and without an eyespot; microtubular flagellar root system with one multilayered structure and usually with one additional, smaller microtubular root; rhizoplasts unknown; interzonal mitotic spindle persistent during cytokinesis; chromosomal microtubules often persistent at telophase; little or no shortening of chromosome-to-mitotic spindle pole microtubules during anaphase except in parenchymatous forms; predominantly freshwater; sexual reproduction always involves the production of a dormant zygote and zygotic meiosis; levels of organization are unicellular, sarcinoid, filamentous, and parenchymatous.

A. Chlorokybales—Characteristics in addition to those of the Charophyceae: Sarcinoid; plasmodesmata absent; zoospores scaly; zoospore release by disintegration of sarcinoid packet.

B. Klebsormidiales—Characteristics in addition to those of the Charophyceae: unbranched filaments without holdfasts; plasmodesmata absent; zoospores naked and released through pore in wall.

  1. Klebsormidiaceae—Characteristics of the order.

C. Zygnematales—Characteristics in addition to those of the Charophyceae: unicells or unbranched filaments without holdfasts; plasmodesmata absent in filamentous forms; flagellated cells not produced; sexual reproduction by conjugation.

D. Coleochaetales—Characteristics in addition to those of the Charophyceae: branched filaments, discoid thalli, or unicells; sheathed setae present; oogamous; motile cells scaly.

  1. Chaetosphaeridiaceae—Eggs are released; filamentous or unicellular; sympodial growth when filamentous.

  2. Coleochaetaceae—Eggs retained in oogonium; branched filaments or discoid thalli; growth by division of terminal cells of filaments or by division of marginal cells in discoid forms; plasmodesmata present; phragmoplast-cell plate develops during cytokinesis, and the microtubules of the phragmoplast proliferate from the persistent interzonal spindle characteristic of the class; reproductive cells sometimes surrounded by sterile cells.

E. Charales—Characteristics in addition to those of the Charophyceae: complex plant body with apical growth and differentiation into nodes and internodal meristematic regions parenchymatous; oogamous; sterile cells surround antheridia and oogonia; plasmodesmata present; male gametes scaly; zoospores not produced; a phragmoplast-cell plate develops during cytokinesis, and the microtubules of the phragmoplast proliferate from the persistent interzonal spindle characteristic of the class.

III. ULVOPHYCEAE

Motile cells, if produced, scaly or naked and with apically attached flagella and near-radial symmetry externally; motile cells biflagellate, quadriflagellate, or multiflagellate; microtubular

(cont.)

Table I (cont.)

flagellar root system cruciate; rhizoplasts often present; arrangement of basal bodies shows counterclockwise orientation; the two larger of the microtubular flagellar roots tend to end in a line between basal bodies whereas the two smaller ones are more offset and attach to the outside of basal bodies; interzonal mitotic spindle persistent during cytokinesis; little shortening of chromosome-to-mitotic spindle pole microtubules during anaphase in many cases; predominantly marine, but a number of genera have freshwater as well as marine species; alternation of generations (sporic meiosis) common, and dormant zygotes are not known; levels of organization are sarcinoid, filamentous, thalloid, parenchymatous, coenocytic, and siphonous.

IV. PLEURASTROPHYCEAE

Motile cells are thecate or naked and more or less flattened; microtubular flagellar root system cruciate; rhizoplasts present although sometimes thin and delicate; arrangement of basal bodies shows counterclockwise orientation; mitotic spindle always metacentric; interzonal mitotic spindle collapses at telophase and a phycoplast develops; chromosome-to-mitotic spindle pole microtubules shorten during anaphase; sexual reproduction not known. The only genus of flagellates assigned to the class is thecate, has four flagella inserted in a flagellar pit, and possesses basal bodies parallel to each other and to the long axis of the cell. Swarmers produced by advanced genera lack a pit and have two flagella that are "bent down" to lie at an angle to the long axis of the cell during forward swimming mode; the two larger microtubular roots of the swarmers pass between the basal bodies while the two smaller ones pass to the outside of the basal bodies; phycoplasts of advanced genera are of special type that includes microtubules that cup the telophase nuclei as well as those that parallel the cleavage plane; plasmodesmata absent. All advanced genera are freshwater; levels of organization are flagellate, coccoid, sarcinoid, and filamentous.

A. Tetraselmidales—Characteristics in addition to those of the Pleurastrophyceae: flagellates only; mitotic spindle develops from rhizoplasts; pyrenoid present, with nuclear or cytoplasmic invaginations; one genus known.

B. Pleurastrales—Characteristics in addition to those of the Pleuarastrophyceae: coccoid, sarcinoid, or filamentous; pyrenoid often indistinct or lacking.

V. CHLOROPHYCEAE

Flagellates and swarmers, if produced, thecate or naked and with apically attached flagella and radial or near-radial symmetry externally; motile cells biflagellate, quadriflagellate, or multiflagellate; microtubular flagellar root system cruciate; rhizoplasts sometimes present in flagellates, but unknown in swarmers; basal bodies arranged so that they show clockwise orientation; the two smaller of the microtubular flagellar roots tend to end in a line between the basal bodies whereas the two larger ones are more offset and attach to outside of basal bodies; interzonal mitotic spindle collapses at telophase and a phycoplast develops; centrioles lie at or just lateral to the spindle poles; spindle never metacentric; chromosome-to-mitotic spindle pole microtubules always shorten during anaphase; predominantly freshwater; sexual reproduction nearly always involves the production of a dormant zygote and zygotic meiosis; levels of organization are flagellate, coccoid, non-motile colonial, motile colonial, sarcinoid, filamentous, and parenchymatous.

A. Chlamydomonadales—Characteristics in addition to those of the Chlorophyceae: flagellates and sedentary forms that possess flagellate characteristics (pseudocilia, a theca, an eyespot).

Table I (cont.)

B. Volvocales—Characteristics in addition to those of the Chlorophyceae: motile colonies.

C. Chlorococcales—Characteristics in addition to those of the Chlorophyceae: non-motile colonies and coccoid unicells that do not possess a theca around the vegetative stage when it is mature; phycoplast often radiates from centrioles if centrioles are present; multiple mitoses often occur before cytokinesis and development of phycoplasts.

D. Sphaeropleales—Characteristics in addition to those of the Chlorophyceae: unbranched filaments in which, during active growth, new walls are deposited inside the old filament wall. The new walls may surround the entire periphery of new cells or they may be "H-shaped" partial walls; uninucleate or multinucleate; plasmodesmata absent.

    1. Sphaeropleaceae—New wall surrounds entire surface of daughter cells; broken old walls may be "H-shaped" in section; pyrenoid with cytoplasmic invaginations; anisogamous or oogamous.

    2. Microsporaceae—New wall around daughter cells "H-shaped" in section to begin with; sexual reproduction unknown.

E. Chlorosarcinales—Characteristics in addition to those of the Chlorophyceae: filaments and sarcinoid forms; vegetative cell division occurs, but plasmodesmata are absent.

F. Chaetophorales—Characteristics in addition to those of the Chlorophyceae: branched or unbranched filaments or occasionally parenchymatous; motile cells biflagellate or quadriflagellate; cytokinesis in vegetative cell division accomplished by a phycoplast-associated cell plate; plasmodesmata present; centrioles do not lie in the plane of the phycoplast but on the opposite sides of the daughter nuclei.

    1. Chaetophoraceae—Branched or unbranched filaments; pyrenoids with appressed thylakoids; hairs or setae, when present, are multicellular; isogamous or anisogamous.

    2. Aphanochaetaceae—Branched filaments; pyrenoids penetrated shallowly by a few thylakoids, and cytoplasm often projects slightly into the pyrenoid; hairs or setae, when present, are usually unicellular; oogamous.

    3. Schizomeridaceae—Mature thallus parenchymatous throughout and unbranched; pyrenoids traversed by several undulating thylakoids; chloroplasts perforate; sexual reproduction unknown.

G. Oedogoniales—Characteristics in addition to those of the Chlorophyceae: branched or unbranched filaments; motile cells with a ring of numerous flagella; distinct type of cytokinesis and cell elongation involving migration of phycoplast-associated cell plate and extension of "wall rings"; plasmodesmata present; plasmodesmata of unique type having a "barrel shape" in longitudinal section and peripheral subunits; oogamous.

---

some genera of the Micromonadophyceae is not yet extensive enough even to attempt some basic interpretations (e.g. whether or not they possess a cruciate flagellar root system).

Earlier attempts to classify green algae on the basis of comparative fine structure and development (Stewart et al., 1973; Stewart and Mattox, 1975) included only a few genera. They were designed to explain progress in the field and were not recommended as general-use classifications. The classifi-

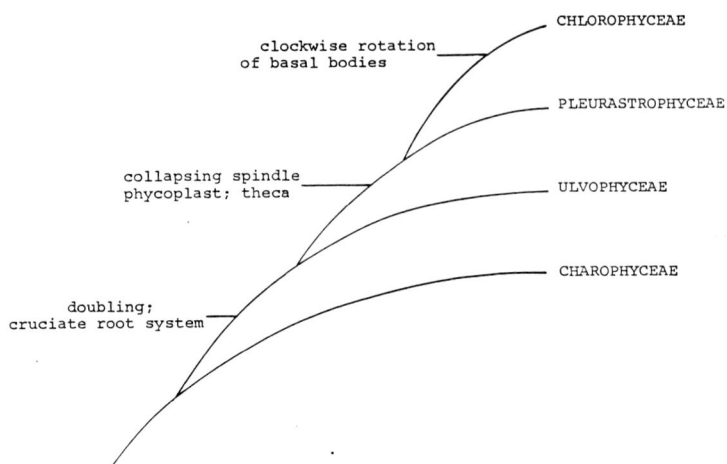

Fig. 5. A phylogenetic "tree" illustrating presumed evolutionary events among some green flagellates that led to differences in the flagellate ancestors of the four algal classes that include advanced green algal genera, Charophyceae, Ulvophyceae, Pleurastrophyceae, and Chlorophyceae. It is assumed that the earliest green flagellates were scale-covered and asymmetrical both externally and in their flagellar root system, which included only one of each kind of root. The swarmers of the Charophyceae are seen as a minimal change from that primitive condition. A doubling of one of the flagella of an early green flagellate, and also of the roots associated with it, gave rise to flagellates with a cruciate root system with two of each kind of root. Flagellates of this type gave rise to the Ulvophyceae with little additional change in organization. Among other flagellates with a cruciate system, the fusion or interweaving of scales gave rise to a theca, and the presence of a theca resulted in selection toward the phycoplast and collapsing interzonal spindle. Flagellates of this type still retained the primitive counterclockwise orientation of basal bodies and gave rise to the Pleurastrophyceae. In some flagellates of that type there was a further evolution from counterclockwise orientation to clockwise orientation in a manner described by O'Kelly and Floyd (1984b), and this gave rise to the flagellate ancestor of the Chlorophyceae. Flagellates of each type are known except a type that would fit in the most primitive position at the beginning of the tree, but there is no reason to think one does not exist (see Fig. 1h–j).

cation of 1975, however, has been used to select green algal genera for comparative biochemical investigation, because it was found to correspond more to the results of those investigations than did older classifications (e.g. Bekheet and Syrett, 1977; Lumsden et al., 1977). Such investigations exemplify the value of an improved classification of green algae. A classification that more closely reflects evolutionary relationships is not merely a theoretical advance for those whose main interest is taxonomy or phylogeny, but also a more useful framework for other fields. It seems that enough information has now been gathered to attempt a comprehensive class-level system for the green algae. That is our aim here, and we do recommend this classification for general use. We believe that the classification presented in this section is soundly based on analysis of characters and character correlations. The foregoing discussion on evolution of green flagellates provides a theoretical foundation for the classification, but the actual structural and developmental features that define the five proposed classes are distinctive enough to stand on their own as indicative of natural grouping, even if some evolutionary events among green flagellates have been misinterpreted. The classification is, of course, not without faults, and some are serious. Nevertheless, we think it is as good as can presently be devised. Its strengths and weaknesses are specifically discussed after presentation of the classification itself. When enough information is available to recommend changes in ordinal and familial classification, they are included in Table I. Orders and some families are given for Charophyceae, Chlorophyceae, and Pleurastrophyceae, but not for Micromonadophyceae and Ulvophyceae. For the latter two classes, other subgrouping into orders and families should continue to be used for the present. However, O'Kelly and Floyd (1983) and Floyd and O'Kelly (1984) are obtaining clear-cut results in their investigations within the Ulvophyceae, and these will doubtless lead eventually to a much more natural and soundly based system of orders and families for that class (see O'Kelly and Floyd, this volume, Chapter 4).

The remainder of this section is devoted to a discussion of each group listed in the classification of Table I. Although some phylogenetic ideas are presented, we wish to emphasize again that in each case an effort was made to base a group on distinct structural and developmental characteristics, and combinations thereof, that might be viewed as delimiting natural taxa without adherence to a specific evolutionary scheme. For example, if the scheme shown in Fig. 5 were strictly followed, a decision might be made to treat the Ulvophyceae, Pleurastrophyceae, and Chlorophyceae as subclasses of a single large class. Another alternative might be the combination of only the Pleurastrophyceae and Chlorophyceae into a single class. Aside from the drawback that those alternative treatments would be based as much on ideas as on facts, there is the practical problem that such classes would be difficult to refer to and discuss in a general way because of the diversity of organisms that would be included.

Available space allows only a few key genera or groups to be mentioned below as being included in a particular class, order, or family. Those mentioned, however, should allow botanists or protistologists to place unlisted genera. It should be understood that many genera that have not been carefully studied will be placed in the classification on the basis of their traditional alignment with genera that have been more thoroughly examined. In a number of cases (e.g. some of the many genera of the Chlorococcales, some flagellates, and some filamentous forms), future study could well result in the transfer of a genus from one group of this classification to another.

1. Micromonadophyceae

This class consists of green flagellates that exhibit primitive characteristics for many of the features used to define the other classes. They are scaly or naked. When naked, they are assumed to have lost scales. That assumption is based on the fact that scales are present in all the classes of green algae and in many "non-green" flagellate groups, and were therefore probably present in the earliest green flagellates. Furthermore, the few species of the Micromonadophyceae that lack scales are among the smallest forms (e.g. *Micromonas, Pedinomonas,* see Fig. 1d, e), which are reduced in other ways as well and whose structure demonstrates that they are very closely related to scaly members of the class (e.g. the similarity of the naked *Micromonas* to the scaly *Mantoniella;* see Fig. 1a and description by Barlow and Cattolico, 1980). Naked members of the class have persistent interzonal mitotic spindles (e.g. *Pedinomonas,* Pickett-Heaps and Ott, 1974), and this feature is strong evidence that they were never thecate since all scaly green flagellates have a persistent interzonal spindle, whereas all thecate flagellates have a collapsing interzonal spindle and a phycoplast.

The characteristics given in Table I clearly define the Micromonadophyceae for practical taxonomic purposes, but since they are primitive characteristics for all the green algae, they fall far short of the ideal for an evolutionary classification. The Micromonadophyceae are a "natural group" in the sense that they doubtless had a common ancestor, but that ancestor apparently gave rise to all the other green algae as well. Any group defined by primitive features only is subject to losing its members to other groups when new information provides the more satisfying and accurate alternative of defining groups on the basis of shared derived characters. It is likely, or even certain, that at least most of the genera of the Micromonadophyceae will prove to be more closely related to green algae of other groups than to each other; examination of Fig. 1a–j will give the reader some idea of the diversity of flagellates in this class. However, we strongly recommend the general use of the class, because there is presently insufficient information to justify more specific alignments, and it is a much clearer treatment than current alternatives. Much of the interest of the Micromonadophyceae lies, of course, in the fact that they represent a pool of primitive genera that are probably similar to forms from which higher green algae sprang and that might eventually yield evidence as to the origin of green algae. The Micromonadophyceae cannot be referred to as the "Prasinophyceae" for reasons to be discussed further in the next section.

Genera representative of those to be included in the class are *Nephroselmis* (=*Heteromastix*), *Micromonas*, *Pyramimonas*, *Mesostigma*, *Monomastix*, *Pterosperma*, *Dolichomastix*, *Pedinomonas*, *Trichloris*, *Cymbomonas*, *Scourfieldia*, etc.

2. Charophyceae

The swarmers of the Charophyceae are notable for their striking dissimilarity to swarmers of other advanced green algae and for their similarities to flagellated male gametes of bryophytes and vascular cryptogams. Although their distinctiveness became clear with the earliest electron microscopic examinations, the recent work of Sluiman (1983) is the first critically detailed reconstruction of a charophycean motile cell. Swarmers of the Charophyceae have a conspicuous and complex flagellar root that is called a multilayered structure (MLS, formerly referred to as the "driergruppe" or "viergruppe" in some land plants). In his work with the zoospore of *Coleochaete pulvinata*, Sluiman not only found some heretofore unsuspected peculiarities, such as the fact that the flagella extend to the right rather than directly out from their point of insertion (see Fig. 2a, drawn from descriptions and illustrations of Sluiman, 1983), but also demonstrated the presence of a second, small microtubular root in addition to the MLS. The small root is probably of rather general occurrence in the Charophyceae, but has been overlooked (see discussion by Sluiman).

The Charophyceae also possess a persistent interzonal mitotic spindle (Fig. 6a, b), a generalized or primitive feature for the green algae. In the most complex charophycean algae, *Coleochaete* and *Chara*, there is a phragmoplast and cell plate that is similar to those structures in land plants and that apparently evolved in part from the proliferation of the persistent interzonal spindle (Pickett-Heaps, 1972; Marchant and Pickett-Heaps, 1973). Although the characteristics of swarmer structure and cell division that define the Charophyceae are primitive for the green algae, there seems to be no reason to doubt that the class is monophyletic. The asymmetry of charophycean swarmers and the presence of only one of each kind of flagellar root were probably features of the ancestral green flagellate, but it also seems likely that the ancestral green flagellate possessed basal bodies and roots not found in the Charophyceae (see section on phylogeny). If so, the loss of those features might be shown to be the derived feature shared by charophycean algae after additional studies of asymmetrical green flagellates. A further indication of monophylety is the fact that characters not

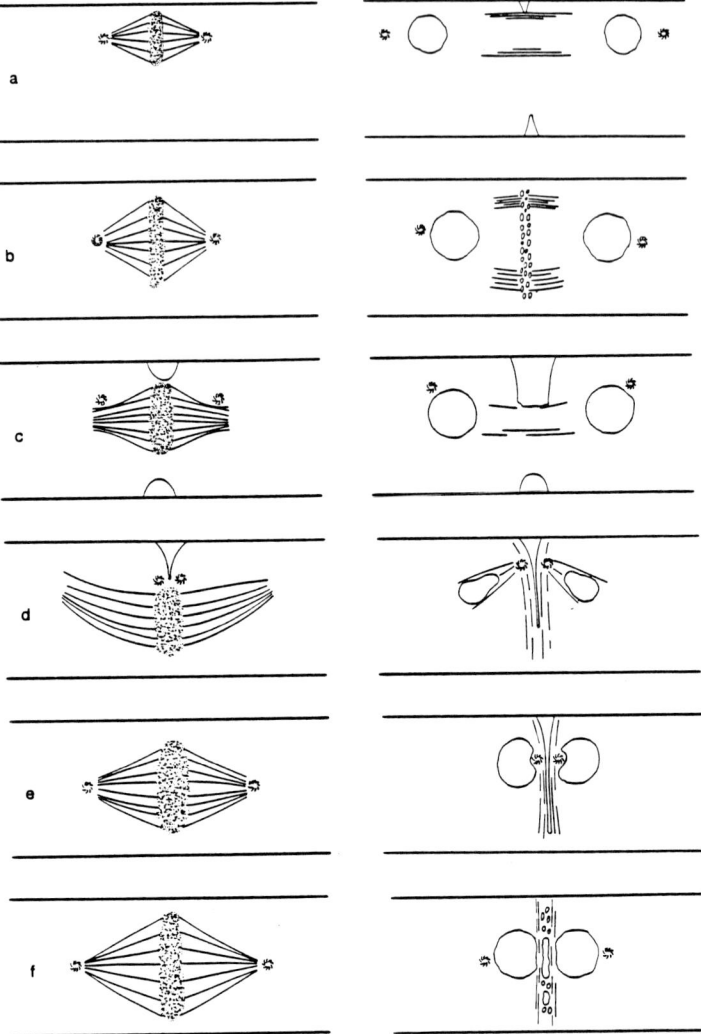

Fig. 6. Metaphase and telophase stages representative of the Charophyceae, Ulvophyceae, Pleurastrophyceae, and Chlorophyceae. (a) Charophyceae: telophase drawing shows persistent interzonal spindle characteristic of orders without cell plate formation and plasmodesmata. (b) Charophyceae: telophase drawing of more advanced orders with plasmodesmata (Coleochaetales and Charales) in which a phragmoplast and cell plate develop from the persistent interzonal spindle. (c) Ulvophyceae: the metaphase drawing shows the usual position of the centrioles lateral to the spindle poles in cellular Ulvophyceae, and the telophase drawing has a cleavage furrowing approaching the persistent interzonal spindle as occurs in cellular genera. (d) Pleurastrophyceae: the centrioles lie in the same plane as the chromosomes at metaphase (a metacentric

originally used to form the present concept of the Charophyceae have turned out to be uniform for the class, thus providing supportive correlations. Examples are the universal lack of eyespots in the Charophyceae, the extension of the flagella from the point of insertion to the right of the cell, and the fact that all the orders, families, and genera appear to be predominantly freshwater organisms.

The concept of the Charophyceae given here is not changed appreciably from that of Stewart and Mattox (1975) except for the inclusion of a few more genera and the establishment of an additional order. Genera to be included are mentioned below with the discussions of orders and families.

(a) Chlorokybales. This order was proposed by Rogers et al. (1980) to include the genus *Chlorokybus*. No additional genera have yet been discovered. *Chlorokybus* is the only known charophyte with sarcinoid organization and is possibly the most primitive genus known for the class.

(b) Klebsormidiales. This order was established in 1975 by Stewart and Mattox. It is construed here to include *Raphidonema, Stichococcus* (Pickett-Heaps, 1974, 1976; Chappell and Floyd, 1981), and *Klebsormidium*. Some species presently included in the genus *Ulothrix* (assigned to the class Ulvophyceae of this classification) will probably have to be transferred to the Klebsormidiales (G. M. Lokhorst, personal communication). Other filamentous genera not yet examined with the electron microscope but that can be placed here on the basis of subapical insertion of flagella and the lack of eyespots are *Hormidiella* and *Microsporopsis* (see Silva, 1982).

(c) Zygnematales. The Zygnematales are a well-established order, and comparative fine structural studies support their recognition as a natural group. The Zygnematales are so clearly charophycean from consideration of other characteristics such as biochemical features and development during cell division that the lack of flagellated cells poses no difficulty in determining the position of the Zygnematales in this classification. The Zyg-

---

spindle); at telophase the interzonal spindle collapses, and (as shown in the drawing) phycoplast microtubules line the cleavage furrow while other microtubules originating at the same place as the phycoplast "cup" the daughter nuclei. (e) Chlorophyceae: the metaphase spindle is more usual than that of the Pleurastrophyceae; after collapse of the interzonal spindle at telophase, the nuclei and associated centrioles rotate so that the centrioles lie in the cleavage plane at a site where phycoplast microtubules originate. (f) Chlorophyceae: this drawing of telophase shows a difference seen in advanced chlorophycean orders with cell plates and plasmodesmata (Chaetophorales and Oedogoniales); the telophase nuclei do not rotate and when centrioles are present, they do not lie in the plane of the phycoplast. Nuclear envelopes are not included in any of the metaphase drawings. Nuclear envelopes are usually present at metaphase in the Chlorophyceae and Ulvophyceae, usually dispersed in the Charophyceae, and present but highly vesiculated in the Pleurastrophyceae.

nematales also appear to be closely related to the Klebsormidiales (Pickett-Heaps, 1972; Rogers et al., 1980). Genera to be included are those of current classifications and can be assigned to families already established.

(d) Coleochaetales. Further discussion of this order is given below with the treatment of the families.

Chaetosphaeridiaceae: A detailed study of growth and reproduction of a species of *Chaetosphaeridium* was made by Thompson (1969), who showed that it is filamentous and that sexual reproduction is oogamous. Growth of the filament is by an unusual "sympodial" pattern. Whether *Chaetosphaeridium* or any other of the genera of this family is unicellular, as usually reported, remains to be seen. Moestrup (1974) has studied the structure of the zoospore of *Chaetosphaeridium*, and it appears to be closely similar to that of *Coleochaete*. Another genus that can be included at present is *Conochaete*, but it has not yet been studied by modern methods.

Coleochaetaceae: It seems desirable, at least for now, to separate the Coleochaetaceae from the Chaetosphaeridiaceae. The extent of similarities or differences between the two families must remain uncertain until the vegetative stages of *Chaetosphaeridium* are studied with the electron microscope. The Coleochaetaceae have received much attention lately because of their apparent close relationship to the ancestry of land plants (see Stewart and Mattox, 1975; Graham, 1982a; Graham and Wilcox, 1983). Graham's work has provided some exciting new information and ideas concerning the phylogenetic position of *Coleochaete*. The recent work of Sluiman (1983) gives detailed information for the zoospore of *Coleochaete pulvinata*. Mitosis and phragmoplast development have been described by Marchant and Pickett-Heaps (1973). *Coleochaete* is presently the only genus assigned to the family.

(e) Charales. Until the earlier informal rearrangement of green algae (Stewart and Mattox, 1975), the Charales had been the only order included in the Charophyceae. Those botanists not involved with comparative fine structure have continued to use the Charophyceae in that restricted sense. Enough information is now available that the continued separation of the Charales in a class distinct from other green algae is not desirable in that it obscures relationships and is contrary to the usual utilization of motile cell structure for the definition of algal classes. In the past it was difficult or impossible to perceive the nature of the kinship between the Charales and simpler green algae, and so the separation of the Charales was prudent. While it is still true that the Charales have attained a vegetative complexity unique in the green algae, the evolutionary beginnings of that complexity are seen in the Coleochaetales, where changes more fundamental than those distinguishing the Charales and Coleochaetales might have occurred. The Charales have apical growth and parenchymatous tissue, but the Coleochaetales can also be parenchymatous (Graham, 1982a), and their terminal

and marginal growth (see Table I) probably reveals the origin of apical growth. Simpler and probably earlier kinds of the covering of reproductive cells by sterile cells, as occurs in the Charales, can be seen in the Coleochaetales. The phragmoplast/cell plate and plasmodesmata are similar to those of embryophytes and occur in *Coleochaete* as well as the Charales. It seems clear that the complexities of the Charales represent elaborations and refinements of more basic evolutionary innovations in the ancestors of the Coleochaetales. Finally, it was always obvious that the Charales must be related in some way to other green algae; the only difference now is that the simpler algae related to the Charales can be identified. There are no greater conceptual difficulties involved in placing the Charales in the same class with the Coleochaetales than in placing the Laminariales or Fucales in the same class with the Ectocarpales.

Current subdivision of the Charales into families and genera has not been affected by electron microscopic studies.

3. Ulvophyceae

This class was first tentatively proposed (Stewart and Mattox, 1978, Fig. 3) when it became clear that a large group of green algae possess a cruciate flagellar root system similar to that of the Chlorophyceae, but have a persistent interzonal mitotic spindle and occasionally scaly zoospores like the Charophyceae. Earlier (Stewart and Mattox, 1975) it had been supposed that the mitotic spindle of these algae collapsed at telophase, as in the Chlorophyceae, even though a phycoplast did not develop. Interpretation was made difficult by the development of a thick cleavage furrow in those studied earliest (Mattox and Stewart, 1974). However, Fig. 14 in that paper showed the impingement of the cleavage furrow on a persistent interzonal spindle, and Sluiman *et al.* (1983) have since shown that another alga of this type, *Ulothrix zonata,* also has a persistent interzonal spindle. When the class name was first used it was spelled "Ulvaphyceae" and that spelling was continued on numerous later occasions. Other workers began to spell the name with an "o". Although a good reason has not been given for the change in spelling, we will use the new spelling here in order to prevent unnecessary confusion.

The main structural features by which the Ulvophyceae differ from the Charophyceae are the striking dissimilarities in the flagellar apparatus of swarmers (compare Fig. 2a with Fig. 2b). The Charophyceae have one each of two different microtubular flagellar roots in an asymmetrical cell with subapically inserted flagella while the Ulvophyceae have two each of two different roots (a cruciate root system, 180° rotational symmetry) in a cell with apically inserted flagella and a near radial symmetry externally. Since comparable differences are also known among flagellates of the Micromonadophyceae, a separate origin of the two classes from different kinds of green flagellates is indicated (see also Fig. 5). Otherwise, the Charophyceae and Ulvophyceae share the apparently primitive features of a scaly covering on swarmers and an interzonal spindle that persists during cytokinesis in cellular forms. A number of ulvophycean genera produce swarmers with scaly coverings (see Mattox and Stewart, 1973), indicating that the class had a scaly ancestor, but many also have naked swarmers, apparently having lost scales. It seems unnecessary to elaborate further on the structural distinctions of the Ulvophyceae since they have been reviewed earlier by Melkonian (1980) and Hoops et al. (1982) and are discussed further by O'Kelly and Floyd (this volume, Chapter 4) and by Roberts (this volume, Chapter 13).

The Ulvophyceae also differ from the Charophyceae and from the other two classes that include advanced genera, Pleurastrophyceae and Chlorophyceae, in being predominantly marine and including many genera with alternation of generations. Reports of alternation of generations in chlorophycean algae seem highly doubtful, with a single possible exception being that of *Draparnaldiopsis* (Singh, 1942). There are some good reasons for believing that the Ulvophyceae had a marine origin while the Charophyceae, Pleurastrophyceae, and Chlorophyceae had freshwater origins. A very large majority of the genera of the Ulvophyceae is either completely marine or has marine as well as freshwater species. Exceptions are few and include *Dichotomosiphon, Trichosarcina,* and the genera in Trentepohliaceae. All filamentous marine green algae or larger green seaweeds studied ultrastructurally have been shown to be ulvophycean; none is known to be charophycean, pleurastrophycean, or chlorophycean. Additionally, no ulvophycean alga is known to produce the thick-walled, dormant zygote often considered typical of sexual reproduction in freshwater green algae. The wide occurrence of alternation of generations in the Ulvophyceae might be due to the possibility that the more stable marine environment fosters the evolution of longer life cycles and alternation of generations. It seems possible that alternation of generations is a primitive condition within the Ulvophyceae, and those that lack it have lost or modified it. An inescapable conclusion is that freshwater ulvophycean algae have invaded fresh water secondarily. An interesting case is the apparent loss of alternation of generations in some freshwater species of *Cladophora* (see Graham, 1982b, and references given there). This suggests that alternation of generations is not particularly advantageous in freshwater habitats.

The matters discussed in the foregoing paragraph are very significant because they show how a concept of the Ulvophyceae that was based on fine structural characteristics has revealed correlations in ecology and life cycles. No previous classification based on levels of organization had ever provided the slightest hint that the marine habitat or alternation of generations might correspond to such a large extent with a class-level treatment of all the green algae.

Some green algae to be included in the Ulvophyceae are all of the coenocytic or siphonous forms except coenocytic members of the Chlorococcales and genera of the Sphaeropleaceae. Also to be included are: marine species of *Ulothrix* and those freshwater species of *Ulothrix* that have codiolum stages or other ulvophycean features; any genus that produces a codiolum stage (Sluiman et al., 1982; Floyd and O'Kelly, 1984); all marine branched filaments that were formerly classified in the Chaetophorales (O'Kelly and Yarish, 1980; O'Kelly and Floyd, 1983; Floyd and O'Kelly, 1984); genera usually included in the Ulvales except for *Schizomeris* (Mattox et al., 1974; Mattox and Stewart, 1974); the Trentepohliaceae (Chapman and Henk, 1983; Chapman, this volume, Chapter 8; Roberts, this volume, Chapter 13); *Pseudendoclonium; Trichosarcina; Pseudendocloniopsis*, a sarcinoid form (unpublished); *Ctenocladus;* and *Smithsoniella* (Sears and Brawley, 1982). Improvements to the classification of the Ulvophyceae are provided by O'Kelly and Floyd (this volume, Chapter 4).

4. Pleurastrophyceae

This class represents a small but very distinct group of green algae whose advanced members possess characteristics that leave little doubt that they had their origin from a different kind of green flagellate than other genera of advanced green algae. In this case, as with chlamydomonad flagellates in the Chlorophyceae, there is a flagellate known that possesses the important features by which the coccoid, sarcinoid, and filamentous Pleurastrophyceae are distinguished. That flagellate, *Tetraselmis,* which is so far unique in possessing both a theca and flagellar scales, is therefore placed in the Pleurastrophyceae with the advanced genera. *Tetraselmis* includes all the species formerly classified in *Platymonas* and *Prasinocladus* (Melkonian, 1979; Norris et al., 1980).

The advanced genera presently assigned to the Pleurastrophyceae are *Pleurastrum,* branched filaments without a holdfast; *Microthamnion,* branched filaments with a holdfast; *Pseudotrebouxia,* sarcinoid, lichenized species; *Trebouxia,* coccoid, usually lichenized forms; and *Friedmannia,* a sarcinoid soil alga. They share with *Tetraselmis* a metacentric mitotic spindle, a collapsing telophase spindle, a phycoplast (see Fig. 6d), and counterclockwise rotation of flagellar basal bodies (Melkonian and Berns, 1983; O'Kelly and Floyd, 1984). The advanced genera, as far as presently known, also share an unusual type of phycoplast in which microtubules cup the telophase nuclei as well as lie along the cleavage plane (Fig. 6d; also see Molnar et al., 1975).

The cupping microtubules might correspond to the fact that telophase nuclei in *Tetraselmis* are attached to rhizoplasts that emanate from the basal bodies as do the phycoplast microtubules (Stewart et al., 1974). The metacentric spindle is not only an unusual feature in the green algae, but, to our knowledge, does not occur anywhere else among protists. In other organisms with centric spindles, the centrioles or basal bodies migrate toward the future spindle poles as the spindle develops, and at metaphase they lie at or just lateral to the spindle poles. However, in genera of the Pleurastrophyceae the basal bodies (*Tetraselmis*, Stewart et al., 1974) or the centrioles (advanced genera, Molnar et al., 1975) remain in their interphase positions during spindle development. As a result, the basal bodies or centrioles lie in a conspicuously unusual position in the same plane as the metaphase chromosomes. Molnar et al. (1975) recommended that this unusual type of arrangement be called "metacentric" and pointed out that the filamentous *Pleurastrum* might be closely related to the flagellate *Tetraselmis* and the coccoid *Trebouxia* on that basis. Later, Deason et al. (1979) showed that the sarcinoid *Friedmannia* is also related to this group. The evolutionary origin of the distinctive metacentric spindle remains obscure, but it seems exceedingly unlikely that it had a multiple origin within the green algae, especially since it occurs in green algae that possess other distinctive features in common.

Melkonian and Berns (1983) have shown that the advanced genera assigned here to the Pleurastrophyceae have zoospores that are very similar to each other and share some distinctive features. The zoospores of *Microthamnion* were described by Watson and Arnott in 1973 (more details are available in Watson, 1971). They showed some peculiarities not then known for other green algae. Melkonian and Berns (1983) have shown that *Friedmannia* and others of this group share those characteristics and that the two flagella of the swarmers lie in the counterclockwise orientation or $^{11}/_5$ positions, unlike all other swarmers of advanced green algae with collapsing telophase spindles and phycoplasts. O'Kelly and Floyd (1984) have additionally reported that *Tetraselmis* shows counterclockwise orientation, and so it is correspondingly unlike all other green flagellates that are known to have collapsing telophase spindles and phycoplasts.

The Pleurastrophyceae are another example, similar to the Ulvophyceae and Charophyceae, where important differences in cell division and the taxonomic conclusions based on them were later shown to correlate with major differences in the structure of the flagellar apparatus and probably with distinctly different lines of evolution from green flagellates to advanced forms. The Pleurastrophyceae are also an especially good example of what many other investigations had shown previously (see Stewart et al., 1973; Stewart and Mattox, 1975; Melkonian and Berns, 1983)—that the different levels of organization or growth habit have arisen in a number of evolutionary lines, and therefore must not be used as taxonomic criteria when better characteristics are available.

In summary, the Pleurastrophyceae share the primitive feature of counterclockwise orientation with the Ulvophyceae, but differ from them in important features of telophase spindle collapse and phycoplast development. Corresponding to that difference is the theca of *Tetraselmis*, whereas the flagellates ancestral to the Charophyceae and Ulvophyceae must have been scaly, as the swarmers of those classes often are. The swarmers of advanced pleurastrophycean algae are all naked and have apparently lost the thecal covering, a common phenomenon among the Chlorophyceae, whose ancestry from a *Chlamydomonas*-like or *Carteria*-like cell is clear. The presence of a flagellar pit in *Tetraselmis* and the lack of it in swarmers of advanced Pleurastrophyceae probably represent a parallel to the loss of pits and the evolution of papillas apparent in other lines.

It is probably significant that the advanced genera of the Pleurastrophyceae also differ from those of the Ulvophyceae in being, as far as is known, completely restricted to freshwater. As pointed out earlier, the flagellar apparatuses of the Ulvophyceae and Pleurastrophyceae are similar in having a cruciate flagellar root system and counterclockwise orientation. As with alternation of generations in the Ulvophyceae, it seems likely that the fine structural differences between the two classes can be correlated with freshwater versus marine habitat and origin. Since one of the main selective advantages of the theca of green flagellates would seem to lie with its osmoregulatory properties, one is led to the conclusion that the theca was more likely to evolve in freshwater forms. Since the evolution of the phycoplast is associated with the evolution of the theca (see section on phylogeny), it is probable that the flagellate ancestor of the Pleurastrophyceae not only differed from that of the Ulvophyceae in having a theca and phycoplast, but also in its preference for fresh water. Since the flagellar apparatus of the Pleurastrophyceae is so similar to that of the Ulvophyceae, it might be suspected that some of the marine genera assigned to the Ulvophyceae on the basis of the structure of the flagellar apparatus might eventually prove to be more properly assigned to the Pleurastrophyceae when their cell division is carefully studied. Matters discussed in this paragraph, however, may well mean that fine structural differences can again be correlated with present ecology and ecological origin as in the discussion of the Ulvophyceae above. It can be discounted that some species of *Tetraselmis* are marine and some freshwater, because a few of the simpler unicellular members of both the freshwater groups Charophyceae and Chlorophyceae are also known to inhabit coastal waters. If the Pleurastrophyceae are basically freshwater, the Ulvophyceae remain as the only group of advanced green algae that are fundamentally marine. Sexual reproduction is unknown for any member of the Pleurastrophyceae, and therefore they provide no relevant information as to the correlation of the types of green algal life cycles with freshwater and marine habitats as mentioned in the earlier discussion of the Ulvophyceae.

The Pleurastrophyceae are a small class, but they are as distinct as the other classes. The use of the Pleurastrophyceae is necessary if the other classes are used as described here. The class should not be rejected because of its size or because of the unfamiliarity of the concept.

(a) Tetraselmidales. This is a new order made necessary by the inclusion of *Tetraselmis* as the sole flagellate genus of the Pleurastrophyceae. The genera *Platymonas* and *Prasinocladus* have been merged into *Tetraselmis*, as mentioned earlier.

(b) Pleurastrales. This is a new order designed to include all advanced genera of the Pleurastrophyceae. It can also be viewed as including a single family, Pleurastraceae. The genera appear to be very closely related even though they range from coccoid through sarcinoid to branched filamentous. *Trebouxia* and *Pseudotrebouxia,* in fact, are so similar to *Pleurastrum* in fine structure of their vegetative cells, and are apparently so closely related to that genus, that a case could be made for including them all in a single genus. *Pleurastrum terrestre* often grows unicellularly, especially on certain media (Tupa, 1974), and filamentous algae are not so strongly filamentous when lichenized (Ahmadjian, 1967). The main difference between the genera appears to be that the tendency toward non-filamentous growth has become fixed in *Trebouxia* and *Pseudotrebouxia.* For those who doubt that coccoid, sarcinoid, and filamentous forms can be that closely related, it is pointed out that a cessation of filamentous growth in a *Pleurastrum* filament could yield both coccoid and sarcinoid forms simultaneously. The reason for that is the fact that some cells in a filament of *Pleurastrum* can become subdivided by cleavages in planes other than those that would lead to growth in length of the filament, resulting in a "sarcinoid cell" within the filament (Tupa, 1974; K. E. Molnar, K. D. Stewart, and K. R. Mattox, unpublished). Other cells in the filament can, at the same time, cleave to form numerous zoospores so that they can be viewed as similar to "coccoid" cells within the filament. A loss of filamentous growth by *Pleurastrum* could obviously lead to unicells that have the potential for either sarcinoid or coccoid growth. This appears to be the reason for the fact that some species formerly classified as *Trebouxia* were found to differ from other species of that genus in being sarcinoid and were transferred from *Trebouxia* to the new genus *Pseudotrebouxia* (Archibald, 1975). With the older classification this also required that *Pseudotrebouxia* be transferred to a different

order (Chlorosarcinales) than that to which *Trebouxia* was assigned (Chlorococcales).

5. Chlorophyceae

The main features by which the Chlorophyceae are defined are the theca (Fig. 2d), the collapsing telophase spindle, the phycoplast (Fig. 6e, f) and clockwise orientation of flagella (Fig. 3d, e). Since all of those features are apparently derived, the Chlorophyceae can be viewed as the most specialized class of the green algae. The distinguishing features are shared by a large number of flagellate genera as well as advanced genera. Some flagellates of the Chlorophyceae and the swarmers of many advanced genera are, of course, naked, but these still possess a phycoplast and can be assumed to have lost the theca secondarily. The important point is that, if a cell covering is present on a chlorophycean flagellate or swarmer, it is of the type referred to here as a theca.

The Chlorophyceae are predominantly freshwater. The few unicellular, planktonic species found in coastal waters are often members of genera that have much greater numbers of freshwater species (e.g. *Chlamydomonas*). The Chlorophyceae whose sexual reproduction is well known produce a dormant zygote and almost always have zygotic meiosis (rare exceptions might occur; see earlier discussion of the Ulvophyceae), features shared with the other large freshwater class Charophyceae, but not with the marine Ulvophyceae.

Whereas all the advanced green algae were once considered to have a *Chlamydomonas*-like or *Carteria*-like ancestor, electron microscopy of the last decade has shown those genera to be specialized rather than primitive. It is nevertheless clear that the advanced genera of one large group, the Chlorophyceae, did have ancestors similar to *Chlamydomonas* or *Carteria*. Some advanced chlorophycean genera have swarmers with a structure very similar to that of *Chlamydomonas* (see Melkonian, 1977), whereas others show varying degrees of difference. The pattern of variation in the Chlorophyceae suggests that various chlorophycean flagellates that had the basic characteristics of the class in common but varied in other details independently gave rise to advanced genera, a pattern not detected in the other green algal classes. In any case, there is not yet enough information to adequately document these independent events in the Chlorophyceae or to base an ordinal and familial system on it. For the present it is suggested that the Chlorophyceae be subdivided into the orders given in Table I and discussed below.

(a) Chlamydomonadales. This order is meant to include all unicellular chlorophycean flagellates. As mentioned at the beginning of this section, all

non-motile chlorophycean genera that possess certain flagellate characteristics are also included in the Chlamydomonadales (e.g. genera included in the Tetrasporales of other classifications). It seems possible that chlorophycean flagellates might be discovered that possess characteristics warranting their inclusion in a separate order, but the treatment here matches the level of current understanding. Representative genera are *Chlamydomonas, Carteria, Pteromonas, Dysmorphococcus, Dunaliella, Asteromonas, Haematococcus, Chlorogonium, Polytoma, Brachiomonas, Phacotus, Tetraspora, Palmella*, etc. Genera that are similar to the coccoid *Nautococcus mammilatus* can also be included in this order because they possess a theca, a flagellate characteristic (see description of the cell covering of *N. mammilatus* in Deason and Schnepf, 1977). Many other coccoid green algae (e.g. *Chlorella* and some species of *Chlorococcum*) have walls more like those of advanced green algae in that they do not have the subunit structure characteristic of a theca [see Deason (1983) for a description of wall structure in coccoid Chlorophyceae].

(b) *Volvocales*. Motile "colonies" are classified in a different order from unicellular flagellates since they often appear to be truly multicellular organisms rather than the colonies they are traditionally referred to as being. Hoops and Floyd (1982a,b, 1983) have also shown considerable ontogenetic reorganization of the flagellar apparatus during development of "colonies." They relate these changes to the fact that flagellar beat patterns in these organisms are different from those of the Chlamydomonadales. The origin of the Volvocales probably involved significant evolutionary changes from the flagellate form and these warrant ordinal status. Genera to be included are *Volvox, Astrephomene, Pandorina, Eudorina, Platydorina, Pyrobotrys*, etc.

(c) *Chlorococcales*. This order is similar to the traditional treatment. It is very likely that some genera will be transferred from the Chlorococcales to other orders or even classes when their fine structural features become known. The Chlorococcales and the Sphaeropleales (below) are phylogenetically greatly removed from algae assigned to the Ulvophyceae, and it is now clear that the origin of those other coenocytes does not lie with the Chlorococcales, as often hypothesized in the past. Families and genera to be included in the Chlorococcales are, for the most part, those of other current classifications. The Chlorellales are not recognized here, because the lack of zoospore production or its failure to be observed is not important enough for recognition at the ordinal level.

(d) *Sphaeropleales*. This order is very distinctive and includes those filamentous Chlorophyceae whose lateral walls at maturity consist of segments rather than a continuous structure, but that otherwise lack the unusual features of the Oedogoniales. The filaments of the Sphaeropleales appear to have a different evolutionary origin from filaments of the Chlorosarcinales and Chaetophorales. The structure of *Microspora* and *Sphaeroplea* suggests (Caceres and Robinson, 1980; see also Pickett-Heaps, 1973; Silva, 1982) that the evolutionary origin of the filament in Sphaeropleales lies with a coccoid cell that produced rows of autospores whose escape was delayed. The new cells of *Sphaeroplea* become completely surrounded by a new wall, and the lateral wall begins to break apart (Caceres and Robinson, 1980). Development in *Microspora* is similar except that each new cell makes only a partial new wall. The new partial wall nevertheless becomes part of the "lateral wall." The structure of the lateral walls in filaments of the Chlorosarcinales and Chaetophorales suggests a different and more direct origin from a green flagellate, as Fritsch (1935) proposed for the evolution of filaments. Development in the latter two orders involves the important difference of growth of the old lateral wall so that segments and broken lateral walls do not occur in a manner similar to those of the Sphaeropleales. The occurrence of H-pieces in walls of some filamentous genera of the Xanthophyceae supports the viewpoint given here since many xanthophytes are coccoid, and flagellate members are not known. It would seem that in that class the evolution of filaments most likely proceeded from a coccoid condition like that proposed for the Sphaeropleales. As Caceres and Robinson (1980) have suggested, these conditions indicate that the closest relatives of *Sphaeroplea* and *Microspora* are probably among the Chlorococcales. Present evidence is not extensive enough to allow a judgement concerning whether or not the Microsporaceae and Sphaeropleaceae had a common or separate origin from chlorococcalean algae, but the best treatment at present appears to be their recognition as separate families in a single order.

Sphaeropleaceae: The characteristics are as discussed above and given in Table I. Genera to be included are *Sphaeroplea; Atractomorpha,* a new anisogamous genus recently described by Hoffman (1983); and *Cylindrocapsa* (as described by Hoffman and Hofman, 1975; Hoffman, 1976). A genus not yet examined electron microscopically but that can be included for the present on features otherwise visible is *Binuclearia* (see Silva, 1982). It seems unlikely to be a matter of chance that all three of the genera assigned to the Sphaeropleaceae that have been studied electron microscopically (*Sphaeroplea, Atractomorpha,* and *Cylindrocapsa*) have cytoplasmic invaginations in their pyrenoids. That is a rare feature in advanced green algae and, to our knowledge, has been reported in only two other cases in advanced Chlorophyceae, the chlorococcalean genus *Ankyra* (Swale and Belcher, 1971) and the Oedogoniales (see Hoffman, 1976).

Microsporaceae: The characteristics are discussed above for the order and given in Table I. Genera to be included are *Microspora* and *Radiofilum.* Another genus with apparent affinities as seen with the light microscope is *Ulothrichopsis* (see Silva, 1982).

(e) Chlorosarcinales. This order includes all filamentous and sarcinoid chlorophycean algae that possess vegetative cell division (desmoschisis; see Bold and Wynne, 1978, p. 127) but lack the plasmodesmata and multicellularity of the Chaetophorales and Oedogoniales. The Sphaeropleales do not have desmoschisis, and the interpretation of the Oedogoniales in that regard is problematical.

At the moment it is still necessary to distinguish this group on the basis of growth habit or vegetative characteristics, because fine structural information is still scarce. Sarcinoid genera and filamentous genera are classified in the same order to avoid compounding the problem by having two orders defined on that basis. Matters considered earlier in the discussion of the Pleurastrales also indicate that the difference between filamentous and sarcinoid green algae can be very minor, at least in some cases. The definition of the Chlorosarcinales and the undesirability of separating filamentous and sarcinoid forms solely on the basis of growth habit were discussed earlier (Rogers *et al.,* 1980). Although it might be true that some of the genera included here in the Chlorosarcinales will prove distinctive enough to be moved to another order, that decision will doubtless be based on characteristics other than growth habit. To be presently included are all sarcinoid genera except *Chlorokybus, Pseudendocloniopsis, Friedmannia,* and *Pseudotrebouxia,* which are distributed among the Charophyceae, Ulvophyceae, and Pleurastrophyceae. A filamentous genus that has been examined with the electron microscope and is to be classified in the Chlorosarcinales is *Gloeotilopsis* (based on unpublished observations of *Gloeotilopsis sterilis*). Filamentous genera that have not been recently studied but that are included on the basis of traditional alignments are *Interfilum, Gloeotila,* etc., which may have to be reassigned when examined with the electron microscope. It is unfortunate that the genus for which this order is named, *Chlorosarcina,* has not been critically studied with the electron microscope. It seems likely to possess all the features used here to define the Chlorosarcinales, but this needs verification.

(f) Chaetophorales. The Chaetophorales are a very distinct order, and they, along with the Oedogoniales, mark the most complex organization attained in the Chlorophyceae. The presence of plasmodesmata, whose evolutionary origin apparently coincided with the origin of cell plate formation in the class, marks the true multicellularity of the order (see Stewart *et al.,* 1973; Stewart and Mattox, 1975).

Multicellularity is also reflected by the presence of specialized cells and, in some cases, localized growth. In many ways the level of organization in the

Chaetophorales parallels that attained by the Coleochaetales in the Charophyceae. Genera to be included are listed with the discussion of families below.

Chaetophoraceae: As pointed out before (Stewart and Mattox, 1975), the genera of this family are a very homogeneous and closely related group. Pyrenoid structure, for example, which is diverse in the green algae, is uniform within the Chaetophoraceae. Other aspects of cell structure in the order are so uniform that a single cell from any of the genera can scarcely be distinguished electron microscopically from that of another. Genera to be included are *Uronema, Stigeoclonium, Chaetophora, Draparnaldia, Draparnaldiopsis,* and *Fritschiella.* Two species presently classified in *Ulothrix, U. belkae* and *U. fimbriata,* are assigned to this family and must be transferred to *Uronema* since the type species of *Ulothrix* is now known to be ulvophycean (see Sluiman *et al.,* 1980, 1983). Some other species presently assigned to *Ulothrix* will also undoubtedly be found to be chlorophycean and chaetophoracean.

Aphanochaetaceae: The only genus included here that has been examined with the electron microscope is *Aphanochaete* (Stewart *et al.,* 1973; Stewart and Mattox, 1975). *Thamniochaete* and *Chaetonema* are also included on the basis of their usual alignment with *Aphanochaete* and their possession of unicellular hairs. *Aphanochaete* has recently been shown to be oogamous (Luther and Hällfors, 1981), contrary to some earlier reports. Oogamy was reported for *Chaetonema* earlier (Meyer, 1930). The differences between this group and the Chaetophoraceae—structure of the pyrenoid, unicellular hairs, and oogamy—seem sufficient for separation at the family level.

Schizomeridaceae: The distinctions of the single genus, *Schizomeris,* are not so clear-cut as the other two families of the order, but separate treatment at least keeps the differences of *Schizomeris* from being forgotten. The unbranched, parenchymatous organization of *Schizomeris* is certainly unusual, even if growth habit is suspect as a useful criterion. The pyrenoid structure and the perforate chloroplasts are also different from the other two families (Mattox *et al.,* 1974). Sexual reproduction is not known. A previous report that the flagellar roots of the zoospores of *Schizomeris* are all of equal size (Birkbeck *et al.,* 1974) has been shown to be erroneous (Floyd and Hoops, 1980). The flagellar apparatus is typical for quadriflagellate zoospores of the Chlorophyceae. G. L. Floyd and H. J. Hoops (personal communication) have also shown that the reason the zoospores of *Schizomeris* sometimes possess more than four flagella is that aberrant zoospores with more than one papilla are occasionally produced.

(g) Oedogoniales. The features of this distinctive order are well known and need no further discussion here except to point out the possibility that the

Oedogoniales are distantly related to the Sphaeropleaceae (see discussion of Sphaeropleales and Sphaeropleaceae above). The genera are *Oedogonium*, *Bulbochaete*, and *Oedocladium*.

STRENGTHS AND WEAKNESSES OF THE CLASSIFICATION

The most important strength of the classification is the degree to which characteristics correlate and predictions can be made. That feature leaves little doubt that the classification is more natural than any previous system. If knowledge of one or two of the characteristics used to define the classes is obtained for a green alga of unknown affinities, it can make and has made possible the prediction of other important features possessed by the alga. When a theca is present on a flagellate or swarmer, it is a reliable indication that a phycoplast will be produced at cytokinesis. A clockwise orientation of flagella also indicates the presence of a phycoplast. Correlations occur with the type of life history and with the marine habitat. Other important correlations could be mentioned and can be understood from the earlier descriptions and discussions, but the ones mentioned are sufficient for the point. The feature that is most difficult to correlate with other characteristics is the level of organization or growth habit of advanced green algae, and it is classifications using those features that this classification is proposed to replace.

The greatest weakness of the classification is the present necessity to place all scaly or naked green flagellates with persistent interzonal spindles in a single class, Micromonadophyceae. Any group, like that one, that must be defined solely on the basis of primitive characteristics is almost certain to contain genera that are more closely related to those of other groups than to each other. It is not a source of confidence that the presence of scales is always correlated with a persistent interzonal spindle, since primitive characteristics usually show more correlation with each other than advanced ones do. The fact that the advanced characteristics of the theca and phycoplast correlate is apparently the result of the phycoplast's being selected for as a result of the restriction of cell elongation by the evolving theca (Mattox and Stewart, 1977). Further improvement of the classification of green algae will most likely result from increased knowledge of the structure of micromonadophycean genera. That knowledge will probably result in many or all of those genera being distributed to the various other classes on the basis of derived rather than primitive features. In a classification like this one, where classes are based on the structure of the flagellated cell, it is not desirable to separate flagellates from advanced genera as is done with the Micromonadophyceae. It was possible on the basis of present information for this classification to include the flagellate *Tetraselmis* in the Pleurastrophyceae and chlamydomonad flagellates in the Chlorophyceae. The goal should be to treat the genera of the Micromonadophyceae, as far as proves possible, in the same manner. It is nevertheless possible or even likely that some lines of evolution in green flagellates did not lead to advanced genera, and these might remain in a group including flagellates only. Significant improvements might be made in the understanding of the evolution of green algae and their classification if some green flagellates of very primitive structure are found and studied carefully. Known flagellates that are not available in culture and therefore have not been studied might prove to be very enlightening (e.g. *Trichloris* and *Cymbomonas;* see Fig. 1h–j).

Other weaknesses of the classification lie with the fact that it is still necessary to use growth habit and level of organization in some cases (e.g. the Chlorococcales and Chlorosarcinales). It does not necessarily follow that all such groups will prove to be polyphyletic, but the likelihood is that most of them will.

COMMENTS ON OTHER CURRENTLY USED GREEN ALGAL CLASSES

Other green algal classes often used or recently proposed are separately discussed below.

1. Bryopsidophyceae

The use of this class is contrary to the traditional use of the characteristics of the motile cell for class-level distinctions. Comparative cytological studies of recent years have shown that siphonous green algae are very closely related to both cellular and coenocytic members of the Ulvophyceae of this classification. It would now be difficult to define the Bryopsidophyceae on the basis of important characteristics comparable or parallel to features used for other classes.

2. Chlamydophyceae

Chlamydophyceae was recently proposed by Ettl (1981) to include chlamydomonad flagellates and other algae that could be interpreted as prolonged stages of the normal development of chlamydomonad flagellates. For example, it was held that genera like *Chlorococcum* represent an extended non-motile stage like the briefer non-motile stage that chlamydomonad flagellates pass through every time cell division occurs. Ettl does not give an adequate reason why the Chlamydophyceae should be separated at the level of class from other green algae that have zoospores or gametes with chlamydomonad or carterial structure. Additionally, the assumption that all the non-motile chlorophycean unicells assigned to the Chlamydophyceae represent evolutionarily arrested stages of chlamydomonad development is incorrect. A mature cell of *Chlorococcum oleofaciens* is surrounded by a cell wall that is distinctly different from a theca, and is therefore unlike a prolonged chlamydomonad developmental stage. The zoospores produced by that species of *Chlorococcum* do have a theca (Miller, 1978), which emphasizes the difference between the flagellated and vegetative stages. Ettl's concept places very closely related algae in different classes. For example, *Dunaliella* is assigned to the Chlorophyceae, not to the Chlaymdophyceae, because it lacks a theca. *Dunaliella* is comparable to *Chlamydomonas* in every other way. The loss of the theca by a flagellate adapted to briny habitats is not surprising and hardly represents a difference important enough to recognize at the level of class. Ettl's assignment of the Chlorellales to the Chlorophyceae and *Chlorococcum* to the Chlamydophyceae is another example of class-level distinctions being based on trivial differences. See Deason (1984) for a detailed critique of Ettl's concept of the classes Chlamydophyceae and Chlorophyceae.

3. Codiolophyceae

Kornmann's (1973) proposal of the Codiolophyceae did not gain the degree of recognition that it deserved. Recent electron microscopic studies have confirmed that genera that possess a codiolum stage are very closely related to each other, as Kornmann argued (see Floyd and O'Kelly, 1984; O'Kelly and Floyd, this volume, Chapter 4). While it is true that codiolum-producing algae are also related to the "non-codiolum" algae classified here in the Ulvophyceae and should be assigned to the Ulvophyceae, Floyd and O'Kelly (1984) have shown that the genera Kornmann assigned to the Codiolophyceae form a distinctive order within the Ulvophyceae. Although Kornmann's proposal had a number of good features and provided an impetus for further research, the present understanding of "codiolum algae" emphasizes the advantage of basing algal classes on the characteristics of the motile cell.

4. Oedogoniophyceae

Although the Oedogoniales possess a number of unusual features, they also possess the phycoplast and collapsing spindle characteristic of the Chlorophyceae. The band connecting the flagellar basal bodies of oedogonialean algae is also conspicuously striated, like those of most Chlorophyceae and unlike those of the Pleurastrophyceae, the other class with a phycoplast. Other features of the Oedogoniales, however distinctive, do not appear to be important enough for class-level recognition. The swarmers of the Oedogoniales apparently do not owe their stephanokont condition to an independent origin from a stephanokont green flagellate. Although the stephanokont zoospores of some Ulvophyceae are doubtless independently evolved in that class and bear no evolutionary connection to those in the Oedogoniales, they do demonstrate that the stephanokont condition can evolve in an alga which still produces gametes with the typical cruciate flagellar root system (Roberts et al., 1980, 1981). The evidence therefore suggests that a filamentous ancestor of the Oedogoniales produced swarmers with a cruciate root system and that the oedogonialean stephanokont flagellar apparatus was not derived directly from a flagellate.

5. Prasinophyceae

Some green flagellates that had features different from *Chlamydomonas* and *Carteria* were separated from the Chlorophyceae by Chadefaud (see Chadefaud, 1977, for his recent comments). When Manton and Parke (e.g. Manton and Parke, 1960) began to examine some of these flagellates by electron microscopy, they were found to have some striking and previously unknown characteristics such as flagellar and body scales. These seemed very different from chlamydomonad flagellates, and a class-level separation was deemed warranted. Christensen (1962) assigned such genera to the Prasinophyceae, presumably basing the name on the genus *Prasinocladus*. Research of the last ten years, however, has shown that the boundary between genera assigned to the Prasinophyceae and other green flagellates is not clear, as mentioned in the section on phylogeny. Prasinophytes cannot now be defined and distinguished on the traditional basis. Scales are known from all evolutionary lines of green algae. The presence of a flagellar pit in a flagellate with a flagellar apparatus distinctly like that of *Carteria* [*Hafniomonas*, see Ettl and Moestrup (1980), and comments in section on phylogeny of this paper], and evidence that the theca of chlamydomonad algae had its origin from a scaly covering completely blur the distinction formerly made between Prasinophyceae and chlamydomonad flagellates. By and large the Prasinophyceae were defined with primitive characteristics, as with the Micromonadophyceae of this classification. In our opinion, however, the Micromonadophyceae is to be preferred because its conceptual basis does not lie with the perception of a large and vacant gap between scaly green flagellates and "volvocalean" flagellates, as does the concept of the Prasinophyceae. Another consideration is the fact that the genus from which the name Prasinophyceae was derived, *Prasinocladus,* is now merged with *Tetraselmis*. *Tetraselmis* is not assigned to the Micromonadophyceae but to the Pleurastrophyceae on the basis of derived characteristics not shared by any other genus ever included in the Prasinophyceae.

6. Zygnematophyceae

The Zygnematales are a very large, important, and distinctive group. It seems likely that the large size of the group and the number of phycologists working with it have been as important as their lack of flagellated cells and conjugatory sexual reproduction in the occasional recognition of the Zygnematales as a class. It is clear that the Zygnematales are related to other algae assigned here to the Charophyceae, and the kinship to the Klebsormidiales is probably very close (see Pickett-Heaps, 1972; Stewart and Mattox, 1975; Rogers *et al.*, 1980). In any case, the relationship of the Zygnematales to all other green algae indicates that its loss of flagellated cells and development of a special reproductive process was a later evolutionary event than the flagellate characteristics used in this classification. Since that is the case, it is our opinion that the special features of the Zygnematales should not be given the same rank in an overall classification.

LATIN DIAGNOSES

Latin diagnoses are given below for the five classes used here. The diagnoses are based on the English descriptions given in Table I. Since the names Tetraselmidales and Pleurastrales have apparently not been used previously, Latin diagnoses are also given for them. English descriptions of the Tetraselmidales and Pleurastrales are also given here in place of the abbreviated descriptions given in Table I.

MICROMONADOPHYCEAE, *classis nova* Flagellatae squamatae vel nudae; fusus mitoticus interzonalis per cytokinesem persistens.

CHAROPHYCEAE, *sensu lato* Cellulae motiles, si procreatae, squamatae vel nudae, asymmetricae flagellis insertis lateraliter vel subapicaliter et dextrorsum ab loco insertionis extensis; cellulae motiles semper biflagellatae et sine stigmate; systema radicum microtubularium flagellorum habens structuram multistratam unam et plerumque unam radicem microtubularem; rhizoplasti ignoti; fusus mitoticus interzonalis per cytokinesem persistens; microtubuli chromosomatum per telophasem saepe persistentes; microtubuli inter chromosomata et polum fusi paulum haudve abbreviati per anaphasem, practer formas parenchymatas; habitantes plerumque aquas dulces; reproductio sexualis zygota dormienti procreata et meiose zygotica; plantae unicellulares vel sarcionoides vel filamentosae vel parenchymaticae.

ULVOPHYCEAE, *classis nova* Cellulae motiles, si procreatae, squamatae vel nudae, flagellis insertis apicaliter, symmetricae radialiter extus; cellulae motiles biflagellatae vel quadriflagellatae vel multiflagellatae; systema radicum microtubularium flagellorum cruciatum; rhizoplasti saepe praesentes; in cellulis biflagellatis corpora basalia, a fronte cellulae visa, posita quasi rotatione contra horologii motum; radicum microtubularium flagellorum duo maiores tendentes terminare linea inter corpora basalia, duo minores plus ablegatae et affixae parti exteriori corporum basilium; fusus mitoticus interzonalis per cytokinesem persistens; microtubuli inter chromosomata et polum fusi saepe paulum abbreviati; plantae marinae pro parte maxima, etsi aliquot genera habent species et aquae dulcis et aquae salinae; alternatio generationum (meiosis sporica) communis, zygotis dormientibus ignotis; plantae sarcinoides vel filamentae vel thalloides vel parenchymaticae vel coenocyticae vel siphonaceae.

PLEURASTROPHYCEAE, *classis nova* Cellulae motiles thecatae vel nudae et plus minusve complanatae; systema radicum microtubularium flagellorum cruciatum; rhizoplasti praesentes etsi interdum tenues subtilesque; fusus mitoticus semper metacentricus; corpora basalia, a fronte cellulae visa, posita quasi rotatione contra horologii motum; fusus mitoticus interzonalis in telophase collabens, phycoplasto tum evoluto; microtubuli inter chromosomata et polum fusi mitotici per anaphasem abbreviati; reproductio sexualis incognita. Genus unicum flagellatorum ad hanc classem assignatum thecatum, quattuor flagellis in foveam insertis, corporibus basalibus parallelibus aliud ad aliud et ad axem longum cellulae. Cellulae motiles generibus provectis procreatae sine fovea; flagella duo, "deorsum flexa" ita ad angulum ab axe longo cellulae prorsum natantis iacentia; in cellulis motilibus, radicum microtubularium maiores inter corpora basalia transientes, du minores ad partem exteriorem corporum basalium; generum provectorum phycoplasti peculiaris forma, ei habentes microtubulos involventes nucleos in telophase et microtubulos parallelos ad planum fissurae; plasmodesmata absentia. Omnia genera provecta aquam dulcem habitant; plantae flagellatae vel coccoides vel sarcinoides vel filamentosae.

CHLOROPHYCEAE, *sensu stricto* Flagellatae et cellulae motiles generum provectorum, si procreatae, thecatae vel nudae; flagella apicaliter inserta; symmetrica radialiter extus; cellulae motiles biflagellatae vel quadriflagellatae vel multiflagellatae; systema radicum microtubularium flagellorum cruciatum; rhizoplasti interdum praesentes in flagellatis, sed incogniti in cellulis motilibus generum provectorum; corpora basalia, a fronte cellulae visa, posita quasi rotatione horologii; radicum microtubularium flagellorum duo minores tendentes terminare linea inter corpora basalia, maiores plus ablegatae et affixae parti exteriori corporum basalium; fusus mitoticus interzonalis in telophase collabens, phycoplasto tum evoluto; centriola ad pola fusi vel vix lateraliter eis; fusus numquam metacentricus; microtubuli inter chromosomata et polum fusi mitotici per anaphasem semper abbreviati; plantae aquam dulcem habitant; reproductio sexualis zygota dormienti procreata meiose zygotica; plantae flagellatae vel coccoides vel sarcinoides vel coloniales immotae vel coloniales motiles vel filamentosa vel parenchymaticae.

TETRASELMIDALES, new order Chlorophytan flagellates with metacentric spindle, phycoplast, and flagellar pit.

TETRASELMIDALES, *ordo novus* Flagellatae virides fusum metacentricum et phycoplastum et foveam flagellarum habentes.

PLEURASTRALES, new order Coccoid, sarcinoid, or filamentous green algae with metacentric spindle and a phycoplast; zoospores biflagellate, flattened; basal bodies in 11 and 5 o'clock positions (counterclockwise rotation) as viewed from the anterior.

PLEURASTRALES, *ordo novus* Algae virides; coccoides vel sarcinoides vel filamentosae; fusum metacentricum et phycoplastum habentes; zoosporae biflagellatae, complanatae, corporibus basalibus positis quasi rotatione contra horologii motum, a fronte cellulae visis.

ACKNOWLEDGEMENTS

We thank Gary Floyd, Michael Melkonian, Charles O'Kelly, Hans Sluiman, Harold Hoops, and Larry Hoffman for valuable discussions and for generously making their unpublished data available to us. We are grateful to Charles Werth for providing the Latin diagnoses. This work was supported by National Science Foundation grant DEB 81-03489.

REFERENCES

Ahmadjian, V. (1967). "The Lichen Symbiosis." Ginn (Blaisdell), Boston, Massachusetts.
Archibald, P. A. (1975). *Trebouxia* de Puymaly (Chlorophyceae, Chlorococcales) and *Pseudotrebouxia* gen.nov. (Chlorophyceae, Chlorosarcinales). *Phycologia* **14**, 125–137.
Barlow, S. B. and Cattolico, R. A. (1980). Fine structure of the scale-covered green flagellate *Mantoniella squamata* (Manton and Parke) Desikachary. *Br. Phycol. J.* **15**, 321–333.
Bekheet, I. A. and Syrett, P. J. (1977). Urea degrading enzymes in green algae. *Br. Phycol. J.* **12**, 137–143.
Birkbeck, T. E., Stewart, K. D. and Mattox, K. R. (1974). The cytology and classification of *Schizomeris leibleinii* (Chlorophyceae). II. The structure of quadriflagellate zoospores. *Phycologia* **13**, 71–79.
Bold, H. C., and Wynne, M. J. (1978). "Introduction to the Algae: Structure and Reproduction." Prentice-Hall, Englewood Cliffs, New Jersey.
Bourrelly, P. (1973). "Les Algues d'eau douce. Initiation à la Systématique. Tome I. Les Algues Vertes" (rev.ed.). Boubée, Paris.
Caceres, E. J. and Robinson, D. G. (1980). Ultrastructural studies on *Sphaeroplea annulina* (Chlorophyceae). Vegetative structure and mitosis. *J. Phycol.* **16**, 313–320.
Chadefaud, M. (1977). Les Prasinophycées. Remarques historiques, critiques et phylogénétiques. *Bull. Soc. Phycol. Fr.* **22**, 1–18.

Chapman, R. L. and Henk, M. C. (1983). Ultrastructure of *Cephaleuros virescens* (Chroolepidaceae; Chlorophyta). IV. Absolute configuration analysis of the cruciate flagellar apparatus and multilayered structures in the pre- and post-release gametes. *Am. J. Bot.* **70**, 1340–1355.

Chappell, D. F. and Floyd, G. L. (1981). Cell division in the weakly filamentous *Raphidonema sessile* (=*Raphidonemopsis sessilis*) (Chlorophyta). *Trans. Am. Microsc. Soc.* **100**, 74–82.

Christensen, T. (1962). Alger. In "Botanik" (T. W. Böcher, M. Lange, and T. Sørensen, eds), Vol. 2, No. 2, pp. 1–178. Munksgaard, Copenhagen.

Deason, T. R. (1983). Cell wall structure and composition as taxonomic characters in the coccoid Chlorophyceae. *J. Phycol.* **19**, 248–251.

Deason, T. R. (1984). A discussion of the classes Chlamydophyceae and Chlorophyceae Ettl 1981 and their subordinate taxa. *Plant Syst. Evol.* (in press).

Deason, T. R. and Schnepf, E. (1977). Fine structure of *Nautococcus mammilatus* (Chlorococcales, Chlorophyceae), a coccoid alga with tomentose cell walls. *J. Phycol.* **13**, 218–224.

Deason, T. R., Ryals, P. E., O'Kelley, J. C. and Bullock, K. W. (1979). Fine structure of mitosis and cleavage in *Friedmannia israelensis* (Chlorophyceae, Chlorosarcinaceae). *J. Phycol.* **15**, 452–457.

Domozych, D. S., Stewart, K. D. and Mattox, K. R. (1981a). Development of the cell wall in *Tetraselmis:* Role of the golgi apparatus and extracellular wall assembly. *J. Cell Sci.* **52**, 351–371.

Domozych, D. S., Stewart, K. D. and Mattox, K. R. (1981b). In vivo cell wall ontogenesis in chlorophycean flagellates. 1. The cell wall, endomembrane system and interphase wall expansion. *Cytobios* **32**, 142–165.

Domozych, D. S., Stewart, K. D. and Mattox, K. R. (1981c). In vivo cell wall ontogenesis in chlorophycean flagellates. 2. Golgi apparatus activation, extracellular assembly and internal cellular controls. *Cytobios* **32**, 167–178.

Domozych, D. S., Stewart, K. D. and Mattox, K. R. (1981d). The role of microtubules during the cytokinetic events of cell wall ontogenesis and papilla development in *Carteria crucifera*. *Protoplasma* **106**, 193–204.

Ettl, H. (1981). Die neüe Klass Chlamydophyceae, eine naturliche Gruppe der Grunalgen (Chlorophyta). *Plant Syst. Evol.* **137**, 107–117.

Ettl, H., and Moestrup, Ø. (1980). Light and electron microscopical studies on *Hafniomonas* gen. nov. (Chlorophyceae, Volvocales), a genus resembling *Pyramimonas*. *Plant Syst. Evol.* **135**, 177–210.

Floyd, G. L. (1978). Mitosis and cytokinesis in *Asteromonas gracilis,* a wall-less green monad. *J. Phycol.* **14**, 440–445.

Floyd, G. L. and Hoops, H. J. (1980). *Schizomeris leibleinii* revisited: Ultrastructure of the flagellar apparatus. *J. Phycol.* **16**, Suppl., 11.

Floyd, G. L. and O'Kelly, C. J. (1984). Motile cell ultrastructure and the circumscription of the orders Ulotrichales and Ulvales (Ulvophyceae, Chlorophyta). *Am. J. Bot.* **71**, 111–120.

Floyd, G. L., Hoops, H. J. and Swanson, J. A. (1980). Fine structure of the zoospore of *Ulothrix belkae* with emphasis on the flagellar apparatus. *Protoplasma* **104**, 17–31.

Fritsch, F. E. (1935). "The Structure and Reproduction of the Algae," Vol. 1. Cambridge Univ. Press, London and New York.

Gayral, P. and Fresnel, J. (1979). *Exanthemachrysis gayraliae* Lepailleur (Prymnesiophyceae, Pavlovales): Ultrastructure et discussion taxinomique. *Protistologica* **15**, 271–282.

Graham, L. E. (1982a). The occurrence, evolution, and phylogenetic significance of parenchyma in *Coleochaete* Breb. (Chlorophyta). *Am. J. Bot.* **69**, 447–454.

Graham, L. E. (1982b). Cytology, ultrastructure, taxonomy, and phylogenetic relationships of Great Lakes filamentous algae. *J. Great Lakes Res.* **8**, 3–9.

Graham, L. E. and Wilcox, L. W. (1983). The occurrence and phylogenetic significance of putative placental transfer cells in the green alga *Coleochaete*. *Am. J. Bot.* **70**, 113–120.

Heath, I. B. (1981). An investigation of protistan phylogeny using a numerical taxonomy (cluster) analysis of mitotic systems. *BioSystems* **14**, 261–270.

Hibberd, D. J. (1976). The ultrastructure and taxonomy of the Chrysophyceae and Prymnesiophyceae (Haptophyceae): A survey with some new observations on the ultrastructure of the Chrysophyceae. *Bot. J. Linn. Soc.* **72**, 55–80.

Hibberd, D. J. (1980). Prymnesiophytes (=Haptophytes). *In* "Phytoflagellates" (E. R. Cox, ed.), Vol. 2, Elsevier/North-Holland, New York.

Hoffman, L. R. (1976). Fine structure of *Cylindrocapsa* zoospores. *Protoplasma* **87**, 191–219.

Hoffman, L. R. (1983). *Atractomorpha echinata* gen. et sp. nov., a new anisogamous member of the Sphaeropleaceae (Chlorophyceae). *J. Phycol.* **19**, 76–86.

Hoffman, L. R. and Hofman, C. S. (1975). Zoospore formation in *Cylindrocapsa*. *Can. J. Bot.* **53**, 439–451.

Hoops, H. J. and Floyd, G. L. (1982a). Ultrastructure of the flagellar apparatus of *Pyrobotrys* (Chlorophyceae). *J. Phycol.* **18**, 455–462.

Hoops, H. J. and Floyd, G. L. (1982b). Ultrastructure and taxonomic position of the rare volvocalean alga *Chlorcorona bohemica*. *J. Phycol.* **18**, 462–466.

Hoops, H. J. and Floyd, G. L. (1983). Ultrastructure and development of the flagellar apparatus and flagellar motion in the colonial green alga *Astrephomene gubernaculifera*. *J. Cell Sci.* **63**, 21–41.

Hoops, H. J., Floyd, G. L. and Swanson, J. A. (1982). Ultrastructure of the biflagellate motile cells of *Ulvaria oxysperma* (Kütz.) Bliding and phylogenetic relationships among ulvaphycean algae. *Am. J. Bot.* **69**, 150–159.

Kornmann, P. (1973). Codiolophyceae, a new class of Chlorophyta. *Helgol. Wiss. Meeresunters.* **25**, 1–13.

Lembi, C. A. (1975). The fine structure of the flagellar apparatus of *Carteria*. *J. Phycol.* **11**, 1–9.

Lumsden, J., Henry, L. and Hall, D. O. (1977). Superoxide dismutase in photosynthetic organisms. *In* "Superoxide and Superoxide Dismutases" (A. M. Michelson, J. M. McCord, and I. Fridovich, eds), pp. 218–235. Academic Press, New York.

Luther, H. and Hällfors, G. (1981). Oogamy in *Aphanochaete* (Chlorophyceae, Chaetophorales). I. Morphology and ecology. *Ann. Bot. Fenn.* **18**, 169–181.

Manton, I. (1964). The possible significance of some details of the flagellar bases in plants. *J. R. Microsc. Soc.* **82**, 279–285.

Manton, I. (1975). Observations on the microanatomy of *Scourfieldia marina* Throndsen and *Scourfieldia caeca* (Korsch.) Belcher et Swale. *Arch. Protistenkd.* **117**, 358–368.

Manton, I. and Ettl, H. (1965). Observations on the fine structure of *Mesostigma viride* Lauterborn. *J. Linn. Soc. London, Bot.* **59**, 175–184.

Manton, I. and Parke, M. (1960). Further observations on small green flagellates with special reference to possible relatives of *Chromulina pusilla* Butcher. *J. Mar. Biol. Assoc. U. K.* **39**, 275–298.

Marano, F. (1976). Étude ultrastructurale de la division chez *Dunaliella*. *J. Microsc. Biol. Cell.* **25**, 279–282.

Marchant, H. J. and Pickett-Heaps, J. D. (1973). Mitosis and cytokinesis in *Coleochaete scutata*. *J. Phycol.* **9**, 461–471.

Mattox, K. R. and Stewart, K. D. (1973). Observations on the zoospores of *Pseudendoclonium basiliense* and *Trichosarcina polymorpha* (Chlorophyceae). *Can. J. Bot.* **51**, 1425–1430.

Mattox, K. R. and Stewart, K. D. (1974). A comparative study of cell division in *Trichosarcina polymorpha* and *Pseudendoclonium basiliense* (Chlorophyceae). *J. Phycol.* **10**, 447–456.

Mattox, K. R. and Stewart, K. D. (1977). Cell division in the scaly green flagellate *Heteromastix angulata* and its bearing on the origin of the Chlorophyceae. *Am. J. Bot.* **64**, 931–945.
Mattox, K. R., Stewart, K. D. and Floyd, G. L. (1974). The cytology and classification of *Schizomeris leibleinii* (Chlorophyceae). I. The vegetative thallus. *Phycologia* **13**, 63–69.
Melkonian, M. (1977). The flagellar root system of zoospores of the green alga *Chlorosarcinopsis* (Chlorosarcinales) as compared with *Chlamydomonas* (Volvocales). *Plant Syst. Evol.* **128**, 79–88.
Melkonian, M. (1979). An ultrastructural study of the flagellate *Tetraselmis cordiformis* Stein (Chlorophyceae) with emphasis on the flagellar apparatus. *Protoplasma* **98**, 139–151.
Melkonian, M. (1980). Ultrastructural aspects of basal body associated fibrous structures in green algae: A critical review. *BioSystems* **12**, 85–104.
Melkonian, M. and Berns, B. (1983). Zoospore ultrastructure in the green alga *Friedmannia israelensis:* An absolute configuration analysis. *Protoplasma* **114**, 67–84.
Meyer, K. (1930). Über den Befruchtungsvorgang bei *Chaetonema irregulare*. *Arch. Protistenkd.* **72**, 147–157.
Mignot, J.-P. (1974). Étude ultrastructurale des *Bicoeca,* protistes flagellés. *Protistologica* **10**, 543–565.
Mignot, J.-P. and Brugerolle, G. (1975). Étude ultrastructurale du flagelle phagotrophe *Colponema loxodes* Stein. *Protistologica* **11**, 547–554.
Miller, D. H. (1978). Cell wall chemistry and ultrastructure of *Chlorococcum oleofaciens* (Chlorophyceae). *J. Phycol.* **14**, 189–194.
Moestrup, Ø. (1974). Ultrastructure of the scale-covered zoospores of the green alga *Chaetosphaeridium,* a possible ancestor of higher plants and bryophytes. *Biol. J. Linn. Soc.* **6**, 111–125.
Moestrup, Ø. and Ettl, H. (1979). A light and electron microscopical study of *Nephroselmis olivacea* (Prasinophyceae). *Opera Bot.* **49**, 1–40.
Moestrup, Ø. and Thomsen, H. A. (1976). Fine structural studies on the flagellate genus *Bicoeca.* I. *Bicoeca maris,* with particular emphasis of the flagellar apparatus. *Protistologica* **12**, 101–120.
Molnar, K. E., Stewart, K. D. and Mattox, K. R. (1975). Cell division in the filamentous alga *Pleurastrum* and its comparison with the unicellular *Platymonas* (Chlorophyceae). *J. Phycol.* **11**, 287–296.
Norris, R. E., Hori, T. and Chihara, M. (1980). Revision of the genus *Tetraselmis* (Class Prasinophyceae). *Bot. Mag.* **93**, 317–339.
O'Kelly, C. J. and Floyd, G. L. (1983). The flagellar apparatus of *Entocladia viridis* motile cells and the taxonomic position of the resurrected family Ulvellaceae (Ulvales, Chlorophyta). *J. Phycol.* **19**, 153–164.
O'Kelly, C. J. and Floyd, G. L. (1984). Flagellar apparatus absolute orientations and the phylogeny of the green algae. *BioSystems* **16**, 227–251.
O'Kelly, C. J. and Yarish, C. (1980). Observations on marine Chaetophoraceae (Chlorophyta). I. Sporangial ontogeny in the type species of *Entocladia* and *Phaeophila. J. Phycol.* **17**, 549–558.
O'Kelly, C. J., Floyd, G. L. and Dube, M. A. (1984). The fine structure of motile cells in the genera *Ulvaria* and *Monostroma,* with special reference to the taxonomic position of *Monostroma oxyspermum* (Ulvophyceae, Chlorophyta). *Plant Syst. Evol.* (in press).
Pickett-Heaps, J. D. (1972). Cell division in *Klebsormidium subtilissimum* (formerly *U. subtilissima*) and its possible phylogenetic significance. *Cytobios* **6**, 167–183.
Pickett-Heaps, J. D. (1973). Cell division and wall structure in *Microspora. New Phytol.* **72**, 347–355.

Pickett-Heaps, J. D. (1974). Cell division in *Stichococcus*. *Br. Phycol. J.* **9**, 63–73.
Pickett-Heaps, J. D. (1976). Cell division in *Raphidonema longiseta*. *Arch. Protistenkd.* **118**, 209–212.
Pickett-Heaps, J. D. and Ott, D. W. (1974). Ultrastructural morphology and cell division in *Pedinomonas*. *Cytobios* **11**, 41–58.
Roberts, K. R., Sluiman, H. J., Stewart, K. D. and Mattox, K. R. (1980). Comparative cytology and taxonomy of the Ulvaphyceae. II. Ulvalean characteristics of the stephanokont flagellar apparatus of *Derbesia tenuissima*. *Protoplasma* **104**, 223–238.
Roberts, K. R., Sluiman, H. J., Stewart, K. D. and Mattox, K. R. (1981). Comparative cytology and taxonomy of the Ulvaphyceae. III. The flagellar apparatus of the anisogametes of *Derbesia tenuissima* (Chlorophyta). *J. Phycol.* **17**, 330–340.
Roberts, K. R., Stewart, K. D. and Mattox, K. R. (1982). Structure of the anisogametes of the green siphon *Pseudobryopsis* sp. (Chlorophyta). *J. Phycol.* **18**, 498–508.
Rogers, C. E., Mattox, K. R. and Stewart, K. D. (1980). The zoospore of *Chlorokybus atmophyticus*, a charophyte with sarcinoid growth habit. *Am. J. Bot.* **67**, 774–783.
Rogers, C. E., Domozych, D. S., Stewart, K. D. and Mattox, K. R. (1981). The flagellar apparatus of *Mesostigma viride* (Prasinophyceae): Multilayered structures in a scaly green monad. *Plant Syst. Evol.* **138**, 247–258.
Schiller, J. (1925). Die planktonischen Vegetationen des adriatischen Meeres. *Arch. Protistenkd* **53**, 59–123.
Sears, J. R., and Brawley, S. H. (1982). *Smithsoniella* gen. nov., a possible evolutionary link between the multicellular and siphonous habits in Ulvophyceae, Chlorophyta. *Am. J. Bot.* **69**, 1450–1461.
Silva, P. C. (1982). Chlorophycota. *In* "Synopsis and Classification of Living Organisms" (S. Parker, ed.), Vol. 1, pp. 133–161. McGraw-Hill, New York.
Singh, R. N. (1942). Reproduction in *Draparnaldiopsis indica* Bharadwaja. *New Phytol.* **4**, 262–273.
Sluiman, H. J. (1983). The flagellar apparatus of the zoospore of the filamentous green alga *Coleochaete pulvinata*: Absolute configuration and phylogenetic significance. *Protoplasma* **115**, 160–175.
Sluiman, H. J., Roberts, K. R., Stewart, K. D. and Mattox, K. R. (1980). Comparative cytology and taxonomy of the Ulvaphyceae. I. The zoospore of *Ulothrix zonata* (Chlorophyta). *J. Phycol.* **16**, 537–545.
Sluiman, H. J., Roberts, K. R., Stewart, K. D., Mattox, K. R. and Lokhorst, G. M. (1982). The flagellar apparatus of the zoospore of *Urospora penicilliformis* (Chlorophyta). *J. Phycol.* **18**, 1–12.
Sluiman, H. J., Roberts, K. R., Stewart, K. D. and Mattox, K. R. (1983). Comparative cytology and taxonomy of the Ulvaphyceae. IV. Mitosis and cytokinesis in *Ulothrix zonata* (Chlorophyta). *Acta bot. neerl.* **32**, 257–269.
Stewart, K. D. and Mattox, K. R. (1975). Comparative cytology, evolution and classification of the green algae, with some consideration of the origin of other organisms with chlorophylls a and b. *Bot. Rev.* **41**, 104–135.
Stewart, K. D. and Mattox, K. R. (1978). Structural evolution in the flagellated cells of green algae and land plants. *BioSystems* **10**, 145–152.
Stewart, K. D. and Mattox, K. R. (1980). Phylogeny of Phytoflagellates. *In* "Phytoflagellates" (E. R. Cox, ed.), Vol 2, pp. 433–462. Elsevier/North-Holland, New York.
Stewart, K. D., Mattox, K. R. and Floyd, G. L. (1973). Mitosis, cytokinesis, the distribution of plasmodesmata, and other cytological characteristics in the Ulotrichales, Ulvales, and Chaetophorales: Phylogenetic and taxonomic considerations. *J. Phycol.* **9**, 128–141.

Stewart, K. D., Mattox, K. R. and Chandler, C. D. (1974). Mitosis and cytokinesis in *Platymonas subcordiformis*, a scaly green monad. *J. Phycol.* **10,** 65–79.

Swale, E. M. F. and Belcher, J. H. (1971). Investigation of a species of *Ankyra* Fott by light and electron microscopy. *Br. Phycol. J.* **6,** 41–50.

Thompson, R. H. (1969). Sexual reproduction in *Chaetosphaeridium globosum* (Nordst.) Klebahn (Chlorophyceae) and description of a species new to science. *J. Phycol.* **5,** 285–292.

Tupa, D. (1974). An investigation of certain chaetophoracean algae. *Beiheifte Z. Nova Hedw.* **46,** 1–155.

Watson, M. W. (1971). Ultrastructure and germination of *Microthamnion* zoospores: Changes in structure and organelles. Ph.D. Dissertation, University of Texas, Austin.

Watson, M. W. and Arnott, H. J. (1973). Ultrastructural morphology of *Microthamnion* zoospores. *J. Phycol.* **9,** 15–29.

3 | Flagellar Apparatus Ultrastructure in Relation to Green Algal Classification*

M. MELKONIAN

Botanisches Institut, Universität Münster, Münster, Federal Republic of Germany

Abstract: Classification of the green algae into larger systematic categories (from order to class level) is for the most part uncertain at the present time. Ultrastructural aspects of green algae have been extensively used for new arrangements in the past ten years, especially at the class level. The flagellated cell plays an important role in these considerations, because it is assumed that structural characters in flagellated cells are evolutionarily conservative. Structurally and biochemically the flagellar apparatus is one of the most complex cell organelles in a flagellated cell. It comprises roughly 300 different polypeptides and exhibits great structural complexity. More than 100 different structural characters have been distinguished in the flagellar apparatus of green algae, and it should theoretically be possible to use these features in green algal classification at all systematic levels. In this review, based on both published information about more than 100 different green algae and on previously unpublished information about another 40 green algae, the different ultrastructural characters of the flagellar apparatus and their structural variability will be discussed. Emphasis is given to structural characters that have not previously been reviewed elsewhere, including the flagellar tip, the flagellar shaft, the transition region and the basal body. The information available on flagellar apparatus ultrastructure will be used to indicate some evolutionary trends in flagellated cells of green algae.

INTRODUCTION

Classification of green algae is performed in different ways and for different purposes. For those investigators interested in the ecology of algae or in applied aspects it may be desirable to use a classification system suitable for identification of green algae in the field. Phycologists interested in natural

---

*Dedicated to Professor Irene Manton *prima inter pares* in her 80th year.

Systematics Association Special Volume No. 27, "Systematics of the Green Algae", edited by D. E. G. Irvine and D. M. John, 1984, pp. 73–120. Academic Press, London and Orlando.
ISBN 0 12 374040 1

Copyright © by the Systematics Association
All rights of reproduction in any form reserved

relationships within green algae will, however, take advantage of a range of different characters (structural, physiological, biochemical, etc.) and use all possible experimental methods to gain more information about the organism in question. It is this latter approach that usually requires laboratory cultures of green algae. Laboratory cultures made available through Culture Collections provide a constant source of information and can be used by later investigators for comparative investigations using new approaches or new methods of study. It is therefore highly desirable that any newly described green algal taxon is made available as a laboratory culture to recognized Culture Collections.

The flagellar apparatus is one of the most complex organelles in a green algal cell both biochemically and structurally. Because of its complexity and fairly wide distribution among all major groups of green algae, it should potentially be a very useful character in the systematics of green algae. This is even more likely since the flagellar apparatus is generally regarded as one of the most ancient organelles in a eucaryotic cell and in such a cell flagella may have evolved even before the acquisition of mitochondria and chloroplasts (see Stewart and Mattox, 1980). The structural complexity of the flagellar apparatus is therefore most likely a reflection of its long evolutionary history. The enormous variability of flagellar apparatus ultrastructure in the green algae could then be viewed as indicating that they might represent one of the most ancient groups of eucaryotic flagellated algae. In recapitulating phylogeny, flagellated reproductive cells produced by non-flagellated vegetative green algae will show some characters in their flagellar apparatus that resemble those of their ancestral flagellates while exhibiting also additional derived characters. By taking into account functional transitions in the flagellar apparatus it is often possible to derive advanced green algae (those green algae exhibiting multicellular organization) from a particular type of ancestral green flagellate, and it is even possible to find present-day flagellates which might be regarded as related to those ancestral types.

It is the purpose of this review to describe the 100 or so structural characters in the flagellar apparatus of green algae and to indicate their structural variability and taxonomic importance. Emphasis will be given to those characters that have not previously been reviewed or for which new and unpublished information is now available. These include the flagellar tip, the flagellar shaft, the transition region and the basal body. The principal lines of evolution within the group will be indicated. An attempt is made to relate the flagellar apparatus in motile cells of advanced green algae to that of present-day green flagellates. Limitations in using flagellar apparatus ultrastructure for green algal classification will be discussed.

THE ULTRASTRUCTURAL CHARACTERS AND THEIR STRUCTURAL VARIABILITY

No terminology and classification of flagellar apparatus ultrastructure in green algae may be unequivocally logical or otherwise totally satisfying, and this may simply be a reflection of the functional unity that the flagellar apparatus represents and which makes separation into distinctive parts extremely difficult. The flagellar apparatus of green algae has been subdivided into three parts: the flagellum, the basal body and associated structures of the basal body (Melkonian, 1982a). The flagellum is covered by the flagellar membrane and contains as its principal structural component the axoneme. The axoneme is embedded in the flagellar matrix and the flagellar membrane is associated with extracellular surface components. Along its longitudinal axis three parts may be distinguished in a flagellum: the flagellar tip, the flagellar shaft and the flagellar transition region, the latter linking external with internal parts of the flagellar apparatus. Each of these three regions of a flagellum contains the above-mentioned components (membrane, axoneme, matrix, surface components), and there is increasing evidence in the green algae that most of these components are different and specialized in the three regions of a flagellum. The basal body is that part of the flagellar apparatus which gives rise to and is continuous with the flagellar axoneme. In a more restricted sense it is a cylinder of microtubule triplets which in a flagellated cell is continuous with the outer microtubule doublets of the axoneme. In practice the delimitation of the basal body from the proximal part of the flagellum (the transition region) is often difficult. Associated structures of the basal body include all structures that are directly attached to the basal body triplets. They usually include the so-called connecting fibres which interlink different basal bodies and flagellar roots—amorphous, fibrillar or microtubular structures which attach at basal bodies and terminate somewhere in the cell. Since the flagellar apparatus of green algae is fully integrated structurally and functionally into the cytoplasm, it has numerous positional and structural links with other organelles and structures in the motile green algal cell, of which some have only comparatively recently been evaluated (for a review, see Melkonian, 1984). These organelles and structures, including the so-called secondary cytoskeletal microtubules which are organized at microtubular flagellar roots, are generally not regarded as an intrinsic part of the flagellar apparatus, although they may be functionally related to it and may have been most important in evolutionary transitions of the flagellar apparatus of green algae (see Melkonian, 1982b).

1. The Flagellar Tip

That the distal part of a flagellum is often structurally specialized has already been noted by light microscopists such as Loeffler (1889). Shadow-cast preparations for electron microscopy revealed a tremendous variation of flagellar tip structure in the algae in general and in green algae in particular. A summary and synthesis of this early ultrastructural information on green algae was provided by Manton (1965), who also indicated the potential usefulness of flagellar tip characters for the systematics and phylogeny of the algae. Later reviews of flagellar tip ultrastructure in the green algae include Norris (1980) on prasinophycean flagellates, Moestrup (1982) on algae in general and Melkonian (1982a) on green algae in particular. It is unfortunate that information on the internal structure of flagellar tips in green algae is lagging behind that provided some two decades ago for the external structure of the flagellar tip. In most of the recent detailed studies on the flagellar apparatus ultrastructure in selected green algae the emphasis is on the basal apparatus and no information on the flagellar tip is provided. Despite these limitations a survey of published information and some unpublished results strongly suggest that the flagellar tip may be a useful character for classifying green algae. It is, however, necessary to take into account that the flagellar tip is a highly dynamic structure that changes considerably in external shape and in internal structure with the flagellar beat cycle.

In the flagellar tip of green algae one can distinguish the flagellar tip surface (including different types of scales, flagellar hairs and other surface coats), the tip membrane and the internal structures of the tip (axoneme, matrix). It is of course not easy to define where the flagellar tip begins and where the flagellar shaft terminates. The flagellar tip is here defined as the region of the flagellum from the termination of the outer axonemal doublets to the actual tip in a non-sliding (straight) flagellum. It should be further mentioned that individual outer axonemal doublets may be of genetically fixed unequal length, thus complicating the definition given above. Preliminary evidence has been presented that in some green algae two opposite outer axonemal doublets may be considerably longer than other outer doublets (Melkonian, 1982a). The tip membrane may be structurally specialized, but it is extremely difficult to visualize a tip membrane specialization, for instance, in freeze–fracture preparations (e.g. McLean et al., 1981). That the tip membrane is functionally most important during mating in some green algae is well established (e.g. Mesland et al., 1980). The internal structures of the flagellar tip in green algae include one structure which links the central pair microtubules (CPM) to the tip membrane (for review of these structures in other eucaryotes see Dentler, 1981). This linking structure consists of several transverse plates and is termed the central microtubule cap (CMC; Dentler and Rosenbaum, 1977). Other internal features of the flagellar tip include: the presence of electron-dense material between the two central pair microtubules, relative length of the CPM in relation to the outer doublets, and the way in which the CPM and the outer axonemal doublets terminate (are the two CPM of equal lengths?, are B-tubules shorter or longer than A-tubules?, what is the individual length of outer doublets?, are A-tubules always attached to the flagellar membrane by a "distal filament"?, etc.). Taken together the above-mentioned structural features of the flagellar tip represent about 15 individual characters, mostly qualitative. For most of these characters there is very limited information about their distribution and variability within the green algae and more comparative research is needed. However, information on some characters (e.g. length of the flagellar tip, occurrence of special flagellar hairs associated with the flagellar tip, and the presence or absence of a CMC) is already proving to be of taxonomic value.

Early electron microscope studies on whole mount preparations with the shadow-cast technique indicated that mainly two types of flagellar tips occurred in the green algae: blunt-ended tips and those with hair-points of variable length (see Manton, 1965; Moestrup, 1982). It later became apparent that the obvious difference between the two types is in the relative length of the CPM to the outer axonemal doublets, i.e. blunt-ended flagella have relatively short CPM relative to outer doublets, whereas flagella with long hair-points are characterized by relatively long CPM (again relative to outer doublets). Since the apparent length relationship between outer doublets and CPM changes during the flagellar beat cycle due to microtubule sliding (see Satir, 1968) both the external shape of the flagellar tip and its internal structure vary accordingly during the flagellar beat cycle. This may be illustrated by two examples.

Pedinomonas species are generally described as having unilaterally, slightly pointed or acute flagellar tips of 0.8–0.9 μm in length (Manton and Parke, 1960; Belcher, 1968b). A reinvestigation of *Pedinomonas tuberculata* (unpublished observations) has shown that such a flagellar tip only occurs when the flagellum is folded back around the cell body (the preferential flagellum orientation in *Pedinomonas;* see Ettl and Manton, 1964) i.e. the flagellum is maximally bent (maximal microtubule sliding). The internal structure of the flagellar tip shows that microtubule sliding has occurred in opposite outer doublets in such a way that one outer doublet extends to the very tip of the flagellum and is of apparently equal length to the CPM, which also extend to the tip membrane. The opposite set of outer doublets abruptly terminates some 0.8 μm from the actual tip of the flagellum,

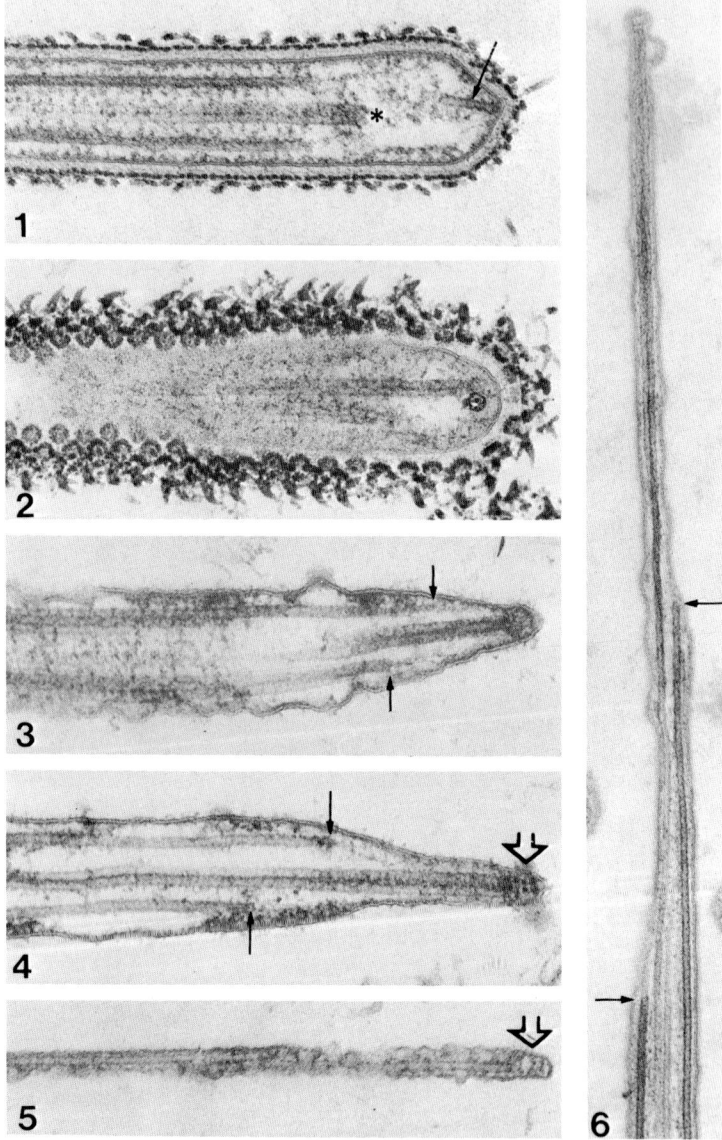

Fig. 1. Longitudinal section through flagellar tip of *Tetraselmis cordiformis* (× 52 000). The cells were rapidly fixed and the preferential flagella orientation was that of two pairs of flagella folded back around the cell body. *Arrow:* single microtubule of one outer doublet projects into the tip. *Asterisk:* the CPM have apparently detached from the flagellar tip membrane.

Fig. 2. Flagellar tip (slightly oblique longitudinal section) of *Nephroselmis olivacea*. (× 52 000). Note the blunt-ended flagellar tip and extension of CPM to the tip membrane.

giving rise to a unilaterally-pointed flagellum. The real length of the flagellar tip (see definition above which relates to a non-sliding flagellum) may be calculated by dividing the length distance between opposite outer doublets by two. This gives the actual length of the tip to be about 0.4 μm in a non-sliding flagellum of *Pedinomonas tuberculata*. This example clearly shows how microtubule sliding may affect both the external shape of a flagellar tip (as seen in shadow-cast preparations) and its internal structure. It is apparent that the effect of microtubule sliding in *Pedinomonas* obscures the fact that *Pedinomonas* has a rather blunt-ended flagellum (definition of a blunt-ended flagellum: relative length of CPM to outer doublets less than 0.5 μm). The effect of microtubule sliding on flagellar tip ultrastructure is most drastic when the relative length of the CPM is very short (extremely blunt-ended flagella) and when great angles of bending occur during the flagellar beat cycle. This is illustrated in the second example.

It was reported that in *Chara* spermatozoids the outer doublets on one side of the flagellar tip terminate before those on the other (indicating microtubule sliding) and the CPM do not reach the tip (Pickett-Heaps, 1968; Moestrup, 1982). In this case a very short CPM (relative to outer doublets) has presumably been detached from the tip membrane by microtubule sliding. A similar observation was made for the blunt-ended flagellar tip of the prasinophycean flagellate *Tetraselmis cordiformis* in maximally bent flagella (Fig. 1). This alga was rapidly fixed and the flagella orientation in most cells was such that the four flagella were bent backwards in pairs along opposite sides of the cell body. When flagella tips of such cells were investigated by thin-section electron microscopy it was most surprising to find that the tips were still blunt-ended, although slightly pointed (see Fig. 1) when compared to straight flagella. Most interesting, however, was the observation that the CPM detached from the tip membrane and that a single microtubule of one of the outer doublets was in contact with the tip membrane and extended this membrane (see Fig. 1). On several occasions it was found that this microtubule was somewhat curved inside the flagellar tip (indicated in Fig. 1 by the microtubule passing out of the plane of section). It is here suggested that this configuration is similar to the one observed in

---

Figs 3, 4. Two longitudinal sections through the flagellar tip of *Chlorosarcinopsis pseudominor* zoospores (× 68 000). *Small arrows* indicate amount of microtubule sliding in opposite outer doublets; *open arrow*: central microtubule cap.

Fig. 5. Flagellar tip (hair-point) of *Friedmannia israelensis* zoospores (× 52 000). *Open arrow*: central microtubule cap.

Fig. 6. Longitudinal section through the flagellar tip region of *Scourfieldia caeca* (× 42 500). Note the enormous amount of microtubule sliding in opposite outer doublets and the relatively long hair-point.

Chara spermatozoids, the only difference being that the flagellum (possibly the membrane or the scales) exerts some resistance to microtubule sliding and thus an elongated flagellar tip is not produced. Due to the force that the maximally slided outer microtubule exerts on the tip membrane the CPM presumably detached from the tip membrane in this species too. It is interesting to note that we have not seen a central microtubule cap linking the distal ends of the CPM to the tip membrane in any prasinophycean flagellate with blunt-ended flagella. Has the CMC evolved in blunt-ended flagella to prevent the periodic detachment of the CPM during the flagellar beat cycle, which could damage the flagellar tip membrane? What happens to the CPM in a flagellate such as *Chlamydomonas reinhardii* that has blunt-ended flagella and a CMC? It is apparently difficult for a flagellate with blunt-ended flagella and a CMC to perform a very powerful and effective stroke of the flagellum. Has this problem been elegantly circumvented by the evolution of flagella tips with hair points, as are now found in all groups of advanced green algae (with the possible exception of the Charophyceae)? In flagella that are either very long or undergo a large amount of bending there is usually a very prominent hair-point, i.e. a relatively long CPM. For example, the peculiar backward-swimming *Scourfieldia caeca* bends the very long flagella backwards along the cell body when resting. In such cells there occurs a tremendous amount of microtubule sliding (1.2 μm) but the flagellar tip is also much longer (Fig. 6). These are several questions for which there is at present no answer, but it is clear that the flagellar tip may be functionally very important and may yield useful information for the gross classification of green algae if studied in more detail.

A survey of flagellar tip lengths in the green algae (over 50 different green algae were evaluated) shows that all flagellates with scaly flagella exhibit blunt-ended flagellar tips. From the foregoing discussion it is clear that flagellar tip lengths can only be accurately measured from thin sections, not from shadow-cast preparations. Most of the information available on flagellar tip lengths, however, comes from shadow-cast preparations, and therefore a certain amount of uncertainty must be attached to the actual lengths which have been measured from published micrographs. To measure flagellar tip lengths accurately it is only necessary to have a median longitudinal section through the flagellar tip in either a non-sliding or a sliding condition. In the non-sliding condition (also in a flagellum in which sliding has occurred but in which the section is perpendicular to the two opposite groups of outer doublets that have slided maximally) the flagellar tip length is simply measured as the distance between the termination of outer doublets (measured from the point where the outer doublet gives rise to a thin distal filament) and the termination of CPM. In a flagellum which exhibits microtubule sliding, the flagellar tip length is measured as the distance between the termination point of the longest outer doublet and the termination of the CPM plus the distance divided by two between the termination points of opposite outer doublets. Interestingly, the flagellar tips of the scaly flagella of the Charophyceae are also blunt-ended (Moestrup, 1970, 1974). The slightly-pointed flagellar tip of *Chaetosphaeridium* zoospores (length 0.8 μm) can be easily accommodated within the blunt-ended flagella assuming that most of its length as measured from shadow-cast preparations is due to microtubule sliding (Moestrup, 1974). The same appears to be true for the scaly flagellate *Mantoniella squamata* in which one report documents a blunt-ended flagellum (Barlow and Cattolico, 1980) whereas another shows a shortly-pointed flagellar tip (again about 0.8 μm long). The latter was interpreted as an artifact (Moestrup, 1982) but may well relate to the external shape of the flagellar tip of a sliding flagellum. In addition to their association with scaly flagella, blunt-ended flagellar tips also occur in a small group of green flagellates which lack flagellar scales. These are *Asteromonas* (Floyd, 1978, judged from SEM micrograph), *Hafniomonas* (Belcher, 1968a), *Chlamydomonas reinhardii* (Ringo, 1967a; Witman *et al.*, 1972; Snell, 1976; Bergman *et al.*, 1975), *Pteromonas* (Belcher and Swale, 1967) and *Furcilla* (Belcher, 1968c). Quite interestingly, two of these flagellates have been assumed on other grounds to bear a relationship to scaly green flagellates, e.g. *Hafniomonas* (Ettl and Moestrup, 1980; Melkonian, 1981a) and *Asteromonas* (Floyd, 1978), both being forms with shallow apical flagellar grooves similar to the deeper grooves found in quadriflagellate prasinophycean taxa. Blunt-ended flagella may also be characteristic for most colonial flagellates related to *Chlamydomonas*.

In three advanced groups of green algae (the Ulvophyceae; the advanced members of the Chlorophyceae, and the Pleurastrophyceae; see Mattox and Stewart, this volume, Chapter 2) hair-points occur in the flagellar tip (relative length of CPM to outer doublets more than 1 μm). Hair-points around 2 μm in length appear to be common in both the Ulvophyceae (Manton and Friedmann, 1960; Evans and Christie, 1970; Hoops *et al.*, 1982) and the Pleurastrophyceae (Watson, 1975, and Fig. 5 for *Friedmannia israelensis*). In most advanced chlorophycean motile cells the hair-point is admittedly slightly shorter (1–2 μm, Figs 3,4). In most of these organisms, however, critical thin section work is still lacking.

The longest hair-points known in the green algae occur in some peculiar very small uni- or biflagellates with naked flagella. These are *Scourfieldia* (Manton, 1975; Melkonian and Preisig, 1982; see Fig. 6), *Micromonas pusilla* (Manton, 1959; Manton and Parke, 1960; Moestrup, 1982), and *Monomastix* (Manton, 1967). In *Monomastix* the flagellar tip is about 5–8 μm long and there is apparently only a single microtubule in the flagellar tip region (Manton, 1967). Partly on the basis of this elongated flagellar tip these flagellates have been grouped into a separate class of green algae, the Loxophyceae *sensu* Christensen (Christensen, 1962; Moestrup, 1982), but from the discussion above it is evident that hair-points have probably evolved in the advanced groups of green algae several times. There is thus no good reason to relate the three genera of flagellates on the basis of their flagellar tip ultrastructure. It is simpler to assume that these flagellates evolved long flagellar tips independently. Such flagellar tips allow extensive flagella bending and consequently microtubule sliding. The loss of scales might have been instrumental in the evolution of hair-points in the different groups of advanced green algae. One of the reasons for the occurrence of blunt-ended flagella in *Chlamydomonas*-type flagellates might lie in the presence of a rather thick flagellar surface coat (e.g. *Pteromonas, Chlamydomonas, Asteromonas*) which may functionally be related to flagellar scales. Could the flagellar surface coat (which in a reduced form also occurs in some "primitive" motile cells of the Ulvophyceae; see Moestrup, 1982, for flagellated cells of *Monostroma grevillei*) represent the remnants of a scaly flagellar covering?

The genus *Pedinomonas,* which has sometimes been aligned with the Loxophyceae (Moestrup, 1982) cannot on the basis of its rather blunt-ended flagella, be accommodated in this class. Other features of the flagellar apparatus ultrastructure support this conclusion (see below).

In some green algae the flagellar tip is extracellularly associated with special types of flagellar hairs (see Moestrup, 1982). These hairs are best studied in negative stain preparations of whole flagella, which also enable the investigator to decide whether they are tubular or non-tubular (Figs 7 and 8). The flagellar hairs associated with the flagellar tip of *Scourfieldia caeca* (Melkonian and Preisig, 1982; see Fig. 8) are non-tubular. Hairs on the flagellar tip of *Mantoniella squamata* (Manton and Parke, 1960; Barlow and Cattolico, 1980) are thinner and twice the length compared to the tubular hair-scales which occur along the flagellar shaft of this species, but apparently they have not been studied in negative stain preparations and a conclusion regarding their tubular or non-tubular nature cannot be drawn at present. In *Spermatozopsis,* which is a *Chlamydomonas*-type green alga, a special set of non-tubular flagellar hairs occurs attached to the flagellar tip membrane (Melkonian and Preisig, 1984). The function of these special flagellar hairs is unknown.

2. *The Flagellar Shaft*

As with the flagellar tip, the flagellum proper or flagellar shaft can be subdivided into the surface, the membrane, the axoneme and the matrix. Whereas features of the flagellar surface (e.g. scales, flagellar hairs and the tomentum) are generally regarded as excellent structural characters in rela-

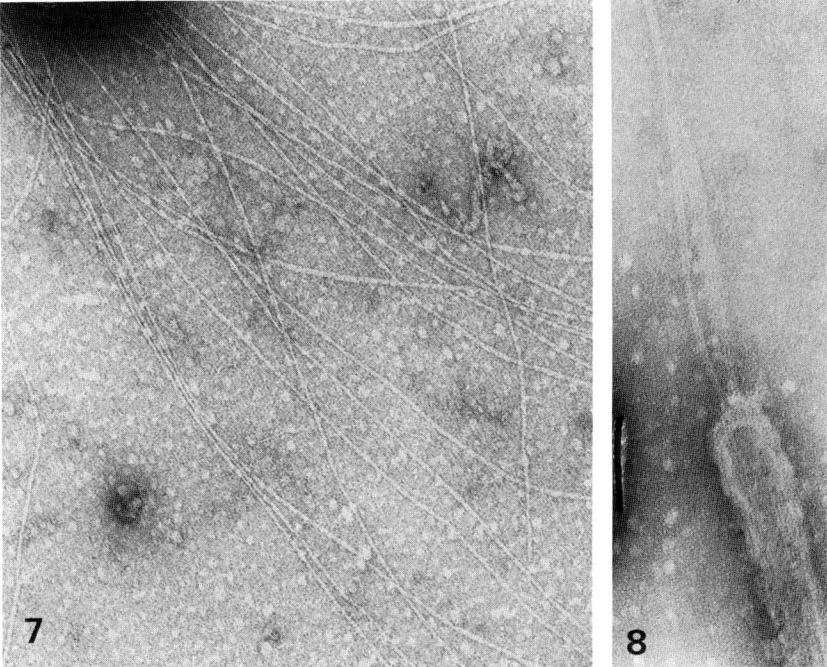

Fig. 7. Negative stain preparation of a flagellum of *Pedinomonas tuberculata* indicating the presence of long non-tubular flagellar hairs associated with the flagellar tip (× 100 000).

Fig. 8. Negative stain preparation of a flagellum of *Scourfieldia caeca* (× 100 000). Special sets of non-tubular flagellar hairs are exclusively associated with the flagellar tip.

tion to green algal classification (see recent reviews by Norris, 1980, and Moestrup, 1982), the internal features of the flagellar shaft, often simply referred to as the "9+2," are generally viewed as "too conservative and too anciently established to be useful" for phylogenetic considerations (Stewart and Mattox, 1980). This short overview intends to show that details of the flagellar axoneme can be very useful for derivation of concepts about the principal evolutionary lines in the green algae.

Since flagellar surface structures in green algae have been adequately reviewed elsewhere (Norris, 1980; Moestrup, 1982) it may be sufficient here just to summarize briefly the major aspects and add some new and unpublished observations.

The presence of flagellar scales is one of the most distinctive features of prasinophycean flagellates. Each genus is usually characterized by its specific complement of flagellar scales, and in some genera (e.g. *Pyramimonas*) it is even possible to distinguish species and strains on the basis of flagellar scale structure (e.g. Pennick et al., 1978). Up to four morphologically-different types of flagellar scales may occur on a single flagellum, usually in separate layers (e.g. *Nephroselmis olivacea;* Moestrup and Ettl, 1979). On the basis of flagellar scale structure three major groups can be distinguished among scaly green flagellates; the first group contains the three genera *Mantoniella*, *Mamiella* (formerly *Nephroselmis gilva sensu* Parke and Rayns, 1964; see discussion in Moestrup, 1983) and *Dolichomastix* (Manton, 1977). These flagellates have a single layer of spider web-like scales on their flagellar surface which is very similar to the scaly covering on the cell body. Only *Mamiella* has flagellar scales which are structurally more different than the two types of cell body scales (Moestrup, 1984). They are characterized by an adnate spine on their dorsal side which points in the direction of the flagellar tip. Similar adnate spines occur as a structural component of the intermediate flagellar scales in some other prasinophycean genera, such as *Pyramimonas* and *Pachysphaera,* and it is quite possible, as already noted by Manton (1977), that both types of flagellar scales with adnate spines are homologous, in which case the *Pyramimonas*-type is probably derived from a *Mamiella*-type. More investigations into this possible relationship may be profitable.

In the other prasinophycean flagellates a new and different type of flagellar scale is present: the pentagonal or square-shaped small scales which are often called "underlayer scales" since when they occur they are always the layer closest to the plasma membrane. If homology exists between the spider web scales of *Mamiella* and the intermediate flagellar and body scales in *Pyramimonas,* then evolutionarily speaking the pentagonal or square-shaped scales have evolved as an underlayer to the spider web scales. Because of their peculiar structure the pentagonal scales can be assumed to have evolved only once. In this case two groups of advanced green algae, namely the Ulvophyceae and the Charophyceae, may be related to ancestral flagellates which contained a layer of pentagonal or square-shaped scales on the cell body and flagella, since in both groups flagellated reproductive cells occur which have this type of scale either on both the cell body and flagella (the Charophyceae) or on only the cell body (the Ulvophyceae). This feature is one of the most conclusive evidences that scaly green flagellates were ancestral to these two classes of advanced green algae.

The third group of prasinophycean flagellates is characterized by an additional flagellar scale type, replacing the spider-web scales, recently designated as "rod-shaped scales with irregular margin" (Melkonian, 1982c). These scales occur as a second layer on top of the pentagonal or square-shaped scales in the genera *Pseudoscourfieldia* (Manton, 1975), *Nephroselmis* (Manton et al., 1965; Moestrup and Ettl, 1979; Moestrup, 1983), and

Tetraselmis (Manton and Parke, 1965; Norris *et al.*, 1980; Melkonian, 1982c). The rod-shaped scales have been shown to occur as "double scales" on the flagellum of *Tetraselmis cordiformis* (Melkonian, 1982c) and two *Nephroselmis* species (Moestrup and Ettl, 1979; Moestrup, 1983). Very interestingly, these "double scales" also occur on the cell body of *Pseudoscourfieldia marina* (Manton, 1975). In *Nephroselmis* there is good evidence that in the more primitive marine species, e.g. *N. rotunda,* the so-called stellate scales which overlie the square-shaped underlayer scales on the cell body are in fact rod-shaped "double scales" in which the two rod-shaped scales are linked to each other at right angles to produce a "new" morphological scale type. All types of intermediate forms may be seen in some of the published micrographs of *Nephroselmis* (=*Heteromastix*) *rotunda* (Manton *et al.*, 1965; M. Melkonian, unpublished observations on the same strain). It is also significant that during morphogenesis of the *Tetraselmis* theca apparently rod-shaped scales, initially termed "stellate" scales, are interconnected by fibrillar material, and finally in a self-assembly process form the characteristic theca of *Tetraselmis* (Manton and Parke, 1965; Domozych *et al.*, 1981; M. Melkonian, unpublished observations). In addition to these types of flagellar scales, all three genera have a type of tubular hair-scale of identical length and substructure (Manton, 1975; Moestrup and Ettl, 1979; Melkonian, 1982c). These hair-scales occur in two rows on opposite sides of a flagellum and at their attachment sites the pentagonal underlayer scales are structurally modified and occur as two rows of square-shaped scales (Melkonian, 1982c, for *Tetraselmis cordiformis;* Moestrup, 1983, for *Nephroselmis pyriformis*). Since these square-shaped underlayer scales are easily recognized in both tangential flagellar sections (Figs 9, 11, and 12) and in cross sections (Fig. 10), it is now possible to state that two rows of square-shaped underlayer scales not only occur in the three genera of prasinophycean flagellates discussed above but also in *Pyramimonas* species [Figs 11 and 12, and are illustrated in the type species of *Pyramimonas, P. tetrarhynchus* (Moestrup and Walne, 1979)]. It may well be universally found that two opposite rows of square-shaped underlayer scales occur in prasinophycean flagellates that contain pentagonal flagellar scales and hair-scales. *Mesostigma,* which lacks hair-scales but has pentagonal underlayer scales, apparently lacks the two rows of square-shaped underlayer scales (unpublished observations).

The conclusion from the above considerations of flagellar scale structure in scaly green flagellates is that three groups can be distinguished, the *Mamiella* group, the *Pseudoscourfieldia* group, and the *Pyramimonas* group (to which the genera *Halosphaera, Pterosperma* and *Pachysphaera* also belong). *Mesostigma* in its flagellar scale structure is somewhat aberrant, but other structural features of the flagellar apparatus indicate that it is related to the

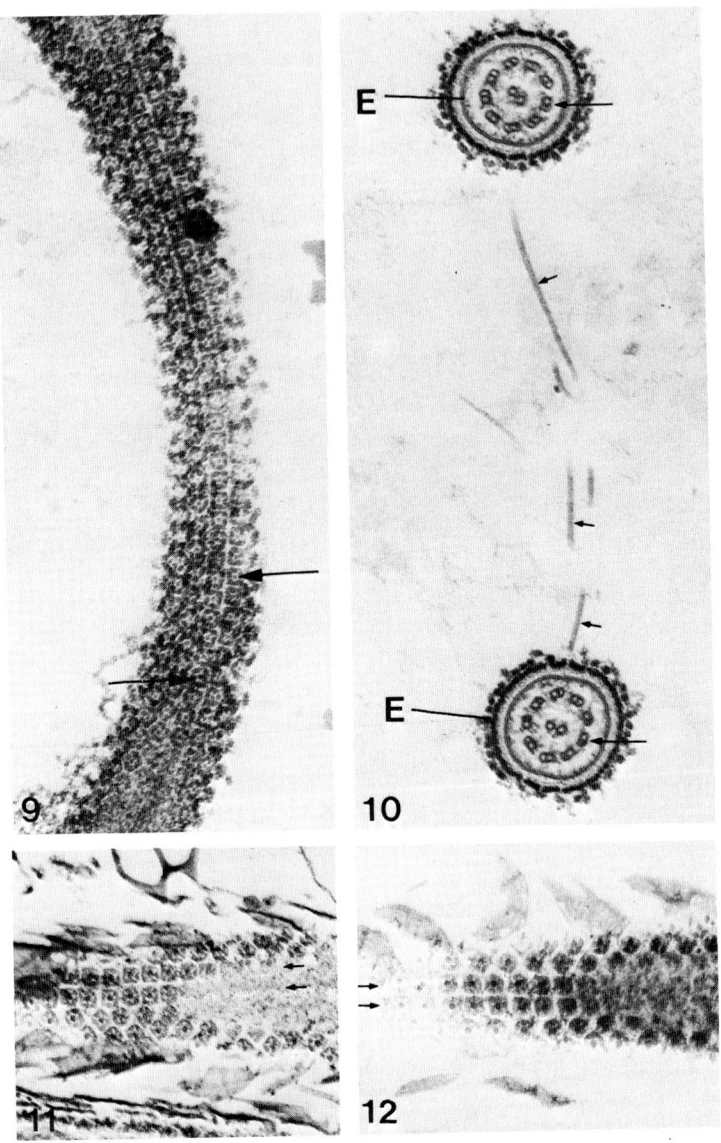

Fig. 9. Tangentially sectioned flagellum of *Tetraselmis cordiformis* (× 51 300). Note that two rows of square-shaped scales occur (*arrows*), and these are located perpendicular to the plane of bending.

Fig. 10. Flagella cross sections from *Tetraselmis cordiformis* (× 51 000). These are two coupled flagella extending along one broad side of the cell. Both flagella have the same absolute orientation indicated by the position of a peculiar outer doublet lacking the outer dynein arm (*long arrows*). E = direction of the effective stroke of the two flagella. Hair-scales (*short arrows*) from the two flagella occur in one plane and distally nearly touch each other.

Figs 11, 12. Tangential flagellar sections in *Pyramimonas parkeae* (Fig. 11) and *Pyramimonas amylifera* (Fig. 12) showing two rows of square-shaped underlayer scales (*arrows*) (× 51 300).

genus *Pyramimonas*. The close relationship on the basis of flagellar scale structure between *Pseudoscourfieldia, Nephroselmis*, and *Tetraselmis* is somewhat surprising since these organisms differ in their mitosis and cytokinesis and in the detailed structure of the flagellar root system. However, they seem to be a monophyletic group which may have been important in the evolution of advanced green algae. It is in this group that the phycoplast presumably originated (possibly closely linked to the occurrence of rod-shaped scales and their transformation into a theca).

Flagellar hairs of a type very different from the tubular hair-scales of the prasinophytes occur on the flagellar surface of other flagellated green algae (see Moestrup, 1982). Most intensively studied are the non-tubular flagellar hairs of *Chlamydomonas reinhardii* (Ringo, 1967a; Witman *et al.*, 1972; Bergman *et al.*, 1975; Snell, 1976). These hairs can be seen in shadow-cast preparations or by negative staining, but are very difficult to see in thin sections. Although their occurrence in two rows on opposite sides of a flagellum has been established by freeze-fracture electron microscopy (Bergman *et al.*, 1975), their relationship to the tubular hair-scales and to the axonemal doublets (see Melkonian, 1982c, for a discussion of these aspects in *Tetraselmis cordiformis*) remains unknown. The width of the non-tubular flagellar hairs of *Chlamydomonas reinhardii* was given as 16 nm by Witman *et al.* (1972), 12–13 nm by Bergman *et al.* (1975) and 15 nm by Snell (1976) from negatively stained preparations. These measurements are not very different from each other and indicate that the non-tubular hairs of *Chlamydomonas* are not thinner than the prasinophycean hair-scales. The reports in the literature that non-tubular flagellar hairs in the green algae are extremely thin might be an artifact of the shadow-cast technique, since the non-tubular flagellar hairs are flattened and may not give the same image in shadow-cast preparations compared to the tubular, cylindrical hair-scale. The poor staining behaviour in thin section of the flagellar hair of *Chlamydomonas* compared to prasinophycean hair-scales is probably related to a different chemical composition of the two flagellar hair types. In accordance with this view the flagellar hair of *Chlamydomonas reinhardii* consists of a single glycoprotein of about 200 000 daltons, whereas preliminary biochemical analysis of hair-scales from *Tetraselmis striata* show that proteins do not form a significant portion of the hair-scale (Melkonian *et al.*, in preparation).

Non-tubular flagellar hairs probably occur in most *Chlamydomonas*-type flagellates (e.g. *Chlamydomonas moewusii*, McLean *et al.*, 1974; *Pteromonas*, Belcher and Swale, 1967) and they are also present in zoospores of *Chlorosarcinopsis dissociata* (unpublished observations). It is suggested that non-tubular flagellar hairs are much more widely distributed among the green algae than previously thought. The thin flagellar hairs of *Pedinomonas* species closely resemble those found in *Chlamydomonas,* as previously pointed out by Witman *et al.* (1972). In *Pedinomonas* they occur in two rows on opposite sides of a flagellum, are of apparently similar length (1 μm) and have the same substructure for *Pedinomonas tuberculata* (see Fig. 7). They are non-tubular in negative stain preparations and flattened, and their width may be given as 5–10 nm depending on whether the flat or the narrow side is exposed. Although similar to the flagellar hairs in *Chlamydomonas,* the hairs in *Pedinomonas tuberculata* are slightly thinner, and conclusions about homology to non-tubular flagellar hairs in other green algae must await biochemical characterization of the *Pedinomonas* flagellar hair. The occurrence of long, thin, non-tubular hairs cannot at present be regarded as an exclusive characteristic of a separate class of green algae, the Loxophyceae *sensu* Christensen (Moestrup, 1982).

The flagellar axoneme is functionally the most important part of a flagellum, since it contains the machinery for microtubule sliding and controlled bending and the motor for these movements, the dynein-ATPases (see Satir and Ojakian, 1979). Structural variations in axonemal structures have mainly been observed and studied in motility mutants of *Chlamydomonas reinhardii* (e.g. Witman *et al.,* 1978). The general structure of the flagellar axoneme was described in detail for *Chlamydomonas reinhardii* (e.g. Ringo, 1967a,b; Hopkins, 1970; Witman *et al.,* 1972, 1978) and will not be repeated here. Attention will be drawn, however, to five structural specializations of the flagellar axoneme in different green algae, and an absolute configuration of the green algal flagellum (see Fig. 19) will be established.

In some recent critical studies on flagellar apparatus ultrastructure of certain green algae, it has consistently been found that one of the outer axonemal doublets lacks the outer dynein arm (Melkonian, 1982a,c; Melkonian and Preisig, 1982; Hoops and Witman, 1983; see also Figs 10, 13, 16, and 17). A critical examination of further flagellated cells of green algae and evaluation of published micrographs of flagellar cross sections have produced clear evidence that this character is universal in the green algae (see

---

Fig. 13. Cross sections through the two flagella of a cell of *Scourfieldia caeca* near their proximal origin. (× 80 000). One axonemal outer doublet *(arrow)* lacks the outer arm in both flagella being distributed on opposite sides of the two flagella.

Fig. 14. Cross section through the flagellum of *Volvox globator* in a region just distal to the basal body triplet origin (× 80 000). *Arrow* indicates prominent A–B-intermicrotubule link.

Fig. 15. Cross sections through the flagellar transition region in the two flagella of a vegetative cell of *Volvox globator* (× 80 000). Each outer doublet in the transition region is linked to the flagellar membrane by prominent electron-dense connectives *(arrow).* The lower flagellar cross section shows the spectacular stellate pattern of this species.

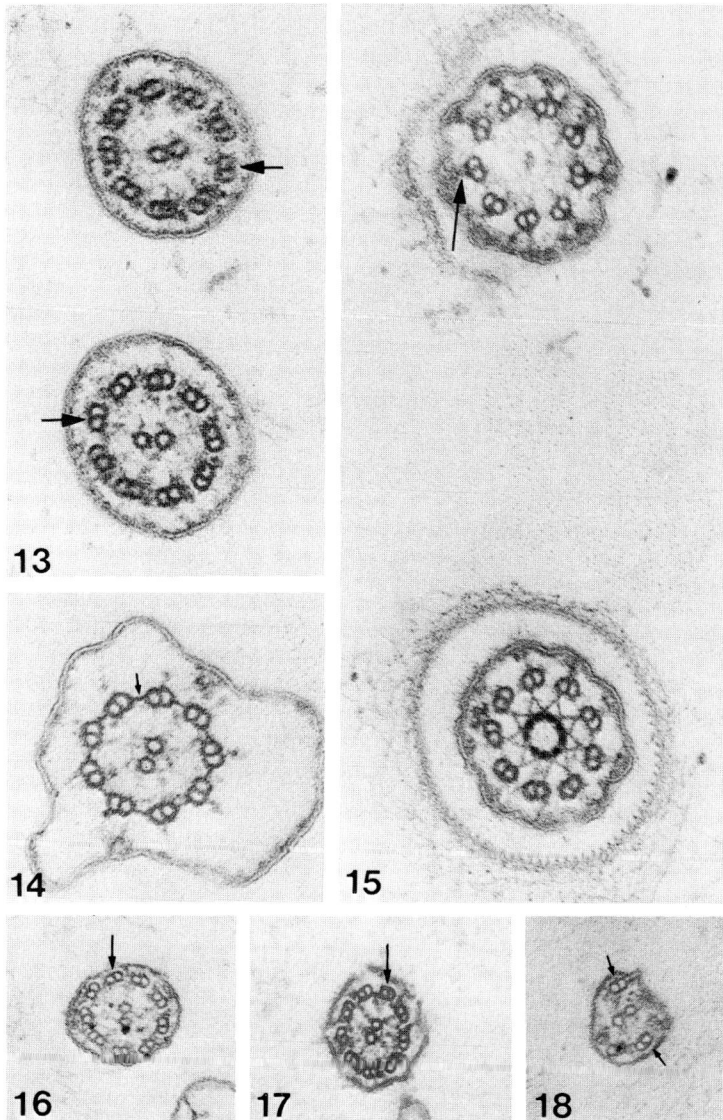

Fig. 16. Flagellar cross section of a zoospore of *Pleurastrum terrestre* with one outer doublet lacking the outer dynein arm (*arrow*) (× 63 000).

Fig. 17. Flagellar cross section through a zoospore of *Fritschiella tuberosa* again with one outer doublet lacking the outer dynein arm (*arrow*) (× 63 000).

Fig. 18. Flagellar tip region in cross section of a zoospore of *Pleurastrum terrestre* (× 67 500). The two *arrows* indicate two outer doublets which are apparently longer than the other outer doublets.

Table I. The occurrence of a single outer axonemal doublet which lacks the outer dynein arm in the green algae.

Alga	Reference
Acetabularia (gamete)	Berger et al. (1975)
Cephaleuros (gamete)	Chapman and Henk (1983)
Chlamydomonas reinhardii (veg. + gamete)	Ringo (1967a,b), Bermann et al. (1975), Witman et al. (1978), Huang et al. (1979), Moestrup (1982)
Chlorosarcinopsis (2 spp.) (zoospore)	Melkonian (1982a)
Dunaliella primolecta	Hyams and Chasey (1974)
Fritschiella tuberosa (zoospore)	This paper
Golenkinia minutissima (spermatozoid)	Moestrup (1972)
Microthamnion kuetzingianum (zoospore)	Watson (1975)
Nephroselmis (3 spp.)	Moestrup and Ettl (1979), Moestrup (1983), M. Melkonian (unpublished observations)
Oedogonium (zoospore)	Pickett-Heaps (1971)
Pleurastrum terrestre (zoospore)	This paper
Scourfieldia caeca	Melkonian and Preisig (1982)
Stigeoclonium (zoospore)	Manton (1964a)
Tetraselmis cordiformis	Melkonian (1982c)
Trentepohlia (gamete?)	Graham and McBride (1975)
Ulva lactuca (male gamete)	This paper
Volvox globator (veg.)	M. Melkonian (unpublished observations)

Table I, which lists 20 species from 17 different genera). Since at present no other eucaryotic flagellated cell is known to exhibit this structural peculiarity, it may provisionally be regarded as one structural character of a green algal flagellum. Although a single axonemal doublet apparently lacking the outer dynein arm is seen in micrographs of green algal flagella prepared as early as 1964 (Manton, 1964a), it seems to have escaped attention until 1979, when Huang et al. (1979) reported its presence in *Chlamydomonas reinhardii*. These authors found this peculiar doublet in virtually all of their flagella cross-sections (only one doublet out of the nine per section) and noted that it had no constant positional relationship with the CPM. At that time it was not certain if this was a particular doublet or if any doublet around the flagellar axoneme could show this structural specialization. A significant advance was made recently when it was shown in *Tetraselmis cordiformis* that there was a second independent structural marker on the axoneme which always had a constant positional relationship to the doublet lacking the outer arm (Melkonian, 1982c). The two opposite rows of hairscales attached through the flagellar membrane to B-tubules of two axonemal doublets, and a constant relationship between these two doublets and the doublet lacking the outer arm was always present (Melkonian,

1982c; see Fig. 10). A constant relationship to the central pair microtubules did not exist in *Tetraselmis*. This is not surprising, since there is evidence that CPM are not straight along the flagellar shaft but occur in a twisted configuration (Melkonian and Preisig, 1982, for *Scourfieldia caeca*). Subsequent studies have shown that the positional relationship between the attachment sites of hair-scales and the single doublet lacking the outer arm also holds true for other scaly flagellates (*Nephroselmis pyriformis*, Moestrup, 1983). Other independent structural markers on axonemal doublets confirm this interpretation. B-tubule spokes in *Chlamydomonas* (see Moestrup, 1982, and discussion below) have a constant positional relationship to the doublet lacking the outer arm (Ringo, 1967b; Jacobs and McVittie, 1970; Witman *et al.*, 1978; Moestrup, 1982; Hoops and Witman, 1983). In ulvophycean flagellated cells B-tubule septations also have a constant positional relationship to this doublet (unpublished observations and Figs 25–27). Finally, the same is apparently true for the peculiar wings in flagella of the Trentepohliales (see micrographs in Graham and McBride, 1975; Chapman and Henk, 1983).

This property of a particular outer axonemal doublet to differ structurally from all other axonemal doublets can now be used to label axonemal doublets unequivocally (see Fig. 19). It is proposed that this terminology should be used in all future analyses of flagellar apparatus ultrastructure in the green algae. In contrast to some previous preliminary doublet labelling (e.g. Melkonian, 1982c) it is here advocated to follow established rules for axonemal doublet labelling in other eucaryotic flagellated or ciliated organisms (Afzelius, 1959; Gibbons and Grimstone, 1960; also review by Satir and Ojakian, 1979) and to label consecutively in the direction of axonemal arm projection, i.e. doublet n has dynein arms that project and interact during sliding with doublet $n + 1$. It is proposed to label the doublet lacking the outer arm as No. 1. This is somewhat arbitrary but is roughly in accordance with doublet labelling in other eucaryotic organisms exhibiting a so-called 5–6-bridge (Satir and Ojakian, 1979), since it has been found (see below, and unpublished observations) that in the green algae the doublet lacking the outer arm is opposite the "5–6-bridge" in ciliated organisms if the direction of the flagellar effective stroke is taken as a reference (see Fig. 19; see also Satir and Ojakian, 1979, for review of the relationship between axonemal structure and direction of the effective stroke in organisms with a flagellum (cilium) containing a "5–6-bridge"). The labelling of doublets proceeds anticlockwise if the flagellum is viewed from the tip and clockwise if viewed from its base (the latter in Fig. 19). In Fig. 19 five structural specializations of the green algal axoneme (including the wings of the Trentepohliales) are incorporated into the absolute configuration of the green algal flagellum. The proposed scheme of labelling in the flagellar axoneme

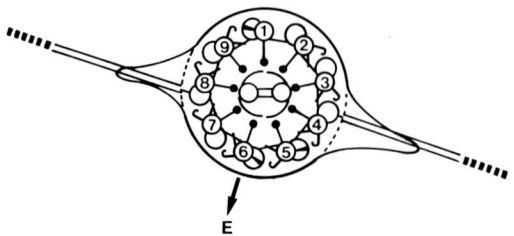

Fig. 19. Schematic presentation of the absolute configuration of the green algal flagellum in cross section. This figure has been designed so as to incorporate structural peculiarities of the green algal axoneme from various algae. The flagellum is oriented as if viewed from base to tip. Individual outer axonemal doublets have been labelled starting with the doublet lacking the outer dynein arm as No. 1 and then proceeding in the direction of dynein arm projection (clockwise). Nexin links interconnecting axonemal doublets have been incorporated into this drawing as well as radial spokes and the CPM sheath. Microtubule septations and spokes occur in the same B-tubules (labelled No. 1, 5 and 6) and only the septations have been drawn. Flagellar hair-scales and the peculiar wing-like flagellar extensions of the Trentepohliales are associated with doublets 4 and 8 and are both perpendicular to the plane of beat of the flagellum (E = direction of the flagellar effective stroke).

of green algae should also be applicable for the transition region and basal body. The relationship between the direction of the flagellar effective stroke (see Fig. 19) and the position of the doublet lacking the outer arm has been investigated in four green algae (*Tetraselmis cordiformis*, *Volvox globator*, *Ulva lactuca* and *Chlorosarcinopsis pseudominor*; M. Melkonian, unpublished observations) by serial-section electron microscopy of cells with known flagellar orientation or in which flagella orientation was established by serial sections. In flagellated green algae which contain more than one flagellum per cell (the great majority) the flagella can have various orientations to each other all of which are functionally important. In biflagellates it is generally found that opposite flagella are rotated with respect to each other by 180°. This is also the apparent reason for the occurrence of the well-known cruciate flagellar root system with socalled 180° rotational symmetry (Floyd et al., 1980). The rotation results in the doublet lacking the outer dynein arm being located on opposite sides of the two flagella (e.g. Fig. 13). Since it is here suggested that this peculiar outer doublet is functionally related to the determination of the flagella beat direction, the 180° rotation of two opposite flagella is apparently a prerequisite for the "breast stroke" flagellar movement which is so typical of the green algae. Recently Hoops and Witman (1983) have studied outer doublet heterogeneity in *Chlamydomonas reinhardii* and have reached similar conclusions. Although opposite orientation of the flagella appears to be necessary for forward movement of the "breast stroke" type, it is not sufficient for this type of movement. *Scourfieldia caeca* is a biflagellate organism that naturally swims backwards with the two flagella undulating and hydrodynamically coupled (Belcher, 1964; Melkonian and Preisig, 1982) though it has flagella with opposite orientation (Fig. 13). Flagella with opposite orientation therefore are not prevented from undertaking the "flagella-type" backwards swimming mode. This conclusion is supported by the fact that most, if not all, green algae that exhibit breast stroke-type flagellar movement are capable also of swimming backwards with the flagella-type movement over short time periods (the so-called avoidance reaction). No change in flagella polarity (rotation) by 180° is likely to occur during the switch from one to another type of movement. It has previously been suggested that the backwards swimming mode in the green algae is evolutionarily the primitive one, since it does not require an effective calcium-ATPase located in the flagellar membrane (Melkonian, 1983) which regulates the internal free calcium ion concentration. It is possible that during evolution the switch from backwards swimming to forward swimming occurred once an effective Ca-ATPase had evolved to keep the internal calcium concentration low in an environment in which it was usually rather high (marine environment). Because of their peculiar flagella orientation (180° rotation of flagella) green algae were then able to use the flagella in a breast stroke-type movement, which probably also favoured evolution of an apical flagellar insertion and allowed individual regulation of flagella in complex behavioural responses (see Melkonian, 1982a, 1983).

Several conclusions may be drawn from Fig. 19, which incorporates data on axonemal structural specializations in several green algae into the absolute configuration of the green algal flagellum:

1. B-tubule septations and B-tubule spokes occur in the same three outer doublets in the Ulvophyceae and *Chlamydomonas*, respectively, and so they are probably homologous. It is therefore no surprise that typical B-tubule septations also occur in flagella of unquestionably chlorophycean taxa, e.g. in the *Stigeoclonium* zoospore (Manton, 1964a). B-tubule septations are therefore not useful for separating the Ulvophyceae from other green algae.

2. Hair-scales of prasinophycean flagellates occur in two rows on opposite sides of a flagellum and are oriented perpendicular to the plane of flagellar beat (see also Figs 9 and 10). The assumption that hair-scales play an important role in the motility of the flagella (Melkonian, 1982c) is thus substantiated, since their orientation indicates that they increase the flagellar diameter considerably. In the coupled flagellar pairs of *Tetraselmis* the two rows of hair-scales are oriented in the same plane, and hair-scales from different flagella touch each other at their tips (Fig. 10). Hair-scales act as a rudder-like structure increasing the efficiency of the flagellar beat, and may

thus compensate for the restriction of microtubule sliding in scaly flagella (see above).

3. The wing-like projection of the flagellar membrane and matrix characteristic of the Trentepohliales (Graham and McBride, 1975; Chapman, 1980, 1981) is also perpendicular to the plane of beat of the flagella, as are the hair-scales. The flagellar wings of the trentepohlialean algae have a similar function in increasing the effective diameter of the flagellum.

3. Transition Region

This is the region of the flagellum between the flagellar shaft and the basal body (as defined by Moestrup, 1982) extending from the proximal termination of the CPM to the origin of the C-tubules of the basal body (Pitelka, 1974). However, this definition should not be taken too strictly since typical transition region structures occur often where the CPM are still present (e.g. the "coiled fibre" of *Pyramimonas* and in some green algae the distal part of the stellate structure), or conversely these structures may extend into the triplet region of the basal body in some green algae (e.g. in *Acetabularia* gametes; Woodcock and Miller, 1973). Transition region structures in the chlorophyll *c*-containing algae were reviewed by Hibberd (1979), but no such review exists with respect to green algae. On the basis of a comparative analysis of the flagellar transition region in about 100 different green algae, such a review is attempted here. The information has been taken from both published micrographs and unpublished data on 40 additional green algae. Quantitative measurements sometimes had to be made on the printed micrographs and some inaccuracies must be taken into account.

The transition region of the green algae contains the well-known stellate structure first discovered in *Polytoma obtusum* by Lang (1963) and then described in more detail by Manton (1964a) for the *Stigeoclonium* zoospore, by Ringo (1967a) for *Chlamydomonas reinhardii*, and by Olson and Kochert (1970) for *Volvox carteri*. A first comparative analysis and synthesis was provided by Manton (1964b) involving four genera of green algae and later

---

Fig. 20. Longitudinal section through the transition region of the flagellum of *Scourfieldia caeca* (× 77 000).

Fig. 21. Cross section through the flagellar transition region of *Scourfieldia caeca* in the position of the transverse septum (× 77 000).

Fig. 22. Longitudinal section through the transition region of the flagellum of *Nephroselmis rotunda* (× 50 000). *Arrow* indicates the additional transverse septum.

Fig. 23. Slightly oblique longitudinal section through the transition region of the flagellum of *Nephroselmis olivacea* (× 57 500). Note the prominent additional transverse septum which is linked to the flagellar membrane (*arrows*).

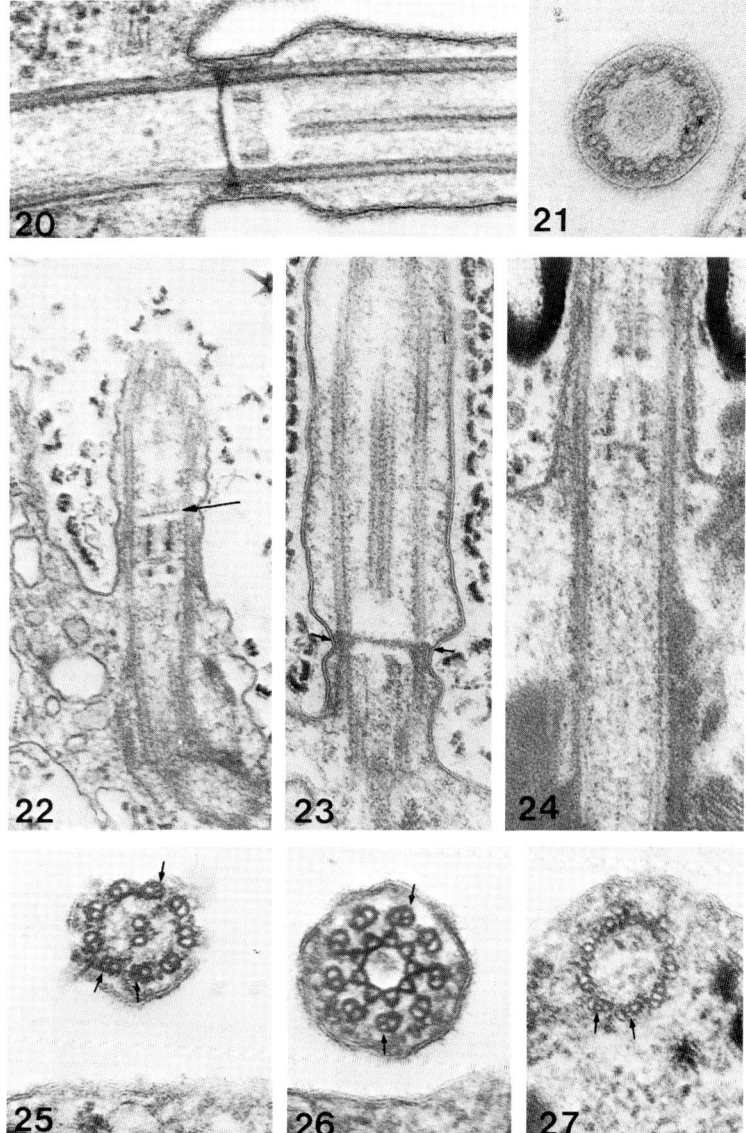

Fig. 24. Longitudinal section through the flagellar transition region of *Tetraselmis cordiformis* (× 70 300).

Figs 25–27. Cross sections through a flagellum, a transition region, and a basal body of male gametes of *Ulva lactuca*. Note the B-tubule septations that always occur in the same doublets or triplets (*arrows*). In Fig. 25 it is likely that the single outer doublet with the B-tubule septation also lacks the outer dynein arm (*top arrow*). Figs 25, 27. × 70 300; Fig. 26 × 70 700.

some aspects of the transition region were reviewed for prasinophycean flagellates by Norris (1980) and for green algae in general by Melkonian (1982a) and Moestrup (1982).

The stellate structure is a cylinder consisting of a number of filaments that connect A-tubules of every second outer axonemal doublet with each other and in cross sections producing the spectacular stellate pattern (Fig. 15). The filaments ascend in a helix with several gyres (Manton, 1964b) with 2 to 3 gyres probably corresponding to one complete circle of interconnections. The number of gyres for each stellate structure can be roughly calculated from longitudinal sections in which the number of electron-dense dots visible represent the number of gyres (e.g. Fig. 32). It is interesting to note that the minimum number of gyres that can be measured in the green algae is 2 to 3. In longitudinal sections through flagella the stellate structure may be continuous or consist of a distal (towards the flagellar tip) and a proximal part (Figs 28–32). Quantitative measurements on the stellate structure include the total length and the relative lengths of the distal part to its proximal part. Very rarely is the stellate structure absent in green algae, and the two known cases (zoospores of *Urospora* species, Sluiman *et al.*, 1982; spermatozoids of *Golenkinia minutissima*, Moestrup, 1972) are presumably secondary losses of this structure.

A transverse septum or diaphragm is commonly present in the flagellar transition region of a great number of green algae in addition to the stellate structure. The positional relationship of this transverse diaphragm to the stellate structure is an important feature in the overall appearance of the transition region (e.g. Fig. 33) and is important in phylogenetic considerations. The transverse septum may be separated from the stellate structure by an electron-translucent space: the septum may either be located proximally to a continuous stellate structure (as in *Scourfieldia,* Fig. 20) or, more rarely, intercalated between a distal and a proximal part of a stellate structure (e.g. *Nephroselmis rotunda,* Fig. 22; and in some primitive ulvophycean algae, e.g. *Monostroma, Ulothrix,* see Fig. 33). More commonly the transverse septum attaches to the distal part of a stellate structure. This occurs most often in the Chlorophyceae. Rarely, the transverse septum may be linked to the proximal part of a stellate structure (e.g. *Tetraselmis* spp., Fig. 24). Finally, a

---

Fig. 28. Longitudinal section through the proximal parts of a flagellum in zoospores of *Chlorosarcinopsis pseudominor* ($\times$ 70 000). *Arrow* indicates that the CPM are oriented in the plane of the flagellar beat and not perpendicular to it.

Fig. 29. Longitudinal section through the transition region in the flagellum of zoospores of *Ulva lactuca* ($\times$ 87 500). *Small arrow* indicates the little-developed transverse septum.

Fig. 30. Longitudinal section through the flagellar transition region of zoospores of *Fritschiella tuberosa* ($\times$ 80 000).

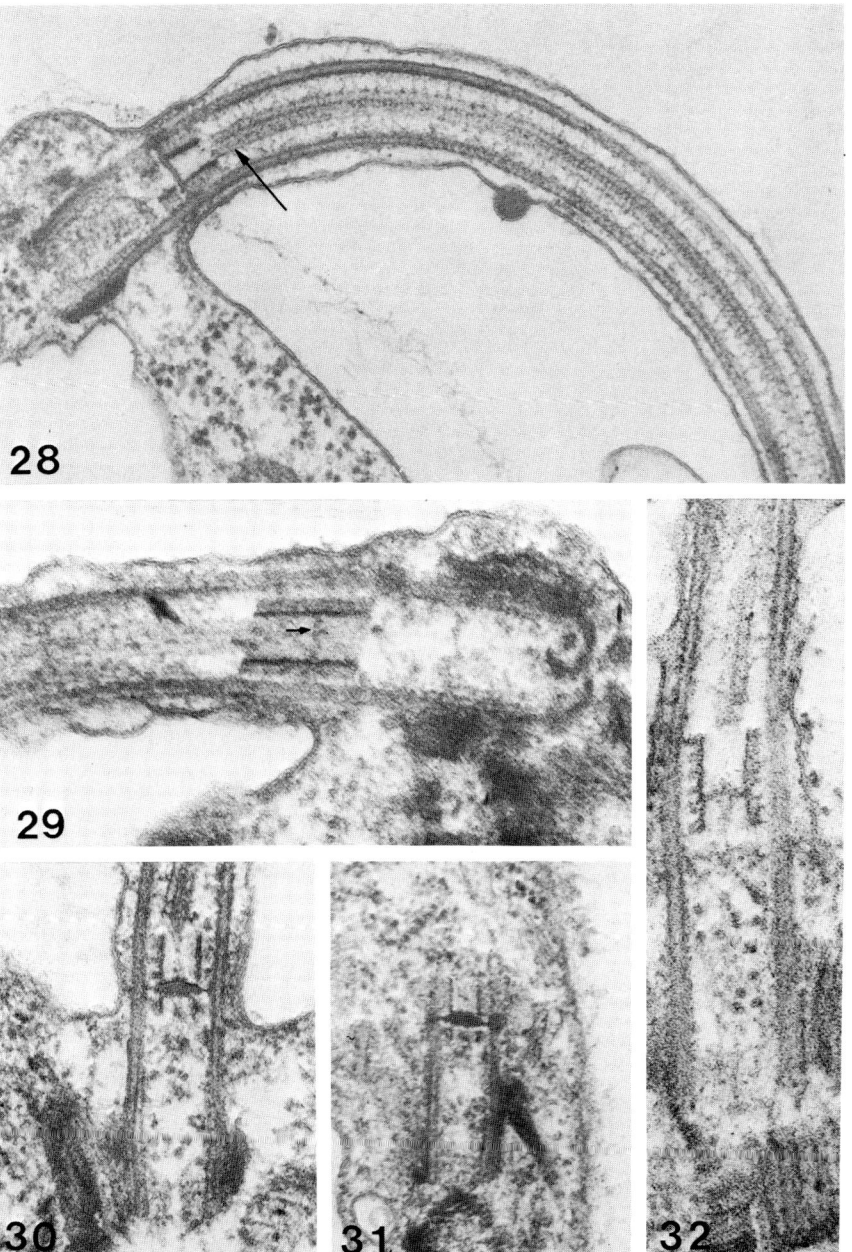

Fig. 31. Similar section to that in Fig. 30, but here the zoospore has retracted its flagella and the transition region can be seen in longitudinal section (× 60 000).

Fig. 32. Longitudinal section through the transition region of a flagellum in *Pedinomonas tuberculata* (× 90 000).

transverse septum may be absent or very reduced (Fig. 29 for *Ulva*), in which case the stellate structure is most often a continuous cylinder; sometimes a distal and a proximal part is present (Fig. 33). Unusual features of the flagellar transition region in green algae include the presence of additional transverse septa (known in *Nephroselmis*—Moestrup and Ettl, 1979; Manton et al., 1965; see Figs 22 and 23; probably in *Pseudoscourfieldia*, Manton, 1975) and the presence of an additional electron-dense cylinder, the so-called coiled fibre or transitional helix. The latter is presently found only in species of the prasinophycean genus *Pyramimonas* (see Norris, 1980; Melkonian, 1982a; Moestrup, 1982). How this structure relates to the transitional helix of chlorophyll c-containing algae (Hibberd, 1979) is unknown, but the structural and positional similarity is unmistakable.

Minor structural characters in the transition region of green algae include the length, shape and substructure of the transverse septum, and the occurrence and development of nine longitudinal filaments on the inner side of the stellate structure. These are associated with the overlapping regions of the filaments of this structure (for examples of these specializations see Figs 15, 20–24, and 28–32). From this list it is apparent that the flagellar transition region offers more than 15 useful structural characters for green algal classification and, with other ultrastructural features of flagellated cells, can be used to test evolutionary trends in the group (see Fig. 33).

The principal types of flagellar transition region structures are depicted in Fig. 33 and will now be briefly discussed.

(a) *The Mantoniella type.* In three genera of green algae (*Mantoniella, Monomastix, Scourfieldia*) the prominent transverse septum is located proximally to the stellate structure and separated from it by a translucent space, i.e. the stellate structure may be viewed as consisting of a distal part only. That this interpretation may be correct can be deduced from the occurrence of a transverse septum of very similar structure in *Nephroselmis rotunda* (Manton et al., 1965) which here separates a longer distal from a very short proximal part. The stellate structure is relatively short in these genera (75–120 nm; Fig. 20 for *Scourfieldia*).

(b) *The Nephroselmis type.* This type occurs in two genera of scaly green flagellates, namely *Nephroselmis* and *Pseudoscourfieldia*. There is some heterogeneity in this type since in the freshwater *Nephroselmis olivacea* the transverse septum separating the distal from the proximal part of the stellate structure is missing or largely reduced, whereas the additional transverse septum occurring distally from the stellate structure is much more elaborate than in the marine *Nephroselmis* species or *Pseudoscourfieldia*. Also, on the basis of flagellar and cell body scale garniture *Nephroselmis olivacea* appears to be more complex and derived. The flagellar transition region in the

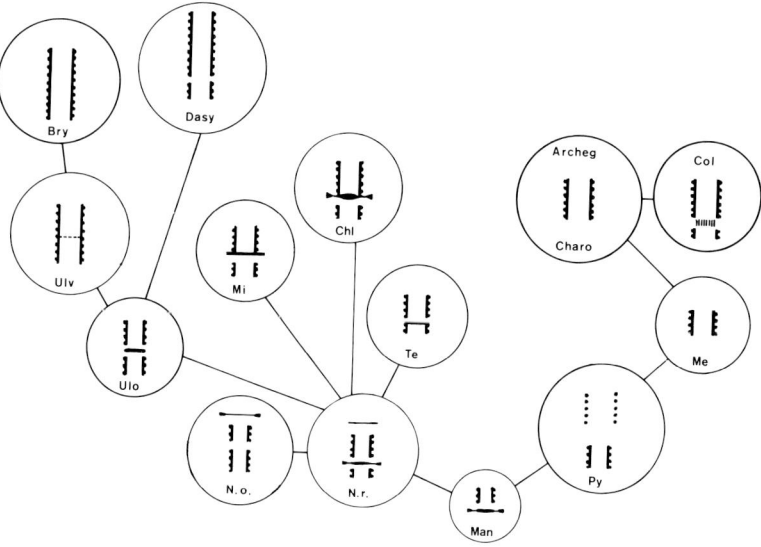

Fig. 33. Schematic presentation of transition region structures of the flagellum in different green algae. The different types of flagellar transition regions have been arranged so as to indicate possible structural and phylogenetic relationships (interlinked by lines). It is assumed that the *Mantoniella* type is the most primitive type of transition region in the green algae. The following symbols have been used: Man: *Mantoniella;* Py: *Pyramimonas;* Me: *Mesostigma;* Col: *Coleochaete;* Charo and Archeg: Charophyceae and archegoniate land plants; N.r.: *Nephroselmis rotunda;* N.o.: *Nephroselmis olivacea;* Te: *Tetraselmis,* Chl: Chlorophyceae; Mi: *Microthamnion;* Ulo: Ulotrichales; Ulv: Ulvales; Bry: Bryopsidales; Dasy: Dasycladales.

marine species of *Nephroselmis* and in *Pseudoscourfieldia* is therefore regarded as more primitive and characteristic for this group (see Fig. 33). It appears that the function of the original transverse septum (between the two parts of the stellate structure), whatever its function may have been, has been taken over by the new, additional transverse septum in *Nephroselmis olivacea;* both septa show an inverse structural development in the two types of *Nephroselmis* species (Figs 22 and 23). The *Nephroselmis* type is then characterized by a longer distal and shorter proximal part to the stellate structure and two transverse septa, the more prominent located between the two parts of the stellate structure but not associated with any of them, and a second transverse septum separated by some 30–50 nm from the distal end of the stellate structure.

(c) *The Tetraselmis type.* This type consists of a stellate structure with a much longer distal part and a little-developed and very short proximal part.

An inconspicuous transverse septum is associated with the proximal part and apparently has no connections to the flagellar membrane (Figs 24 and 33).

(d) *The Chlorophyceae type.* This is the best studied type of transition region found to date in about 40 different green algae. The stellate structure consists of a longer distal part and a shorter proximal part (a very common ratio between their lengths is 2–3:1); intercalated between both parts is a prominent transverse septum which is most often attached to the distal part of the stellate structure and never to the proximal part alone. It may, however, be developed to such an extent that it extends from the distal to the proximal part filling the space between both parts. The transverse septum attaches to the flagellar membrane and is often centrally dilated (e.g. Figs 30 and 31). An internal ring of nine longitudinal filaments may also occur quite frequently in this transition region type (see Fig. 15). The length of the stellate structure (including the transverse septum and intervening translucent space) is usually 140–160 nm. There is a tendency for shorter stellate structures in colonial Volvocales (100–140 nm) and in some chlorococcalean algae (110–140 nm) while one of the shortest stellate structures exhibiting both distal and proximal parts is that of the zoospore of *Sorastrum* (Marchant, 1974). In the latter case the length ratio between distal and proximal part is 1:1 indicating that the distal part of the stellate structure is reduced in these zoospores. In the Oedogoniales the stellate structure is relatively long (150–200 nm; Pickett-Heaps, 1971; Hoffman, 1973; Markowitz, 1978).

(e) *The Ulotrichales type.* In the Ulvophyceae four types of flagellar transition regions may be distinguished (see Fig. 33). The Ulotrichales type appears to be the most primitive and more related to both the Chlorophycean type and the *Nephroselmis* type. It consists of a stellate structure with a distal and proximal part and an intercalated transverse septum. The distal part of the stellate structure is only slightly longer than the proximal part (ratio: 1.3–1.6:1) and the transverse septum is either not associated with any of the two parts or may be associated with the proximal part. It is not found associated with the distal part (Jonsson and Chesnoy, 1974; Moestrup, 1978; Sluiman *et al.*, 1980; Hoops *et al.*, 1982; Floyd and O'Kelly, 1984). The length of the stellate structure is intermediate (between 150 and 170 nm).

(f) *The Ulvales type.* This type can be derived from the Ulotrichales type by assuming that the transverse septum became reduced and the distal and proximal parts of the stellate structure merged to form a continuous and rather long cylinder (see Fig. 33). That this interpretation may be correct is indicated by the presence of a very faint transverse septum occurring in most Ulvales studied (e.g. Fig. 29). The ratio between the distal and proximal parts of the stellate structure is similar to that of the Ulotrichales (1–2:1). The stellate structure in the Ulvales type is relatively long (175–240 nm). Three genera exhibit this type of transition region: *Ulva, Enteromorpha* and *Blidingia* (Evans and Christie, 1970; Swanson and Floyd, 1978; Melkonian, 1979, 1980a,b). It should be mentioned that the stellate structure is slightly shorter in the male gametes compared to the female gametes in *Ulva lactuca* (150 nm compared to 180 nm), a tendency that is even more pronounced in the next transition region type.

(g) *The Bryopsidales type.* This type can be derived from the Ulvales type by the complete disappearance of the transverse septum and further elongation of the continuous stellate structure. In this group the stellate structures may reach enormous lengths (190–280 nm) and conspicuous differences in length occur between male and female gametes in at least three different genera: *Bryopsis lyngbyei* (230 nm against 280 nm; Melkonian, 1981, and unpublished observations), *Pseudobryopsis* (130 nm against 190 nm; Roberts *et al.,* 1982) and *Derbesia* (140 nm against 200 nm; Roberts *et al.,* 1981). In all cases the male gamete has the shorter stellate structure.

(h) *The Dasycladales type.* At present the transition region is only documented for *Acetabularia* (Berger *et al.,* 1975; Herth *et al.,* 1981). The stellate structure is very long (300 nm according to Herth *et al.,* 1981) and apparently consists of both a distal and proximal part with the distal part much longer than the proximal. Sometimes electron-dense material can be seen between both parts, but it is not clear if this is remnants of a transverse diaphragm. The Dasycladales type may be derived from either the *Bryopsis* type or more probably directly from the Ulotrichales type.

(i) *The Pedinomonas type.* This type is not illustrated in Fig. 33 since its relationship to the transition region of other green algae must be regarded as uncertain at present. The *Pedinomonas* type of transition region (see Fig. 32) is structurally most similar to the Ulvales type. It consists of a continuous stellate structure (in *Pedinomonas minor* the proximal part is apparently little developed; see Ettl and Manton, 1964) with both distal and proximal parts separated from each other by a transverse septum which does not interrupt the continuous stellate structure and therefore is not linked to the flagellar membrane. It has been found in three species of *Pedinomonas* (Ettl and Manton, 1964; Sweeney, 1976; this paper).

(j) *The Pyramimonas type.* This type is most readily derived from a *Mantoniella* type by disappearance of the transverse septum. If this interpretation is correct then in the *Pyramimonas* type only the distal part of the stellate structure would be present. In accordance with this view the stellate structure in most *Pyramimonas* species is relatively short (100–120 nm) with the possible exception of the octoflagellate *Pyramimonas amylifera* (200 nm; see

Manton, 1966; Moestrup and Thomsen, 1974; Norris and Pearson, 1975; Melkonian, 1981b, and unpublished observations). All studied species of *Pyramimonas* contain an additional "coiled fibre."

(k) *The Mesostigma type*. This type can be derived from the *Pyramimonas* type by the loss of the "coiled fibre". It is obvious that this type also appears to be present in most Charophyceae and archegoniate sperm cells, although there is insufficient information on the flagellar transition region in these organisms to permit final conclusions. In some bryophyte sperms the stellate structure may be totally missing, whereas it is present in other bryophytes (Carothers and Duckett, 1979). In most Charophyceae and in the archegoniate sperm cells, the transition region consists of a stellate structure which forms a continuous cylinder and is not differentiated into distal and proximal parts. In addition a transverse septum is lacking. An apparent exception to this are the two closely-related genera *Chaetosphaeridium* (Moestrup, 1974) and *Coleochaete* (Sluiman, 1983). In *Chaetosphaeridium* a continuous stellate structure is present, but in addition there occur three peculiar transverse septa, two at the ends of the stellate structure and one in the middle (Moestrup, 1974). In *Coleochaete* zoospores the stellate structure consists of a distal and proximal part with an intercalated transverse septum (Sluiman, 1983). Interestingly, in both genera the structure of the transverse septum is so unlike that of the other transverse septa described above (not a continuous plate but composed of transversely striated material) that it is best at present to regard it in these two genera as representing yet another independent origin of this structure (as in *Nephroselmis*). If this interpretation is true then the transition region of the flagellated cells in *Chaetosphaeridium* and *Coleochaete* is not so different from that of other members of the Charophyceae and the archegoniate land plants to which they are phylogenetically related.

Algae not included in the above flagellar transition region types are *Friedmannia* (Pleurastrophyceae) and members of the Trentepohliales. This is partly because more information is needed on their flagellar transition region and partly because some members seem to fit into one or the other of the above-mentioned types. *Microthamnion* zoospores (Watson, 1975), for example, have a Chlorophyceae-type flagellar transition region but differ from it in some minor aspects (see Fig. 33). Also, the Trentepohliales (Graham and McBride, 1975; Chapman, 1981) contain in their motile cells a transition region which conforms to the Bryopsidales type.

In Fig. 33 the different structural types of flagellar transition regions that occur in the green algae are grouped together and interlinked so as to reflect structural and possibly evolutionary relationships. The direction of evolution of course can be read in different ways. There is no absolute way of determining a primitive and a derived condition solely on the basis of transition region ultrastructure, especially since the function of the transition region in the algae is still largely a matter of speculation (see Moestrup, 1982). If, however, other ultrastructural data of the flagellar apparatus and of the cell are taken into consideration and the main groups are arranged so as to suggest natural relationships, then the transition region types fit nicely into this scheme (Fig. 33) and most transitions between one type and another are logical one-step transitions.

4. Basal Body Length

The ultrastructure of basal bodies is fairly homogeneous within the green algae, though detailed comparative investigations have yet to be performed and so the range of variation is unknown. Discussion is therefore limited to data that are available about the lengths of basal bodies in about 40 different genera of green algae. It should be mentioned that basal body lengths have been determined from longitudinal sections through basal bodies and from the proximal termination of the transition region (either the stellate structure or a transverse plate) to the proximal end of the basal body. This may strictly speaking not be the actual length of the basal body, which was defined as the triplet region of the axoneme. Nonetheless, the procedure chosen here appears to be the most acceptable since the origin of the triplets is difficult to determine from longitudinal sections and it is known that some basal bodies have triplets of considerably unequal length (Melkonian and Berns, 1983, for the *Friedmannia* zoospore).

In scaly green flagellates basal bodies are generally very long irrespective of cellular dimensions: in tiny *Mantoniella* the length varies between 650 and 800 nm (Salisbury and Floyd, 1978; Barlow and Cattolico, 1981), and in the rather larger sized *Pyramimonas amylifera* the length is given as 740–900 nm (Manton, 1966; Woods and Triemer, 1981; unpublished observations). Not a single scaly green flagellate has a basal body length of less than 500 nm. It may therefore be concluded that the ancestral green algae had very long basal bodies and conversely relatively short transition regions. Some small naked flagellates also have longer basal bodies (e.g. *Scourfieldia*, 600–640 nm), whereas others have relatively short basal bodies (*Pedinomonas* species, 300–400 nm). This character in *Pedinomonas* species would seem to be derived rather than primitive. The few data on basal body length available for the Charophyceae indicate that, as with scaly green flagellates, they are relatively long. Thus in *Chaetosphaeridium* zoospores the basal body length may reach 650 nm. All other groups of advanced green algae have much shorter basal bodies (300–400 nm in the Chlorophyceae and the same length range for the Ulvophyceae and most Pleurastrophyceae). Genera of advanced green algae that are usually regarded as being more primitive have slightly longer basal bodies, e.g. species of *Carteria* and *Hafniomonas* have basal bodies of 600 nm length (Ettl and Moestrup, 1980; Domozych *et al.*, 1981; unpublished observations on *Carteria obtusa*). Similarly the basal bodies of *Microthamnion* and trentepohlialean-type zoospores are about 550 nm long and 500 nm long, respectively; *Microthamnion* may be regarded as an unspecialized member of the Pleurastrophyceae. Only in the class Ulvophyceae do the motile cells not have longer basal bodies even in the basic order Ulotrichales. In the Chlorophyceae two evolutionary side-lines have either shorter (the colonial Volvacales, e.g. 230 nm in *Chlorcorona;* see Hoops and Floyd, 1982) or longer basal bodies (the Oedogoniales, 500–900 nm).

Basal body lengths appear to be useful systematic characters.

5. Basal Body–Associated Structures

Structures associated with the basal bodies of green algae (connecting fibres, flagellar roots) have been intensively studied over the past few years and a number of detailed reviews on these structures exist (Moestrup, 1978, 1982; Melkonian, 1980a, 1982b). It might therefore be sufficient to outline only some major aspects of these structures and then to treat the absolute configuration of the basal parts of the flagellar apparatus in some detail.

Connecting fibres are fibrillar or amorphous structures (usually banded or plate-like) that link different basal bodies within one flagellar apparatus with each other (Melkonian, 1980a). They are not present in uniflagellate green algae with a second non-functional basal body. The function of connecting fibres is still largely unknown, although some recent progress has been made. Wright *et al.* (1983) have described a mutant of *Chlamydomonas reinhardii* with a deficient distal striated connecting fibre. Using this mutant Hoops *et al.* (1984) showed that flagellar waveform, frequency and beat synchrony are similar to those of wild-type cells, indicating that in *Chlamydomonas reinhardii* the distal connecting fibre is probably not required for initiation or synchronization of flagellar motion. However, the orientation of the flagella was impaired so that effective strokes occurred in virtually any direction. The authors concluded that the distal striated fibre may be important in establishing or maintaining the correct rotational orientation of the basal bodies to ensure effective cellular movement. Melkonian and Preisig (1984), using the naked biflagellate *Spermatozopsis,* demonstrated that basal body reorientation occurs when cells switch from the forward to the backward swimming mode and that this is paralleled by an extreme change in length of the distal connecting fibre. They suggested that in this organism the distal fibre may be contractile and involved in basal body reorientation. Structurally the connecting fibres are very diverse in the green algae and different types have been distinguished (e.g. capping plates, distal striated connecting fibres, synistosomes; see Melkonian, 1980a). The structure, number and relative position of the connecting fibres in the flagellar apparatus are regarded as important systematic characters.

Flagellar roots in the green algae may be either microtubular or fibrillar (see Moestrup 1978, 1982; Melkonian, 1980a). The microtubular roots occur in basic numbers of two per basal body originating from and attaching to specific basal body triplets. In biflagellate cells one of the basal bodies may lack microtubular flagellar roots, in which case the whole microtubular root system consists of two roots only (apparently in all Charophyceae; Sluiman, 1983) and in the Mamiellaceae (a group of small scaly flagellates; Moestrup, 1984). If microtubular roots develop from the second basal body which has opposite orientation to the first basal body, these are formed in an arrangement that has been called 180° rotational symmetry (Floyd et al., 1980) and simply indicate that the triplet containing the root template for root type X in the second basal body is now opposite (rotated by 180°) to the corresponding triplet in the first basal body. In most green algae, therefore, a cruciate microtubular flagellar root system evolved containing two types of microtubular roots which Moestrup (1978) termed the X-2-X-2 cruciate flagellar root system. In contrast to an earlier assumption (Moestrup, 1978) that the two-stranded microtubular root in the green algae is structurally the most conservative and shows virtually no variation, the most recent analysis indicates that two-stranded roots are not very common in prasinophycean flagellates: they are lacking in *Nephroselmis* (Moestrup and Ettl, 1979), in some *Pyramimonas* species (unpublished observations), in *Mesostigma* (Rogers et al., 1981; unpublished observations) and possibly in *Mamiella* (Moestrup, 1984) In the Trentepohliales they do not occur either (Chapman and Henk, 1983). From recent studies in *Mamiella gilva* (Moestrup, 1984) and *Mesostigma viride* (Melkonian, in preparation), it is apparent that one root type contains microtubules in two rows and exhibits configurational changes (see Melkonian, 1978). From these structural features and its absolute configuration with respect to its basal body, it is concluded that this root type corresponds to the X-root of a cruciate flagellar root system. The other root type (which in *Mesostigma* is proximally associated with a multilayered structure) contains from three (*Mamiella*) to six microtubules (*Mesostigma*). In this root type no configurational changes of root microtubules occur and its absolute configuration with respect to its basal body shows that this type is homologous to the two-stranded microtubular root of the usual X-2-X-2 cruciate root system. Very interestingly, in both *Mamiella* and *Mesostigma* this root type proximally originates at the basal body with two microtubules to which are added after a small distance one (*Mamiella*) to five (*Mesostigma*) additional microtubules. One may therefore conclude that this root type is already potentially two-stranded. If the additional microtubules (which should be regarded as intrinsic root microtubules and not as secondary cytoskeletal microtubules; see Melkonian et al., 1980) were to be lost during evolution, two-stranded microtubular roots would develop and these might have evolved several times independently from a multimembered flagellar root. That this may have occurred in a close relative of *Mamiella* is indicated since *Mantoniella squamata* contains two microtubular roots with four and two microtubules each.

In *Nephroselmis* (Moestrup and Ettl, 1979) the second basal body bears only one microtubular root, which structurally and on the basis of its absolute configuration with respect to its basal body is homologous to the multilayered structure–containing roots in *Mesostigma* and to the two-stranded roots in the usual X-2-X-2 root system. All these interrelationships between different flagellar root systems in prasinophycean flagellates are shown in Fig. 34. The absolute orientations of basal bodies and flagellar

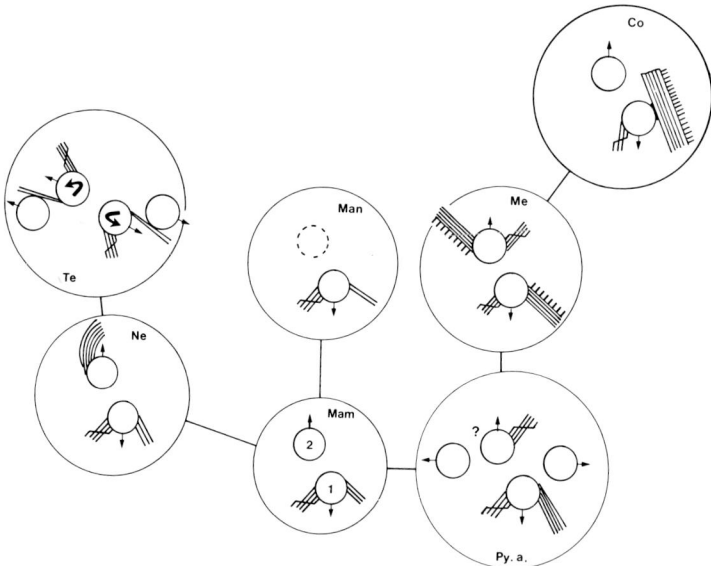

Fig. 34. Schematic presentation of evolutionary relationships within the flagellar apparatus of some green algae with scaly flagella. Basal bodies No. 1 and No. 2 are labelled. The *arrows* give the direction of the flagellar effective stroke. All types of flagellar apparatus have been arranged to reveal the absolute configuration (they are viewed from the flagellar tip). The abbreviations are: Mam: *Mamiella;* Man: *Mantoniella;* Py.a.: *Pyramimonas amylifera;* Me: *Mesostigma;* Co: *Coleochaete;* Ne: *Nephroselmis olivacea;* Te: *Tetraselmis.* In *Tetraselmis* the two basal bodies, 1 and 2, have rotated around their axes in the anticlockwise direction (*curved arrows*).

roots are indicated as well as the presumptive direction of the flagellar effective strokes. The linkages between the different flagellar apparatuses reflect structural similarities and possible natural relationships.

In addition to microtubular flagellar roots most green algal flagellated cells contain fibrillar roots of two types (termed system I and system II fibrous roots by Micalef and Gayral, 1972; more fully described in Melkonian, 1980a). Floyd *et al.* (1980) use a different terminology, namely rhizoplasts (system II fibres) and striated microtubule–associated components (system I fibres). From some early but very detailed studies on the system I fibre of the *Oedogonium* zoospore (Hoffman, 1970), it can be deduced that this component is also actually a fibre since it contains longitudinally-oriented rows of subunits. Functionally the best studied system is the system II fibrous root (the rhizoplast) of *Tetraselmis* (Salisbury and Floyd, 1978; Robenek and Melkonian, 1979; Salisbury, 1983). It appears that biflagellate cells of green algae may either have one, two or even four system II fibres per flagellar apparatus. In the biflagellate or uniflagellate scaly green flagellates only one system II fibre occurs in the flagellar apparatus (in *Mamiella, Mantoniella,* and *Nephroselmis*). It is not known to which of the two basal bodies this root attaches, but it is most likely that the basal body termed No. 1 in Fig. 34 gives rise to the system II fibre of *Mantoniella* (Barlow and Cattolico, 1980). In *Nephroselmis olivacea* the single system II fibre branches proximally to attach at both basal bodies (Moestrup and Ettl, 1979). Two system II fibres occur in the quadriflagellate prasinophycean cells and the number of system II fibres may even be increased in the octoflagellate species of *Pyramimonas*. System II fibres occur in three groups of advanced green algae, the Ulvophyceae, the Pleurastrophyceae (although reduced in the more derived genera; Melkonian and Berns, 1983) and the Chlorophyceae (in this class limited to flagellates). In the Charophyceae system II fibres are apparently absent, while in *Mesostigma,* the only prasinophycean flagellate which has been shown to contain a typical multilayered structure in its flagellar apparatus (Rogers *et al.,* 1981), two very thin fibrillar bands (each consisting of only two to three filaments) have been detected which link the basal bodies to the chloroplast and they originate from the lateral sides of the basal bodies (unpublished observations). It is not known if these fibres are homologous to the system II fibres of the related genus *Pyramimonas*.

System I fibres occur associated with flagellar root microtubules (see Melkonian, 1980a). They are present in most, if not all, members of the Chlorophyceae, and in this group often the fibre occupies a characteristic position running parallel to and overlying the two-stranded microtubular root. In a few well-studied organisms this fibre has been shown to be continuous between opposite two-stranded microtubular roots (see Brown *et al.*, 1976; Goodenough and Weiss, 1978; Melkonian, 1978). System I fibres have also frequently been observed in the Ulvophyceae, apparently underlying the two-stranded roots, but there is no evidence that in this group these fibres occur overlying these roots (Melkonian, 1979, 1980b). A continuous system I fibre between opposite two-stranded roots in the Ulvophyceae would also not be possible because of the peculiar 11/5 o'clock displacement of the basal bodies in ulvophycean-type cells. The evidence that system I fibres are also associated with the *X*-type roots must still be regarded as non-conclusive. The only prasinophycean flagellate that has been reported to contain a system I fibre is *Nephroselmis olivacea* (Moestrup and Ettl, 1979).

In *Nephroselmis olivacea* the system I fibre is associated with a three-stranded root of basal body 1 which is homologous to the two-stranded root of the cruciate flagellar root system of the Ulvophyceae and Chlorophyceae, i.e. this system I fibre is attached to the same root type as in the two groups of advanced green algae in which system I fibres are present. System I fibres have not so far been detected in the Trentepohliales and in the Pleurastrophyceae. In summary, the occurrence and location of system I fibres in the flagellar apparatus may be of considerable importance in the systematics and phylogeny of this group.

Floyd *et al.* (1980) have shown that it is important to establish the absolute configuration of the flagellar apparatus in a green algal cell in order to distinguish between mirror image arrangements. This type of approach has led to some of the most valuable data for green algal systematics in the past four years (Floyd *et al.*, 1980; Melkonian, 1981b; Hoops *et al.*, 1982; Roberts *et al.*, 1982; Melkonian and Berns, 1983; Sluiman, 1983). The main conclusions are that there is only one actual configuration in the flagellar apparatus of a green algal motile cell and that the mirror image arrangement does not occur. The obvious reason for this is that the basal body exhibits a structural polarity; it has a distal and a proximal end which are different from each other. In cross sections through basal bodies this is exemplified by the imbrication of the basal body triplets: when a basal body is viewed from its distal end (from the flagellar tip) the imbrication of the triplets is clockwise (the dynein arms of the axonemal doublets project counterclockwise); if the basal body is viewed from its proximal end the imbrication of the basal body triplets is anticlockwise. It is now found that there is a constant positional relationship between the two flagellar roots of a basal body and the principal connecting fibre that links the two basal bodies (termed Nos 1 and 2 in Fig. 34) with opposite orientation to each other. The connecting fibre is attached to two or three basal body triplets which are located between the two triplets to which the flagellar roots attach. If the basal body is viewed from its distal end (Fig. 34), and is oriented in such a way that the connecting fibre attachments point to the top of the drawing (basal body No. 1 in Fig. 34), one root is located on the right side of the connecting fibre and the other on the left. This absolute orientation of the microtubular flagellar roots presumably remained unchanged during the evolution of all major groups of green algae (see Figs 34 and 35). It is now evident that a definition of the two different root types in the green algae should be based on their absolute orientation and attachment to specific basal body triplets (which may ultimately be labelled once the doublet labelling proposed in this paper has been transferred to the basal body) instead of being based on structural aspects of the two root types, which have undergone considerable evolutionary change (see Figs 34 and 35). For the purpose of this review one root type will be described as the right root, the other as the left root (in the above sense). The right root in most green algae is a two-stranded root (in the old terminology of the X-2-X-2 root system *sensu* Moestrup, 1978). It may, however, have more than two microtubules in a single row. When it contains more than two microtubules it is usually, or perhaps invariably, proximally associated with a multilayered structure (MLS). The MLS (possibly on basal body 1) of *Mesostigma* is therefore homologous to the MLS in the Charophyceae and to the MLS in sperm cells of archegoniate higher plants. It is also homologous to the two-stranded root of the X-2-X-2 microtubular root system (Fig. 34). The left root always exhibits root tubule configurational changes and at least proximally occurs with microtubules arranged in two rows (Fig. 34). This is also the case in the zoospores of *Trentepohlia,* in which the right root has remnants of an MLS (Chapman and Henk, 1983). The left root of basal body 1 is retained in the Charophyceae as a 2-over-1 root (Sluiman, 1983).

The method of absolute configuration analysis has enabled the investigator to distinguish individual basal bodies and flagellar roots, and has therefore led to sounder and much more logical ideas about the evolutionary relationships between the major green algal groups and between green flagellates and advanced groups of green algae (Figs 34 and 35).

THE FLAGELLAR APPARATUS IN GREEN ALGAL SYSTEMATICS

This section will be limited to a few examples in which the flagellar apparatus ultrastructure has been supportive or even decisive in taxonomic rearrangements within the green algae.

The discovery of scales on flagellar and cell surfaces in a number of green flagellates led to the recognition that these flagellates cannot be placed into families and orders together with naked or walled volvocalean-type cells, hence the erection of the green algal class Prasinophyceae (see Norris, 1980).

Although it is now evident that some prasinophycean flagellates have clear affinities to certain advanced groups of green algae (see following section), they nevertheless share a sufficiently large number of common characters to qualify as a distinctive taxonomic group.

The flagellar apparatus ultrastructure of *Klebsormidium flaccidum* (Marchant et al., 1973) has shown that this organism belongs to the Charophyceae and confirms the findings of earlier ultrastructural studies on its cell division. In a similar way the chlorosarcinalean *Chlorokybus atmophyticus* was transferred to the Charophyceae on the basis of its flagellar apparatus and motile cell ultrastructure (Rogers et al., 1980). *Ulothrix zonata* was assigned to the Ulvophyceae and was removed from the Chlorophyceae on the basis of its flagellar apparatus ultrastructure (Sluiman et al., 1980). Recently it has been established that a number of green algae previously classified with the Chlorophyceae have to be incorporated into a separate group and for these organisms Mattox and Stewart (this volume, Chapter 2) propose the new class name Pleurastrophyceae *sensu* Mattox and Stewart. This proposal is largely based on the ultrastructure of the flagellar apparatus in the flagellated cells of these organisms (Melkonian and Berns, 1983). The following genera of advanced green algae were assigned to the Pleurastrophyceae: *Trebouxia* (formerly of the Chlorococcales), *Pseudotrebouxia* and *Friedmannia* (formerly of the Chlorosarcinales) and *Pleurastrum* and *Microthamnion* (previously placed in the Chaetophorales).

From these examples it may be seen that the flagellar apparatus ultrastructure will be important for green algal classification in the future in those groups (orders) which have previously been defined mainly on the basis of growth characteristics and morphology (e.g. the Chlorococcales, the Chlorosarcinales, the Chaetophorales and the Ulotrichales). Here the most significant taxonomic transfers are to be expected. In other groups of green algae the flagellar apparatus is likely to support previous classifications and may be significant in the more precise circumscription of green algal taxa (e.g., O'Kelly and Floyd, this volume, Chapter 4). It is also likely that the flagellar apparatus ultrastructure is increasingly used to delimit taxa at lower systematic levels (species, genus, family). When used at lower systematic levels, however, functional modifications may interfere with a detailed evaluation of structural data and could lead to incorrect interpretations.

The flagellar apparatus ultrastructure is unlikely to play a major role in routine green algal taxonomy because of difficulties preparing material. It is hoped that taxonomists will make available in culture collections the material they have investigated for further study by specialists using ultrastructural or biochemical techniques. The value of investigating the ultrastructure of the flagellar apparatus not only relates to its importance as a character in green algal classification, and in phylogenetic considerations, but also relates to the function of this very complex cell organelle.

PHYLOGENETIC CONCLUSIONS

The phylogenetic conclusions of this paper (summarized in Figs 33–35) are based mainly on flagellar apparatus ultrastructure (including the flagellar tip, the flagellar shaft, the transition region, the basal body and basal body–associated structures, with emphasis on absolute configurations), but included are such features as the general motile cell ultrastructure (including the type of cell body scale present) and mitotic and cytokinetic mechanisms.

I have assumed that the most ancient green alga was also structurally the most simple. This is not self-evident, because it may equally well be assumed that the most ancient green alga was a structurally very complex flagellate (Mattox and Stewart, this volume, Chapter 2). The latter view is based on analogies with some other algal classes. Zooflagellates with a complex internal structure are known which show striking structural similarities to members of these algal classes (Stewart and Mattox, 1980; Mattox and Stewart, this volume, Chapter 2). It must, however, be mentioned that no zooflagellate is known to date which can be related to a green flagellate. Even if one accepts the endosymbiotic origin of the green algal chloroplast, there is no need to postulate that the ancestral colourless flagellate must have been structurally very complex. Mattox and Stewart (this volume, Chapter 2) very interestingly reach the conclusion that during evolution of the green algae a single flagellum with two different types of flagellar roots had doubled to give rise to a biflagellate cell. Such a biflagellate cell is here considered at the base of the evolutionary radiation in the green algae (related to present-day *Mamiella* or *Dolichomastix*; see Figs 33 and 34) due to its overall simplicity of motile cell ultrastructure. Mattox and Stewart (this volume, Chapter 2), however, assume that before this stage was reached this simple flagellate must have been highly complex (with three or four flagella). This complex zooflagellate then lost all but one of its flagella.

While this sequence of evolution is possible it would appear to be not very probable. It is easier to assume that a simple uni- or biflagellate cell was primarily simple and not derived from a much more complex cell for which there is no evidence at present. It is also more likely that identical flagella are more primitive in general than dissimilar and structurally specialized flagella. The following evolutionary sequence, which is highly speculative, may be envisaged: a simple uniflagellate cell, which swam with a flagella-type movement (as in present-day animal sperms and some zoosporic fungi), took up a green prokaryote at the cell end opposite to the flagellar insertion, which is the end that leads the cell during movement. (This is a type of uptake behaviour that can be observed in various present-day eucaryotes.) The next evolutionary step would have been the formation of an additional basal body that exhibited opposite orientation to the mature basal body (basal body 2 in previous terminology; Fig. 34). Both basal bodies replicated during mitosis. Although structurally a copy of the first basal body, this second basal body did not develop, for instance, flagellar roots, because this was apparently not necessary as long as basal body 1 (which had two types of microtubular flagellar roots) replicated during mitosis and the daughter basal body (with the potential to form the two flagellar roots) was distributed together with basal body 2 into one of the daughter cells. It is suggested that basal body 2 only grew out flagellar roots when it became necessary to ensure the correct positioning of certain cell organelles in the vicinity of this basal body. An example of such a situation may be seen in present-day *Nephroselmis olivacea* (Moestrup and Ettl, 1979). Basal body 2 has only a single microtubular root (a right root, and therefore homologous to the opposite three-stranded root; see Fig. 34), which is probably involved in positioning of the eyespot (Moestrup and Ettl, 1979). In *Mamiella* the eyespot is located on that side of the cell which is opposite the flagellar insertion and in the future cleavage plane of the cell during mitosis, and the eyespot may simply be replicated and transferred to each daughter cell. In *Nephroselmis*, however, this is not possible because the eyespot is located in the anterior part of the cell below the shorter of the two flagella. To ensure distribution of two eyespots to the daughter cells, a second eyespot must form *de novo* on the opposite side of the cell. It is likely that the newly developed microtubular root performs an organizing function in this process.

It is possible that a cruciate flagellar root system arose in a way similar to the peculiar root system of *Nephroselmis*. This ensured the proper distribution of cell organelles (especially eyespots, contractile vacuoles, etc.) which were not located in the plane of cell division and could therefore not be simply replicated by division. In the Charophyceae the basal bodies are still in their primitive parallel orientation as in the prasinophycean ancestors. The right root developed a multilayered structure or elaborated an initial one that may have been barely recognisable. One of the unsolved questions is, did the Charophyceae always have only two microtubular flagellar roots (as assumed by Mattox and Stewart, this volume, Chapter 2) and then possibly originate from a *Mamiella*-type flagellate, or did they have four flagellar roots (perhaps a cruciate flagellar root system as in *Mesostigma*, the only known prasinophycean flagellate that contains an MLS) and two of these disappeared during evolution, possibly due to the distinctly subapical insertion of the two flagella? While a decision on these two alternatives is not possible at present, the second hypothesis is favoured here because of similarities in the flagellar transition region structure between *Mesostigma* and the Charophyceae, the presence of two contractile vacuoles in *Mesostigma*, which may indicate derivation from a quadriflagellate prasinophyte, and the surprising similarity of the underlayer scales of the Charophyceae and *Mesostigma* (maple leaf type).

The other classes of advanced green algae have flagella with either opposite flagella orientation (flagella form an angle of 180° to each other; the Ulvophyceae, the Pleurastrophyceae, the Trentepohliales) or flagella that form various angles with each other (the Chlorophyceae *sensu* Stewart and Mattox). The hypothetical derivation of two advanced groups of green algae, namely the Chlorophyceae and the Pleurastrophyceae, is shown in Fig. 35. A hypothetical biflagellate scaly alga similar to *Pseudoscourfieldia* and having developed a cruciate flagellar root system formed a theca by fusion of scales (which might have been similar to the rod-shaped scales of *Tetraselmis*) and a phycoplast. In the line of evolution leading to the Pleurastrophyceae a switch in flagella behaviour occurred in this flagellate resulting in the evolution of forward swimming. Both flagella were then operating in a breast stroke type of movement and were beating in a plane that had already been present in the flagellate (see arrows that indicate the direction of the flagellar effective stroke in flagella with opposite orientation, Fig. 35). From Fig. 35 it can be seen that such a folding down of the two flagella in the direction of the effective stroke would lead to basal body overlap (as observed in the Pleurastrophyceae and in the Ulvophyceae) and to the peculiar disposition of the left root, which is seen to take a path perpendicular to the basal body axis. Also it may be seen that an 11/5 o'clock displacement of the basal bodies results during this folding down mechanism. This is apparently due to both the characteristic direction of the

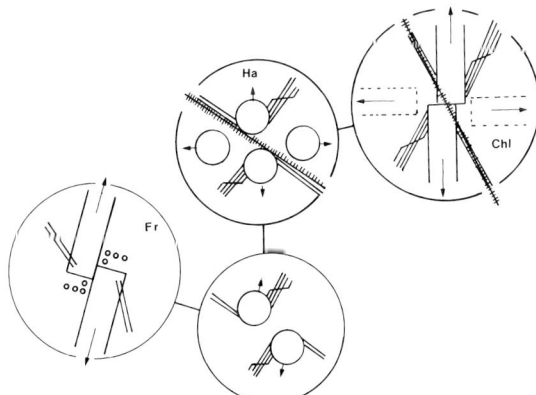

Fig. 35. Schematic presentation of evolutionary relationships among the flagellar apparatus of some advanced green algae. From a hypothetical ancestral type both a *Friedmannia*-type (Fr) flagellar apparatus and a chlorophycean-type flagellar apparatus (Ha: *Hafniomonas;* Chl: Chlorophyceae) have evolved. For further details see text.

effective stroke in the two flagella and the displacement of the two basal bodies in the ancestral flagellate. It should be mentioned that the term "11/5 o'clock" orientation of basal bodies should only be applied to basal body orientations in advanced green algae, since in prasinophytes the basal bodies are parallel and the actual orientation of the flagella is dependent on the direction of the effective stroke. This is most clearly demonstrated in *Tetraselmis* (Fig. 34). Here basal bodies 1 and 2 have apparently rotated around their axis (in the anticlockwise direction) to enable the cell to use the four flagella as two coupled pairs. If a folding down mechanism should occur in such an organism it would result in the formation of a basal body orientation which may be described as slightly 1/7 o'clock. If, therefore, the actual orientation of the flagellar effective stroke is not known in a prasinophycean flagellate the relative orientation of the basal bodies may lead to wrong conclusions. *Tetraselmis*, for instance, cannot be described as showing an 11/5 o'clock orientation of its basal bodies.

In the evolution of the Ulvophyceae and the Trentepohliales it is assumed that a similar folding down mechanism has occurred independently in a scaly green flagellate. In the Ulvophyceae this flagellate had probably not yet evolved a phycoplast, but may have been structurally similar to the phycoplast-containing flagellate that gave rise to the Pleurastrophyceae and the Chlorophyceae.

The evolution of the Chlorophyceae has possibly taken place by a slightly different mechanism. It is assumed that a quadriflagellate phycoplast-containing cell was ancestral in the Chlorophyceae, and that basal bodies 1 and 2 shifted into a 1/7 o'clock orientation and simultaneously also rotated around their axis by 40–80° into the anticlockwise direction (Fig. 35). It may be that the system I fibrous root that overlies opposite two-stranded roots in the Chlorophyceae was instrumental in the proposed basal body movements. This system I fibre occurs in the relatively "primitive" quadriflagellate *Hafniomonas*, which still has largely parallel basal bodies. How the folding down of opposite flagella has occurred in the Chlorophyceae is not known, but the papilla formation may be closely related to this process (as suggested by Mattox and Stewart, this volume, Chapter 2).

ACKNOWLEDGEMENTS

I would like to express thanks to Barbara Surek for considerable help in the preparation of the manuscript. Mrs. B. Berns has helped during various stages of the experimental work. This work is supported by a grant from the Deutsche Forschungsgemeinschaft (Me 658/2-1).

REFERENCES

Afzelius, B. A. (1959). Electron microscopy of the sperm tail. Results obtained with a new fixative. *J. biophys. biochem. Cytol.* **5,** 269–278.
Barlow, S. B. and Cattolico, R. A. (1980). Fine structure of the scale-covered green flagellate *Mantoniella squamata* (Manton and Parke) Desikachary. *Br. phycol. J.* **15,** 321–333.
Barlow, S. B. and Cattolico, R. A. (1981). Mitosis and cytokinesis in the Prasinophyceae. I. *Mantoniella squamata* (Manton and Parke) Desikachary. *Am. J. Bot.* **68,** 606–615.
Belcher, J. H. (1964). Further notes on *Scourfieldia caeca*. *Br. Phycol. Bull.* **2,** 371–373.
Belcher, J. H. (1968a). A study of *Pyramimonas reticulata* Korshikov (Prasinophyceae) in culture. *Nova Hedw.* **15,** 179–190.
Belcher, J. H. (1968b). A morphological study of *Pedinomonas major* Korshikov. *Nova Hedw.* **16,** 131–140.
Belcher, J. H. (1968c). The fine structure of *Furcilla stigmatophora* (Skuja) Korshikov. *Arch. Mikrobiol.* **60,** 84–94.
Belcher, J. H. and Swale, E. M. F. (1967). Observations on *Pteromonas tenuis* sp. nov. and *P. angulosa* (Carter) Lemmermann (Chlorophyceae, Volvocales) by light and electron microscopy. *Nova Hedw.* **13,** 353–359.
Berger, S., Herth, W., Franke, W. W., Falk, H., Spring, H. and Schweiger, H. G. (1975). Morphology of the nucleocytoplasmic interactions during the development of *Acetabularia* cells. II. The generative phase. *Protoplasma* **84,** 223–256.
Bergman, K., Goodenough, U. W., Goodenough, D. A., Jawith, J. and Martin, H. (1975). Gametic differentiation in *Chlamydomonas reinhardtii*. II. Flagellar membranes and the agglutination reaction. *J. Cell Biol.* **67,** 606–622.
Brown, D. L., Massalski, A. and Patenaude, R. (1976). Organization of the flagellar apparatus and associated cytoplasmic microtubules in the quadriflagellate alga *Polytomella agilis*. *J. Cell Biol.* **69,** 106–125.
Carothers, Z. B. and Duckett, J. G. (1979). Spermatogenesis in the systematics and phylogeny of the Hepaticae. *In* "Bryophyte Systematics" (G. C. S. Clarke and J. G. Duckett, eds), pp. 425–446. Academic Press, London.
Chapman, R L, (1980). Ultrastructure of *Cephaleuros virescens* (Chroolepidaceae; Chlorophyta). II. Gametes. *Am. J. Bot.* **67,** 10–17.
Chapman, R. L. (1981). Ultrastructure of *Cephaleuros virescens* (Chroolepidaceae; Chlorophyta). III. Zoospores. *Am. J. Bot.* **68,** 544–556.
Chapman, R. L. and Henk, M. C. (1983). Ultrastructure of *Cephaleuros virescens* (Chroolepidaceae, Chlorophyta). IV. Absolute configuration analysis of the cruciate flagellar apparatus and multi-layered structures in prerelease and postrelease gametes *Am. J. Bot.* **70,** 1340–1355.
Christensen, T, (1962). Alger *In* "Botanik" (T. W. Böcher, M. Lange, and T. Sørensen, eds), Vol. 2. No. 2, pp. 1–178. Munksgaard, Copenhagen
Dentler, W. L. (1981). Microtubule—membrane interactions in cilia and flagella. *Int. Rev. Cytol.* **72,** 1–47.
Dentler, W. L. and Rosenbaum, J. L. (1977). Flagellar elongation and shortening in *Chlamydomonas*. III. Structures attached to the tips of flagellar microtubules and their relationship to the directionality of flagellar microtubule assembly. *J. Cell Biol.* **74,** 747–759.
Domozych, D. S., Stewart, K. D. and Mattox, K. R. (1981). Development of the cell wall in *Tetraselmis*: Role of the Golgi apparatus and extracellular wall assembly. *J. Cell Sci.* **52,** 351–371.

Ettl, H. and Manton, I. (1964). Die feinere Struktur von *Pedinomonas minor* Korschikoff. *Nova Hedw.* **8**, 421–451.
Ettl, H. and Moestrup, Ø. (1980). Light and electron microscopical studies on *Hafniomonas* gen. nov. (Chlorophyceae, Volvocales), a genus resembling *Pyramimonas* (Prasinophyceae). *Plant Syst. Evol.* **135**, 177–210.
Evans, L. V. and Christie, A. O. (1970). Studies on the shipfouling alga *Enteromorpha*. I. Aspects of the fine-structure and biochemistry of swimming and newly settled zoospores. *Ann. Bot. (London)* [N.S.] **34**, 451–466.
Floyd, G. L. (1978). Mitosis and cytokinesis in *Asteromonas gracilis,* a wall-less green monad. *J. Phycol.* **14**, 440–445.
Floyd, G. L., Hoops, H. J. and Swanson, J. A. (1980). Fine structure of the zoospore of *Ulothrix belkae* with emphasis on the flagellar apparatus. *Protoplasma* **104**, 17–31.
Floyd, G. L. and O'Kelly, C. J. (1984). Motile cell ultrastructure and the circumscription of the orders Ulotrichales and Ulvales (Ulvophyceae, Chlorophyta). *Am. J. Bot.* **71**, 111–120.
Gibbons, I. R. and Grimstone, A. V. (1960). On flagellar structure in certain flagellates. *J. biophys. biochem. Cytol.* **7**, 697–716.
Goodenough, U. W. and Weiss, R. L. (1978). Interrelationships between microtubules, a striated fiber, and the gametic mating structure of *Chlamydomonas reinhardi*. *J. Cell Biol.* **76**, 430–438.
Graham, L. E. and McBride, G. E. (1975). The ultrastructure of multilayered structures associated with flagellar bases in motile cells of *Trentepohlia aurea*. *J. Phycol.* **11**, 86–96.
Herth, W., Heck, B. and Koop, H. U. (1981). The flagellar root system in the gamete of *Acetabularia mediterranea*. *Protoplasma* **109**, 257–269.
Hibberd, D. J. (1979). The structure and phylogenetic significance of the flagellar transition region in the chlorophyll c-containing algae. *BioSystems* **11**, 243–261.
Hoffman, L. R. (1970). Observations on the fine structure of *Oedogonium*. VI. The striated component of the compound flagellar "roots" of *O. cardiacum*. *Can. J. Bot.* **48**, 189–196.
Hoffman, L. R. (1973). Fertilization in *Oedogonium*. I. Plasmogamy. *J. Phycol.* **9**, 62–84.
Hoops, H. J. and Floyd, G. L. (1982). Ultrastructure and taxonomic position of the rare volvocalean alga, *Chlorcorona bohemica*. *J. Phycol.* **18**, 462–466.
Hoops, H. J. and Witman, G. B. (1983). Outer doublet heterogeneity reveals structural polarity related to beat direction in *Chlamydomonas* flagella. *J. Cell Biol.* **97**, 902–908.
Hoops, H. J., Floyd, G. L. and Swanson, J. A. (1982). Ultrastructure of the biflagellate motile cells of *Ulvaria oxysperma* (Kütz.) Bliding and phylogenetic relationships among Ulvophycean algae. *Am. J. Bot.* **69**, 150–159.
Hoops, H. J., Wright, R. L., Jarvik, J. W. and Witman, G. B. (1984). Flagellar waveform and rotational orientation in a *Chlamydomonas* mutant lacking normal striated fibers. *J. Cell Biol.* **98**, 818–824.
Hopkins, J. M. (1970). Subsidiary components of the flagella of *Chlamydomonas reinhardtii*. *J. Cell Sci.* **7**, 823–839.
Huang, B., Piperno, G. and Luck, D. J. L. (1979). Paralyzed flagella mutants of *Chlamydomonas reinhardii* defective for axonemal doublet microtubule arms. *J. biol. Chem.* **254**, 3091–3099.
Jacobs, M. and McVittie, A. (1970). Identification of the flagellar proteins of *Chlamydomonas reinhardii*. *Exp. Cell Res.* **63**, 53–61.
Jonsson, S. and Chesnoy, L. (1974). Étude ultrastructurale de l'incorporation des axonèmes flagellaires dans les zygotes du *Monostroma grevillei* (Thuret) Wittr., Chlorophycée marine. *C. r. hebd. Seanc. Acad. Sci., Ser. D.* **278**, 1557–1560.
Lang, N. J. (1963). An additional ultrastructural component of flagella. *J. Cell Biol.* **19**, 631–634.

Loeffler, F. (1889). Eine neue Methode zum Färben der Mikroorganismen, im besonderen ihrer Wimperhaare und Geisseln. *Zentralbl. Bakteriol., Parasitenkd. Infektionskr.* **6,** 209–224.

McLean, R. J., Laurendi, C. J. and Brown, R. M., Jr. (1974). The relationship of gamone to the mating reaction in *Chlamydomonas moewusii. Proc. natl Acad. Sci. U.S.A.* **71,** 2610–2613.

McLean, R. J., Katz, K. R., Sedita, N. J., Menoff, A. L., Laurendi, C. J. and Brown, R. M. (1981). Dynamics of Concanavalin A binding sites on *Chlamydomonas moewusii* flagellar membranes. *Ber. Dtsch. bot. Ges.* **94,** 387–400.

Manton, I. (1959). Electron microscopical observations on a very small flagellate: The problem of *Chromulina pusilla* Butcher. *J. mar. biol. Assoc. U. K.* **38,** 319–333.

Manton, I. (1964a). Observations on the fine structure of the zoospore and young germling of *Stigeoclonium. J. exp. Bot.* **15,** 399–411.

Manton, I. (1964b). The possible significance of some details of flagellar bases in plants. *J. R. microsc. Soc.* **82,** 279–285.

Manton, I. (1965). Some phyletic implications of flagellar structure in plants. *Adv. Bot. Res.* **2,** 1–34.

Manton, I. (1966). Observations on scale production in *Pyramimonas amylifera* Conrad. *J. Cell Sci.* **1,** 429–438.

Manton, I. (1967). Electron microscopical observations on a clone of *Monomastix* Scherffel in culture. *Nova Hedw.* **14,** 1–11.

Manton, I. (1975). Observations on the microanatomy of *Scourfieldia marina* Throndsen and *Scourfieldia caeca* (Korsch.) Belcher et Swale. *Arch. Protistenkd.* **117,** 358–368.

Manton, I. (1977). *Dolichomastix* (Prasinophyceae) from arctic Canada, Alaska and South Africa: A new genus of flagellates with scaly flagella. *Phycologia* **16,** 427–438.

Manton, I. and Friedmann, I. (1960). Gametes, fertilization, and zygote development in *Prasiola stipitata* Suhr. *Nova Hedw.* **1,** 443–462.

Manton, I. and Parke, M. (1960). Further observations on small green flagellates with special reference to possible relatives of *Chromulina pusilla* Butcher. *J. mar. biol. Assoc. U.K.* **39,** 275–298.

Manton, I. and Parke, M. (1965). Observations on the fine structure of two species of *Platymonas* with special reference to flagellar scales and the mode of origin of the theca. *J. mar. biol. Assoc. U.K.* **45,** 743–754.

Manton, I., Rayns, D. G., Ettl, H. and Parke, M. (1965). Further observations on green flagellates with scaly flagella: The genus *Heteromastix* Korshikov. *J. mar. biol. Assoc. U.K.* **45,** 241–255.

Marchant, H. J. (1974). Mitosis, cytokinesis and colony formation in the green alga *Sorastrum. J. Phycol.* **10,** 107–120.

Marchant, H. J., Pickett-Heaps, J. D. and Jacobs, K. (1973). An ultrastructural study of zoosporogenesis and the mature zoospore of *Klebsormidium flaccidum. Cytobios* **8,** 95–107.

Markowitz, M. M. (1978). Fine structure of the zoospore of *Oedocladium carolinianum* (Chlorophyta) with special reference to the flagellar apparatus. *J. Phycol.* **14,** 289–302.

Melkonian, M. (1978). Structure and significance of cruciate flagellar root systems in green algae: Comparative investigations in species of *Chlorosarcinopsis* (Chlorosarcinales). *Plant Syst. Evol.* **130,** 265–292.

Melkonian, M. (1979). Structure and significance of cruciate flagellar root systems in green algae: Zoospores of *Ulva lactuca* L. (Ulvales, Chlorophyceae). *Helgol. wiss. Meeresunters.* **32,** 425–435.

Melkonian, M. (1980a). Ultrastructural aspects of basal body associated fibrous structures in green algae: A critical review. *BioSystems* **12,** 85–103.

Melkonian, M. (1980b). Flagellar roots, mating structure and gametic fusion in the green alga *Ulva lactuca* (Ulvales). *J. Cell Sci.* **46,** 149–169.

Melkonian, M. (1981a). Structure and significance of cruciate flagellar root systems in green algae: Female gametes of *Bryopsis lyngbyei* (Bryopsidales). *Helgol. wiss. Meeresunters.* **34**, 355–369.
Melkonian, M. (1981b). The flagellar apparatus of the scaly green flagellate *Pyramimonas obovata:* Absolute configuration. *Protoplasma* **108**, 341–355.
Melkonian, M. (1982a). The functional analysis of the flagellar apparatus in green algae. In "Prokaryotic and Eukaryotic Flagella" (W. B. Amos and J. G. Duckett, eds.), pp. 591–608. Cambridge Univ. Press, London and New York.
Melkonian, M. (1982b). Structural and evolutionary aspects of the flagellar apparatus in green algae and land plants. *Taxon* **31**, 255–265.
Melkonian, M. (1982c). Effect of divalent cations on flagellar scales in the green flagellate *Tetraselmis cordiformis. Protoplasma* **111**, 221–233.
Melkonian, M. (1983). Functional and phylogenetic aspects of the basal apparatus in algal cells. *J. submicrosc. Cytol.* **15**, 121–125.
Melkonian, M. (1984). Flagellar root-mediated interactions between the flagellar apparatus and cell organelles in green algae. In "Compartments in Algal Cells and their Interaction" (W. Wiessner, D. Robinson, and R. C. Starr, eds), pp. 96–108. Springer Verlag, Berlin-Heidelberg.
Melkonian, M. and Berns, B. (1983). Zoospore ultrastructure in the green alga *Friedmannia israelensis:* An absolute configuration analysis. *Protoplasma* **114**, 67–84.
Melkonian, M. and Preisig, H. R. (1982). Twist of central pair microtubules in the flagellum of the green flagellate *Scourfieldia caeca. Cell Biol. Int. Rep.* **6**, 269–277.
Melkonian, M. and Preisig, H. E. (1984). Ultrastructure of the flagellar apparatus in the green flagellate *Spermatozopsis similis. Plant Syst. Evol.* (in press).
Melkonian, M., Kröger, K.-H. and Marquardt, K.-G. (1980). Cell shape and microtubules in zoospores of the green alga *Chlorosarcinopsis gelatinosa* (Chlorosarcinales): Effects of low temperature. *Protoplasma* **104**, 283–293.
Mesland, D. A. M., Hoffman, J. L., Caligor, E. and Goodenough, U. W. (1980). Flagellar tip activation stimulated by membrane adhesions in *Chlamydomonas* gametes. *J. Cell Biol.* **84**, 599–617.
Micalef, H. and Gayral, P. (1972). Quelques aspects de l'infrastructure des cellules végétatives et des cellules reproductrices d'*Ulva lactuca* L. (Chlorophyceae). *J. Microsc. (Paris)* **13**, 417–428.
Moestrup, Ø. (1970). The fine structure of mature spermatozoids of *Chara corallina* with special reference to microtubules and scales. *Planta* **93**, 295–308.
Moestrup, Ø. (1972). Observations on the fine structure of spermatozoids and vegetative cells of the green alga *Golenkinia. Br. phycol. J.* **7**, 169–183.
Moestrup, Ø. (1974). Ultrastructure of the scale-covered zoospores of the green alga *Chaetosphaeridium,* a possible ancestor of the higher plants and bryophytes. *Biol. J. Linn. Soc.* **6**, 111–125.
Moestrup, Ø. (1978). On the phylogenetic validity of the flagellar apparatus in green algae and other chlorophyll a and b containing plants. *BioSystems* **10**, 117–144.
Moestrup, Ø. (1982). Flagellar structure in algae: A review with new observations particularly on the Chrysophyceae, Phaeophyceae (Fucophyceae), Euglenophyceae and *Reckertia. Phycologia* **21**, 427–528.
Moestrup, Ø. (1983). Further studies on *Nephroselmis* and its allies (Prasinophyceae). II. The question of the genus *Bipedinomonas. Nord. J. Bot.* **3**, 609–627.
Moestrup, Ø. (1984). Further studies on *Nephroselmis* and its allies (Prasinophyceae). II. *Mamiella* gen. nov. (Mamiellaceae fam. nov.). *Nord. J. Bot.* **4**, 109–121.

Moestrup, Ø. and Ettl, H. (1979). A light and electron microscopical study of the flagellate *Nephroselmis olivacea* (Prasinophyceae). *Opera bot.* **49,** 1–39.

Moestrup, Ø. and Thomsen, H. A. (1974). An ultrastructural study of the flagellate *Pyramimonas orientalis* with particular emphasis on Golgi apparatus activity and the flagellar apparatus. *Protoplasma* **81,** 247–269.

Moestrup, Ø. and Walne, P. (1979). Studies on scale morphogenesis in the Golgi apparatus of *Pyramimonas etrarhynchus* (Prasinophyceae). *J. Cell Sci.* **36,** 437–459.

Norris, R. E. (1980). Prasinophytes. *In* "Phytoflagellates" (E. R. Cox, ed.), Vol. 2, pp. 85–145. Elsevier/North-Holland, New York.

Norris, R. E. and Pearson, B. R. (1975). Fine structure of *Pyramimonas parkeae,* sp. nov. (Chlorophyta, Prasinophyceae). *Arch. Protistenkd.* **117,** 192–213.

Norris, R. E., Hori, T. and Chihara, M. (1980). Revision of the genus *Tetraselmis* (Class Prasinophyceae). *Bot. Mag.* **93,** 317–339.

Olson, L. W. and Kochert, G. (1970). Ultrastructure of *Volvox carteri.* II. The Kinetosome. *Arch. Mikrobiol.* **74,** 31–40.

Parke, M. and Rayns, D. G. (1964). Studies on marine flagellates. VII. *Nephroselmis gilva* sp. nov. and some allied forms. *J. mar. biol. Assoc. U.K.* **44,** 209–217.

Pennick, N. C., Clarke, K. J. and Belcher, J. H. (1978). Studies on the external morphology of *Pyramimonas.* 1. *P. orientalis* and its allies in culture. *Arch. Protistenkd.* **120,** 304–311.

Pickett-Heaps, J. D. (1968). Ultrastructure and differentiation in *Chara (fibrosa).* IV. Spermatogenesis. *Aust. J. biol. Sci.* **21,** 255–274.

Pickett-Heaps, J. D. (1971). The autonomy of the centriole: Fact or fallacy? *Cytobios* **3,** 205–214.

Pitelka, D. R. (1974). Basal bodies and root structures. *In* "Cilia and Flagella" (M. A. Sleigh, ed.), pp. 437–469. Academic Press, London.

Ringo, D. L. (1967a). Flagellar motion and fine structure of the flagellar apparatus in *Chlamydomonas. J. Cell Biol.* **33,** 543–571.

Ringo, D. L. (1967b). The arrangement of subunits in flagellar fibers. *J. Ultrastruct. Res.* **17,** 266–277.

Robenek, H. and Melkonian, M. (1979). Rhizoplast-membrane associations in the flagellate *Tetraselmis cordiformis* Stein (Chlorophyceae) revealed by freeze-etching and thin sections. *Arch. Protistenkd.* **122,** 340–351.

Roberts, K. R., Stewart, K. D. and Mattox, K. R. (1981). The flagellar apparatus of *Chilomonas paramecium* (Cryptophyceae) and its comparison with certain zooflagellates. *J. Phycol.* **17,** 159–167.

Roberts, K. R., Stewart, K. D. and Mattox, K. R. (1982). Structure of the anisogametes of the green siphon *Pseudobryopsis* sp. (Chlorophyta). *J. Phycol.* **18,** 498–508.

Rogers, C. E., Mattox, K. R. and Stewart, K. D. (1980). The zoospore of *Chlorokybus atmophyticus,* a charophyte with sarcinoid growth habit. *Am. J. Bot.* **67,** 774–783.

Rogers, C. E., Domozych, D. S., Stewart, K. D. and Mattox, K. R. (1981). The flagellar apparatus of *Mesostigma viride* (Prasinophyceae): Multilayered structures in a scaly green flagellate. *Plant Syst. Evol.* **138,** 247–258.

Salisbury, J. L. (1983). Contractile flagellar roots: The role of calcium. *J. submicrosc. Cytol.* **15,** 105–110.

Salisbury, J. L. and Floyd, G. L. (1978). Calcium-induced contraction of the rhizoplast of a quadriflagellate green alga. *Science* **202,** 975–977.

Salisbury, J. L., Swanson, J. A., Floyd, G. L., Hall, R. and Maihle, N. J. (1981). Ultrastructure of the flagellar apparatus of the green alga *Tetraselmis subcordiformis.* With special consideration given to the function of the rhizoplast and the rhizanchora. *Protoplasma* **107,** 1–11.

Satir, P. (1968). Studies on cilia. III. Further studies on the cilium tip and a "sliding filament" model of ciliary motility. *J. Cell Biol.* **39,** 77–94.

Satir, P. and Ojakian, G. K. (1979). Plant cilia. *Encycl. Plant Physiol., New Ser.* **7,** 224–249.

Sluiman, H. J. (1983). The flagellar apparatus of the zoospore of the filamentous green alga *Coleochaete pulvinata:* absolute configuration and phylogenetic significance. *Protoplasma* **115,** 160–175.

Sluiman, H. J., Roberts, K. R., Stewart, K. D. and Mattox, K. R. (1980). Comparative cytology and taxonomy of the Ulvaphyceae. I. The zoospore of *Ulothrix zonata* (Chlorophyta). *J. Phycol.* **16,** 537–545.

Sluiman, H. J., Roberts, K. R., Stewart, K. D., Mattox, K. R. and Lokhorst, G. M. (1982). The flagellar apparatus of the zoospore of *Urospora penicilliformis* (Chlorophyta). *J. Phycol.* **18,** 1–12.

Snell, W. J. (1976). Mating in *Chlamydomonas:* A system for the study of specific cell adhesion. I. Ultrastructural and electrophoretic analysis of flagellar surface components involved in adhesion. *J. Cell Biol.* **68,** 48–69.

Stewart, K. D. and Mattox, K. R. (1978). Structural evolution in the flagellated cells of green algae and land plants. *BioSystems* **10,** 145–152.

Stewart, K. D. and Mattox, K. R. (1980). Phylogeny of phytoflagellates. *In* "Phytoflagellates" (E. R. Cox, ed.), Vol. 2, pp. 433–462. Elsevier/North-Holland, New York.

Swanson, J. A. and Floyd, G. L. (1978). Fine structure of the zoospores and thallus of *Blidingia minima. Trans. Am. microsc. Soc.* **97,** 549–558.

Sweeney, B. M. (1976). *Pedinomonas noctilucae* (Prasinophyceae), the flagellate symbiotic in *Noctiluca* (Dinophyceae) in Southeast Asia. *J. Phycol.* **12,** 460–464.

Watson, M. W. (1975). Flagellar apparatus, eyespot and behaviour of *Microthamnion kuetzingianum* (Chlorophyceae) zoospores. *J. Phycol.* **11,** 439–448.

Witman, G. B., Carlson, K., Berliner, J. and Rosenbaum, J. L. (1972). *Chlamydomonas* flagella. I. Isolation and electrophoretic analysis of microtubules, matrix, membranes, and mastigonemes. *J. Cell Biol.* **54,** 507–539.

Witman, G. B., Plummer, J. and Sander, G. (1978). *Chlamydomonas* flagellar mutants lacking radial spokes and central tubules. *J. Cell Biol.* **76,** 729–747.

Woodcock, C. L. F. and Miller, G. J. (1973). Ultrastructural features of the life cycle of *Acetabularia mediterranea.* I. Gametogenesis. *Protoplasma* **77,** 313–329.

Woods, J. K. and Triemer, R. E. (1981). Mitosis in the octaflagellate prasinophyte, *Pyramimonas amylifera* (Chlorophyta). *J. Phycol.* **17,** 81–90.

Wright, R. L., Chojnacki, B. and Jarvik, J. W. (1983). Abnormal basal body number, location and orientation in a striated fiber defective mutant of *Chlamydomonas reinhardtii. J. Cell Biol.* **96,** 1697–1707.

4 | Correlations among Patterns of Sporangial Structure and Development, Life Histories, and Ultrastructural Features in the Ulvophyceae

C. J. O'KELLY* and G. L. FLOYD

Department of Botany, Ohio State University, Columbus, Ohio, USA

Abstract: A number of the ultrastructural, reproductive, and biochemical features of critically examined green algae permit the development of classification schemes that supersede those based primarily on thallus morphology. Within the Ulvophyceae, we recognize the orders Ulotrichales, Ulvales, Siphonocladales (incorporating the Cladophorales), Dasycladales, and Caulerpales, based on the phase alternation present in the sexual life history, several zoosporangial and gametangial features (including the cleavage pattern present, the structure of the exit aperture, and the presence or absence of either a vesicle or a plug), and variations in a number of flagellar apparatus features, especially the shapes of proximal sheaths and terminal caps and the presence or absence of "winged" microtubules in rootlets. Additional supporting evidence is presented from, for example, studies of cell wall structural polysaccharides, chloroplast pigment arrays, and chromosome numbers. We expand the concepts of the Ulotrichales and Ulvales to include algae having filamentous, foliose, and (in the Ulotrichales) hemisiphonous vegetative morphologies, and demonstrate parallel evolution of thallus form both within and among the orders. Trends toward the modification or loss of certain features may reflect the course of evolution in the Ulvophyceae. Similar investigations, we suggest, should result in more "natural" ordinal circumscriptions in all green algal classes, and should provide much new information on the phylogenetic relationships of these algae.

INTRODUCTION

Even with the large amount of information that is now available on the ultrastructural features of green algae, systematic treatments based on ultra-

---

*Present address: Department of Botany, La Trobe University, Bundoora, Victoria 3083, Australia.

Systematics Association Special Volume No. 27, "Systematics of the Green Algae", edited by D. E. G. Irvine and D. M. John, 1984, pp. 121–156. Academic Press, London and Orlando.
ISBN 0 12 374040 1   Copyright © by the Systematics Association
                              All rights of reproduction in any form reserved

structural data apparently have not been accepted by the larger community of phycologists and plant taxonomists, and appear in few textbooks except in marginal or parenthetical notations. Perhaps this is due, in part, to the relatively rapid progress in the field of green algal ultrastructural systematics (see Mattox and Stewart, this volume, Chapter 2; Melkonian, this volume, Chapter 3). More likely, we feel, other systematists have been unwilling to discard long-held beliefs that thallus morphology in the green algae is of great phylogenetic significance, even in the face of the stiffest challenges from ultrastructural information. Bold and Wynne (1978, p. 245) succinctly identified the problem when they stated that the charophycean orders Klebsormidiales, Zygnematales, Coleochaetales, and Charales are "rather strange bedfellows on the basis of traditionally recognized criteria!" Because ultrastructural work has left several orders in disarray, especially the Chaetophorales, Chlorosarcinales, Ulotrichales, and Ulvales, and few attempts to reconstruct these orders along criteria consistent with ultrastructural findings have appeared since the treatment of certain groups by Stewart and Mattox (1975; see also Silva, 1982; Floyd and O'Kelly, 1984; Mattox and Stewart, this volume, Chapter 2), it is perhaps understandable that general acceptance of ultrastructural systematics has been delayed. That the transmission electron microscope might be the only tool capable of providing phylogenetically reliable morphological and anatomical information on green algae also may have provoked a not unjustifiable sense of horror among those not trained in (or budgeted for!) its use.

If the structures associated with flagellar apparatuses and with the mitotic and cytokinetic processes are as evolutionarily conservative as we think they are, then, we reasoned, it should be possible to identify certain of these fine structural features that are consistent within natural orders, and to find other features (e.g. structural, reproductive, biochemical) that correlate well with the ultrastructural features and do not necessarily correlate well with patterns of thallus morphology. Our studies on the features of certain filamentous organisms usually referred to the "marine Chaetophoraceae" (O'Kelly, 1983; O'Kelly and Floyd, 1983) drew us to this problem, and suggested to us our course of action. We subsequently expanded our investigations to include most of the Ulvophyceae, looking for correlations among ultrastructural features, especially those of the motile cells, with which we have been most directly concerned (Hoops et al., 1982; O'Kelly and Floyd, 1983), and other features such as sexual life history patterns, zoosporangial and gametangial structure and development, and selected biochemical attributes. We report our results to date in this paper. We shall consider the following questions:

1. What orders are referable to the Ulvophyceae, and how are these orders best circumscribed?

2. What features are consistently present in the Ulvophyceae, and what trends of possible evolutionary significance may be discerned?
3. To what extent may the findings of our study on the Ulvophyceae be useful in systematic and phylogenetic treatments of the other classes within the Chlorophyta?

THE ULVOPHYCEAE: ORDINAL CIRCUMSCRIPTIONS

1. The Ulvophyceae Defined

We used the following ultrastructural criteria to determine if an alga under investigation belonged to the Ulvophyceae. Since these features have been described in some detail by O'Kelly and Floyd (1983, 1984), Melkonian (this volume, Chapter 3), and Mattox and Stewart (this volume, Chapter 2), we will not discuss them here. In the flagellar apparatuses of motile cells, excluding stephanokont zoospores, we looked for 180° rotational symmetry, counterclockwise absolute orientation, basal bodies arranged into upper and lower pairs with the proximal ends of the upper pair overlapping and both pairs arranged perpendicular to the long axis of the cell, cruciately arranged rootlets showing an X-2-X-2 pattern, and the absence of multilayered structures. We have found, in addition, that striated bands connecting rootlets to basal bodies (Fig. 1) and striated distal fibers (see Fig. 7) may be found in most ulvophycean motile cells (Floyd and O'Kelly, 1984). During mitosis, we expected to see a closed, centric, persistent spindle, and at cytokinesis a centripetal furrow associated with neither phycoplast nor phragmoplast microtubules. Although the known features of trentepohlialean motile cells are consistent with those outlined above for the Ulvophyceae, they differ from all the other taxa included here by possessing structures vaguely like multilayered structures (MLS) associated with two of the four rootlets, and plasmodesmata in the cross walls between vegetative cells. Pending a more complete understanding of the phylogenetic significance of these and other anomalous features in this order (see O'Kelly and Floyd, 1984; Chapman, this volume, Chapter 8; Roberts, this volume, Chapter 13), we will exclude the Trentepohliales from our discussion.

2. The Ulotrichales

The traditional concept of the Ulotrichales has persisted to the present despite the discovery that, from species having the characteristic filamentous, uniseriate, unbranched morphology, genera referable to the Charophyceae, the Chlorophyceae, and the Ulvophyceae can be recognized (Stewart and Mattox, 1975; Sluiman et al., 1980a). Among algal morphologists there has been considerable discussion of the scope of the order,

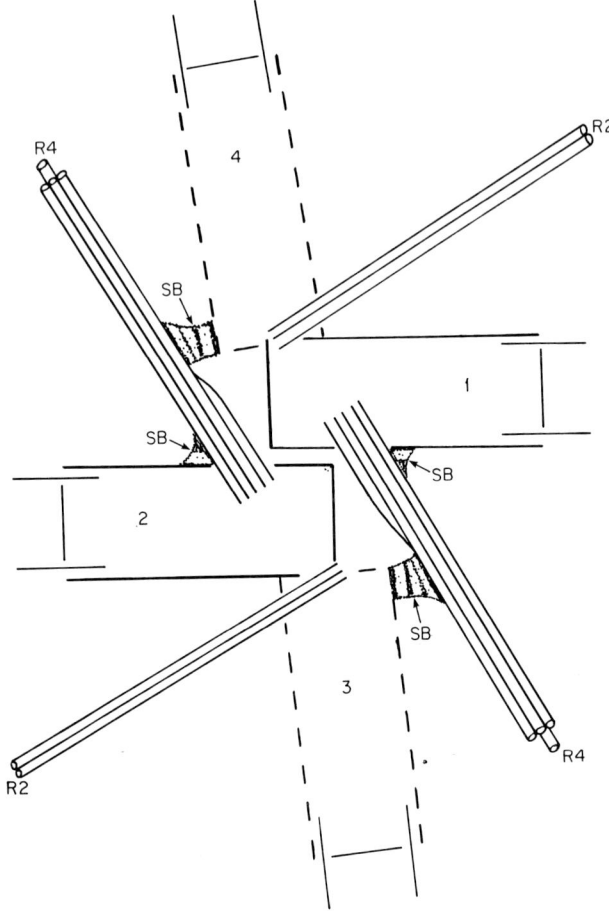

Fig. 1. Generalized diagrammatic illustration of the flagellar apparatus in an ulvophycean quadriflagellate zoospore, as seen from the anterior end of the cell. The distal fiber and several other components are not shown. Note the presence of 180° rotational symmetry. The basal bodies are organized into an upper pair (1, 2) and a lower pair (3, 4, *dashed lines*). The lower pair is absent in biflagellate motile cells. The basal bodies and microtubular rootlets (R2, R4) show counterclockwise absolute orientation; that is, they are rotated in the counterclockwise direction relative to a strictly cruciate arrangement of these components (see O'Kelly and Floyd, 1983). Striated bands (SB) connect members of both basal body pairs to the adjacent four-membered rootlet.

Table I. The Ulotrichales.

"*Eugomontia* group"	"*Chlorocystis* group"
Eugomontia	*Chlorocystis*
Pseudopringsheimia sensu Perrot	*Halochlorococcum*
Pseudendoclonium sensu Vischer	"*Acrosiphonia* group"
"*Monostroma* group"	*Acrosiphonia*
Gomontia	*Chlorothrix*
Monostroma	*Pseudendoclonium sensu* Yarish
"*Ulothrix* group"	*Spongomorpha*
Capsosiphon	*Urospora*
Gayralia	
Trichosarcina	
Ulothrix	

but only Kornmann (1963, 1965, 1973; see also Tanner, 1981), has seriously challenged its traditional foundation. Kornmann disregarded the vegetative morphology of the gametophytic thallus as a reliable feature, emphasizing instead life history patterns and the morphology and development of the sporophytic generation. Our investigations (Floyd and O'Kelly, 1984; O'Kelly and Floyd, 1984) have confirmed and extended Kornmann's treatment (Table I), although we continue to recognize this group at the ordinal level, not at the class level as proposed by Kornmann in 1973.

(a) Motile cell features: Ulotrichalean genera produce quadriflagellate zoospores and biflagellate gametes that usually possess typically ulvophycean flagellar apparatus features. The discovery of striated distal fibers in several genera (Floyd and O'Kelly, 1984; O'Kelly *et al.*, 1984) is worth emphasizing. Exceptionally, flagellar apparatuses having a V-shaped configuration occur in *Pseudendoclonium* as defined by Vischer in 1933 (see also Tupa, 1974) and in *Ulothrix* (O'Kelly and Floyd, 1984). Other minor anomalies occur in *Urospora* motile cells (see Floyd and O'Kelly, 1984). The motile cell ultrastructural features diagnostic of the Ulotrichales include the presence of body scales (except in the "*Acrosiphonia* group") and the structures of terminal caps and proximal sheaths. Terminal caps in this order are more or less prominent electron-dense flaps located on the anterior surface of basal bodies 1 and 2, often near the insertions of the two-membered rootlets, and they fold over and cover a small part of the proximal end of the basal body (Sluiman *et al.*, 1980a; Hoops *et al.*, 1982; Floyd and O'Kelly, 1984; O'Kelly *et al.*, 1984; Fig. 2). Proximal sheaths occur beneath all basal bodies in ulotrichalean motile cells, but those associated with the upper basal body pair are wedge-shaped, narrow proximally and broadening distally (Fig. 3), while those associated with the lower basal body pair (except in *Urospora*)

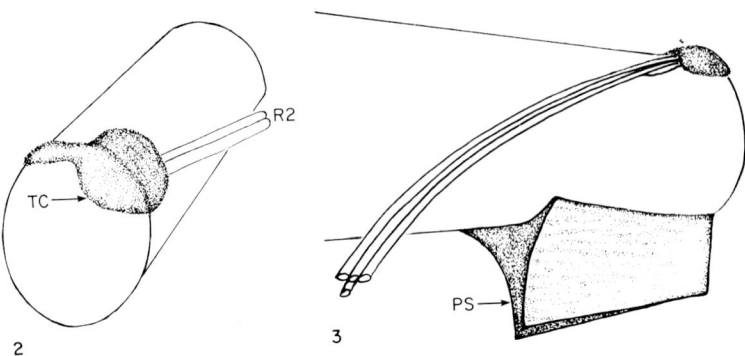

Figs 2, 3. Diagrammatic representations of motile cell ultrastructural features typical of the Ulotrichales. Fig. 2. The simple, overlapping terminal cap (TC). Note the apparent continuity of the terminal cap with the electron-dense material present near the insertion of the two-membered rootlet (R2). Fig. 3. The wedge-shaped proximal sheath (PS) typically found subtending the members of the upper basal body pair.

are half-cylindrical structures appressed to the posterior surface of the basal bodies (Floyd and O'Kelly, 1984; O'Kelly et al., 1984).

(b) *Mitosis and cytokinesis:* Relatively few ultrastructural examinations of nuclear and cellular division exist for ulotrichalean algae. For the most part the typical ulvophycean configuration is present during mitosis, but the fate of the spindle is difficult to discern. In several uninucleate-celled genera the spindle is impinged upon by the precocious cleavage furrow in early telophase, (see Sluiman et al., 1983). Other authors are silent on this point. The cytokinetic mechanism in uninucleate-celled genera is typically ulvophycean. Whether the furrow is always precocious is not known, although light microscope studies (see Fritsch, 1935; Jónsson, 1962; Hudson and Waaland, 1974) indicate a close temporal association between the processes of mitosis and cytokinesis in all genera studied, regardless of whether the cells are uninucleate or multinucleate. A variation occurs in the hemisiphonous genera *Acrosiphonia* (Hudson and Waaland, 1974) and *Urospora* (Lokhorst and Star, 1983) in which a ring or band of microtubules is associated with the leading edge of the cleavage furrow.

(c) *Sexual life history:* Virtually all the genera that have been examined with the electron microscope and for which sexual life histories are known have an alternation of heteromorphic phases in which the gametophyte is multicellular and the sporophyte is unicellular (Floyd and O'Kelly, 1984; see Table I). The two genera in the "*Chlorocystis* group" are exceptional in that both life history phases are unicellular (Kornmann and Sahling, 1983). Chromosomal alternation usually parallels the morphological alternation, with meiosis occurring during the first nuclear divisions in the sporophyte

(Gross, 1931; Jónsson, 1968, 1970), although parthenogenetic development of gametes may lead to a morphological alternation presumably without chromosomal alternation (Jónsson, 1968; Kornmann, 1970; Lokhorst, 1978; Lokhorst and Trask, 1981). There are three exceptional life history patterns. In *Eugomontia sacculata* Kornm. the gametophytic and sporophytic vegetative thalli are nearly identical (Kornmann, 1960). The zoosporangia formed by the sporophytes, however, are structurally and cytologically identical to the single-celled sporophytes found in other genera (Kornmann, 1960; Wilkinson and Burrows, 1970; Floyd and O'Kelly, 1984). Another species, described by Perrot (1969) and identified (incorrectly) as *Pseudopringsheimia confluens* (Rosenv.) Wille, has a similar life history. In this species, however, the gametophyte is a discoid, pseudoparenchymatous epiphyte, while the sporophyte is a shell-boring, branching filament. Curiously, all cells but the apical cells in the sporophytes died soon after formation in Perrot's cultures. The apical cells eventually develop into sporangia having features similar to the sporangia of *Eugomontia* and in which meiosis occurs. In some species of *Acrosiphonia,* on the other hand, the sporophyte produces the gametophytic thallus directly rather than by zoospore formation, and the club-shaped structure typical of sporophytes in other ulotrichalean species, including other members of *Acrosiphonia,* may be reduced or eliminated (Kornmann, 1970).

(d) Sporangial and gametangial structure and development: In ulotrichalean gametophytes, gametangia and zoosporangia having apparently identical structure and development are produced in cells that show little apparent precleavage differentiation from the vegetative state at the light microscope level. In most of the uninucleate-celled genera a series of sequential cleavages produces the motile cells (e.g. Gross, 1931; Jónsson, 1962), while in the hemisiphonous genera simultaneous or progressive cleavage prevails (Jónsson, 1962). According to Gross (1931), however, the gametophytes of some species of *Ulothrix* produce swarmers by simultaneous cleavage of a multinucleate mother cell. The exit aperture through which the motile cells escape is an undifferentiated pore or slit (Jónsson, 1962; Tatewaki, 1969; Lokhorst and Vroman, 1972, 1974a,b; Lokhorst, 1978; Lokhorst and Trask, 1981; Kornmann and Sahling, 1983; Floyd and O'Kelly, 1984), except in the larger species of *Urospora,* where the exit apertures may be associated with a papilliform protuberance (Jónsson, 1962; Lokhorst and Trask, 1981), and in *Acrosiphonia,* which has operculate exit apertures (Jónsson, 1962). *Gayralia* zoosporangia have no exit aperture, since the zoospores, enclosed within a vesicle, are released by wall dissolution (Gayral, 1964, 1965; Bliding, 1968). A vesicle surrounds the motile cells in most genera examined (Jónsson, 1962; Bliding, 1963, 1968; Tatewaki, 1969; Lokhorst and Vroman, 1972, 1974a,b; Lokhorst, 1978; Lokhorst and Trask, 1981; Kornmann and Sahl-

ing, 1983; Floyd and O'Kelly, 1984). This vesicle may or may not be released from the sporangium or gametangium along with the motile cells (Floyd and O'Kelly, 1984). At least in gametangia of *Monostroma grevillei* (Thur.) Wittr. and *Urospora gregaria* (A. Br.) Floyd & O'Kelly, and zoosporangia of *Gayralia oxysperma* (Kütz.) Vinogr., this vesicle is a more or less electron-dense wall layer that is added to the existing wall layers during motile cell development and is separated from the rest of the wall during motile cell release (unpublished observations; Fig. 4). Events during zoosporangial development in sporophytes are not as well understood, but are clearly different in several ways from events in gametophytes. The zoosporangial mother cell, whether or not it is the only cell of the sporophyte, undergoes considerable enlargement before zoosporogenesis commences. Multiplication of nuclei occurs in the absence of cytokinesis in the sporophytes of *Acrosiphonia* (Jónsson, 1970) and *Monostroma* (Jónsson, 1968), and in the sporangia of *Eugomontia* (Floyd and O'Kelly, 1984) and *Pseudopringsheimia sensu* Perrot (Perrot, 1969). *Eugomontia* zoosporangia subsequently undergo simultaneous cleavage (Floyd and O'Kelly, 1984), and this cleavage pattern may also be present in *Acrosiphonia* (Jónsson, 1970)

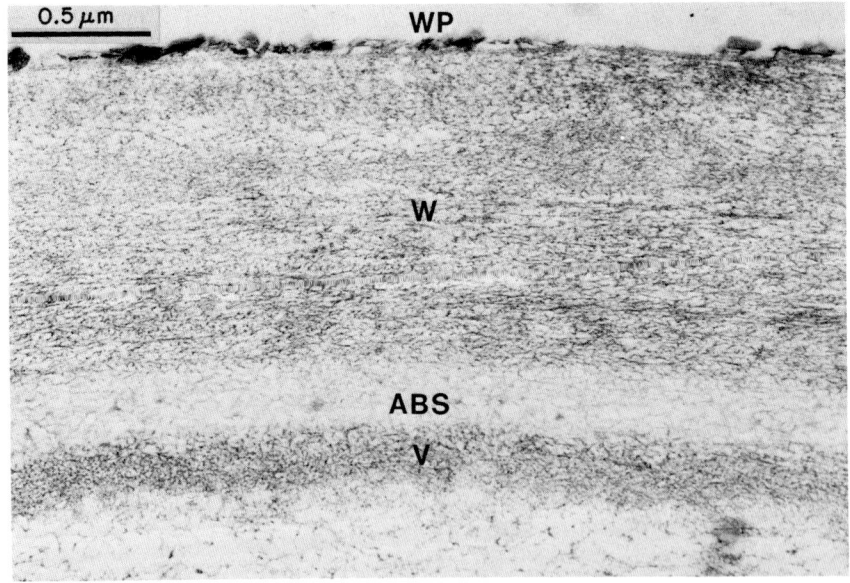

Fig. 4. The cell wall of a mature female gametangium of *Urospora gregaria*, as seen in cross section through the electron microscope. The vesicle (V) is separated from the rest of the wall (W) by an electron-lucent zone (ABS) that may represent an abscission zone. Electron-dense particles (WP) appear on the surface of the wall.

and *Monostroma* (Jónsson, 1968), while sequential cleavage patterns may be the rule in *Spongomorpha* (Jónsson, 1962) and *Ulothrix* (Dodel, 1876). The exit aperture is undifferentiated at the light microscope level except in shell-boring species, where an exit tube may be present (Kornmann, 1973). No vesicles have been observed.

(e) Cell wall structural polysaccharides: The structure and chemical composition of cell wall structural polysaccharides have proven to be useful in taxonomic treatments of multinucleate-celled green algal genera (see below), and the cell walls of *Acrosiphonia* and *Urospora* have been examined in this context. Parker (1970) reviewed the literature to that time (see also Jónsson, 1962), and new information on the *Urospora* cell wall has since appeared (Bourne et al., 1974; Carlberg and Percival, 1977). Interpretations of the published results vary, but three generalizations may be made that are sufficient for our purposes. First, the microfibrils in *Acrosiphonia* and *Urospora* motile cells are more or less randomly arranged. Second, these microfibrils do not yield the x-ray spectrum of "native" cellulose under any of the conditions tested. Third, these microfibrils yield glucose and several other sugars on hydrolysis. These cell wall features have provided strong evidence in support of the separation of *Acrosiphonia* and *Urospora* from other multinucleate-celled green algal groups (Jónsson, 1962; Parker, 1970). Domozych et al. (1980), on the other hand, found apparently pure cellulose in the cell walls of a species of *Pseudendoclonium sensu* Vischer. We are unaware of any other comparable examinations of uninucleate-celled ulotrichalean genera.

(f) Siphonaxanthin and siphonein: Following the suggestion by O'Kelly (1982a) that siphonaxanthin is a primitive feature in green algae that has been preserved in those ulvophycean algae that inhabit relatively deep waters, we would expect to find siphonaxanthin or siphonein in at least some ulotrichalean algae. Most members of Ulotrichales, however, inhabit shallow water or soil, hence it is not surprising that siphonaxanthin has not been found in the few previously examined species (Goodwin, 1974). *Eugomontia sacculata* is a subtidal species (Wilkinson and Burrows, 1970, 1972), and our preliminary chromatographic analysis of its chloroplast pigments reveals the presence of an orange xanthophyll. The mobility of this pigment from *Eugomontia* is close to, but not identical with, that of siphonein isolated from *Codium fragile* ssp. *tomentosoides,* and upon saponification its mobility is identical to that of siphonaxanthin from several sources. The name "siphonein" is actually attached to a class of esterified siphonaxanthins; the esterifying molecule may be any one of a number of fatty acids (see Goodwin, 1974). Therefore we suggest, in recognition of the analytical work that remains to be done on the *Eugomontia* pigment, that this orange xanthophyll represents a form, perhaps previously undescribed, of siphonein. Should

this be confirmed, *Eugomontia* would be only the second genus of green algae known to possess siphonein but not siphonaxanthin, the other genus being the caulerpalean *Dichotomosiphon* (Goodwin, 1974).

Even though the Ulotrichales demonstrates considerable morphological diversity, our analysis of its features reveals a well-circumscribed group of closely related organisms with typically ulvophycean features, supporting our interpretation of this group as an order within the Ulvophyceae and not as a separate green algal class (Floyd and O'Kelly, 1984). The following features are diagnostic of the order:

1. Simple, overlapping terminal caps.

2. Wedge-shaped proximal sheaths subtending the basal bodies that form the upper pair, with those subtending the lower pair usually of a different structure.

3. Sexual life history an alternation of heteromorphic phases, or if isomorphic then with the sporophyte-borne zoosporangia closely resembling the sporophytes of other genera in structure and development.

4. Different structure of, and developmental sequences in, gametophyte-borne gametangia and zoosporangia *versus* sporophyte-borne zoosporangia.

5. Exit aperture of gametophyte-borne gametangia and zoosporangia a simple, unelevated hole, or if more complex then not resembling those in other orders.

6. A vesicle composed of modified wall materials surrounding the motile cells produced by gametophytes, and usually released with them.

Within this order, five groups of genera may be recognized on an informal basis pending further investigations of the taxa involved (see Table I). The *Eugomontia* group consists of those genera having a filamentous or pseudoparenchymatous, "chaetophoralean" vegetative morphology, and either a multicellular sporophyte, or body scales on the motile cells that have a distinctive substructure, or both. In the *Monostroma* group the gametophytic thallus is a pseudoparenchymatous disc at least during early development, the sporophyte is unicellular (as it is in the remaining groups), and scales have a distinctive substructure. In the *Ulothrix* group the earliest stage in the development of the gametophyte is an erect, uniseriate, unbranched filament, and scales, if present (they may be absent from some marine species of *Ulothrix*), lack a distinctive substructure. The two genera of the *Chlorocystis* group have coccoid, unicellular gametophytes. The sexual life history in these algae has only recently been determined (Kornmann and Sahling, 1983), and our preliminary "unpublished" investigations on the ultrastructure of their motile cells have confirmed their status as members of the Ulotrichales, with possible close affinities to *Ulothrix*. The *Acrosiphonia* group includes genera having filamentous or hemisiphonous gametophytic thallus forms that may start development with either erect or prostrate filaments. Scales are absent from the motile cells. A unique and consistent pyrenoid structure is also typical of this last-named group (Lokhorst and Trask, 1981; Berger-Perrot and Thomas, 1982; unpublished observations).

3. The Ulvales

When the Ulvales has been recognized as an order distinct from the Ulotrichales in traditional treatments, it has included within it only those species having parenchymatous, foliose, or tubular thalli composed of uninucleate cells. This concept has persisted to the present despite demonstrations that certain genera referable to this order on a morphological basis, for example, *Schizomeris* (Mattox *et al.*, 1974), belong neither to the Ulvales nor to the Ulvophyceae, while other genera lacking parenchymatous vegetative morphology are in fact closely related to *Ulva* (O'Kelly, 1983; O'Kelly and Floyd, 1983; Floyd and O'Kelly, 1984). Our circumscription of the Ulvales (Floyd and O'Kelly, 1984; Table II) includes filamentous as well as foliose forms, illustrating the broadened scope of the order that we have adopted to accommodate the results from current research on included organisms.

(a) Motile cell features: The quadriflagellate zoospores and biflagellate gametes that are typical of the Ulvales possess (by definition) ulvophycean ultrastructural features. Most recently, we have documented striated distal fibers in several ulvalean species (Stuessy *et al.*, 1983; Floyd and O'Kelly, 1984; O'Kelly *et al.*, 1984) although the distal fiber and its striated region are rudimentary in *Entocladia* (O'Kelly and Floyd, 1983). As in all ulvophycean orders except the Ulotrichales, body scales are absent. Two features are diagnostic. The terminal caps that are present on the basal bodies of the upper pair are bilobed and cover the entire proximal end of the associated basal bodies (Fig. 5). Terminal caps are usually absent from the lower basal

Table II. The Ulvales.

Ulvellaceae	Ulvaceae
Acrochaete	*Chloropelta*
Endophyton	*Enteromorpha*
Entocladia	*Ulva*
Ochlochaete	*Ulvaria*
Pilinella	"*Pseudendoclonium* group"
Pringsheimiella	*Blidingia*
Pseudopringsheimia	*Pseudendoclonium*
Ulvella	
Incertae sedis	
Kornmannia	
Percursaria	

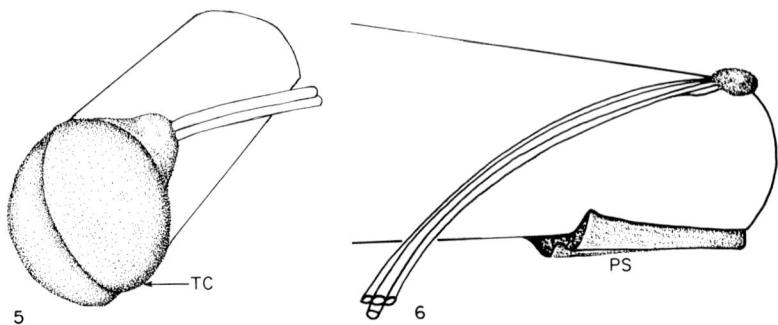

Figs 5, 6. Diagrammatic representations of motile cell ultrastructural features typical of the Ulvales. Fig. 5. The bilobed terminal cap (TC). Fig. 6. The proximal sheath (PS).

body pair in ulvalean zoospores, but in *Entocladia* (O'Kelly and Floyd, 1983) simple overlapping terminal caps may be found in this position. Proximal sheaths subtending all basal bodies are composed of two equal subunits, triangular in cross section, that narrow and finally join together proximally (Fig. 6).

(b) Mitosis and cytokinesis: The ultrastructural aspects of nuclear and cellular division have been examined in *Enteromorpha* (McArthur and Moss, 1978) and in *Ulva* (Løvlie and Bråten, 1970; Bråten and Nordby, 1973). As discussed by Mattox and Stewart (1974) and Stewart and Mattox (1975, 1978), there is little to distinguish mitosis and cytokinesis in these two genera from mitosis and cytokinesis in the ulotrichalean genera that they studied.

(c) Sexual life history: An alternation of isomorphic phases is the typical sexual life history pattern in the Ulvales (Floyd and O'Kelly, 1984). The single exception occurs in the enigmatic genus *Kornmannia*, in which Yamada and Tatewaki (1965) have reported a heteromorphic alternation between a blade-like sporophyte and a discoid gametophyte.

(d) Sporangial and gametangial structure and development: Except for the reduction division that occurs during the formation of zoospores involved in the sexual life history, the structures and developmental events in zoosporangia are identical to those of gametangia, and there are no distinctions in these structures and events between sporophytes and gametophytes. With the apparent exception of certain subgenera (e.g. *Gemina*) of *Ulva*, zoosporangia and gametangia may form from any vegetative cell. The formation of an exit papilla, along with certain cytoplasmic events, marks the initial differentiation of the mother cells (Nordby, 1974; O'Kelly and Yarish, 1980; O'Kelly, 1982b, 1983). In addition, an electron-dense wall layer, or capsule, is produced in the genera examined (Bråten and Løvlie,

1968; McArthur and Moss, 1979; O'Kelly and Yarish, 1980). The ultrastructural aspect of the capsule may be remarkably similar to that of the vesicle observed in certain ulotrichalean genera (see above). Subsequent cleavage is sequential in all cases examined (Nordby, 1974; O'Kelly and Yarish, 1980, 1981; O'Kelly, 1982b, 1983). The capsule apparently remains attached to the rest of the wall. *Percursaria percursa* (Ag.) Bory (but not *P. dawsonii* Hollenb., unpublished observations) is exceptional in that a vesicle reportedly does surround the released motile cells (Kornmann, 1956).

(e) Cell wall structural polysaccharides: Examinations of cell wall structural polysaccharides in the Ulvales have been few, restricted to the genera *Enteromorpha* and *Ulva*. Cell walls from these two genera have, in the hands of most workers, failed to yield the x-ray diagram for "native" cellulose, and several hydrolysates from extracted microfibrils have yielded other sugars, especially xylose, in addition to glucose (Kreger, 1962; Siegel and Siegel, 1973). Dennis and Preston (1961; see also Mackie and Preston, 1974) isolated rodlets from *Ulva* walls that had an x-ray diffraction pattern similar to that of native cellulose and yielded only glucose on hydrolysis. Kreger (1962), however, suggested that similar rodlets produced the x-ray diagram of a reconstituted cellulose, not native cellulose. Domozych *et al.* (1980) have found several similarities between the cell wall features of *Ulva* and those of *Pseudendoclonium sensu* Vischer (Ulotrichales).

(f) Siphonaxanthin and siphonein: Yokohama (1981) and O'Kelly (1982a) have demonstrated the presence of siphonaxanthin in two species of *Ulva* and in several filamentous species, respectively. Only subtidal species have the pigment. Siphonein is absent.

The diagnostic features of the Ulvales include:
1. Bilobed terminal caps.
2. Proximal sheaths consisting of two equal subunits and having the same structure on all basal bodies.
3. Sexual life history an alternation of isomorphic phases.
4. Gametangia and zoosporangia identical in structure and development, lined with a capsule consisting of electron-dense wall material, and producing motile cells by sequential cleavages.
5. Exit aperture of gametangia and zoosporangia papillate, producing a more or less rounded pore after motile cell release.
6. A vesicle not usually surrounding the motile cells.

Our findings clearly document the separateness of the orders Ulotrichales and Ulvales, settling a long-standing taxonomic question. While the range of morphological form is not as great in the Ulvales as in the Ulotrichales, the correlated features we have identified call for the inclusion of filamentous as well as parenchymatous forms into the Ulvales. Within this order, we recognize two families, as well as some allied genera of uncertain taxonomic position (see Table II). Filamentous or pseudoparenchymatous, "chaetophoralean" thalli, as well as the absence of rhizoplasts from motile cells, typify the Ulvellaceae (O'Kelly and Floyd, 1983), while parenchymatous thalli and the presence of rhizoplasts in specific arrays, depending on the type of motile cell being considered (Melkonian, 1979, 1980; Floyd and O'Kelly, 1984), identify the Ulvaceae. We have placed *Pseudendoclonium sensu* Nielsen (Nielsen, 1980; in our opinion the correct concept of this genus) and *Blidingia* together in a "*Pseudendoclonium* group" based on our preliminary observations of their motile cells (Swanson and Floyd, 1978; unpublished observations). *Percursaria* and *Kornmannia* are referred to the Ulvales with some reservations. The principal anomalous feature of *P. percursa* is the reported presence of a vesicle surrounding its motile cells. This may not prove to be a formidable objection, however. Because the ultrastructural appearance of the ulvalean capsule is very similar to that of the ulotrichalean vesicle, it may be that the one has been derived from the other, and it is conceivable, therefore, that the capsule layer may regain or have retained its function as a vesicle in certain genera. The zoosporangia of *P. dawsonii* have typically ulvacean structural and developmental features, at least as observed at the light microscope level (unpublished observations), and we suggest that this unusual species may be referable to *Enteromorpha*. *Kornmannia*, on the other hand, is more problematical. Zoosporangial and, apparently, gametangial structure and development are of the ulvalean type (Yamada and Tatewaki, 1965; Bliding, 1968; unpublished observations), and the zoospores have bilobed terminal caps (unpublished observations). Early thallus development resembles that of *Blidingia*, and the two genera also share the feature of the absence of eyespots from the zoospores (Bliding, 1968). However, the sexual life history pattern (Yamada and Tatewaki, 1965) and the ultrastructural features of the pyrenoid (Hori, 1973) are unlike anything found elsewhere in the Ulvales.

4. The Siphonocladales

The taxonomic history of the Siphonocladales is fraught with controversy. Much of the discussion has centered on whether or not the Cladophoraceae should be separated into its own order (see Fritsch, 1946; Jónsson, 1962; Hoek, 1981, this volume, Chapter 5). Relevant ultrastructural data are relatively few, and the ulvophycean affinity of the Siphonocladales in general (Sluiman *et al.*, 1980b) and the Cladophoraceae in particular (Floyd, 1981; Taylor *et al.*, 1982) has only recently been suggested. Despite these problems, our analysis suggests to us, at least in a preliminary way, that the differences between the Cladophoraceae and other siphonocladalean algae have been overemphasized, and that only one order need be recognized

Table III. Critically examined siphonocladalean taxa.

Chaetosiphonaceae	Anadyomenaceae
Blastophysa	*Anadyomene*
Chaetosiphon	*Microdictyon*
Arnoldiellaceae	Siphonocladaceae
Basicladia	*Siphonocladus*
Cladophoraceae	Valoniaceae
Chaetomorpha	*Dictyosphaeria*
Cladophora	*Valonia*
Rhizoclonium	

(Table III; Hoek, 1981, this volume, Chapter 5). Hoek (1981) has suggested that the combined order should bear the name Cladophorales, according to the nomenclatural principle of priority. We have chosen to retain Siphonocladales, however, because the latter name has been used in this broad sense for a number of years (e.g. Oltmanns, 1904; Chapman and Chapman, 1973), while the former has generally retained the narrow application intended by Fritsch (1935). Since strict application of the principle of priority is not required at the ordinal level (Art. 11.4 of the *International Code of Botanical Nomenclature;* Voss, 1983), we are free to make this choice.

(a) Motile cell features: Most siphonocladalean genera that have been carefully examined produce quadriflagellate zoospores and biflagellate gametes. Ultrastructural information is available from members of the Cladophoraceae (Floyd, 1981; Taylor *et al.*, 1982; Floyd *et al.*, in preparation), the Chaetosiphonaceae, which has recently been allied with the Cladophoraceae (O'Kelly and Floyd, 1981; unpublished observations), and, in a very preliminary way, the Valoniaceae (Hori and Enomoto, 1978a,b; unpublished observations). Typically ulvophycean features prevail throughout, including prominently striated distal fibers (Fig. 7). Terminal caps, however, are entirely absent. Proximal sheaths are much-reduced half-cylindrical structures, and appear to be entirely absent from a few species. The basal bodies themselves are unusual, at least in the Cladophoraceae and Chaetosiphonaceae, in that the microtubules are arranged in nine doublets through most of the length of the basal body instead of the standard nine triplets (Floyd *et al.*, in preparation). In all cells in which the microtubular rootlets have been observed, a "flattening" is present: the rootlets diverge at a small angle (less than 20°) from the basal bodies, and adjacent rootlets may be nearly or quite parallel to one another as they extend posteriorly into the cell (Floyd *et al.*, in preparation). A coarsely striated component appears above the two-membered rootlets, starting near their insertion into the flagellar apparatus and extending posteriorly a variable distance (depending on the species) into

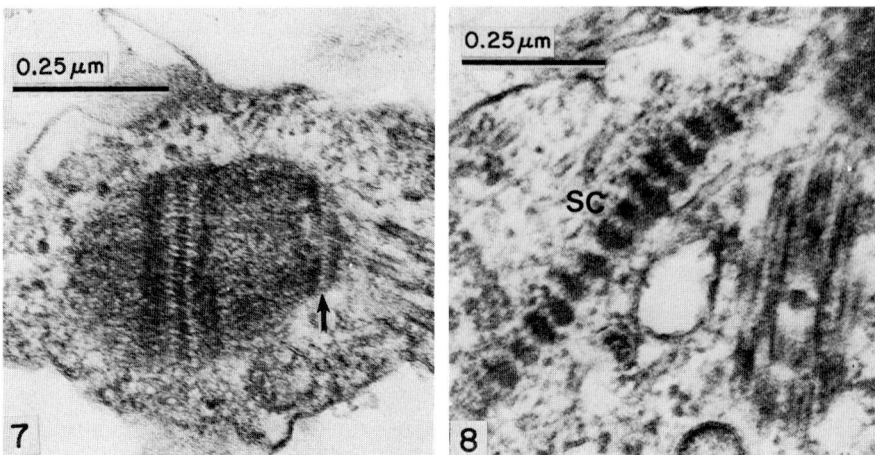

Figs 7, 8. Electron micrographs of motile cell ultrastructural features typical of the Siphonocladales. Fig. 7. The distal fiber, as seen in the biflagellate gamete of *Chaetosiphon* sp. Note the prominently striated central region, and an additional striated area (*arrow*) near the point of attachment of the distal fiber to a basal body. Fig. 8. The coarsely striated rootlet-associated component (SC), seen here associated with a two-membered rootlet (out of the plane of section) in the biflagellate zoospores of *Cladophora glomerata*.

the cell (Fig. 8). Additional electron-dense material that may represent striated microtubule-associated components of the type seen in the Ulotrichales and Ulvales may also be present. A single "wing" structure also appears, connecting the lowermost microtubule in the multistranded rootlets to one of the uppermost microtubules. This structure is identical to that seen in *Batophora*, a member of the Dasycladales (see Roberts, this volume, Chapter 13). At least one rhizoplast occurs in *Valonia* zoospores, whereas in the cladophoracean and chaetosiphonacean genera examined, rhizoplasts are absent. In some cases, unstriated structures may be observed (unpublished observations) attaching to the undersides of basal bodies and accompanying the multistranded rootlets posteriorly into the cell in much the same way as rhizoplasts.

(b) Mitosis and cytokinesis: In the Siphonocladales, nuclear and cytoplasmic division are separate, unassociated events (see, e.g., Fritsch, 1946; Jónsson, 1962; McDonald and Pickett-Heaps, 1976). Ultrastructural aspects of mitosis and cytokinesis have been studied in *Cladophora* (McDonald and Pickett-Heaps, 1976; Scott and Bullock, 1976), *Dictyosphaeria* (Hori and Enomoto, 1978a,b), and *Valonia* (Hori and Enomoto, 1978c). Ulvophycean features are present throughout; the distinctive spindle arrangement at

telophase is especially noteworthy. In contrast to the situation in the basal bodies, the microtubules in the centrioles of *Cladophora* demonstrate the typical nine-triplet arrangement (Scott and Bullock, 1976). Cytoplasmic microtubules may accompany the cleavage furrow in *Cladophora* (McDonald and Pickett-Heaps, 1976).

(c) Sexual life history: An alternation of isomorphic phases appears to be the sexual life history pattern most commonly encountered in the Siphonocladales. This pattern is generally regarded as typical of the Cladophoraceae; the reports of a haplobiontic life history with meiosis at gametogenesis in *Cladophora glomerata* (e.g., Schussnig, 1954) await independent confirmation (see Shyam, 1980). We have evidence that an isomorphic life history pattern is also present in the chaetosiphonacean genus *Chaetosiphon* and the arnoldiellacean genus *Basicladia* (O'Kelly and Floyd, 1981; unpublished observations). This life history pattern is known also in the anadyomenacean genera *Anadyomene* and *Microdictyon* (Iyengar and Ramanathan, 1940, 1941; Mayhoub, 1975) and the valoniacean genus *Dictyosphaeria* (Enomoto and Okuda, 1977, 1981), and may also occur in *Valonia* (Chihara, 1959). Cytological investigations of *Siphonocladus* (Jónsson and Puiseux-Dao, 1959) and *Valonia* (Schussnig, 1938) suggesting the presence of a haplobiontic life history with meiosis at gametogenesis require confirmation.

(d) Sporangial and gametangial structure and development: Zoosporangia are structurally and developmentally identical to gametangia in all siphonocladalean genera for which data exist, except for the occurrence of meiosis in one or the other, usually (see above) the zoosporangia. They may form from any vegetative cell, which develops one or several more or less prominent exit papillae but otherwise undergoes little change in shape. Simultaneous cleavage of the multinucleate mother cell to produce the swarmers appears to be universal and, eventually, swarmers are released without a vesicle through one or more of the exit apertures. Few if any differences have been observed in exit aperture structure between the Cladophoraceae and other siphonocladalean genera (Jónsson and Puiseux-Dao, 1959; unpublished observations). In several genera, we (unpublished observations) have noticed that swarmer release is immediately preceded by the extrusion of a plug of mucilaginous material from the exit papilla (Figs 9 and 10).

(e) Cell wall structural polysaccharides: The siphonocladalean genera that have been examined possess cell walls that produce the x-ray diffraction pattern typical of "native" cellulose. The microfibrils in these walls are laid down in nonrandom, parallel arrays, and they yield only glucose on hydrolysis (see Parker, 1970).

(f) Siphonaxanthin and siphonein: Several siphonocladalean algae possess siphonaxanthin, and one species of *Cladophora* has been reported to contain

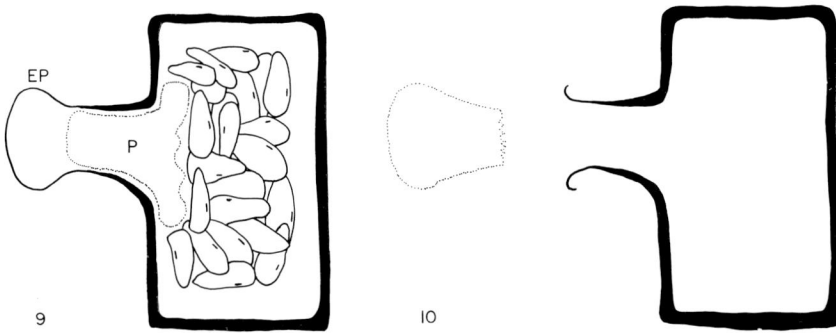

Figs 9, 10. Diagrammatic illustrations of plug formation and discharge in siphonocladalean zoosporangia and gametangia. The diagrams most closely resemble structures present in *Chaetosiphon* spp. In other siphonocladalean genera, the exit papilla is not nearly so pronounced. Fig. 9. A mature zoosporangium (or gametangium) prior to discharge of motile cells. The plug (P) is clearly visible within the lumen of the exit papilla (EP). Fig. 10. After motile cell release. The apex of the papilla has ruptured or has dissolved away, and the plug is jettisoned along with the swarmers.

siphonein as well (Goodwin, 1974). As in other Ulvophyceae, those species living in subtidal habitats are most likely to have one or both pigments (Yokohama, 1981).

We suggest that the following features are diagnostic of the Siphonocladales:

1. Terminal caps absent, and proximal sheaths reduced or absent.
2. "Flattening" of the flagellar apparatus.
3. Coarsely striated components present above the two-membered rootlets, and a "wing" in the multistranded rootlets.
4. Mitosis and cytokinesis distinctly separate processes.
5. Sexual life history usually an alternation of isomorphic phases.
6. Zoosporangia identical to gametangia in structure and development, with motile cell release through an exit papilla that has released a plug of mucilaginous material.
7. No vesicle surrounding the released motile cells.
8. Cellulosic cell walls, with microfibrils arranged in parallel arrays.

Of all these diagnostic features, none permits the separation of the Cladophoraceae from the rest of the Siphonocladales. Many workers have used "segregative division" to separate the two taxa at the ordinal level, but the mechanisms and structures associated with segregative division remain poorly understood and, in our opinion, it is equally possible that segregative division is little more than an extreme expression of the separation between the processes of mitosis and cytokinesis that is found in all members of the order as we have circumscribed it here. For additional discussion of this topic, see Hoek (1981, this volume, Chapter 5). Some other workers have made use of a chromosomal feature of possible taxonomic significance. Godward (1966) and Sarma (1982), among others, have observed a polyploid series in cladophoracean chromosome numbers, with $x = 6$ (other workers have proposed $x = 4, 8,$ or 12). Species of *Basicladia* and *Chaetosiphon* also fall within this polyploid series (O'Kelly and Floyd, 1981; unpublished observations). In other siphonocladalean families, however, no polyploid series is readily apparent as yet (Godward, 1966; Sarma, 1982). In our opinion, this feature may also reflect phylogenetic advancement within the order, rather than a clear-cut distinction between two orders. A better understanding of segregative division, and of the range of chromosome numbers in the more advanced siphonocladalean families, would help to support or dispute our suggestions.

The families that we have recognized as belonging to the Siphonocladales (see Table III) are based on morphological and cytological criteria that have been supported by our investigations to date.

5. The Dasycladales

The status of the morphologically and cytologically distinctive Dasycladales as a separate order within the green algae is now generally accepted. That this order has ulvophycean affinities has only recently been established, however (Roberts *et al.*, 1982), and some of the ultrastructural features of its motile cells have been regarded as problematical (Herth *et al.*, 1981). Our findings support the morphological concept of the order, confirm its placement as a fairly isolated member of the Ulvophyceae (see also Roberts *et al.*, 1982; Roberts, this volume, Chapter 13), and suggest that it may be most closely allied with the Siphonocladales.

(a) Motile cell features: Dasycladalean genera produce only biflagellate motile cells, which usually function as gametes. The flagellar apparatuses in the gametes of *Acetabularia* (Herth *et al.*, 1981; see also Roberts *et al.*, 1982; O'Kelly and Floyd, 1983) and *Batophora* (Roberts, this volume, Chapter 13) are flattened, as are those of siphonocladalean motile cells, but in other respects typically ulvophycean features are found, notably the prominently striated distal fiber and the single "wing" structure in the multimembered rootlets. Terminal caps have not been demonstrated, and proximal sheaths are rudimentary. Coarsely striated components appear below all four rootlets in *Acetabularia,* while rhizoplasts are found in *Batophora.* Also in *Batophora,* a "wing" structure identical to that found in the Siphonocladales appears in the multistranded rootlets (Roberts, this volume, Chapter 13).

(b) Mitosis and cytokinesis: Nuclear division has been examined at the ultrastructural level by Berger *et al.* (1975) in *Acetabularia* and by Liddle *et al.* (1976) in *Batophora*. In addition, Koop (1978, 1979) has observed mitosis in secondary nuclei of living *Acetabularia* cells. Typically ulvophycean features, especially the distinctive spindle arrangement at telophase, appear to be present, except that centrioles have not yet been observed in association with dividing nuclei.

(c) Sexual life history: At least in *Acetabularia* (Koop *et al.*, 1977, 1979; Koop, 1978, 1979) the primary nucleus undergoes spindle-associated, presumably meiotic division, hence the life history is a modification of a haplobiontic pattern in which zygotic meiosis may be long delayed. Liddle *et al.* (1976) have suggested a similar life history for *Batophora*. Earlier notions that the primary nucleus divides amitotically and that meiosis occurs at gametogenesis, which still appear in many recent texts, have not been confirmed.

(d) Other features: Koop (1975a,b, 1979) has summarized gametangial development in *Acetabularia*. Cell wall structural polysaccharides are composed mostly of mannan (see Parker, 1970) except in the gametangial cyst of *Acetabularia*, where they are cellulosic (Herth *et al.*, 1975). Neither siphonaxanthin nor siphonein has been found in dasycladalean algae (Goodwin, 1974), which are usually inhabitants of shallow marine or brackish water.

Diagnostic features of the Dasycladales, in addition to the distinctive vegetative morphology, include the following:

1. Quadriflagellate cells absent.
2. Flattening of the flagellar apparatus.
3. Absence of terminal caps.
4. Rudimentary proximal sheaths.
5. Sexual life history haplobiontic with zygotic meiosis, the meiotic division often delayed for a considerable period.
6. Gametangial cysts operculate, the gametes formed by simultaneous cleavage.
7. Cell wall structural polysaccharide a mannan, except for cellulose in the gametangial cyst.

Although several of these features, especially the unique vegetative morphology, the life history, the specialized gametangial structures and the mannan-containing vegetative cell walls, all emphasize the isolated position of the Dasycladales, other features, such as the flattened aspect presented by the flagellar apparatus, the structures of the striated distal fiber, the absence of terminal caps, and perhaps also the cellulosic gametangial cyst walls, suggest a closer relationship of this order to the Siphonocladales than to any other ulvophycean group.

6. The Caulerpales

The order Caulerpales has had a long and complex taxonomic and nomenclatural history. The present usage is that of Bold and Wynne (1978). Some recent authors (see, e.g., Round, 1973) have suggested that this group be elevated to class rank and subdivided into as many as five orders. The ulvophycean affinity of the Caulerpales has been suggested (Roberts et al., 1982; O'Kelly and Floyd, 1984), but relevant ultrastructural information is still lacking for many of the included families. Our analysis tends to confirm the status of the Caulerpales as a distinctive group of organisms separate from other members of the Ulvophyceae, and suggests in a preliminary way that only one order need be recognized (Table IV).

(a) Motile cell features: With the single exception of the genus *Ostreobium* (Kornmann and Sahling, 1980), quadriflagellate motile cells are absent from the Caulerpales. Stephanokont zoospores are now known for many members of the Bryopsidaceae (Bold and Wynne, 1978; Okuda et al., 1979), but in the remaining genera only biflagellate motile cells, often functioning as gametes, have been observed. Typically ulvophycean features have been found in most of the biflagellate motile cells examined with the electron microscope, although the 3-5-3-5 rootlet configuration observed in *Derbesia* female gametes is exceptional (Roberts et al., 1981). Distal fiber structure in male gametes is very similar from one genus to the next (Hori, 1977; Gori, 1979; Roberts et al., 1981, 1982), while that in female gametes may be very different from that in the male gametes in the same species and show considerable variation from genus to genus (Melkonian, 1981; Roberts et al., 1981, 1982; Greuel et al., 1982; unpublished observations). A central striated region, which is considerably less distinct than in other ulvophycean algae, has been clearly demonstrated only in *Pseudobryopsis* gametes (Roberts et al., 1982). Simple, overlapping terminal caps, which have subtle differences in structure from those found in the Ulotrichales, and rudimentary proximal

Table IV. Critically examined caulerpalean taxa.

Ostreobiaceae	Udoteaceae
Ostreobium	*Halimeda*
Bryopsidaceae	*Udotea*
Bryopsis	Caulerpaceae
Derbesia	*Caulerpa*
Pseudobryopsis	Dichotomosiphonaceae
Codiaceae	*Dichotomosiphon*
Codium	

sheaths occur in most genera examined. Rootlets having "winged" microtubules appear in female gametes (Melkonian, 1981; Roberts et al., 1981, 1982; Greuel et al., 1982; unpublished observations), but have been seen less often in male gametes (Moestrup and Hoffman, 1975; Roberts et al., 1982). Rhizoplasts have never been observed, although rootlet-associated striated bands comparable to those in other ulvophycean orders have sometimes been interpreted as rhizoplasts (Roberts et al., 1981, 1982). The structure of the putative mating apparatus in female gametes is also consistent in, and unique to, the Caulerpales, so far as is known (Melkonian, 1981; Roberts et al., 1982). Roberts et al. (1980) and Hori and Kobara (1982) have discussed the ultrastructural features of stephanokont zoospores in, respectively, *Derbesia* and *Pseudobryopsis*.

(b) Mitosis and cytokinesis: Typically ulvophycean aspects of nuclear division have been observed in *Bryopsis* (Burr and West, 1970) and *Caulerpa* (Hori, 1981), although centrioles may or may not be present at the poles. Cytokinesis during gametogenesis in *Bryopsis* is also of an ulvophycean nature (Burr and West, 1970).

(c) Sexual life history: Two types of sexual life history have been reported in the Caulerpales, an alternation of heteromorphic phases in the Bryopsidaceae (Hoek et al., 1972; Mayhoub, 1974; Okuda et al., 1979), and a haplobiontic life history with meiosis presumably at gametogenesis in the remainder (e.g., Borden and Stein, 1969; Meinesz, 1972a,b). Cytological confirmations of these life histories are rare, however (Williams, 1925; Schussnig, 1939; Neumann, 1974), and, in one isolate of *Derbesia,* the morphological alternation occurs in the absence of gamete fusion and without apparent change in the nuclear DNA level (Schnetter et al., 1980). The zygote nucleus in several Caulerpales may become enlarged, and therefore may superficially resemble the primary nucleus in dasycladalean genera (Neumann, 1974). However, caulerpalean "primary nuclei" are clearly distinguishable from dasycladalean primary nuclei at the ultrastructural level (Burr and West, 1971) and do not appear to be the site of meiosis (Neumann, 1974).

(d) Sporangial and gametangial structure and development: Caulerpalean motile cells form in structures that may or may not be differentiated from the rest of the thallus. In the Bryopsidaceae zoosporangia are structurally distinct from gametangia (Bold and Wynne, 1978; Okuda et al., 1979). Throughout the order, simultaneous cleavage results in the formation of swarmers, which are then released through one or more exit papilla, perhaps with the simultaneous release of mucilaginous material (Williams, 1925; Borden and Stein, 1969; Neumann, 1974; Wheeler and Page, 1974).

(e) Cell wall structural polysaccharides: As reviewed by Parker (1970), structural microfibrils in caulerpalean cell walls may be composed either of mannan or xylan. In the Bryopsidaceae sporophyte cell walls possess man-

nan, while gametophyte cell walls have a xylan that may be associated with some cellulose (Frei and Preston, 1968; Parker, 1970; Huizing and Rietema, 1975).

(f) Siphonaxanthin and siphonein: Both siphonaxanthin and siphonein are consistently present in caulerpalean algae, with the exception of *Dichotomosiphon,* which has only siphonein, and one species of *Caulerpa* in which both pigments are lacking (Goodwin, 1974). The presence of these orange xanthophylls is independent of the habitat of the alga (Yokohama, 1981).

Diagnostic features of the Caulerpales, in addition to the coenocytic morphology, include:

1. Quadriflagellate motile cells absent, except in *Ostreobium*.
2. Simple, overlapping terminal caps and rudimentary proximal sheaths present.
3. Modified distal fibers, especially so in female gametes, with the central striated region indistinct or lacking.
4. "Winged" rootlet microtubules, at least in female gametes.
5. Unique structure of the mating body in female gametes.
6. Sexual life history either an alternation of heteromorphic phases or haplobiontic with gametic meiosis.
7. Motile cells produced in more or less differentiated structures by simultaneous cleavages, and released without a vesicle through one or more exit papillas; zoosporangia when present structurally distinct from gametangia.
8. Cell wall structural polysaccharides of mannan, xylan, or possibly xylan plus cellulose, never cellulose alone.
9. Habitat-independent presence of both siphonaxanthin and siphonein (siphonein alone in *Dichotomosiphon*).

The observed similarities in motile cell structure, especially the flagellar apparatuses of male gametes and perhaps also the mating bodies in female gametes, suggest to us that the Caulerpales forms a close-knit group of organisms recognizable as a single order. Features previously used to subdivide the Caulerpales, such as life history patterns, the presence of oogamy, and the presence of heteroplastidy, seem to us to be more useful at the family level (see Table IV). Other features that may eventually prove useful at infraordinal taxonomic levels include the structural variations in the distal fibers of female gametes, the distribution of mannan versus xylan in cell walls (see Parker, 1970), and variations in the nucleus–microbody associations that are apparently present (Roth and Friedmann, 1980). *Ostreobium* is clearly distinct from all other Caulerpales because of its quadriflagellate zoospores, but the family to which it is usually referred, the Phyllosiphonaceae, has been transferred to the Xanthophyceae (Bourrelly, 1973), hence our informal referral of the former genus to the Ostreobiaceae, following Silva (1982).

THE ULVOPHYCEAE: PHYLOGENETIC CONSIDERATIONS

1. The Ulvophyceae Redefined

The present work confirms our previous suggestion (O'Kelly and Floyd, 1983) that most of the flagellar apparatus features previously used to characterize the Ulvophyceae (see Hoops *et al.*, 1982) are not stable enough to be used at the class level. Only the counterclockwise absolute orientation of flagellar apparatus components is typical of all Ulvophyceae [distal fibers may be absent from some members of the Ulvales (unpublished observations)], and not even this feature is the exclusive property of the class. As we have noted (O'Kelly and Floyd, 1984), counterclockwise absolute orientation is present in all green algae except those belonging to the Chlorophyceae, and is also present in at least one other group of algae, the Glaucophyceae; and in one group of phycoplast-containing green algae ("*Pleurastrum* lineage" of O'Kelly and Floyd, 1984; Pleurastrophyceae of Mattox and Stewart, this volume, Chapter 2; see also Melkonian, 1982; Melkonian and Berns, 1983), the critical flagellar apparatus features (counterclockwise absolute orientation, striated distal fiber, striated bands connecting rootlets to basal bodies) are identical to those of many Ulvophyceae. Hence, no flagellar apparatus features may be used to define the Ulvophyceae except in combination with mitotic and cytokinetic features.

Instead, we have seen that selected flagellar apparatus features provide for the recognition of five discrete groups within the Ulvophyceae, which we have recognized at the ordinal level, and that these flagellar apparatus features are supported by a battery of additional features (see Table V). Taken together, these correlated features allow for a classification of the Ulvophyceae (Fig. 11) in which anatomical (ultrastructural), biochemical, and reproductive features take precedence, and vegetative morphology is reduced to a subordinate role, as is the case in most higher plant classifications.

2. Evolutionary Trends within the Ulvophyceae

We have also observed trends within and among these ulvophycean orders that we believe may have phylogenetic significance (see O'Kelly and Floyd, 1984). Some of these trends include:

(a) The elimination of quadriflagellate motile cells. Loss of quadriflagellate motile cells, whether absolute or by their presumed conversion to the stephanokont condition, seems to occur in tandem with advancement in mor-

Table V. Some diagnostic features of ulvophycean orders.

Orders	Quadriflagellate zoospores	Scales	Terminal cap	Proximal sheath	Life history[*]	Sporangial exit aperture	Cell wall	Pigment[†]
Ulotrichales	present	present[‡]	simple overlapping	wedge-shaped (upper pair); half-cylindrical (lower pair)	D[h§]	irregular pore	not defined	sn[∥]?
Ulvales	present	absent	bilobed	two equal subunits	D[i]	papilla	not defined	sx[∥]
Siphonocladales	present	absent	none	rudimentary or absent	D[i]	papilla with plug	cellulose	sx[∥]
Dasycladales	absent	absent	none	rudimentary	H[h]	operculum	mannan or cellulose	none
Caulerpales	absent[#]	absent	simple overlapping	rudimentary	d[h] or H[d]	papilla	mannan or xylan	sx, sn[**]

[*] Abbreviations after Bold and Wynne (1978).
[†] sn = siphonein; sx = siphonaxanthin; none = neither siphonaxanthin nor siphonein present.
[‡] Absent from the "Acrosiphonia group."
[§] D[i] in Eugomontia.
[∥] Presence of the pigment dependent on habitat.
[#] Present in Ostreobium.
[**] Presence of the pigment independent of habitat.

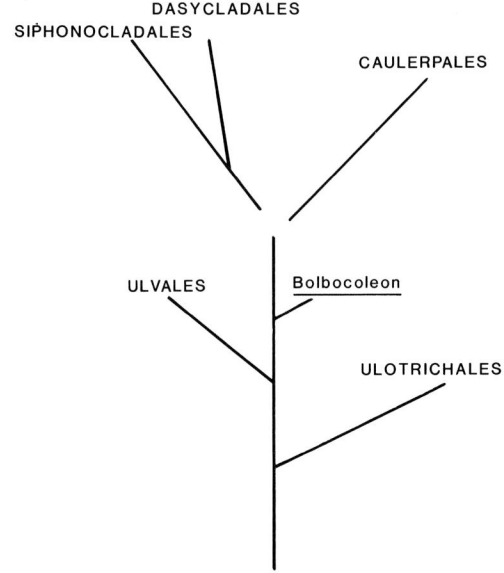

Fig. 11. Suggested phylogenetic relationships among the orders of the Ulvophyceae. For details, see text.

phological features and the loss of an isomorphic life history. The trend is most obvious in the Caulerpales, the Dasycladales, and in the ulotrichalean genus *Acrosiphonia*.

(b) Reduction, modification, or loss of motile cell components. Body scales, proximal sheaths, terminal caps, and distal fibers bearing a prominently striated central region appear to us to be primitive features, hence reduction, modification, or loss of these features indicates derived states. Most of these trends are clear only when orders are compared to one another, but a clear trend toward the reduction and eventual loss of body scales appears in the Ulotrichales.

(c) Development of basal body orientation perpendicular to the long axis of the cell during forward swimming. As we have discussed elsewhere (O'Kelly and Floyd, 1984), one of the more important events in flagellar apparatus evolution in the green algae was the exsertion of the ancestrally pit-borne flagellar apparatus into an apical papilla, with the consequent rearrangement of the basal bodies from the primitive parallel arrangement to the derived arrangement perpendicular to the cell's long axis. V-shaped basal body arrangements in certain ulotrichalean motile cells may be relicts of this evolutionary tendency in the Ulvophyceae, and indicate, along with the presence of body scales and other features, that this order is the most primitive known in the class (Floyd and O'Kelly, 1984).

(d) Loss of the isomorphic life history. We have argued (Floyd and O'Kelly, 1984) that an alternation of isomorphic phases is the primitive life history in the Ulvophyceae, using the essentially isomorphic life history and other primitive features of *Eugomontia* as evidence. Upon accepting this premise, clear trends toward the reduction and eventual loss of one phase, occurring independently in the Ulotrichales, the Siphonocladales–Dasycladales group, and the Caulerpales, may be identified.

(e) Evolution of increasingly complex zoosporangial and gametangial structure and development. Trends toward increasingly complex structural and developmental events, including the evolution of differentiated from undifferentiated exit apertures, and of simultaneous from sequential cleavage patterns, occur independently in the Ulotrichales, and may also be observed in the other ulvophycean orders when considered as a group. In addition, a vesicle of modified wall-like material surrounding the motile cells appears to be a primitive feature, while the presence of a "plug" or other mucilaginous or membranous, nonvesicular material during motile cell release seems to be derived.

(f) Cell walls. As Parker (1970) and Domozych *et al.* (1980) have pointed out, the potential usefulness of cell wall data in phylogenetic studies of the green algae is just beginning to be realized, and interpretation of the fragmentary information on ulvophycean cell wall features (especially in ulotrichalean and ulvalean taxa) remains particularly difficult. A trend toward increasing structural and chemical definition of cell wall microfibrils may be discerned, however. Domozych *et al.* (1980) have underscored the close overall chemical similarity between the cell walls of *Pseudendoclonium sensu* Vischer (Ulotrichales) and *Ulva* (Ulvales), and it may be supposed that these walls represent an intermediate cell wall type between an ancestral, amorphous type (see Domozych *et al.,* 1980) and the derived, well-organized walls found in the Caulerpales, Dasycladales, and Siphonocladales. Why cellulose is elaborated as the primary crystalline component of the cell wall in some cases, while mannan or xylan appears in others, has yet to be explained, as does the change in wall composition between life history phases of individual species in the Caulerpales, Dasycladales, and Ulotrichales.

3. The Parallel Evolution of Thallus Morphology

We conclude from our analysis of these correlated features that superficially similar thallus morphologies have evolved independently in several different ulvophycean lines. For example, differences in motile cell fine structure and early gametophyte development suggest to us that parenchymatous, blade-like or tubular thalli evolved at least twice in the Ulotrichales and at least twice more in the Ulvales. Filamentous, "chaetophoralean" morphologies appear to be most primitive, while hemisiphonous and siphonous morphologies are most derived. Multinucleate-celled thalli have evolved at least twice, once independently in the Ulotrichales and again in the line or lines leading to the Siphonocladales, Dasycladales, and Caulerpales. In the latter case, we suggest that multinucleate-celled thalli evolved from ancestors having "chaetophoralean" morphology as well as gametangia and zoosporangia with exit apertures that are differentiated into papillas and in which a plug is present and a vesicle is absent. These ancestors probably were most closely related to those that gave rise to the present-day Ulvales. Extant genera having suggestive features include the two closely related genera *Bolbocoleon* (Nielsen, 1979; see Fig. 11) and *Sporocladus* (unpublished observations), and the genus *Phaeophila* (O'Kelly and Yarish, 1980). The multinucleate-celled condition may have occurred either by increase of the number of nuclei in cells, as apparently has occurred in the Ulotrichales and that may be indicated by events during zoosporogenesis in *Phaeophila* (O'Kelly and Yarish, 1980), or by the loss of cross walls between cells, as may be occurring in the genus *Smithsoniella* (Brawley and Sears, 1982; Sears and Brawley, 1982). The sizeable differences in diagnostic features between the Siphonocladales and Dasycladales, on the one hand, and the Caulerpales on the other, suggest that these two groups diverged fairly early. The fossil record, in fact, places this divergence no later than the early Cambrian period, about 6.0×10^8 years ago (Tappan, 1980). It remains an open question whether this divergence took place before or after the development of the multinucleate-celled condition in these lines.

Incidentally, we note that the zoosporangia in at least some Trentepohliales possess both an apical plug of sorts and a vesicle, and that the vesicle is composed of cytoplasmic, not wall-like, materials (Graham and McBride, 1978). On the basis of this information, as well as evidence we have cited here and elsewhere (O'Kelly and Floyd, 1984), we reiterate our belief that this order occupies an isolated taxonomic position, and that any attempt to place it in the Ulvophyceae or any other currently recognized green algal class remains premature.

APPLICATION OF CORRELATED FEATURES ANALYSIS TO OTHER GREEN ALGAL CLASSES

We can summarize the results of our analysis of the Ulvophyceae as follows:

1. Flagellar apparatus ultrastructural features do not necessarily provide, by themselves, a sufficiently sound basis for class-level taxonomic treatments; they must be used together with, at least, ultrastructural features of mitosis and cytokinesis.

2. Certain motile cell ultrastructural features correlate well with other ultrastructural, reproductive, and biochemical features and permit the identification of clear-cut groups within the classes, here recognized as taxa at

the ordinal level, in which morphological traits of vegetative thalli are of secondary importance.

3. Trends among these correlated features are of considerable value in determining phylogenetic relationships and the direction of evolution within the higher taxa under investigation.

We can predict that analysis of similar features will permit the identification of natural orders and the establishment of plausible evolutionary tendencies in other classes of green algae as well. In particular, because the Ulvophyceae and the Chlorophyceae have been considered to be closely related (Hoops et al., 1982), we can suggest that many of the features that are indicative of primitiveness in the former class will also be present in primitive members of the latter. Our recent investigations of members of the Chlorophyceae in general, and of the order Tetrasporales (as defined on the basis of vegetative morphology) in particular, have so far confirmed the accuracy of our suggestions. We have demonstrated (O'Kelly and Floyd, 1984) that the clockwise absolute orientation of flagellar apparatus components now known to be typical of most Chlorophyceae (Melkonian, this volume, Chapter 3; Mattox and Stewart, this volume, Chapter 2) is unique to this class, and therefore it has most likely developed from an ancestor having counterclockwise absolute orientation by clockwise displacement of components. The flagellar apparatus of a primitive chlorophycean motile cell, therefore, should have a nearly cruciate arrangement of flagellar apparatus components, reflecting the most probable evolutionary intermediate. If this motile cell also possessed primitive features similar to those in the Ulvophyceae, it would have four flagella (two in gametes) and body scales, the basal body arrangement would be V-shaped, and prominent proximal sheaths and terminal caps would be present. Precisely these features, minus the terminal caps, occur in the quadriflagellate zoospores of *Chaetopeltis* (O'Kelly and Floyd, 1984), and very similar features, minus the scales in this case, appear in the quadriflagellate zoospores of a blade-forming alga that we have tentatively referred to the genus *Phyllogloea* (unpublished observations). Both these genera are commonly referred to the Tetrasporales (Bourrelly, 1973; Wujek and Chelune, 1975), but the zoospores and gametes of *Tetraspora,* which are always biflagellate (a derived feature in the Ulvophyceae!), have little in common with those of *Chaetopeltis* and (?)*Phyllogloea,* having instead many features in common with *Chlamydomonas* (O'Kelly and Floyd, 1984). We have, in addition, demonstrated a close inverse correlation between the presence of quadriflagellate motile cells and the amount of clockwise displacement of basal bodies in the flagellar apparatus in several chlorophycean genera (O'Kelly and Floyd, 1984). Interestingly, it is only among those genera having quadriflagellate zoospores and little or no clockwise basal body displacement (specifically, the Chaetophorales *sensu* Stewart and Mattox, 1975) in which isomorphic or heteromorphic sexual life history patterns have been consistently recorded (Islam, 1963; Bold and Wynne, 1978). These observations, as we have explained in more detail elsewhere (O'Kelly and Floyd, 1984), at once strongly suggest that the Tetrasporales, as traditionally conceived, is a heterogeneous assemblage, and that phylogenetic treatments that derive chlorophycean algae from *Chlamydomonas*-like ancestors may be in error. Further work in these areas should prove to be most rewarding.

NOTE ADDED IN PROOF: Since this volume went to press, three articles have come to our attention that have a direct bearing on the matters discussed.

1. A species of *Entocladia* is now known to possess both siphonaxanthin and siphonein [Goldberg, W. M., Makemson, J. C., and Colley, S. B. (1984). *Biol. Bull.* **166,** 368–383.]; hence, our generalization (p. 133) that siphonein is absent in members of the Ulvales can no longer be considered absolute.

2. The Japanese workers T. Hirayama and T. Hori have examined the flagellar apparatus of *Chaetomorpha spiralis* Okamura quadriflagellate zoospores. The features they found are, for the most part, identical to those that we have described here for the Siphonocladales (p. 135), and they have correctly noted that their organism is referable to the Ulvophyceae. Their report, at the proof stage as this is written (we thank Dr Hori for graciously providing us with copy of the manuscript), is to appear in *Botanica Marina*.

3. A fairly comprehensive account of the flagellar apparatus features in the isogametes of *Batophora* (Dasycladales) is given by K. R. Roberts, K. D. Stewart, and K. R. Mattox, 1984 (*J. Phycol.* **20,** 183–191).

ACKNOWLEDGEMENTS

This research was funded by NSF grants DEB-7911777 (GLF) and DEB-8200361 (CJO and GLF), and an Ohio State University Postdoctoral Fellowship (CJO).

REFERENCES

Berger, S., Herth, W., Franke, W. W., Falk, H., Spring, H. and Schweiger, H. G. (1975). Morphology of the nucleo-cytoplasmic interactions during the development of *Acetabularia* cells. II. The generative phase. *Protoplasma* **84,** 223–256.

Berger-Perrot, Y. and Thomas, J. C. (1982). Étude ultrastructurale comparée du pyrénoïde et des parois dans les genres *Ulothrix, Chlorothrix,* et *Urospora. Phycologia* **21,** 355–369.

Bliding, C. (1963). A critical survey of European taxa in Ulvales. Part I. *Capsosiphon, Percursaria, Blidingia, Enteromorpha. Opera Bot.* **8**(3), 1–160.

Bliding, C. (1968). A critical survey of European taxa in Ulvales. Part II. *Ulva, Monostroma, Kornmannia. Bot. Not.* **121,** 535–629.

Bold, H. C. and Wynne, M. J. (1978). "Introduction to the Algae: Structure and Reproduction." Prentice-Hall, Englewood Cliffs, New Jersey.

Borden, C. A. and Stein, J. R. (1969). Reproduction and early development in *Codium fragile* (Suringar) Hariot (Chlorophyceae). *Phycologia* **8**, 91–99.

Bourne, E. J., Megarry, M. L. and Percival, E. (1974). The carbohydrates of the green seaweed *Urospora*. *J. Carbohydr., Nucleosides Nucleotides* **3**, 235–264.

Bourrelly, P. (1973). "Les Algues d'eau douce. Initiation à la Systématique. Tome I. Les Algues Vertes" (rev. ed.). Boubée, Paris.

Bråten, T. and Løvlie, A. (1968). On the ultrastructure of vegetative and sporulating cells of the multicellular green alga *Ulva mutabilis* Føyn. *Nytt Mag. Bot.* **15**, 209–219.

Bråten, T. and Nordby, Ø. (1973). Ultrastructure of meiosis and centriole behaviour in *Ulva mutabilis* Føyn. *J. Cell Sci.* **13**, 69–81.

Brawley, S. H. and Sears, J. R. (1982). Septal plugs in a green alga. *Am. J. Bot.* **69**, 455–463.

Burr, F. A. and West, J. A. (1970). Light and electron microscope observations on the vegetative and reproductive structures of *Bryopsis hypnoides*. *Phycologia* **9**, 17–37.

Burr, F. A. and West, J. A. (1971). Comparative ultrastructure of the primary nucleus in *Bryopsis* and *Acetabularia*. *J. Phycol.* **7**, 108–113.

Carlberg, G. E. and Percival, E. (1977). The carbohydrates of the green seaweeds *Urospora wormskioldii* and *Codiolum pusillum*. *Carbohydr. Res.* **57**, 223–234.

Chapman, V. J. and Chapman, D. J. (1973). "The Algae," 2nd ed. Macmillan, London.

Chihara, M. (1959). Studies of the life history of the green algae in the warm seas around Japan. 9. Supplementary note on the life history of *Valonia macrophysa* Kütz. *J. Jpn. Bot.* **34**, 257–266.

Dennis, D. T. and Preston, R. D. (1961). Constitution of cellulose microfibrils. *Nature (London)* **191**, 667–668.

Dodel, A. (1876). Die Kraushaaralge, *Ulothrix zonata*. Ihre geschlechtliche und ungeschlechtliche Fortpflanzung. *Jahrb. wiss. Bot.* **10**, 417–550.

Domozych, D. S., Stewart, K. D. and Mattox, K. R. (1980). The comparative aspects of cell wall chemistry in the green algae (Chlorophyta). *J. Mol. Evol.* **15**, 1–12.

Enomoto, S. and Okuda, K. (1977). On the life history of *Dictyosphaeria cavernosa*. *J. Phycol.* **13**, Suppl., 20.

Enomoto, S. and Okuda, K. (1981). Culture studies of *Dictyosphaeria* (Chlorophyceae, Siphonocladales). I. Life history and morphogenesis of *Dictyosphaeria cavernosa*. *Jpn. J. Phycol.* **29**, 225–236.

Floyd, G. L. (1981). Ultrastructure of motile cells from two species of *Cladophora*. *J. Phycol.* **17**, Suppl., 11.

Floyd, G. L. and O'Kelly, C. J. (1984). Motile cell ultrastructure and the circumscription of the orders Ulotrichales and Ulvales (Ulvophyceae, Chlorophyta). *Am. J. Bot.* **71**, 111–120.

Frei, E. and Preston, R. D. (1968). Non-cellulosic structural polysaccharides in algal cell walls. III. Mannan in siphoneous green algae. *Proc. R. Soc. London. Ser. B.* **169**, 127–145.

Fritsch, F. E. (1935). "The Structure and Reproduction of the Algae." Vol. 1. Cambridge Univ. Press, London and New York.

Fritsch, F. E. (1946). The status of the Siphonocladales. *J. Indian bot. Soc., M. O. P. Iyengar Commem. Vol.* pp. 29–50.

Gayral, P. (1964). Sur le démembrement de l'actuel genre *Monostroma* Thuret (Chlorophycées, Ulotrichales s. 1.). *C. r. hebd. Seanc. Acad. Sci., Ser. D* **258**, 2149–2152.

Gayral, P. (1965). *Monostroma* Thuret; *Ulvaria* Ruprecht emend. Gayral; *Ulvopsis* Gayral (Chlorophycées, Ulotrichales): Structure, reproduction, cycles, position systématique. *Rev. gen. Bot.* **72**, 627–638.

Godward, M. B. E. (1966). "The Chromosomes of the Algae." Edward Arnold, London.

Goodwin, T. W. (1974). Carotenoids and biliproteins. *In* "Algal Physiology and Biochemistry" (W. D. P. Stewart, ed.), pp. 176–205. Blackwell, Oxford.

Gori, P. (1979). Ultrastructure of the spermatozoid in *Halimeda tuna* (Chlorophyceae). *Gamete Res.* **2**, 345–355.

Graham, L. E. and McBride, G. E. (1978). Mitosis and cytokinesis in sessile sporangium of *Trentepohlia aurea* (Chlorophyceae). *J. Phycol.* **14**, 132–137.

Greuel, B. G., Floyd, G. L. and O'Kelly, C. J. (1982). Ultrastructural studies on the biflagellate motile cell of *Codium fragile* ssp. *tomentosoides* (Caulerpales, Chlorophyta). *Ohio J. Sci.* **82**(2), 17.

Gross, I. (1931). Entwicklungsgeschichte, Phasenwechsel und Sexualität bei der Gattung *Ulothrix*. *Arch. Protistenkd.* **73**, 206–234.

Herth, W., Kuppel, A. and Franke, W. W. (1975). Cellulose in *Acetabularia* cyst walls. *J. Ultrastruct. Res.* **50**, 289–292.

Herth, W., Heck, B. and Koop, H.-U. (1981). The flagellar root system in the gamete of *Acetabularia mediteranea*. *Protoplasma* **109**, 257–269.

Hoek, C. van den (1981). Chlorophyta: Morphology and classification. In "The Biology of Seaweeds (C. S. Lobban and M. J. Wynne, eds), pp. 86–132. Blackwell, Oxford.

Hoek, C. van den, Cortel-Breeman, A. M., Rietema, H. and Wanders, J. B. W. (1972). L'interpretation de données obtenues, par des cultures unialgales, sur les cycles évolutifs des algues. Quelques examples tirés des recherches conduites au laboratoire de Gronigue. *Mem. Soc. bot. Fr.* pp. 45–66.

Hoops, H. J., Floyd, G. L. and Swanson, J. A. (1982). Ultrastructure of the biflagellate motile cells of *Ulvaria oxysperma* (Kütz.) Bliding and phylogenetic relationships among ulvaphycean algae. *Am. J. Bot.* **69**, 150–159.

Hori, T. (1973). Comparative studies of pyrenoid ultrastructure in algae of the *Monostroma* complex. *J. Phycol.* **9**, 190–199.

Hori, T. (1977). Electron microscope observations on the flagellar apparatus of *Bryopsis maxima* (Chlorophyceae). *J. Phycol.* **13**, 238–243.

Hori, T. (1981). Ultrastructural studies on nuclear division during gametogenesis in *Caulerpa* (Chlorophyceae). *Jpn. J. Phycol.* **29**, 163–170.

Hori, T. and Enomoto, S. (1978a). Developmental cytology of *Dictyosphaeria cavernosa*. I. Light and electron microscope observations on cytoplasmic cleavage in zooid formation. *Bot. mar.* **21**, 401–408.

Hori, T. and Enomoto, S. (1978b). Developmental cytology of *Dictyosphaeria cavernosa*. II. Nuclear division during zooid formation. *Botanica mar.* **21**, 477–481.

Hori, T. and Enomoto, S. (1978c). Electron microscope observations on the nuclear division in *Valonia ventricosa* (Chlorophyceae, Siphonocladales). *Phycologia* **17**, 133–142.

Hori, T. and Kobara, T. (1982). Ultrastructure of the flagellar apparatus in the stephanokont zoospores of *Pseudobryopsis hainanensis*. *Jpn. J. Phycol.* **30**, 31–39.

Hudson, P. R. and Waaland, J. R. (1974). Ultrastructure of mitosis and cytokinesis in the multinucleate green alga *Acrosiphonia*. *J. Cell Biol.* **62**, 274–294.

Huizing, H. J. and Rietema, H. (1975). Xylan and mannan as cell wall constituents of different stages in the life-histories of some siphoneous green algae. *Br. phycol. J.* **10**, 13–16.

Islam, A. K. M. N. (1963). A revision of the genus *Stigeoclonium*. *Beih. Nova Hedw.* **10**, 1–164.

Iyengar, M. O. P. and Ramanathan, K. R. (1940). On the reproduction of *Anadyomene stellata* (Wulf.) Ag. *J. Indian bot. Soc.* **19**, 175–176.

Iyengar, M. O. P. and Ramanathan, K. R. (1941). On the life-history and cytology of *Microdictyon tenuis* (Ag.) Decsne. *J. Indian bot. Soc.* **20**, 157–159.

Jónsson, S. (1962). Recherches sur des Cladophoracées marines. Structure, reproduction, cycles comparés, conséquences systématiques. *Ann. Sci. nat., Bot. Biol. Veg.* [12] **3**, 27–265.

Jónsson, S. (1968). Sur le cycle ontogénique et chromosomique du *Monostroma grevillei* (Thur.) Wittr. de Roscoff. *C. r. hebd. Seanc. Acad. Sci., Ser.* D **267**, 402–405.

Jónsson, S. (1970). Localisation de la meiose dans le cycle de l'*Acrosiphonia spinescens* (Kütz.) Kjellm. (Acrosiphoniacées). *C. r. hebd. Seanc. Acad. Sci., Ser.* D **271**, 1859–1861.

Jónsson, S. and Puiseux-Dao, S. (1959). Observations morphologiques et caryologiques relatives à la reproduction chez le *Siphonocladus pusillus,* (Kütz.) Hauck, Siphonocladacées, en culture. *C. r. hebd. Seanc. Acad. Sci., Ser.* D **249,** 1383–1385.

Koop, H.-U. (1975a). Germination of cysts in *Acetabularia mediteranea. Protoplasma* **84,** 137–146.

Koop, H.-U. (1975b). Multinuclear stages of the life cycle of *Acetabularia mediteranea. Protoplasma* **86,** 351–362.

Koop, H.-U. (1978). Development of *Acetabularia* (Dasycladales). *Inst. wiss. Film* **C1298.**

Koop, H.-U. (1979). The life cycle of *Acetabularia* (Dasycladales, Chlorophyceae): A compilation of evidence for meiosis in the primary nucleus. *Protoplasma* **100,** 353–366.

Koop, H.-U., Heunert, H. H. and Schmid, R. (1977). Division of the primary nucleus of *Acetabularia. Protoplasma* **93,** 131–134.

Koop, H.-U., Schmid, R., Heunert, H. and Spring, H. (1979). Spindle formation and division of the giant primary nucleus of *Acetabularia* (Chlorophyta, Dasycladales). *Differentiation* **14,** 135–146.

Kornmann, P. (1956). Zur Morphologie und Entwicklung von *Percursaria percursa. Helgol. wiss. Meeresunters.* **5,** 259–272.

Kornmann, P. (1960). Die heterogene Gattung *Gomontia.* II. Der fädige Anteil, *Eugomontia sacculata* nov. gen. nov. spec. *Helgol. wiss. Meeresunters.* **8,** 195–202.

Kornmann, P. (1963). Die Ulotrichales, neu geordnet auf der Grundlage entwicklungsgeschichtlicher Befunde. *Phycologia* **3,** 60–68.

Kornmann, P. (1965). Ontogenie und Lebenszyklus der Ulotrichales in phylogenetischer Sicht. *Phycologia* **4,** 163–172.

Kornmann, P. (1970). Phylogenetische Beziehungen in der Grünalgengattung *Acrosiphonia. Helgol. Wiss. Meeresunters.* **21,** 292–304.

Kornmann, P. (1973). Codiolophyceae, a new class of Chlorophyta. *Helgol. Meeresunters.* **25,** 1–13.

Kornmann, P. and Sahling, P.-H. (1980). *Ostreobium quekettii* (Codiales, Chlorophyta). *Helgol. Meeresunters.* **34,** 115–122.

Kornmann, P. and Sahling, P.-H. (1983). Meeresalgen von Helgoland: Erganzung. *Helgol. Meeresunters.* **36,** 1–65.

Kreger, D. R. (1962). Cell walls. *In* "Physiology and Biochemistry of Algae" (R. A. Lewin, ed.), pp. 315–335. Academic Press, New York.

Liddle, L., Berger, S. and Schweiger, H.-G. (1976). Ultrastructure during development of the nucleus of *Batophora oerstedii* (Chlorophyta; Dasycladaceae). *J. Phycol.* **12,** 261–272.

Lokhorst, G. M. (1978). Taxonomic studies on the marine and brackish-water species of *Ulothrix* (Ulotricales, Chlorophyceae) in western Europe. *Blumea* **24,** 191–299.

Lokhorst, G. M. and Star, W. (1983). Fine structure of mitosis and cytokinesis in *Urospora* (Acrosiphoniales, Chlorophyta). *Protoplasma* **117,** 142–153.

Lokhorst, G. M. and Trask, B. J. (1981). Taxonomic studies on *Urospora* (Acrosiphoniales, Chlorophyceae) in western Europe. *Acta bot. neerl.* **30,** 353–431.

Lokhorst, G. M. and Vroman, M. (1972). Taxonomic study of three freshwater *Ulothrix* species. *Acta bot. neerl.* **21,** 449–480.

Lokhorst, G. M. and Vroman, M. (1974a). Taxonomic studies on the genus *Ulothrix* (Ulotrichales, Chlorophyceae). II. *Acta bot. neerl.* **23,** 369–398.

Lokhorst, G. M. and Vroman, M. (1974b). Taxonomic study on the genus *Ulothrix* (Ulotrichales, Chlorophyceae). III. *Acta bot. neerl.* **23,** 561–602.

Løvlie, A. and Bråten, T. (1970). On mitosis in the multicellular alga *Ulva mutabilis* Føyn. *J. Cell Sci.* **6,** 109–129.

McArthur, D. M. and Moss, B. L. (1978). Ultrastructural studies of vegetative cells, mitosis and cell division in *Enteromorpha intestinalis* (L.) Link. *Br. phycol. J.* **13,** 255–267.

McArthur, D. M. and Moss, B. L. (1979). Gametogenesis and gamete structure of *Enteromorpha intestinalis* (L.) Link. *Br. phycol. J.* **14**, 43–57.

McDonald, K. and Pickett-Heaps, J. D. (1976). Ultrastructure and differentiation in *Cladophora glomerata*. I. Cell division. *Am. J. Bot.* **63**, 592–601.

Mackie, W. and Preston, R. D. (1974). Cell wall and intercellular region polysaccharides. In "Algal Physiology and Biochemistry" (W. D. P. Stewart, ed.), pp. 40–85. Blackwell, Oxford.

Mattox, K. R. and Stewart, K. D. (1974). A comparative study of cell division in *Trichosarcina polymorpha* and *Pseudendoclonium basiliense* (Chlorophyceae). *J. Phycol.* **10**, 447–456.

Mattox, K. R., Stewart, K. D. and Floyd, G. L. (1974). The cytology and classification of *Schizomeris leibleinii* (Chlorophyceae). I. The vegetative thallus. *Phycologia* **13**, 63–69.

Mayhoub, H. (1974). Reproduction sexuée et cycle du développement de *Pseudobryopsis myura* (Ag.) Berthold (Chlorophycées, Codiales). *C. r. hebd. Seanc. Acad. Sci., Ser.* D **278**, 867–870.

Mayhoub, H. (1975). Reproduction sexuée et cycle du développement de l'*Anadyomene stellata* (Wulf.) Ag. de la Méditerranée orientale. *C. r. hebd. Seanc. Acad. Sci., Ser.* D **280**, 587–590.

Meinesz, A. (1972a). Sur la croissance et le développement du *Penicillus capitatus* Lamarck forma *mediterranea* (Decaisne) P. et H. Huvé (Caulerpale, Udoteacée). *C. r. hebd. Seanc. Acad. Sci., Ser.* D **269**, 667–669.

Meinesz, A. (1972b). Sur le cycle de l'*Udotea petiolata* (Turra) Boergesen (Caulerpale, Udoteacée). *C. r. hebd. Seanc. Acad. Sci., Ser.* D **275**, 1975–1977.

Melkonian, M. (1979). Structure and significance of cruciate flagellar root systems in green algae: Zoospores of *Ulva lactuca* (Ulvales, Chlorophyceae). *Helgol. wiss. Meeresunters.* **32**, 425–435.

Melkonian, M. (1980). Flagellar roots, mating structure and gametic fusion in the green alga *Ulva lactuca* (Ulvales). *J. Cell Sci.* **46**, 149–169.

Melkonian, M. (1981). Structure and significance of cruciate flagellar root systems in green algae: Female gametes of *Bryopsis lyngbyei* (Bryopsidales). *Helgol. Meeresunters.* **34**, 355–369.

Melkonian, M. (1982). Structural and evolutionary aspects of the flagellar apparatus in green algae and land plants. *Taxon* **31**, 255–265.

Melkonian, M. and Berns, B. (1983). Zoospore ultrastructure in the green alga *Friedmannia israelensis:* An absolute configuration analysis. *Protoplasma* **114**, 67–84.

Moestrup, Ø. and Hoffman, L. R. (1975). A study of the spermatozoids of *Dichotomosiphon tuberosus* (Chlorophyceae). *J. Phycol.* **11**, 225–235.

Neumann, K. (1974). Zur Entwicklungsgeschichte und Systematik der siphonalen Grünalgen *Derbesia* und *Bryopsis*. *Botanica mar.* **17**, 176–185.

Nielsen, R. (1979). Culture studies on the type species of *Acrochaete, Bolbocoleon,* and *Entocladia* (Chaetophoraceae, Chlorophyceae). *Bot. Not.* **132**, 441–449.

Nielsen, R. (1980). A comparative study of five marine Chaetophoraceae. *Br. phycol. J.* **15**, 131–138.

Nordby, Ø. (1974). Light microscopy of meiotic zoosporogenesis and mitotic gametogenesis in *Ulva mutabilis* Føyn. *J. Cell Sci.* **15**, 443–445.

O'Kelly, C. J. (1982a). Chloroplast pigments in selected marine Chaetophoraceae and Chaetosiphonaceae (Chlorophyta): The occurrence and significance of siphonaxanthin. *Botanica mar.* **25**, 133–137.

O'Kelly, C. J. (1982b). Observations on marine Chaetophoraceae (Chlorophyta). III. The structure, reproduction, and life history of *Endophyton ramosum*. *Phycologia* **21**, 247–257.

O'Kelly, C. J. (1983). Observations on marine Chaetophoraceae (Chlorophyta). IV. The structure, reproduction, and life history of *Acrochaete geniculata* (Gardner) comb. nov. *Phycologia* **22**, 13–21.

O'Kelly, C. J. and Floyd, G. L. (1981). The taxonomic position of *Chaetosiphon* and *Wittrockiella* (Chaetosiphonaceae, Chlorophyta). *J. Phycol.* **17**, Suppl., 14.

O'Kelly, C. J. and Floyd, G. L. (1983). The flagellar apparatus of *Entocladia viridis* motile cells, and the taxonomic position of the resurrected family Ulvellaceae (Ulvales, Chlorophyta). *J. Phycol.* **19**, 153–164.

O'Kelly, C. J. and Floyd, G. L. (1984). Flagellar apparatus absolute orientations and the phylogeny of the green algae. *BioSystems* **16**, 227–251.

O'Kelly, C. J. and Yarish, C. (1980). Observations on marine Chaetophoraceae (Chlorophyta). I. Sporangial ontogeny in the type species of *Entocladia* and *Phaeophila*. *J. Phycol.* **16**, 549–558.

O'Kelly, C. J. and Yarish, C. (1981). Observations on marine Chaetophoraceae (Chlorophyta). II. On the circumscription of the genus *Entocladia* Reinke. *Phycologia* **20**, 32–45.

O'Kelly, C. J., Floyd, G. L. and Dube, M. A. (1984). The fine structure of motile cells in the genera *Ulvaria* and *Monostroma*, with special reference to the taxonomic position of *Monostroma oxyspermum* (Ulvophyceae, Chlorophyta). *Plant Syst. Evol.* **144**, 179–199.

Okuda, K., Enomoto, S. and Tatewaki, M. (1979). Life history of *Pseudobryopsis* sp. (Codiales, Chlorophyta). *Jpn. J. Phycol.* **27**, 7–16.

Oltmanns, F. (1904). "Morphologie und Biologie der Algen," Vol. 1. Fischer, Jena.

Parker, B. C. (1970). Significance of cell wall chemistry to phylogeny in the algae. *Ann. N. Y. Acad. Sci.* **175**, 417–428.

Perrot, Y. (1969). Sur le cycle ontogénique et chromosomique du *Pseudopringsheimia confluens* (Rosenv.) Wille. *C. r. hebd. Seanc. Acad. Sci., Ser. D* **268**, 279–282.

Roberts, K. R., Sluiman, H. J., Stewart, K. D. and Mattox, K. R. (1980). Comparative cytology and taxonomy of the Ulvaphyceae. II. Ulvalean characteristics of the stephanokont flagellar apparatus of *Derbesia tenuissima* (Chlorophyta). *Protoplasma* **104**, 223–238.

Roberts, K. R., Sluiman, H. J., Stewart, K. D. and Mattox, K. R. (1981). Comparative cytology and taxonomy of the Ulvaphyceae. III. The flagellar apparatuses of the anisogametes of *Derbesia tenuissima* (Chlorophyta). *J. Phycol.* **17**, 330–340.

Roberts, K. R., Stewart, K. D. and Mattox, K. R. (1982). Structure of the anisogametes of the green siphon *Pseudobryopsis* sp. (Chlorophyta). *J. Phycol.* **18**, 498–508.

Roth, W. C. and Friedmann, E. I. (1980). Taxonomic significance of nucleus-microbody associations, segregated nucleoli and other nuclear features in siphonous green algae. *J. Phycol.* **16**, 449–464.

Round, F. E. (1973). "The Biology of the Algae," 2nd ed. Edward Arnold, London.

Sarma, Y. S. R. K. (1982). Chromosome numbers in algae. *Nucleus* **25**, 66–108.

Schnetter, R., Mohr, B., Bula-Meyer, G. and Seibold, G. (1980). Ecology, life history and nucleus DNA contents of *Derbesia tenuissima* from the Caribbean coast of Colombia. *Proc. Int. Seaweed Symp.* **10**, 357–362.

Schussnig, B. (1938). Der Kernphasenwechsel von *Valonia utricularis* (Roth) Ag. *Planta* **28**, 43–59.

Schussnig, B. (1939). Ein Beitrag zur Entwicklungsgeschichte von *Caulerpa prolifera*. *Bot. Not.* **92**, 75–96.

Schussnig, B. (1954). Gonidiogenese, Gametogenese und Meiose bei *Cladophora glomerata* (L.) Kuetzing. *Arch. Protistenkd.* **100**, 287–322.

Scott, J. L. and Bullock, K. W. (1976). Ultrastructure of cell division in *Cladophora*. Pregametangial cell division in the haploid generation of *Cladophora flexuosa*. *Can. J. Bot.* **54**, 1546–1560.

Sears, J. R. and Brawley, S. H. (1982). *Smithsoniella* gen. nov., a possible evolutionary link between the multicellular and siphonous habits in the Ulvophyceae, Chlorophyta. *Am. J. Bot.* **69**, 1450–1461.

Shyam, R. (1980). On the life cycle, cytology and taxonomy of *Cladophora callicoma* from India. *Am. J. Bot.* **67**, 619–624.

Siegel, B. I. and Siegel, S. M. (1973). The chemical composition of algal cell walls. *CRC Crit. Rev. Microbiol.* **3,** 1–26.
Silva, P. C. (1982). Chlorophyceae. *In* "Synopsis and Classification of Living Organisms" (S. B. Parker, ed.), Vol. 1, pp. 133–161. McGraw-Hill, New York.
Sluiman, H. J., Roberts, K. R., Stewart, K. D. and Mattox, K. R. (1980a). Comparative cytology and taxonomy of the Ulvaphyceae. I. The zoospore of *Ulothrix zonata* (Chlorophyta). *J. Phycol.* **16,** 537–545.
Sluiman, H. J., Stewart, K. D. and Mattox, K. R. (1980b). Moderne opvattingen over de fylogenie van groenwieren en landplanten. *Vakbl. Biol.* **60,** 204–212.
Sluiman, H. J., Roberts, K. R., Stewart, K. D. and Mattox, K. R. (1983). Comparative cytology and taxonomy of the Ulvophyceae. IV. Mitosis and cytokinesis in *Ulothrix* (Chlorophyta). *Acta bot. neerl.* **32,** 257–269.
Stewart, K. D. and Mattox, K. R. (1975). Comparative cytology, evolution and classification of the green algae, with some consideration of the origin of other organisms with chlorophylls a and b. *Bot. Rev.* **41,** 104–135.
Stewart, K. D. and Mattox, K. R. (1978). Structural evolution in the flagellated cells of green algae and land plants. *BioSystems* **10,** 145–152.
Stuessy, C. L., Floyd, G. L. and O'Kelly, C. J. (1983). Fine structure of the zoospores of an *Enteromorpha* species (Ulvales, Chlorophyta) collected from fresh water. *Br. phycol. J.* **18,** 249–257.
Swanson, J. A. and Floyd, G. L. (1978). Fine structure of the zoospores and thallus of *Blidingia minima. Trans. Am. microsc. Soc.* **97,** 549–558.
Tanner, C. E. (1981). Chlorophyta: Life histories. *In* "The Biology of Seaweeds" (C. S. Lobban and M. J. Wynne, eds), pp. 218–247. Blackwell, Oxford.
Tappan, H. (1980). "The Paleobiology of Plant Protists." Freeman, San Francisco, California.
Tatewaki, M. (1969). Culture studies on the life history of some species of the genus *Monostroma. Sci. Pap. Inst. algol. Res. Hokkaido Univ.* **6,** 1–56.
Taylor, M. G., O'Kelly, C. J. and Floyd, G. L. (1982). Ultrastructural investigations on the biflagellate motile cells of three species of Cladophorales (Ulvophyceae, Chlorophyta) and their systematic implications. *Ohio J. Sci.* **82**(2), 21.
Tupa, D. D. (1974). An investigation of certain chaetophoralean algae. *Beih. Nova Hedw.* **46,** 1–250.
Vischer, W. (1933). Über einige kritische Gattungen und die Systematik der Chaetophorales. *Beih. bot. Zentralbl.* **51,** 1–100.
Voss, E. G., ed. (1983). "International Code of Botanical Nomenclature." Bohn, Scheltema & Holkema, Utrecht and Antwerp.
Wheeler, A. E. and Page, J. Z. (1974). The ultrastructure of *Derbesia tenuissima* (De Notaris) Crouan. I. Organization of the gametophyte protoplast, gametangium, and gametangial pore. *J. Phycol.* **10,** 336–352.
Wilkinson, M. and Burrows, E. M. (1970). *Eugomontia sacculata* Kornm. in Britain and North America. *Br. phycol. J.* **5,** 235–238.
Wilkinson, M. and Burrows, E. M. (1972). An experimental taxonomic study of the algae confused under the name *Gomontia polyrhiza. J. mar. biol. Assoc. U. K.* **52,** 49–57.
Williams, M. M. (1925). Cytology of the gametangia of *Codium tomentosum. Proc. Linn. Soc. N.S.W.* **50,** 91–111.
Wujek, D. E. and Chelune, P. (1975). The taxonomic position of *Chaetopeltis. Br. phycol. J.* **10,** 265–268.
Yamada, Y. and Tatewaki, M. (1965). New findings on the life history of *Monostroma zostericola* Tilden. *Sci. Pap. Inst. algol. Res. Hokkaido Univ.* **5,** 105–117.
Yokohama, Y. (1981). Distribution of the green light-absorbing pigments siphonaxanthin and siphonein in marine green algae. *Botanica mar.* **24,** 637–640.

5 | The Systematics of the Cladophorales

C. VAN DEN HOEK

*Marine Botany Research Group, Biological Centre,
University of Gröningen, Haren, The Netherlands*

Abstract: The well-defined chlorophycean order Cladophorales is characterized by: the siphonocladous organizational level; the numerous angular chloroplasts forming mostly a parietal reticulum; the bilenticular pyrenoids in many chloroplasts; the tightly appressed, often paired thylakoids; the crystalline cellulose-I cell walls with a crossed fibrillar pattern; and the isomorphic diplohaplontic life history of sexually-reproducing species. Details of the cruciate zoid type and of the closed mitosis are possibly characteristic of the order. However, these features have been investigated for only very few representatives. The heterogeneous genus *Cladophora* comprises 12 sections characterized by diverging architectures. Other genera of Cladophorales can be derived from at least eight of these architectural types by one or several comparatively simple morphological transformations. For instance, in this way *Microdictyon* can be derived from section Boodleoides, *Struvea* (via *Cladophoropsis*) from section Repentes, *Valonia* and *Chamaedoris* (via *Ernodesmis*) from section Longiarticulatae, and *Anadyomene* from section Longiarticulatae. Phylogenetically, we may possibly interpret the sections of *Cladophora* as the result of the first phylogenetic radiation of the primeval cladophoralean chlorophyte (which possibly resembled the almost unicellular cladophoralean *Bryobesia*). The more complicated genera can be visualized as further specialisations of the rather simple architectures of these sections. These morphological affinities support the opinion that the above and other, related, specialized genera belong to the Cladophorales, and not to the order Siphonocladales. The Siphonocladales cannot be separated from the Cladophorales on the basis of segregative cell division, as only two genera (*Siphonocladus, Dictyosphaeria*) of the 12 ranged in some taxonomies under Siphonocladales have this type of cell division, and they are related to other genera of the order on other grounds. "Segregative cell division" in e.g. *Cladophoropsis* and *Valonia* may be a wounding reaction and not the normal mode of cell division, which does not principally differ from cell division in *Cladophora*. *Sphaeroplea* is often placed in the order Cladophorales on account of its "siphonocladous organizational level". However, in all other respects including the type of mitosis and type of zoid it differs from the Cladophorales, and it should be placed in the separate order Sphaeropleales.

Systematics Association Special Volume No. 27, "Systematics of the Green Algae", edited by D. E. G. Irvine and D. M. John, 1984, pp. 157–178. Academic Press, London and Orlando.
ISBN 0 12 374040 1

Copyright © by the Systematics Association
All rights of reproduction in any form reserved

THE ORDER CLADOPHORALES

The chlorophycean order Cladophorales is well defined and exhibits the following distinct set of characters (Jónsson, 1962; Hoek, 1963, 1978, 1981, 1982a:

(a) The organisational level is siphonocladous, which means that the uniseriate, branched or unbranched filamentous plants are composed of multinucleate cells.

(b) The cells contain numerous discoid angular chloroplasts united in a parietal reticulum or a more or less closed layer, but they may extend into the internal meshes of the protoplasmic foam. In the deceivingly similar order Acrosiphoniales the multinucleate cells each contain one parietal reticulate chloroplast in the form of a cylinder which is open at both ends.

(c) Many chloroplasts contain one bilenticular pyrenoid that is divided into two hemispheres by a single thylakoid (exceptionally into more portions by several thylakoids). Each hemisphere is covered by a bowl-shaped starch grain. This pyrenoid structure, which can be observed under the light microscope, has been studied more precisely by electron microscopy in species of the genera *Cladophora, Chaetomorpha, Rhizoclonium, Valonia, Valoniopsis, Cladophoropsis, Boergesenia, Anadyomene, Microdictyon, Boodlea, Chamaedoris* and *Dictyosphaeria* (Hori and Ueda, 1967, 1975; Chan et al., 1978). In the order Acrosiphoniales the chloroplast contains polypyramidal pyrenoids, i.e. with radiating tubular thylakoid extensions and covered by many radiating starch grains (Jónsson, 1962; Hoek, 1963, 1978; Hori and Ueda, 1967, 1975; Wik-Sjöstedt and Nordquist, 1970; Chan et al., 1978).

(d) The chloroplasts are filled with tightly appressed (often paired) thylakoids (Hori and Ueda, 1967, 1975).

(e) The main wall polysaccharide is a highly crystalline cellulose I, forming numerous lamellae of microfibrils in a crossed fibrillar pattern which is visible under the light microscope. The presence of crystalline cellulose I has been demonstrated (by x-ray analysis and chemical and cytochemical methods) in the cell walls of *Cladophora, Pithophora, Chaetomorpha, Rhizoclonium, Apjohnia, Dictyosphaeria,* and *Siphonocladus* (for reviews, see Mackie and Preston, 1974; Preston, 1974; McCandless, 1981). In *Pithophora* chitin is present, in addition to cellulose I, in the outer wall and in the cross walls (Pearlmutter and Lembi, 1980). In the morphologically-similar Acrosiphoniales the cell walls lack crystalline cellulose I and a crossed fibrillar pattern (Jónsson, 1962), and they are composed of mannan, xylan, glucan and rhamnan in varying proportions (Bachman et al., 1976; Carlberg and Percival, 1977).

(f) The life history of the species with sexual reproduction is isomorphic diplohaplontic with gametophytes producing biflagellate isogametes (or slightly differing anisogametes), and sporophytes producing quadriflagellate meiospores. Cultural as well as karyological evidence is available for species of the genera *Cladophora, Chaetomorpha, Rhizoclonium, Anadyomene, Microdictyon,* and cultural evidence only for species of *Valonia* and *Dictyosphaeria* (see Hoek, 1981, and Tanner, 1981, for reviews; Enomoto and Okuda, 1981, for *Dictyophaeria*). For a number of species only asexual reproduction is known (mostly by biflagellate zoospores, in some species only by vegetative fragmentation). Older reports (Schussnig, 1928, 1954) on diplontic *Cladophora glomerata* with meiotic gametogenesis need confirmation (see Köhler, 1956, for criticism of Schussnig's report, and Hoek, 1963, pp. 11–12, for a critical review of older literature). In the asexually reproducing *C. sericea* var. *biflagellata,* prior to the formation of biflagellate zoospores, the nuclei show diploidization followed by meiosis (Wik-Sjöstedt, 1970). Acrosiphoniales have a haplontic life history with a *Codiolum*-like hypnozygote producing quadriflagellate meiospores. This hypnozygote can also be interpreted as a strongly reduced unicellular sporophyte (see Hoek, 1981; Tanner, 1981).

Two additional features, the ultrastructure of the zoids and the ultrastructure of mitosis, have been investigated for only very few representatives of the order and hence their value for defining the order is at present still uncertain.

1. Ultrastructure of the Zoids

The zoids of Cladophorales are of the cruciate (and not the unilateral) type, as are the large majority of green algal zoids. In the cruciate zoid type the flagella are anchored in the cell by four cruciately-arranged microtubular roots when the zoid is viewed from its anterior end. In one pair of opposite roots each root consists of two microtubules, and the other pair of a varying number of microtubules (in the Cladophorales three microtubules have been observed in a *Chaetomorpha* species and four in two species of *Cladophora;* see Floyd, 1981). This cruciate zoid is approximately bilaterally symmetrical when viewed from the side. This character has become important in recent theories about the phylogeny of green algae and higher plants (see reviews by Moestrup, 1978, 1982; Sluiman *et al.,* 1980a,b; Melkonian, 1981, 1982).

Two subtypes have been recently distinguished in the cruciate zoid type. Subtype (a) has, apart from the four microtubular roots, one to four fibrous roots descending into the cell and exhibiting cross-striations with a repetitive unit of more than 80 nm (system II fibrous roots of Melkonian, 1980a). In some cases narrow fibrous roots with a periodicity of 25–35 nm are closely associated with the two-stranded microtubular roots (Melkonian's "system I" fibrous roots). Moreover, subtype (a) has non-striated or faintly striated "fibres" or "caps" anteriorly connecting the flagellar basal bodies. According to Roberts *et al.* (1982) biflagellate zoids of subtype (a) have overlapping basal bodies. Subtype (b) cruciate zoids lack "system II" fibrous roots, and the fibres connecting anteriorly the basal bodies have a distinct cross-striation; biflagellate zoids of subtype (b) do not have overlapping basal bodies.

Subtype (a) has been ascribed by Roberts *et al.* (1982) to representatives of the heterogeneous (Hoek, 1981) order Ulotrichales (*Ulothrix,* Sluiman *et al.,* 1980b), as well as the order Ulvales (*Ulva,* Melkonian, 1979, 1980b; *Ulvaria,* Hoops *et al.,* 1982), Acrosiphoniales (*Urospora,* Lokhorst and Trask, 1981), Caulerpales (*Bryopsis,* Melkonian, 1981; *Pseudobryopsis,* Roberts *et al.,* 1982, and Hori and Kobara, 1982; *Derbesia,* Roberts *et al.,* 1980, 1981; *Halimeda,* Gori, 1979), Dasycladales (*Acetabularia,* Herth *et al.,* 1981), and Cladophorales (*Dictyosphaeria,* Hori and Enomoto, 1978b).

Roberts and his associates (Sluiman *et al.,* 1980a,b; Roberts *et al.,* 1980, 1981, 1982) claimed that cruciate zoids of subtype (a) consistently characterize, apart from *Ulothrix* and some other "Ulotrichales", the order Ulvales and the "green siphons" (including Acrosiphoniales, Cladophorales, Caulerpales, and Dasycladales). This assemblage they proposed to place under the new class Ulvophyceae proposed by Stewart and Mattox (1978). Moestrup (1982), however, pointed to the great morphological diversity in the structures of the green algal flagellar apparatus, and this does not support the concept of two, internally homologous subtypes of the cruciate zoid type. It seems quite likely that more than two subtypes of the cruciate zoid type can be distinguished. With regard to Cladophorales, actually very little is known about the detailed structure of the flagellar apparatus. According to Hori and Enomoto (1978b) the cruciate zoids of *Dictyosphaeria* have microtubular roots and a cross-striated fibre connecting anteriorly the basal bodies (see also discussion in Hori and Kobara, 1982); this accords with subtype (b) rather than with (a). The same is true for the dasycladalean flagellar apparatuses in *Acetabularia* (Herth *et al.,* 1981) and *Batophora* (Roberts *et al.,* 1982), which also have a cross-striated connecting fibre. Distinction of a class Ulvophyceae (including Cladophorales) seems therefore somewhat premature. At the present state of our knowledge it seems preferable to use all this fascinating new information concerning the ultrastructure of green algal zoids to construct hypothetical evolutionary lineages rather than to disrupt the current formal system in which the green algal classes function as primary operational groups of a high hierarchical level (Silva, 1980; Hoek, 1981).

Subtype (b) cruciate zoids have been described for various representatives of the orders Volvocales, Chlorococcales, and the heterogeneous order Ulotrichales, which would constitute the class Chlorophyceae *sensu stricto* (for a review, see Melkonian, 1980a).

Subtypes (a) and (b) of the cruciate zoid represent possibly two of several archaic evolutionary lineages within the green algae which radiated very early in the evolution of the chlorophytes with cruciate zoids. The unilateral zoid type represents, in this context, possibly another very old evolutionary green algal lineage (Charophyceae *sensu lato*) which has given rise to the higher land plants. In the unilateral zoid the unilaterally attached flagella are anchored in the cell by one broad band of microtubules; near the flagellar bases this microtubular band is part of a typical "multilayered structure" (Pickett-Heaps and Marchant, 1972; Stewart and Mattox, 1975, 1978; Hoek, 1978; Moestrup, 1978; Melkonian, 1980a; Stebbins and Hill, 1980).

2. *Ultrastructure of Mitosis*

During mitosis the long telophase spindle is persistent, as is the nuclear envelope, at least in the few Cladophorales for which this feature has been investigated (*Cladophora,* Scott and Bullock, 1976, and McDonald and Pickett-Heaps, 1976; *Dictyosphaeria,* Hori and Enomoto, 1978b; *Valonia,* Hori and Enomoto, 1978c; for a review, see Brawley and Wetherbee, 1981, pp. 283–284). Moreover, there are several special features: the mitoses are centric, the chromosomes bear kinetochores, spindle microtubules converge to centrioles outside the nuclear envelope, which lacks polar fenestrae, and the chromosomes separate asynchronously.

The mitoses of the numerous nuclei in cells of Cladophorales are not immediately followed by cytokinesis. In the Acrosiphoniales simultaneous mitoses take place in the future plane of cell cleavage and are immediately followed by cytokinesis (Kornmann, 1965, 1966; Hudson and Waaland, 1974). Vegetative cytokinesis in Cladophorales is often a diaphragm-like invagination with which microtubules are possibly associated (G. M. Lokhorst, personal communication). Preceding zoid formation the parietal layer of cytoplasm is partitioned by a ramifying system of cleavage furrows, with which no microtubules have been observed to be associated. The membranes of these furrows are partially Golgi-derived (Hori and Enomoto, 1978a).

In recent speculations about the significance of the type of mitosis–cytokinesis for the phylogeny of green algae and higher land plants, the long persistent telophase spindle with persistent nuclear envelope is considered as primitive (Stewart and Mattox, 1975; Hoek, 1981).

VARIOUS ARCHITECTURAL TYPES WITHIN THE GENUS *CLADOPHORA*

The genus *Cladophora* is a rather heterogeneous assemblage of species, as appears from my revisions of this genus for Europe, for American coasts of the North Atlantic Ocean, and for the coasts of southern Australia (Hoek,

1963, 1982a; Hoek and Womersley, 1984). Within the genus one can distinguish at least 12 different architectural types, which I have treated in these revisions as sections. Such morphological heterogeneity may seem strange as *Cladophora* comprises, in principle, rather simply structured, monosiphonous branched filaments.

I shall now present eight of these 12 architectural types, because they are particularly relevant to the following discussion. Figure 1 illustrates these eight architectural types (or morphotypes), corresponding with the eight following sections (Hoek, 1963, 1982a).

1. Section *Affines* Brand (Fig. 1A): The long *Rhizoclonium*-like filaments grow by frequent intercalary divisions. The few branches, which are concentrated in the basal region, are laterally inserted by steep basal walls. The cells are short (length/width ratio mostly about 1 to 2).

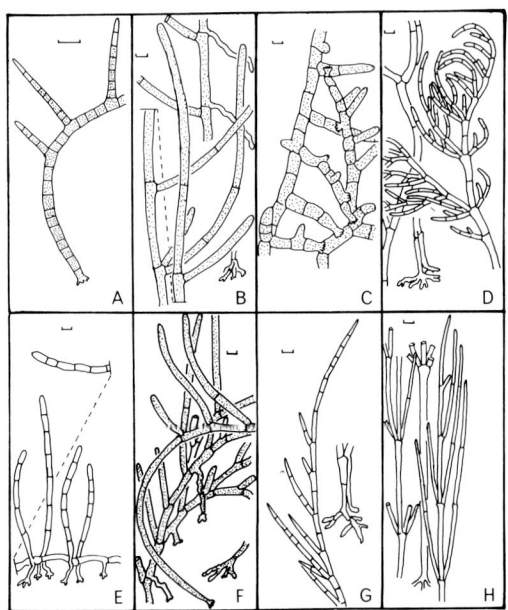

Fig. 1. Semidiagrammatic figure of eight architectural types in the genus *Cladophora*, each representing one section. (A) Section Affines Brand (*Cladophora pachyderma* (Kjellm.) Brand); (B) Section Repentes Kütz. (*C. coelothrix* Kütz.); (C) Section Boodleoides van den Hoek (*C. liebetruthii* Grunn.); (D) Section Glomeratae Kütz. (*C. dalmatica* Kütz.); (E) Section Dorsiventrales van den Hoek (*C. intertexta* Collins); (F) Section Aegagropila (Kütz.) Hansg. [*C. catenata* (L.) Kütz. emend. van den Hoek]; (G) Section Rupestres Kütz. [*C. sericea* (Huds.) Kütz.]; (H) Section Longiarticulatae Hamel [*C. pellucida* (Huds.) Kütz.]. After Hoek (1982a) with modifications. Scales to Fig.: A–D, G, 100 μm; H, 200 μm; E, F, 500 μm.

2. Section *Repentes* Kütz. (Fig. 1B): The long-celled filaments grow mainly by divisions of conspicuous cylindrical apical cells. The branches are laterally inserted by steep basal walls. Rhizoids arise from basal cell poles in the lower as well as the upper plant parts.

3. Section *Aegagropila* (Kütz.) Hansg. (fig. 1F): This section differs from the section *Repentes* by its thick cell walls, and by an easy inversion of polarity.

4. Section *Dorsiventrales* van den Hoek (Fig. 1E): The dorsiventral stolon-like filaments grow by divisions of apical cells. One or two erect filaments, and one to three rhizoids may arise from the proximal, ventral side of a stolon cell.

5. Section *Boodleoides* van den Hoek (Fig. 1C): The short-celled filaments grow mainly by intercalary divisions. The branches are laterally inserted. Rhizoids on apical and intercalary cells grow attached to other filaments of the thallus which thus forms a three-dimensional net-like structure.

6. Section *Rupestres* Kütz. (Fig. 1G): The main filaments often bear unilateral rows of branches and branchlets of different ages, the shorter (younger) ones intercalated between longer (older) ones. Growth is mainly by intercalary divisions. Branches are apically inserted on parent cells (with oblique to almost horizontal basal walls). Rhizoids are only formed from cells near the base.

7. Section *Glomeratae* Kütz. (Fig. 1D): The main filaments bear from their tips downwards rows of increasingly older and hence longer branches (acropetal organization). Growth mainly by divisions of apical cells, but towards the base intercalary growth becomes often important. Branches are apically inserted on parent cells (with oblique to almost horizontal basal walls). Rhizoids are only formed from cells near the base.

8. Section *Longiarticulatae* Hamel (Fig. 1H): The organization is almost strictly acropetal, and growth takes place by divisions of apical cells, followed by cell elongation. Intercalary divisions are rare. Mostly the basal cell is very elongate.

MORPHOLOGICAL AFFINITIES OF *CLADOPHORA* WITH OTHER GENERA OF CLADOPHORALES

The morphological heterogeneity within the genus *Cladophora* is underlined by the fact that the various above-treated architectural types can be placed in comparative morphological series with different genera in the Cladophorales. These genera represent further specialisations of the architectural types within *Cladophora*. These specializations can each be conceived as the result

of one or several of the following hypothetical "morphological transformations" of one of the architectural *Cladophora* types:
1. Planification of the thallus (tendency to blade formation).
2. Interweaving of filaments by hapteroid rhizoids in upper plant parts to strengthen blades or other structures.
3. Lateral coalescence of filaments to strengthen blades.
4. Replacement of succedaneous initiation of branches on one node by simultaneous or nearly simultaneous initiation of branches.
5. Increase of the number of branches per node.
6. Inflation of cells to form swollen vesicles.
7. Differentiation between axis and branches.
8. Reduction or suppression of branching.

I shall now treat a number of examples of such "morphological transformations". In Fig. 2 the morphological relationships between sections of the genus *Cladophora* and other genera of the Cladophorales are illustrated. The central box (A) contains the following sections ("architectural types") of Cladophora: Affines (I), Boodleoides (VI), Aegagropila (IV), Repentes (III), Dorsiventrales (V) and Longiarticulatae (XI). Other sections have been omitted for clarity. Around box A (*Cladophora*) a number of boxes represent various other Cladophoralean genera, namely *Chaetomorpha* (C), *Rhizoclonium* (B), *Wittrockiella* (D), *Rhizoclonium*(?) *grande* (E), *Microdictyon* (F), *Cladophoropsis* (G), *Boodlea* (H), *Struvea* (I), *Chamaedoris* (J), *Ernodesmis* (K), *Valonia* (L), *Anadyomene* (M), and *Bryobesia* (N). Open connections between boxes symbolize morphological affinities.

"Reduction of branching" (morphological transformation 8) leads from section Boodleoides (VI), over section Affines and the genus *Rhizoclonium* to the genus *Chaetomorpha*. Boodleoides and Affines have normal and rhizoidal, laterally-inserted branches; *Rhizoclonium* has only laterally-inserted rhizoidal branches or no branches. *Chaetomorpha* lacks branches entirely. Boodleoides has apical and intercalary cell divisions. Affines, *Rhizoclonium* and *Chaetomorpha* (almost) have only intercalary divisions. As we have no evidence that the branched condition is phylogenetically more primitive than the unbranched condition, one can also conceive the series in a reverse order.

Three morphological transformations (1, 2, 5; see above) lead from the section Boodleoides (VI) to the genus *Microdictyon*. Young *Microdictyon* plants closely resemble *Cladophora* (Fig. 2, No. 5). Adult *Microdictyon* thalli are flat, blade-like networks.

A graded morphological series leads from the section Repentes of *Cladophora* (III), over the genera *Cladophoropsis* (G) and *Boodlea* (H), towards the genus *Struvea* (I). *Cladophoropsis* differs from the section Repentes only by the prolonged or definite postponement of the formation of a cross wall at

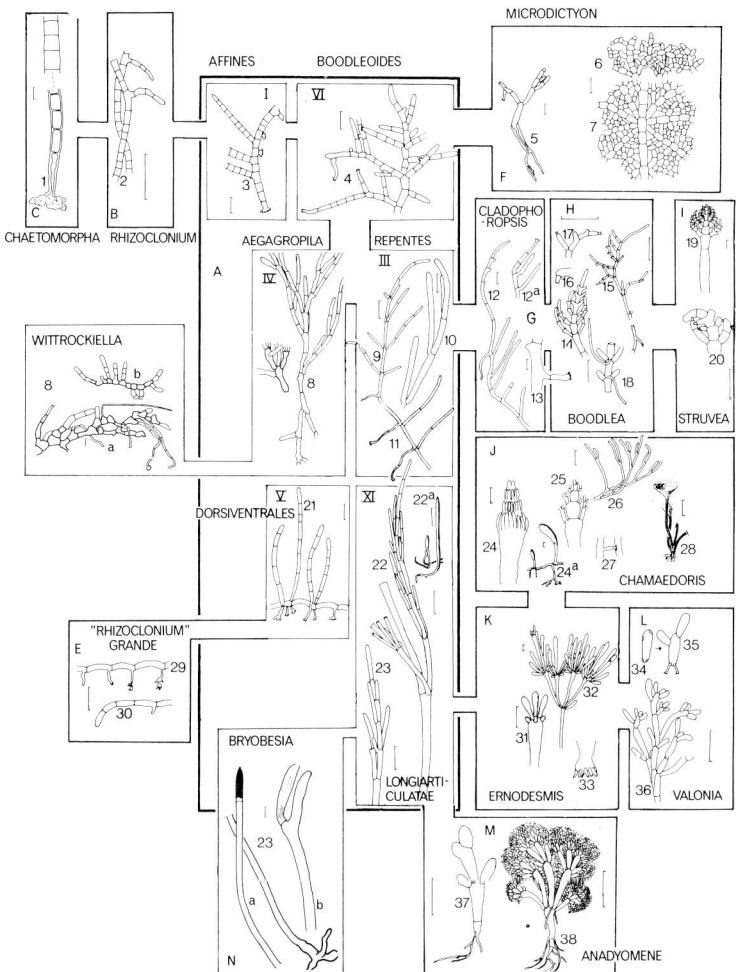

Fig. 2. Morphological relationships between sections of the genus *Cladophora* and other genera of Cladophorales. Central box A: the genus *Cladophora* containing the boxes I (Section Affines), VI (Section Boodleoides), IV (Section Aegagropila), III (Section Repentes), V (Section Dorsiventrales) and XI (Section Longiarticulatae). Box B: *Rhizoclonium*. Box C: *Chaetomorpha*. Box D: *Wittrockiella*. Box E: "*Rhizoclonium*" *grande* Børgesen (1935). Box F: *Microdictyon*. Box G: *Cladophoropsis*. Box H: *Boodlea*. Box I: *Struvea*. Box J: *Chamaedoris*. Box K: *Ernodesmis*. Box L: *Valonia*. Box M: *Anadyomene*. Box N: *Bryobesia* Weber-van Bosse (see Hoek, 1983). Figures in boxes: 1. *Chaetomorpha aerea*; 2. *Rhizoclonium riparium*; 3. *Cladophora incompta*; 4. *Cladophora liebetruthii*; 5. Young plant of *Microdictyon*; 6. Growing margin of *Microdictyon*; 7. Central part of *Microdictyon* blade; 8. *Cladophora aegagropila*; 8a,b. *Wittrockiella salina* Chapman (see Hoek et al., 1984); 9–11. *Cladophora coelothrix*; 12 and 13. *Cladophoropsis*; 14–18. *Boodlea*; 19 and 20. *Struvea*; 21. *Cladophora intertexta*; 22 and 23. *Cladophora pellucida*; 23a. *Bryobesia johannae* Weber-van Bosse (see Hoek, 1983); 24–28. *Chamaedoris peniculum*; 29 and 30. "*Rhizoclonium*" *grande*; 31–33. *Ernodesmis*; 34, 35. *Valonia*; 36. *Valonia aegagropila*; 37. Young plant of *Anadyomene*; 38. Small plant of *Anadyomene*. Modified after Hoek (1982a). Scales to Fig.: 1–7, 8a, 10, 13, 14, 18 and 37=200 μm; 8, 9, 11, 12, 15, 17, 19–26, 29–32 and 38=1 mm; 28 and 36=1 cm.

the point of insertion of the laterals. This postponement may also occur in species of *Cladophora*, though not as consistently as in *Cladophoropsis*. *Cladophoropsis* resembles the section Repentes in having long cylindrical cells and having rhizoids sprouting from intercalary cells. *Boodlea*, and still more *Struvea*, can be derived from the section Repentes and from *Cladophoropsis* by four morphological transformations (1, 2, 4, 7; see above).

The adult thalli of *Boodlea*, but still more so of *Struvea*, consist of blades (strengthened by interweaving) borne by stipes which are very pronounced in *Struvea*. In Repentes and *Cladophoropsis* intercalary rhizoids may incidentally attach to filaments of the same thallus. In *Boodlea* and *Struvea* well-organized interweaving takes place by small terminal rhizoidal cells while intercalary cells may produce rhizoids comparable to those in *Cladophoropsis* and Repentes. Both *Boodlea* and *Struvea* resemble *Cladophoropsis* in the postponement of the formation of a cross wall at the point of insertion of the laterals. The morphology of *Boodlea* is quite variable. Some plants have predominantly unilateral branching and these may be confused with *Cladophoropsis* (Fig. 2, Nos 15 and 17). Other plants consist of an axis bearing (almost) equally developed opposite branches, and these resemble *Struvea* (Fig. 2, No. 14).

The section Longiarticulatae (XI) can be considered as the centre of four comparative morphological series. In the first place the genus *Ernodesmis* can be derived from this section, via *Apjohnia* (Fig. 2, No. 32; see also Fig. 3) by two morphological transformations (5, 6; see above). In *Valonia* (Fig. 2, box L) inflation of the cells has led to the large, macroscopic vesicles. *Ernodesmis* and *Apjohnia* resemble Longiarticulatae in their strict acropetal organization. In Longiarticulatae a main filament grows by elongation of the apical cell and its subsequent division into a new apical cell and a subapical cell. Each side branch starts as a papilliform protuberance below a cross wall and is cut off from the main filament by a basal wall (Figs 3A and B). In *Apjohnia* and *Ernodesmis* there is no difference in the mode of growth of the main filament and side branches. Here a number of papilliform branch initials are formed on the tip of the inflated apical segment, which are cut off from this segment very early in *Ernodesmis*, and much later in *Apjohnia* (Figs 3C and D). It is impossible, or almost so, to indicate a branch which forms a main filament with lower segments of the thallus in these two genera having a radiating fan-like appearance. Some branches, however, are more developed than others. *Valonia aegagropila* (Fig. 2, No. 36) still has an acropetal organization, which is absent, however, in other *Valonia* species. The branch-initials in *Valonia* are not papilliform but lens-shaped, and they are cut off from the periphery of the parent cell by concave walls (Fig. 2, Nos 34, 35; Fig. 3E). Thus, the mode of branching in Longiarticulatae,

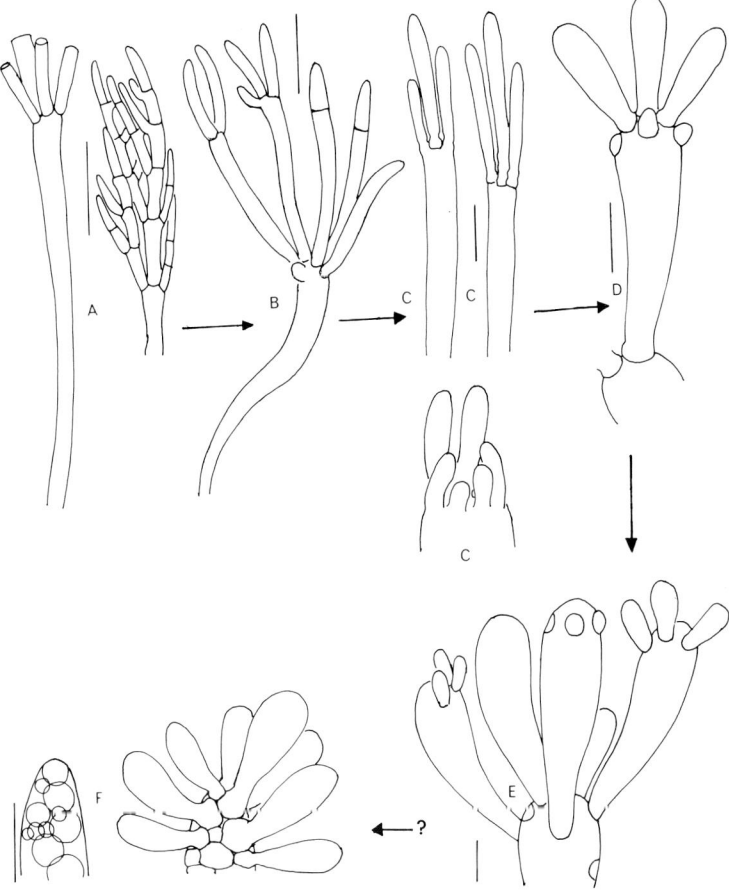

Fig. 3. Comparative morphological series in which *Valonia* (E) is derived from *Cladophora prolifera* (A), over *C. retroflexa* (B), *Apjohnia laetevirens* (C), and *Ernodesmis verticillata* (D). Perhaps segregative cell division in *Siphonocladus tropicus* (F) can be derived from cell division by lens-cells in *Valonia* (E). Scales: 1 mm. (A and B) Redrawn after Hoek (1963, 1982a). (C and D) After Papenfuss and Chihara (1975). (E) After Oltmanns (1922). (F) After Egerod (1952).

Apjohnia, Ernodesmis and *Valonia* can be placed in a graded morphological series (Fig. 3).

In *Apjohnia* and *Ernodesmis* small papilliform initials at the basal cell poles grow into small unicellular rhizoids which anchor the parent cell to the underlying cell. Comparable rhizoidal cells attach *Valonia* to the substrate (Fig. 2, Nos 33 and 35). They can also be compared to the small rhizoidal extensions of cells in the basal region of *Cladophora prolifera* plants; these, however, mostly retain protoplasmic continuity with the parent cell. *C. prolifera* is a typical representative of the Longiarticulatae (Fig. 3A and B).

A second morphological series is formed by Longiarticulatae, *Apjohnia, Ernodesmis,* and *Chamaedoris. Chamaedoris* (Fig. 2, Box J) can be derived from Longiarticulatae via *Apjohnia* and *Ernodesmis* by invoking two additional morphological transformations, namely strong differentiation between axis and laterals and interweaving. In *Chamaedoris* the axis grows by divisions of apical cells; the cut off segments grow enormously in length and width. A whorl of almost equal papilliform branch initials is formed on each node (Fig. 2, Nos 24 and 25). These initials grow into *Cladophoropsis*-like branched filaments. As the older laterals are shed, only a terminal cup-shaped tuft of laterals is present, which is strengthened by "interweaving" by small, intercalary rhizoids.

A third morphological series consists of Longiarticulatae and *Anadyomene* (Fig. 2, Box M). *Anadyomene* can be derived from Longiarticulatae by invoking four morphological transformations (1, 3, 5, 6; see above). Actually *Anadyomene* forms a flat, pseudoparenchymatous, blade-like thallus on the basis of the architecture of Longiarticulatae. Young *Anadyomene* plants (Fig. 2, No. 37) much resemble young Longiarticulatae plants.

A fourth, in this case reductional, morphological series consists of Longiarticulatae and *Bryobesia* (Fig. 2, Box N). This genus was recently shown to belong to the Cladophorales and not, as formerly thought, to the Caulerpales (Hoek, 1983). *Bryobesia* can be derived from Longiarticulatae by suppression of branching (morphological transformation 8) and of growth. An adult *Bryobesia* plant is unicellular and cylindrical and it cuts off by a cross wall one apical zoidangium (Fig. 2, Box N). This stage much resembles a young plant in the Longiarticulatae after its first apical division. Thus *Bryobesia* gives the impression of being a rudimentary *Cladophora,* and it can be either interpreted as a primitive cladophoralean at the base of cladophoralean morphological radiation, or as a highly-reduced and hence a highly-derived cladophoralean.

The genus *Wittrockiella* (Fig. 2, Box D) shows a high morphological affinity with the section Aegagropila. It shares all characters of this section, but it has one in addition, namely the possession of one or two cylindrical non-septate hairs on intercalary cells. However, these hairs are often lacking. The filaments are shorter-celled than in the section Aegagropila, but as in this section the plants are either compact and mat-like, or they may form under certain conditions smooth, hollow spheres (Hoek *et al.,* 1984).

The genus *Willeella* (or the section Willeella of *Cladophora;* cf. Hoek, 1982a) can be derived from the section Rupestres (e.g. *C. rupestris,* or *C. sericea*) by morphological transformations 1 and 4.

Finally, the curious dorsiventral *"Rhizoclonium" grande* Børgesen (1935) can be derived from the section Dorsiventrales by suppression of branching (morphological transformation 8; Fig. 2, Box E). Other *Rhizoclonium* species (Fig. 2, No. 2) do not have a dorsiventral organization.

DISCUSSION

The above-treated ten comparative morphological series connecting sections within the genus *Cladophora* with 14 other cladophoralean genera stress the morphological heterogeneity of the genus *Cladophora*. Actually the genus *Cladophora* does not fulfill the requirement that all species of the same genus should be more closely related to each other than to any species in another cladophoralean genus. In practice the relationship can here be conceived only as morphological affinity. For instance, *C. incompta* (Fig. 2, No. 3) is more closely related to *Rhizoclonium* (Fig. 2, No. 2) than to *C. pellucida* (or *prolifera*) (Fig. 2, Box XI) which, in its turn, is more closely related to *Anadyomene* (Fig. 2, Box M).

One could propose two solutions to this problem of the ill-defined genus *Cladophora:* to merge all genera of Cladophorales into one genus, *Cladophora,* or to raise all sections of *Cladophora* to the genus level. Both solutions are unsatisfactory; the first because a number of distinct genera such as *Microdictyon, Chamaedoris, Ernodesmis,* and *Anadyomene* would be submerged in an amorphous mass genus, and the second because the sections of *Cladophora* are not sharply separated from one another but form a reticulate morphological continuum. The best possible solution is therefore to maintain *Cladophora* and the other genera of Cladophorales as they are now generally conceived. Perhaps exceptions should be made for a few genera which are very near to *Cladophora,* such as *Cladophoropsis* and *Willeella* (see Hoek, 1982a); these could be merged with *Cladophora.*

Phylogenetically, we could possibly interpret the different architectural types in the sections of *Cladophora* as the result of the first phylogenetic radiation of the primeval cladophoralean chlorophyte. The almost unicellular genus *Bryobesia* is possibly nearest to this ancestral type, although it can also be interpreted as the result of extreme reduction (the small size being an adaptation to extreme grazing pressure in the coral reefs where this genus occurs). The more complicated architectural types as realized by genera like *Microdictyon, Struvea, Chamaedoris, Valonia* and *Anadyomene* could be interpreted as further specializations of the relatively simple architectural types found in *Cladophora*. According to this hypothesis, these genera are younger than the sections of *Cladophora*. Unfortunately there are no fossil records to test this hypothesis and we do not have data from which to infer the age of the Cladophorales* and its subgroups. However, the distribution patterns of some species may suggest at least their minimum ages.

On the species level, various tropical Cladophorales have a strikingly disjunct distribution, as they are restricted to the west sides of both the Pacific and the Atlantic Oceans. This type of distribution is shared with other tropical algal species. Examples include *Cladophora catenata, Valonia aegagropila, Valonia ocellata, Valonia ventricosa,* and *Dictyosphaeria cavernosa*. *Cladophora catenata* (=*fuliginosa*) is a very typical species of the section Aegagropila, whose thick walls harbour the characteristic fungus *Blodgettiomyces borneti*. This is true for Caribbean material as well as for material from Queensland, Northern Australia (National Herbarium of Victoria, Melbourne, no. 597628, Endeavour River, Queensland; as *Blodgettia australis* J. Ag. in the exsiccata "Algae Muellerianae Curante J.G. Agardh distributae"). These highly disjunct distributions can probably be interpreted as remnants of a formerly continuous Tethyan distribution area (Hoek, 1982b). If this is correct it would mean that these species already existed when Africa and Eurasia collided in the Miocene, thus causing the closure of the eastern end of the Tethyan Ocean. They would consequently be at least 18 million years old. The origin of the genera and of the sections of *Cladophora* must therefore lie in a much more distant past.

Because of the lack of fossil evidence the present phylogenetic reflections are, in all sobriety, only comparative morphological exercises with extant plants, as are, for that matter, virtually all reflections on green algal phylogeny including those on the ultrastructure of mitosis and flagellate cells. However, these comparative morphological exercises in Cladophorales indicate that more complex and specialized genera such as *Microdictyon, Struvea, Chamaedoris, Valonia* and *Anadyomene* can be derived from architectural types within the genus *Cladophora*. This derivation supports the inclusion of these genera in one and the same order, the Cladophorales, particularly as they show distinct similarities in cell wall structure, pyrenoid structure, chloroplast structure, life history type and (probably) in details of the mitosis.

In some taxonomies (Egerod, 1952; Smith, 1955; Christensen, 1962; Bold and Wynne, 1978; Hoek, 1978; Lee, 1980; Tanner, 1981) the above and other specialized genera [*Microdictyon, Cladophoropsis, Boodlea, Struvea, Pseudostruvea* (Egerod, 1975), *Chamaedoris, Apjohnia, Ernodesmis, Valonia, Valoniopsis* (Børgesen, 1934), *Siphonocladus, Boergesenia, Dictyosphaeria, Anadyomene*] are all or in part placed in the order Siphonocladales, whilst

---

*Cladophorales Haeckel (1894) has priority over Siphonocladales (Blackman & Tansley) Oltmanns (1904); see Papenfuss and Chihara (1975).

Cladophora and other genera with a simpler, less specialized construction are ranged under the Cladophorales [*Cladophora, Basicladia* [as a section of *Cladophora* in Hoek, 1963], *Wittrockiella, Pithophora, Arnoldiella, Gemmiphora, Chaetonella, Chaetocladiella, Cladophorella, Cladostroma, Dermatophyton*; see Bourrelly, 1966, and Vinogradova *et al.*, 1980, for recent reviews of freshwater cladophoralean genera]. Cladophorales and Siphonocladales are merged with each other in most recent taxonomies (Chadefaud, 1960; Jónsson, 1962; Bourrelly, 1966; Fott, 1971; Chapman and Chapman, 1973; Feldmann, 1978; Ettl, 1980; Hoek, 1981). In some of these the Acrosiphoniales are still incorrectly ranged under Cladophorales (Chadefaud, 1960; Fott, 1971; Chapman and Chapman, 1973; Ettl, 1980).

Siphonocladales, if kept apart, are thought to differ from Cladophorales by a peculiar method of cell division and branching termed "segregative cell division," that is, cleavage of the protoplast into walled and rounded portions which later expand into new cells and even branches (Børgesen, 1905, 1913). However, in only two genera, *Dictyosphaeria* and *Siphonocladus,* has well-organized segregative cell division been indisputably demonstrated. A detailed description of this process in cultures of *Dictyosphaeria* has been recently provided by Enomoto *et al.* (1982). The network leading to segregative cell division much resembles the network leading to the formation of swarmers in *Dictyosphaeria,* but in the latter case the colour changes from green to yellowish green (Enomoto and Okuda, 1981). Comparable networks are formed preceding sporulation in several *Cladophora* species, e.g. *C. catenata* and *C. prolifera* (Hoek, 1963, 1982a), and in *Valonia* (Hori and Enomoto, 1978c). These observations suggest that organized segregative cell division may be derived from the cellular processes leading to the formation of swarmers.

Segregative cell division has also been described and illustrated for various other specialized genera of Cladophorales, namely *Cladophoropsis, Boodlea, Struvea, Chamaedoris, Ernodesmis,* and *Valonia* (Børgesen, 1913; Egerod, 1952). In *Ernodesmis* and *Valonia* the formation of lenticular branch initials is considered as a special case of segregative cell division. Diaphragm-like cross walls of the *Cladophora* type have been described for *Anadyomene* (Enomoto and Hirose, 1970) and *Microdictyon* (Enomoto and Hirose, 1971).

A comparative morphological series of the modes of branching in the section Longiarticulatae of *Cladophora* (*C. prolifera, C. retroflexa*) and in *Apjohnia, Ernodesmis* and *Valonia* (see Fig. 3), however, suggests that these modes do not fundamentally differ from one another. In *Cladophora* a side branch starts as a papilliform protuberance with accumulated protoplasm and chloroplasts, and this gradually expands into the young cylindrical side branch. A basal cross wall cuts the branch from the axis (Figs 3A and B). In *C. retroflexa* up to five side branches may grow from the subapical cell, thus resembling the apical whorls in *Apjohnia* (Fig. 3C) and *Ernodesmis* (Fig. 3D). In *Apjohnia* the formation of basal branch walls takes place rather late. In *Ernodesmis*, however, the papilliform branches are cut off very early by a curved basal wall from the parent cell (Fig. 3D). In this species the whorled branches have an aspect much resembling the type of branching in *Apjohnia* and *Cladophora retroflexa*. Finally, in *Valonia* peripheral accumulations of protoplasm with chloroplasts are cut off by curved walls from the parent cell, and only after that do they grow into protuberances which develop into side branches (Fig. 3E). It seems even possible to place real segretative cell division as shown by *Siphonocladus* (Fig. 3F) at the extreme end of the morphological series. Here masses of accumulated protoplasm, nuclei and chloroplasts are entirely surrounded by curved spherical walls, and after that grow into protuberances.

A rarely observed mode of cell division in *Cladophoropsis*, much resembling segregative cell division, has been illustrated by Børgesen (1913; Fig. 4E and F). The contents of one long cell are divided into separate portions with rounded ends; these portions later expand into new cells separated from one another by straight cross walls where these portions are pressed against each other. In *Struvea* Børgesen observed a comparable phenomenon, but his observations are somewhat problematic as they are based on ethanol-preserved material in which protoplasts possibly contracted from active cleavage furrows. Apparently extensive comparative investigations on the mode of branch formation and cell division in living material are now required.

Recently LaClaire (1982) demonstrated that wounding (puncturing) of *Cladophoropsis* cells results in a rapid constriction of the protoplasm of the cell into several segments which pinch apart and pull away from each other within several minutes. About one hour after wounding the ends of the segments become smooth and rounded and form new wall material (Fig. 4A–D). This phase corresponds with Børgesen's illustration of segregative cell division in *Cladophoropsis*. The segments after that expand and eventually fill the entire parent cell wall (120 h after wounding). This wound healing process in *Cladophorosis* certainly resembles segregative cell division, but differs from it in various details. Comparable wound healing processes probably occur in some *Cladophora* species, for instance in *C. coelothrix*, a species which much resembles *Cladophoropsis* (see Figs 68, 70 and 72 in Hoek, 1963, where short-celled filaments in cultures of this long-celled species are probably the result of wounding).

According to LaClaire (1982) wounding reactions in various specialized Cladophorales differ in several respects from one another. For example, in *Ernodesmis* and two species of *Struvea* the protoplast after puncturing contracts but does not divide. A cell wall is synthesized on the naked end. After

5. Cladophorales

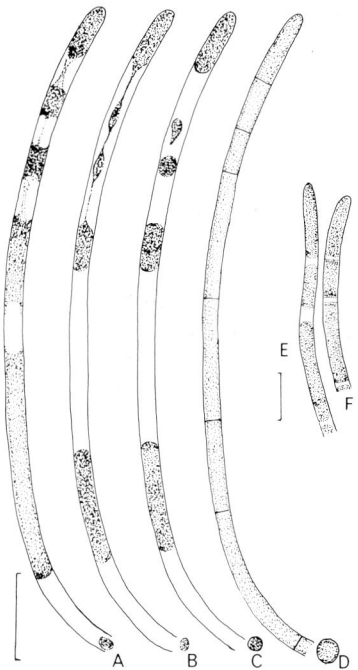

Fig. 4. (A–D) Wound healing in *Cladophoropsis*. Time after wounding: (A) 2 min; (B) 10 min; (C) 60 min; (D) 120 h. The protoplast breaks into segments which later expand into new cells (a process much resembling segregative cell division). (A–D) is based on LaClaire (1982). (E, F) Stages in presumptive segregative cell division in *Cladophoropsis* as illustrated by Børgesen (1913). Possibly these figures represent stages in the wound healing process. Scale: 1 mm.

that this single protoplast expands and grows into a new plant. Papenfuss and Chihara (1975), however, reported and illustrated the occasional occurrence of spherical, walled protoplasts in *Ernodesmis* cells. In *Siphonocladus* the contraction after wounding into one protoplast is extreme, here the wounding reaction differs strikingly from the segregative cell division in this genus. In *Valonia* and *Boergesenia* wounding results in the contraction of the protoplast into hundreds of tiny spheres, each of which can grow into a new individual [see also Preston (1974, p. 193) for this wounding reaction in *Valonia*]. The wounding reaction in *Boergesenia* and *Valonia* is very similar to segregative cell division in *Siphonocladus* and *Dictyosphaeria*, but it differs greatly from their normal processes of cell division and branching. In *Boodlea*, *Chamaedoris* and a third species of *Struvea* the wound was closed by a plug as in the Caulerpales (LaClaire, 1982).

My provisional conclusion is that wounding reactions, which probably occur quite regularly in nature (see, for *Valonia*, Børgesen, 1913, 1940, and remarks by Papenfuss and Chihara, 1975), superficially resemble segregative cell division in *Cladophoropsis, Valonia, Boergesenia,* and possibly also in some other Cladophorales (see also Enomoto and Hirose, 1972), but their normal modes of cell division and branching are in principle comparable to those in *Cladophora*. Genuine segregative cell division is distinct in *Siphonocladus* and *Dictyosphaeria*.

I would like to stress the point that aspects of the above comparative morphological and cytological reflections leading to the conclusion that Cladophorales and Siphonocladales belong to one and the same order have been put forward already in 1938 by J. Feldmann and in 1962 by Jónsson (1962).

In some taxonomies the freshwater genus *Sphaeroplea* is placed in the order Cladophorales (or Siphonocladales) only on account of its siphonocladous organisational level (Pascher, 1931; Fott, 1971; Bold and Wynne, 1978; Lee, 1980). Vinogradova *et al.* (1980) also placed *Sphaeroplea* near the Cladophorales as the order Sphaeropleales in the hierarchically-inflated class Siphonocladophyceae. Apart from being composed of multinucleate cells, *Sphaeroplea* lacks all other features which are here considered characteristic of the order Cladophorales. On the other hand it has a unique vegetative architecture (unbranched filaments of multinucleate cells containing rings of cytoplasm, each with an annular parietal chloroplast) and sexual reproduction (a unique type of oogamy). Therefore, it has been placed, on account of light microscopic features only, in the order Sphaeropleales by various authors (Chadefaud, 1960; Bourrelly, 1966; Feldmann, 1978; Ettl, 1980; Hoek, 1981), and in the order Ulotrichales by some others (Fritsch, 1935; Smith, 1955; Christensen, 1962; Chapman and Chapman, 1973). The special position of *Sphaeroplea* and hence the concept of the order Sphaeropleales is supported by recent electron microscopical investigations (Cáceras and Robinson, 1980, 1981). Mitosis is closed and the phycoplast is associated with the separation of daughter nuclei during late telophase (not with cytokinesis). Only during formation of male biflagellate gametes do cleavage furrows pass through the phycoplasts (for a discussion of types of mitosis–cytokinesis, see Hoek, 1981). These gametes belong to a peculiar variant of the cruciate zoid type, with the four microtubular roots each composed of about eight microtubules, with a distal non-striated fibre connecting the basal bodies and on top of this fibre a unique "apical cone". This zoid type does not conform to one of the two above-treated subtypes of the cruciate zoid type, and it consequently stresses the special position of the order Sphaeropleales.

REFERENCES

Bachmann, P., Kornmann, P. and Zetsche, K. (1976). Regulation der Entwicklung und des Stoffwechsels der Grünalge *Urospora* durch die Temperatur. *Planta* **128,** 241–245.

Bold, H. C. and Wynne, M. J. (1978). "Introduction to the Algae. Structure and Reproduction." Prentice-Hall, Englewood Cliffs, New Jersey.

Børgesen, F. (1905). Contributions à la connaissance du genre *Siphonocladus* Schmitz. *Overs. K. dan. Vidensk. Selsk. Fosh.* No. 3, pp. 259–291.

Børgesen, F. (1913). The marine algae of the Danish West Indies. Part I. Chlorophyceae. *Dan. bot. Ark.* **1**(4), 1–158.

Børgesen, F. (1934). Some marine algae from the northern part of the Arabian Sea with remarks on their geographical distribution. *Biol. Medd.—K. dan Vidensk. Selsk.,* **11**(6), 1–72.

Børgesen, F. (1935). A list of marine algae from Bombay. *Biol. Medd.—K. dan. Vidensk. Selsk.* **12**(2), 1–64.

Børgesen, F. (1940). Some marine algae from Mauritius. I. Chlorophyceae. *Biol. Medd.—K. dan. Vidensk. Selsk.* **15**(4), 1–81.

Bourrelly, P. (1966). "Les Algues d'eau douce. Initiation à la Systématique. Tome I. Les Algues Vertes." Boubée, Paris.

Brawley, S. H. and Wetherbee, L. (1981). Cytology and ultrastructure. *In* "The Biology of Seaweeds" (C. S. Lobban and M. J. Wynne, eds), pp. 248–299. Blackwell, Oxford.

Cáceras, E. J. and Robinson, D. G. (1980). Ultrastructural studies on *Sphaeroplea annulina* (Chlorophyceae). Vegetative structure and mitosis. *J. Phycol.* **16,** 313–320.

Cáceras, E. J. and Robinson, D. G. (1981). Ultrastructural studies on *Sphaeroplea annulina* (Chlorophyceae). II. Spermatogenesis and male gamete structure. *J. Phycol.* **17,** 173–180.

Carlberg, G. E. and Percival, E. (1977). The carbohydrates of the green seaweeds *Urospora wormskioldii* and *Codiolum pusillum*. *Carbohydr. Res.* **57,** 223–234.

Chadefaud, M. (1960). Les végétaux non vasculaires. Cryptogamie. *In* "Traité de Botanique, Systématique" (M. Chadefaud and L. Emberger, eds), Vol. I, pp. 1–1018. Masson, Paris.

Chan, K.-Y., Wong, S. L. L. and Wong, M. H. (1978). Observations on *Chaetomorpha brachygona* Harv. (Chlorophyta, Cladophorales). 1. Ultrastructure of the vegetative cells. *Phycologia* **17,** 419–429.

Chapman, V. J. and Chapman, D. J. (1973). "The Algae." Macmillan, London.

Christensen, T. (1962). Alger. *In* "Botanik" (T. W. Böcher, M. Lange, and T. Sørensen, eds.), Vol. 2, No. 2, pp. 1–178. Munksgaard, Copenhagen.

Egerod, L. E. (1952). An analysis of the siphonous Chlorophycophyta, with special reference to the Siphonocladales, Siphonales, and Dasycladales of Hawaii. *Univ. Calif., Publ. Bot.* **25,** 325–454.

Egerod, L. E. (1975). Marine algae of the Andaman sea coast of Thailand: Chlorophyceae. *Bot. Mar.* **18,** 41–66.

Enomoto, S. and Hirose, H. (1970). On the life history of *Anadyomene wrightii* with special reference to the reproduction, development and cytological sequences. *Bot. Mag.* **83,** 270–280.

Enomoto, S. and Hirose, H. (1971). On the septum formation of *Microdictyon okamurai* Setchell. *Bull. Jpn. Soc. Phycol.* **19,** 90–93.

Enomoto, S. and Hirose, H. (1972). Culture studies on artificially induced aplanospores and their development in the marine alga *Boergesenia forbesii* (Harvey) Feldmann (Chlorophyceae, Siphonocladales). *Phycologia* **11,** 119–122.

Enomoto, S. and Okuda, K. (1981). Culture studies of *Dictyosphaeria* (Chlorophyceae, Si-

phonocladales). I. Life history and morphogenesis of *Dictyosphaeria cavernosa*. *Jpn. J. Phycol.* **29**, 225–236.
Enomoto, S., Hori, T. and Okuda, K. (1982). Culture studies of *Dictyosphaeria* (Chlorophyceae, Siphonocladales). II. Morphological analysis of segregative cell divisions in *Dictyosphaeria cavernosa*. *Jpn. J. Phycol.* **30**, 103–112.
Ettl, H. (1980). "Grundriss der allgemeinen Algologie." Fischer, Jena.
Feldmann, J. (1938). Sur la classification de l'ordre des Siphonocladales. *Rev. gen. Bot.* **50**, 571–597.
Feldmann, J. (1978). Algae. *In* "Précis de Botanique" (H. des Abbayes, M. Chadefaud, J. Feldmann, Y. de Ferré, H. Gaussen, P.-P. Grassé, and A. R. Prévot, eds), Vol. I, pp. 95–320. Masson, Paris.
Floyd, G. L. (1981). Ultrastructure of motile cells from two species of *Cladophora*. *J. Phycol.* **17**, Suppl., 11.
Fott, B. (1971). "Algenkunde," 2nd ed. Fischer, Jena.
Fritsch, F. E. (1935). "The Structure and Reproduction of the Algae," Vol. 1. Cambridge Univ. Press. London and New York.
Gori, P. (1979). Ultrastructure of the spermatozoid in *Halimeda tuna* (Chlorophyceae). *Gamete Res.* **2**, 345–355.
Herth, W., Heck, B. and Koop, H. U. (1981). The flagellar root system in the gamete of *Acetabularia mediterranea*. *Protoplasma* **109**, 257–269.
Hoek, C. van den (1963). "Revision of the European species of *Cladophora*." Brill, Leiden (reprint Koeltz Königstein, 1976).
Hoek, C. van den (1978). (Unter Mitwirkung von H. M. Jahns.) "Algen. Einführung in die Phycologie." Thieme, Stuttgart.
Hoek, C. van den (1981). Chlorophyta: morphology and classification. *In* "The Biology of Seaweeds" (C. S. Lobban and M. J. Wynne, eds), pp. 86–132. Blackwell, Oxford.
Hoek, C. van den (1982a). A taxonomic revision of the American species of *Cladophora* (Chlorophyceae) in the North Atlantic Ocean and their geographic distribution. *Verh. K. ned. Akad. Wet., Afd. Natuurkd., Reeks 2* **78**, 1–236.
Hoek, C. van den (1982b). The distribution of benthic marine algae in relation to the temperature regulation of their life histories. *Biol. J. Linn. Soc.* **18**, 81–144.
Hoek, C. van den (1983). *Bryobesia johannae*, a small tropical representative of the order Cladophorales (Chlorophyceae). *J. Phycol.* **19**, 116–118.
Hoek, C. van den, Ducker, S. C. and Womersly, H. B. S. (1984). *Wittrockiella salina* Chapman (Cladophorales, Chlorophyceae), a mat and ball forming alga. *Phycologia* **23** 39–46.
Hoek, C. van den and Womersley, H. B. S. (1984). Genus *Cladophora*. Kützing 1843. *In* "The Marine Benthic Flora of Southern Australia. Part 1" (H. B. S. Womersley, ed), pp. 185–213. D. J. Woolman, Government Printer, South Australia.
Hoops, H. J., Floyd, G. L. and Swanson, J. A. (1982). Ultrastructure of the biflagellate motile cells of *Ulvaria oxysperma* (Kütz.) Bliding and phylogenetic relationships among ulvaphycean algae. *Am. J. Bot.* **69**, 150–159.
Hori, T. and Enomoto, S. (1978a). Developmental cytology of *Dictyosphaeria cavernosa* I. Light and electron microscope observations on cytoplasmic cleavage in zoid formation. *Bot. mar.* **21**, 401–408.
Hori, T. and Enomoto, S. (1978b). Developmental cytology of *Dictyosphaeria cavernosa*. II. Nuclear division during zooid formation. *Bot. mar.* **21**, 477–481.
Hori, T. and Enomoto, S. (1978c). Electron microscope observations on the nuclear division in *Valonia ventricosa* (Chlorophyceae, Siphonocladales). *Phycologia* **17**, 133–142.
Hori, T. and Kobara, T. (1982). Ultrastructure of the flagellar apparatus in the stephanokont zoospores of *Pseudobryopsis hainanensis* (Chlorophyceae). *Jap. J. Phycol.* **30**, 31–39.

Hori, T. and Ueda, R. (1967). Electron microscope studies on the fine structure of plastids in siphonous green algae with special reference to their phylogenetic relationships. *Sci. Rep. Tokyo Kyoiku Daigaku, Sect. B.* **12,** 225–244.

Hori, T. and Ueda, R. (1975). The fine structure of algal chloroplasts and algal phylogeny. In "Advance of Phycology in Japan" (J. Tokida and H. Hirose, eds), pp. 11–42. Fischer, Jena.

Hudson, P. R. and Waaland, J. R. (1974). Ultrastructure of mitosis and cytokinesis in the multinucleate green alga *Acrosiphonia*. *J. Cell Biol.* **62,** 274–294.

Jónsson, S. (1962). Recherches sur les Cladophoracées marines. Thèse, Université de Paris, F. Masson, Paris.

Köhler, K. (1956). Entwicklungsgeschichte, Geschlechtsbestimmung und Befruchtung bei *Chaetomorpha*. *Arch. Protistenkd.* **101,** 223–268.

Kornmann, P. (1965). Zur Analyse des Wachstums und des Aufbaus von *Acrosiphonia*. *Helgol. Wiss. Meeresunters.* **12,** 219–238.

Kornmann, P. (1966). Wachstum und Zellteilung bei *Urospora*. *Helgol. wiss. Meeresunters.* **13,** 73–83.

LaClaire, J. W., II (1982). Cytomorphological aspects of wound healing in selected Siphonocladales (Chlorophyceae). *J. Phycol.* **18,** 379–384.

Lee, R. E. (1980). "Phycology." Cambridge Univ. Press, London and New York.

Lokhorst, G. M. and Trask, B. J. (1981). Taxonomic studies on *Urospora* (Acrosiphoniales, Chlorophyceae) in Western Europe. *Acta bot. neerl.* **30,** 353–431.

McCandless, E. L. (1981). Polysaccharides of the seaweeds. In "The Biology of Seaweeds" (C. S. Lobban and M. J. Wynne, Eds), pp. 559–588. Blackwell, Oxford.

McDonald, K. L. and Pickett-Heaps, J. D. (1976). Ultrastructure and differentiation in *Cladophora glomerata*. I. Cell division. *Am. J. Bot.* **63,** 592–601.

Mackie, W. and Preston, R. D. (1974). Cell wall and intercellular region polysaccharides. In "Algal Physiology and Biochemistry" (W. D. P. Stewart, ed.), pp. 40–85. Blackwell, Oxford.

Melkonian, M. (1979). Structure and significance of cruciate flagellar root systems in green algae: zoospores of *Ulva lactuca* (Ulvales, Chlorophyceae). *Helgol. wiss. Meeresunters.* **32,** 425–435.

Melkonian, M. (1980a). Ultrastructural aspects of basal body associated fibrous structure in green algae: A critical review. *BioSystems* **12,** 85–104.

Melkonian, M. (1980b). Flagellar roots, mating structure and gametic fusion in the green alga *Ulva lactuca* (Ulvales). *J. Cell Sci.* **46,** 149–169.

Melkonian, M. (1981). Structure and significance of cruciate flagellar root systems in green algae: female gametes of *Bryopsis lyngbyaei* (Bryopsidales). *Helgol. Meeresunters.* **34,** 355–370.

Melkonian, M. (1982). Systematics and evolution of the algae. *Prog. Bot.* **44,** 315–344.

Moestrup, Ø. (1978). On the phylogenetic validity of the flagellar apparatus in green algae and other chlorophyll a and b containing plants. *BioSystems* **10,** 117–144.

Moestrup, Ø. (1982). Flagellar structure in algae: A review, with new observations particularly on the Chrysophyceae, Phaeophyceae (Fucophyceae), Euglenophyceae and *Reckertia*. *Phycologia* **21,** 427–528.

Oltmanns, F. (1922). "Morphologie und Biologie der Algen." Fischer, Jena.

Papenfuss, G. F. and Chihara, M. (1975). The morphology and systematic position of the green algae *Ernodesmis* and *Apjohnia*. *Phycologia* **14,** 309–316.

Pascher, A. (1931). Systematische Übersicht über die mit Flagellaten in Zusammenhang stehenden Algenreihen und Versuch einer Einreihung dieser Algenstämme in die Stämme des Pflanzenreiches. *Beih. bot. Zentralbl., Abt. 2* **48,** 317–332.

Pearlmutter, N. L. and Lembi, C. A. (1980). Structure and composition of *Pithophora oedogonia* (Chlorophyta) cell walls. *J. Phycol.* **16**, 602–616.
Pickett-Heaps, J. D. and Marchant, H. J. (1972). The phylogeny of the green algae: a new proposal. *Cytobios* **6**, 255–264.
Preston, R. D. (1974). "The Physical Biology of Plant Cell Walls." Chapman & Hall, London.
Roberts, K. R., Sluiman, H. J., Stewart, K. D. and Mattox, K. R. (1980). Comparative cytology and taxonomy of the Ulvaphyceae. II. Ulvalean characteristics of the stephanokont flagellar apparatus of *Derbesia tenuissima*. *Protoplasma* **104**, 223–238.
Roberts, K. R., Sluiman, H. J., Stewart, K. D. and Mattox, K. R. (1981). Comparative cytology and taxonomy of the Ulvaphyceae. III. The flagellar apparatus of the anisogametes of *Derbesia tenuissima* (Chlorophyta). *J. Phycol.* **17**, 330–340.
Roberts, K. R., Stewart, K. D. and Mattox, K. R. (1982). Structure of the anisogametes of the green siphon *Pseudobryopsis* sp. (Chlorophyta). *J. Phycol.* **18**, 498–508.
Schussnig, B. (1928). Die Reduktionsteilung bei *Cladophora glomerata*. *Österr. bot. Z.* **77**, 62–67.
Schussnig, B. (1938). Der Kernphasenwechsel von *Valonia utricularis* (Roth) Ag. *Planta* **28**, 43–59.
Schussnig, B. (1954). Gonidiogenese, Gametogenese und Meiose bei *Cladophora glomerata*. *Arch. Protistenkd.* **100**, 287–332.
Scott, J. H. and Bullock, K. W. (1976). Ultrastructure of cell division in *Cladophora*. Pregametangial cell division in the haploid generation of *Cladophora flexuosa*. *Can. J. Bot.* **54**, 1546–1560.
Silva, P. C. (1980). Names of classes and families of living algae. *Regnum veg.* **103**, 1–156.
Sluiman, H. J., Mattox, K. R. and Stewart, K. D. (1980a). Moderne opvattingen over de fylogenie van groenwieren en landplanten. *Vakbl. Biol.* **60**, 204–212.
Sluiman, H. J., Roberts, K. R., Stewart, K. D. and Mattox, K. R. (1980b). Comparative cytology and taxonomy of the Ulvaphyceae. I. The zoospore of *Ulothrix zonata* (Chlorophyta). *J. Phycol.* **16**, 537–545.
Smith, G. M. (1955). "Cryptogamic Botany," 2nd ed., Vol. I, McGraw-Hill, New York.
Stebbins, G. L. and Hill, G. J. (1980). Did multicellular plants invade the land? *Am. Nat.* **115**, 342–353.
Stewart, K. D. and Mattox, K. R. (1975). Comparative cytology, evolution and classification of the green algae with some consideration of the origin of other organisms with chlorophyll a and b. *Bot. Rev.* **41**, 104–135.
Stewart, K. D. and Mattox, K. R. (1978). Structure and evolution in the flagellate cells of green algae and land plants. *BioSystems* **10**, 145–152.
Tanner, C. E. (1981). Chlorophyta: Life histories. *In* "The Biology of Seaweeds" (C. S. Lobban and M. J. Wynne, eds), pp. 218–247. Blackwell, Oxford.
Vinogradova, K. L., Gollerbach, M. M., Zayer, L. M. and Sdobnikova, N. V. (1980). "Opredel'itel' presnovodnych vodoroslej SSSR 13. Zel'enye vodorosli-Chlorophyta; Krasnye vodorosli-Rhodophyta; Burye vodorosli-Phaeophyta." Nauka, Leningrad.
Wik-Sjöstedt, A. (1970). Cytogenetic investigations in *Cladophora*. *Hereditas* **66**, 233–262.
Wik-Sjöstedt, A. and Nordquist, T. (1970). Preliminary observations on the ultrastructure of some *Cladophora* and *Spongomorpha* species. *Bot. mar.* **13**, 6–12.

6 | Current Ideas on Classification of the Ulotrichales Borzi

G. M. LOKHORST

Rijksherbarium, Leiden, The Netherlands

Abstract: Since Kützing's introduction of the family Ulothricheae in 1843 the delimitation of this group as a taxonomic unit has continuously been a subject for discussion. In an historical review, the most important classifications proposed by various authorities will be discussed. In the last century, before the establishment of the ordinal name Ulotrichales by Borzi in 1895, confusion concerning the circumscription of this order resulted from the meagre generic and species descriptions. Moreover, distinction of ulotrichalean genera was mainly based on variable morphological traits. In the first half of the present century, a gradual involvement of life history characteristics apparently led to a better understanding of the systematics of the Ulotrichales. Modern ultrastructural studies, however, have demonstrated the highly heterogeneous nature of the traditional Ulotrichales when classified mainly on the basis of morphological features. In the present review the taxonomic significance of traditional and modern features will be evaluated. Typical ulotrichalean patterns will be discussed concerning the basic architecture of the individual vegetative cells, the mechanism of mitosis–cytokinesis and aspects of the flagellar apparatus of motile reproductive cells. It is concluded that features like the type of life history and the organizational level of the thallus are too variable within the Ulotrichales (as conceived here) to serve as highly useful taxonomic traits at ordinal level.

HISTORICAL REVIEW

The history of the Ulotrichales begins with the establishment by Kützing (1843) of the family Ulothricheae. Apart from the genus *Ulothrix* (with three freshwater species) only the newly-established *Stigeoclonium* (with five species) was included in this family. In Kützing's *Species Algarum* (1849), the number of genera comprising the Ulothricheae was increased to nine. It is curious to note that, in addition to the classification of the genera *Microtham*-

---

Systematics Association Special Volume No. 27, "Systematics of the Green Algae", edited by D. E. G. Irvine and D. M. John, 1984, pp. 179–206. Academic Press, London and Orlando.
ISBN 0 12 374040 1

Copyright © by the Systematics Association
All rights of reproduction in any form reserved

nion, Schizomeris and *Draparnaldia* in the Ulothricheae, the red algae *Bangia* and *Goniotrichum* were included as well. This classification illustrates that nineteenth century ulotrichalean treatments were essentially based on purely morphological traits, leading to the recognition of very unnatural groups. Acceptance of the Ulothricheae *sensu* Kützing was not widespread. For example, Harvey (1860) maintained the name Confervaceae to encompass the green algal genera *Cladophora, Chaetomorpha* and *Oedogonium,* in addition to the ulotrichalean genera *Ulothrix, Hormotrichum, Chaetophora,* and *Draparnaldia.* In the last decades of the past century, however, a more definite delimitation of the Ulothricheae gradually crystallised, as can be noticed from the descriptions given by Rabenhorst (1863, 1868), Hansgirg (1886), and De Toni (1889). Cellular features like the position and shape of the chloroplast started to play a more prominent diagnostic role. Red algae were no longer encountered in ulotrichalean classification systems. Borzi (1895) also referred to the shape of the chloroplast when he introduced the new name Ulotrichales. Together with blue-green algae, branched green filamentous algae were classified in the order Chaetophoroideae by Greville as early as 1824. In later years, this name seemed to be restricted to just *Chaetophora* and branched red algae (Kützing, 1843, 1849). Harvey (1860) and Rabenhorst (1863) ignored the name Chaetophoroideae. A few years later, however, Rabenhorst (1868) separated the unbranched forms as Ulotrichaceae, while the branched genera like *Stigeoclonium, Draparnaldia* and *Chaetophora* were assigned as a clear-cut group to the Chaetophoraceae, from a modern viewpoint an advanced step. After 1868 the classification of branched and unbranched filamentous algae underwent all kinds of variations, i.e. phycologists treated them separately or merged them together at the family or ordinal level. For example, Kirchner (1878) treated them as the Ulotrichinae and the Chaetophoreae (as a subgroup of the Cladophorinae) in the family Confervaceae. Hansgirg (1886) recognized the Chaetophoraceae with a subdivision into the subfamilies Ulotricheae and Chaetophoreae, respectively. Related groups like the Coleochaetaceae and Ulvaceae were ranked at the family level by Hansgirg.

Hauck (1885) and Reinbold (1891) avoided a questionable separation between branched and unbranched marine ulotrichalean algae. These authors used the name Confervaceae to comprise both unbranched genera like *Chaetomorpha, Ulothrix* and *Rhizoclonium,* and branched genera like *Cladophora, Entocladia* and *Bulbocoleon.* By keeping the families Ulvaceae and Confervaceae separate on the basis of growth habit, Hauck and Reinbold apparently considered the affinity between branched and unbranched forms closer than between (un-)branched and foliose ones.

By recommending the use of the name Chaetophorales instead of Confervales at class level, Wille's (1890) concepts of the Chaetophorales became very wide. His inclusion of the oogamous Oedogoniaceae makes the Chaetophorales *sensu* Wille an unnatural, heterogeneous assemblage. West (1904) essentially followed Wille's classification but he removed the Oedogoniaceae from the Chaetophorales.

In his standard work, Heering (1914) retained the name Ulotrichales. Apart from branched and unbranched filamentous plants, he also assigned thalloid plants to this order as the Ulvaceae and the Blastosporaceae (*Prasiola*). Heering especially emphasised the significance of different types of zoospores for a better understanding of the phylogeny and taxonomy of the Ulotrichales. In this respect he referred to Pascher (1914), who divided the Ulotrichales into the Tetrakontae and Dikontae. From a phylogenetic point of view, the quadriflagellate macrozoospore was considered to be closest to the hypothetical flagellate ancestor of this group. The occurrence of three types of reproductive cells found in *Ulothrix zonata* tempted Heering to speculate on evolutionary lines starting from this alga and leading, amongst others, to the Cylindrocapsaceae.

Fritsch (in West and Fritsch, 1927) stressed the heterotrichous organizational level in the Chaetophorales. As a consequence of this, he conceived the Ulotrichales as embracing unbranched ulotrichalean genera only. On the other hand, he enlarged the Ulotrichales in an unnatural way by the inclusion of the Cladophorales and Sphaeropleales. Fritsch stated that one might disagree with the merger of the Cladophorales with the Ulotrichales but there is, however, a close resemblance in reproductive methods in many genera of both groups which favours his classification system.

Pascher (1931) followed Fritsch's separation of the Ulotrichales and Chaetophorales, but both groups were assigned to the subclass Ulotrichinae, which, together with the Microsporinae and Oedogoniinae, constitute the class Ulotrichineae. Pascher's classification of the ulotrichalean algae is unnatural. He apparently did not know how to judge the taxonomic value of the un-ulotrichalean feature of oogamy in the Oedogoniinae, particularly in view of his treatment of conjugalean algae (which show an equally un-ulotrichalean method of reproduction) as a division (the Conjugatae).

In contrast to the last-mentioned authors, Papenfuss (1955) reduced the Chaetophorales (together with the Ulvales, Microsporales and Cylindrocapsales) to subgroups in a broadly conceived order Ulotrichales. He argued that in some genera of the Chaetophorales either the typical erect or the prostrate system may be absent or poorly developed. Moreover, he regarded heterotrichy as a process which evolved independently in a number of different groups which include simple algae as well as advanced types. Therefore Papenfuss found it arbitrary to use this character for the delimitation of major groups. Consequently, he rejected the autonomous status of the order Ulvales since the presence of a parenchymous growth habit in the Ulvales does not distinguish them essentially from other ulotrichalean algae, such as, for example, the heterotrichous *Fritschiella*. Furthermore, since the typical ulvacean alternation of generations is also found in *Draparnaldiopsis* (Singh, 1945), Papenfuss considered this type of life history no longer a bar to inclusion in the Ulotrichales. The Ulotrichales in the sense of Fott (1971) comprise five suborders including the sarcinoid Chlorosarcinineae, which, in his opinion, probably constitute the transitional stage between coccalean and filamentous algae. Hindak (1962) was one of the first to consider genera that show a chlorococcalean thallus but have vegetative cell division and not the formation of autospores like *Elakatothrix, Diplosphaera, Raphidonema* and *Chlorosarcina* as primitive Ulotrichales.

On the basis of a close similarity in chromosome morphology and numbers as found in *Ulothrix, Caespitella, Stigeoclonium* and *Draparnaldia* (Sarma, 1958, 1963; Abbas and Godward, 1964), Godward (1966) favoured inclusion of the Chaetophorales with the Ulotrichales.

Kornmann (1965) conceived the Ulotrichales as a small order, characterised by a common diplohaplontic life history with heteromorphic alternation of generations and a similar ontogenetic development of the germinating zoospore. The Ulotrichales *sensu* Kornmann include the Ulotrichaceae, Codiolaceae (*Urospora*) and Monostromaceae. However, Kornmann's treatment of the Ulotrichales ignores the taxonomic problems that arise if one tries to classify the remaining heterogeneous assemblage of ulotrichalean algae, which predominantly occur in freshwater habitats. In 1973, Kornmann elevated the *Codiolum*-producing algae even to class level, thereby restricting the Ulotrichales to the genus *Ulothrix* only, a unique event.

Round (1963) followed Fritsch (1935) in assuming that the difference between unbranched and heterotrichous thalli is sufficiently fundamental to warrant a segregation of the Ulotrichales and Chaetophorales. On the other hand, he criticised Fritsch for not distinguishing between simple thalli and thalloid expanses, which would have led to a separation of Ulvales and Ulotrichales.

Ramanathan (1964) finished his comprehensive historical account of the Ulotrichales with a conveniently arranged enumeration of the features used for delimiting and subdividing this order. Ramanathan used the type of organizational level of the thallus as the main argument for characterising the Ulotrichales, Chaetophorales and Ulvales. Strangely enough, he regarded the Sphaeropleineae as a suborder of the Ulotrichales. In his elaboration of Heering's (1914) flora, Printz (1964) merged nearly all the traditional ulotrichalean genera into a broadly conceived order the Chaetophorales, exclusive of groups like *Sphaeroplea, Pleurococcus* and *Chlorosarcina*. It is especially to Printz's merit that he gave additional information on the distribution of the species in natural habitats as well as on their phytogeography.

In my opinion, Christensen's (1966) treatment of the Ulotrichales is unsatisfactory because of the inclusion of the multinucleate families Codiolaceae and Acrosiphoniaceae. Proposed as a practical treatment, Bourrelly (1966) divided the Chlorophyta into four classes, one of which constitutes the Ulotrichophycées. Bourrelly was criticised by Round (1971) for the artificial manner in which siphonous algae and the Oedogoniales were included in his class Ulotrichophycées.

Considering submicroscopical data pertaining to the mechanism of mitosis and cytokinesis and the distribution of plasmodesmata, Stewart *et al.* (1973) redivided the Chaetophorales *sensu* Printz into four orders, of which the one constituting the Ulotrichales included *Ulva* and *Klebsormidium*. Stewart *et al.* (1973) were the first to postulate that fundamental differences in ultrastructure clearly demonstrate that exclusive use of the type of growth habit to delimit orders and families in the green algae has led to unnatural juxtapositions. In an epoch-making paper, Stewart and Mattox (1975), mainly based on comparative ultrastructural and biochemical studies, provisionally reclassified the green algae into two classes, and later (1978) into three classes. As a consequence of this, traditional ulotrichalean algae like *Klebsormidium* were assigned to the class Charophyceae, *Ulothrix zonata* to the Ulvophyceae and *Uronema* to the Chlorophyceae. From now on, recognition of the traditional Ulotrichales as a distinct taxonomic entity was completely abandoned.

Hoek (1981) also argued for a subdivision of the Chlorophyta into the classes Chlorophyceae and Charophyceae. Despite the fact that the architecture of the flagellar apparatus of motile reproductive cells, the type of mitosis and cytokinesis and the type of life history have recently become important for the taxonomic understandings of the Chlorophyta, Hoek still adhered to the organizational level. He recognised the Klebsormidiales and the Coleochaetales apart from the Ulotrichales (including the Chlorosarcinales and Chaetophorales) as autonomous orders in the Chlorophyceae.

Silva (1982) essentially followed Stewart and Mattox's system and made an extensive correlation of recently obtained ultrastructural data with certain traditional characteristics; Silva stressed that the ulotrichine line is polyphyletic, having evolved more than once. He erected the new order Ctenocladales to accommodate branched algae like *Pseudendoclonium, Trichosarcina* and *Gongrosira* that exhibit ulvophycean ultrastructural features. The Ulotrichales *sensu* Silva are restricted to those green algae showing an unbranched growth habit, and which are composed of uninucleate cells containing a parietal laminate or cylindrical chloroplast with variously shaped pyrenoids. Ultrastructural features include the collapse of the interzonal spindle at telophase, absence of plasmodesmata, and cytokinesis that involves a precocious cleavage not associated with a phycoplast. The life history is described as diplobiontic with a heteromorphic alternation of generations.

It can be concluded that the taxonomy of the Ulotrichales developed explosively certainly during the last decennium due to the utilisation of electron microscopy in comparative taxonomic studies. In the following section most of the important diagnostic criteria will be evaluated in an attempt to delimit the Ulotrichales *anno* 1983.

DIAGNOSIS OF THE ULOTRICHALES

Thalli sarcinoid, uniseriate (un-)branched, multiseriate or expanded to form monostromatic blades or hollow tubes or distromatic sheets.

Cells fundamentally cylindrical in shape, uninucleate in vegetative stage, possessing one parietal chloroplast, laminate, (in-)completely girdle-shaped or cup-like often with projecting strands.

Chloroplast with one to several pyrenoids surrounded by a sheath of starch grains.

Cell wall continuous, sometimes lamellated.

Plasmodesmata absent.

Attachment by simple or rhizoid-like basal cells.

At mitosis, chromosome movement to nuclear poles mainly by elongation of interzonal spindle microtubules which are persistent at telophase.

Cytokinesis by ingrowth of the plasmalemma, parental wall involved with the formation of daughter cell walls (desmoschisis).

Asexual reproduction by bi/quadriflagellate swarmers or by aplanospores, arising in ordinary cells.

Sexual reproduction by biflagellate gametes, fusion iso/anisogamous.

Life history varying from haplobiontic to isomorphic or heteromorphic diplobiontic.

Flagella apically inserted, flagellar apparatus cruciate.

Basal bodies showing 180° rotational symmetry, arranged relative to each other in a 11/5 o'clock configuration.

Distal connecting fibre not striated, rhizoplasts present.

Root formula x-2-x-2 (x=4).

Examples include: *Friedmannia* (Chlorosarcinales *sensu* Groover and Bold, 1969); *Ulothrix* (Lokhorst, 1974, 1978); *Ulva, Monostroma, Ulvaria, Kornmannia* and *Enteromorpha* (Bliding, 1963, 1968); *Trichosarcina* (Nichols and

Bold, 1965); *Pseudendoclonium, Gongrosira, Entocladia* and *Endophyton* (Ctenocladales *sensu* Silva, 1982).

Excluded are the Microsporales, Sphaeropleales, Cylindrocapsales, Prasiolales, Klebsormidiales, and the chlorophycean phyletic line in the Chaetophorales, Acrosiphoniales and Cladophorales.

THE SIGNIFICANCE OF TRADITIONAL AND MODERN CHARACTERISTICS FOR THE DELIMITATION OF THE ULOTRICHALES

1. General Architecture of the Vegetative Cell

The presence of one parietal chloroplast containing one or more pyrenoids and one nucleus per cell are considered as fundamental characters for the Ulotrichales, as conceived here. Thus, the multinucleate orders Acrosiphoniales and Cladophorales are excluded. Considering the diverse types of advancements in green algae, it is emphasized that a multiplication of the number of nuclei means an important step towards a detailed refinement of the whole metabolism in an individual cell. In this respect, the uninucleate feature is regarded as primitive and the multinucleate as derived, and hence these characteristics are taxonomically meaningful. However, the use of the uninucleate character of the individual cells may raise difficulties and, in occurring cases, demand for compromises. For example, Berger-Perrot (1980) found three marine uninucleate species having features of both *Ulothrix* and *Urospora*. Without challenging the main principles of the traditional Acrosiphoniales, Berger-Perrot preferred to classify these intermediate species in the genus *Urospora* (Acrosiphoniales) based on the posteriorly-pointed zoospores and the micro-anatomy of the pyrenoid.

The multinucleate *Sphaeroplea,* previously often conceived as a member of the Ulotrichales (e.g. Ramanathan, 1964) has to be removed to a separate order, the Sphaeropleales. Apart from showing advanced anisogamy or oogamy and a peculiar cellular organisation including multiple chloroplast rings per cell, ultrastructural studies on *Sphaeroplea* (Cáceres and Robinson, 1980, 1981) argue for the recognition of a separate order Sphaeropleales too, despite similarities in mitosis, cytokinesis and thallus organization with the Chlorococcales.

The genera *Cylindrocapsa* and *Prasiola,* previously often recognised as ulotrichalean algae (e.g. Fott, 1971; Bold and Wynne, 1978) are excluded from the Ulotrichales as conceived here. Both genera possess an axillary chloroplast which often appears stellate. Therefore, in view of their oogamous mode of sexual reproduction and fundamentally different thallus organization, Silva (1982) is prepared to elevate them to autonomous higher taxa, the Cylindrocapsaceae and Prasiolales, respectively. The Ulotrichales are here restricted exclusively to forms that possess one chloroplast per cell (see Fig. 2D–F). It extends as a plate with lobed margins (*Gongrosira*, Tupa, 1974; *Pseudendoclonium*, Yarish, 1975), or is shaped as an (in-)complete girdle (*Ulothrix*, Lokhorst, 1974, 1978) or a cup (*Friedmannia*, Chantanachat and Bold, 1962). The chloroplast morphology itself, however, cannot be considered as a constant diagnostic feature since (adverse) external conditions usually immediately affect its morphology without changing its parietal position (e.g., *Ulothrix*, Lokhorst, 1974, 1978).

The ulotrichalean chloroplast is provided with one or more pyrenoids, which are completely embedded in it and are normally surrounded by a cap of starch grains. In both *Ulothrix* (Lokhorst and Star, 1980) and *Monostroma* (Hori, 1973) the micro-anatomy of the pyrenoid shows a remarkable variability and hence can be used as an additional species characteristic. Hori's findings of different pyrenoid types varying with the alternation of generations in *Monostroma nutidum* and *M. angicava* is surprising, and in my opinion needs yet to be confirmed. In general, following cleavage and subsequent formation of zoospores or gametes, each is equipped with the same basic type of pyrenoid as shown by the parental generation. Compare in this respect the pyrenoid of the *Ulothrix zonata* zoospore (Sluiman *et al.*, 1980b) with that of the interphase cell (Lokhorst and Star, 1980) or the different stages in *Ulva* (Bråten and Løvlie, 1968; Bråten, 1971). Pyrenoids are not always so variable in micro-anatomy. Mattox and Stewart (1974) already noted an almost identical pyrenoid type in the genera *Trichosarcina*, *Pseudendoclonium*, *Ulva*, *Enteromorpha* and *Percursaria*. Our findings (G. M. Lokhorst and W. Star, unpublished) on *Pseudendoclonium printzii*, *P. prostratum* and *P. akinetum* revealed the genus *Pseudendoclonium* as being a clear-cut group in having one basic type of pyrenoid showing a core bisected by one strand of chloroplast thylakoids.

The Microsporales are excluded from the Ulotrichales firstly, because of the presence of a basically reticulate chloroplast which, according to the author's experience, can be broken up into segments, and secondly, by the lack of pyrenoids. Moreover, the ultrastructure of the *Microspora* cell wall (Pickett-Heaps, 1973, 1975) is so essentially different from that of the Ulotrichales that the recognition of a separate order, the Microsporales, seems to be inevitable. The cell wall development in the Ulotrichales does not result in segments appearing like H-pieces in optical section, but in a continuous layer, which may become conspicuously lamellated in older growth stages (e.g. *Ulothrix zonata*, Lokhorst, 1974).

2. The Organizational Level of the Thallus

In the past much taxonomic value has been attributed to the organizational level of the growth habit, which, depending on the author's preference, has led to a broad or restricted view on the classification of the Ulotrichales (see Historical Review).

By merging the Ulvales with some Chaetophorales, Ctenocladales and Chlorosarcinales into the Ulotrichales as conceived here, the latter group comprises sarcinoid forms (*Friedmannia*, Chantanachat and Bold, 1962), unbranched filaments (*Ulothrix*, Lokhorst, 1974, 1978), heterotrichous plants (*Pseudendoclonium*, Tupa, 1974), pluriseriate forms (*Trichosarcina*, Nichols and Bold, 1965), monostromatic blades (*Monostroma*, Bliding, 1968), tube-like habits (*Enteromorpha*, Bliding, 1963) and distromatic blades (*Ulva*, Bliding, 1968). Thus, the significance traditionally attributed to the growth habit has been reduced here for defining the Ulotrichales. Not only at ordinal rank but also at lower ranks this criterion cannot be taken too strictly. In the genus *Ulothrix*, for example, variation in growth habit has been detected. In the freshwater *Ulothrix zonata* and in the estuarine *U. implexa* false and true branching have been found (Lokhorst, 1974). In the estuarine *U. palusalsa* biseriate cell rows may occur and its palmelloid stage is essentially similar to the growth habit of the rather gelatinous *Monostroma bullosum;* compare in this respect Fig. 14f in Lokhorst (1978) and Fig. 5 in Kornmann (1965). Filaments of the marine *Ulothrix flacca* are often coalescent, giving the impression of a multiseriate thallus. In *U. implexa* apical cells may behave like the basal cell, and upon making contact with hard substrates they are able to differentiate into basal cell like rhizoids, after which the filament may assume a loop-shaped habit. A comparable diversity in growth habit is also demonstrated in, for example, *Enteromorpha* (Koeman and Hoek, 1982), including the unbranched *Enteromorpha torta* and the profusely branched *E. radiata* and *E. ahlneriana*.

Nevertheless, early variations in the ontogenetic pattern of the zoospore and its germling, resulting ultimately in a monostromatic layer in the *Monostroma* complex or a distromatic layer in the *Ulva* complex, have taxonomic significance. Hence, in connection with reproductive data, the *Monostroma* complex is broken up into several genera (*Monostroma sensu stricto*, Bliding, 1968; *Gayralia* and *Protomonostroma*, Vinogradova, 1969, 1974, and Gayral, 1971; *Kornmannia*, Bliding, 1968; *Ulvaria*, Bliding, 1963). The *Ulva* complex is divided into *Ulva sensu stricto* (Bliding, 1968; Koeman and Hoek, 1981) and *Chloropelta* (Tanner, 1980). The latter would show the first major departure of a distromatic ulvaceous alga from the ontogenetic type of development exemplified by *Ulva*.

With respect to the type of cytokinesis, desmoschisis or eleutheroschisis (terminology after Groover and Bold, 1969), there has been discussion on the taxonomic boundary between the Ulotrichales and Chlorococcales. Desmoschisis includes vegetative cell division showing a significant role of the parental wall in the formation of daughter cell walls. This type of cell division seems to be characteristic of filamentous and parenchymous algae and higher plants. In contrast, the true Chlorococcales only have the ability to undergo eleutheroschisis, resulting in the production of naked protoplasts (zoospores), or of autospores which become surrounded by completely new walls that are not intimately associated with the parental cell wall. Strangely enough, Fott (1959, 1971) considered cell division in the Ulotrichales and higher organisms as the formation of autospores. However, in Fott's definition of vegetative cell division, the role of the parental cell wall is not as much stressed as is the persistence of the arisen transverse cell wall. He referred to *Cylindrocapsa* as a model for the ulotrichalean growth pattern. However, because of the method of wall formation by daughter cells and certain ultrastructural features pertaining to mitosis and cytokinesis, it is clear that *Cylindrocapsa* is not ulotrichalean but instead has to be considered as a filamentous chlorococcalean alga (H. J. Sluiman, unpublished).

I support Silva's (1982) contention that *Binuclearia* and certain species of *Radiofilum* probably also have to be considered as filamentous members of the Chlorococcales. The same applies to *Ulothrix mucosa* (Lokhorst, 1974), which shows another peculiar phenomenon: when the filaments have become reproductive hardly any difference can be detected between vegetative and spore-producing cells (especially those producing only one zoospore per cell). The only difference is the appearance of a conspicuous stigma. In my opinion this is a phenomenon similar to that found in autospore formation. On the other hand, when evaluating Pickett-Heaps' ultrastructural findings on the chlorococcalean alga *Kirchneriella,* Hoek and Jahns (1978) postulated that no essential difference exists in the cell division mechanism leading to the formation of either autospores or vegetative daughter cells. Hence, in their opinion, the taxonomic boundary between the orders Chlorococcales and Ulotrichales becomes artificial. In support of the ideas of Tschermak-Woess (1982), it is preferred here to use the formation of autospores, occurring without the involvement of the parental wall, as the basis for keeping the Chlorococcales apart from the ulotrichalean members of the Chlorosarcinales (e.g. *Friedmannia*).

Our comparative studies (G. M. Lokhorst, E. J. P. Delbecque, and J. J. M. van Ingen, unpublished) on the growth habit of the freshwater ulotrichalean *Monostroma bullosum* and the estuarine tetrasporalean *Pseudotetraspora marina* have indicated that both species share a spectrum of corresponding features like the arrangement of the cells in groups of two to four separated by a gelatinous wall substance, a sac-like thallus anchored by a prostrate disc, a laminate parietal chloroplast and a similar type of pyrenoid. However, *Monostroma bullosum* consists of a monostromatic sheet (Fig. 1B), while in *Pseudotetraspora marina* the cells are arranged in three-dimensional packets (Fig. 1A). The question raised here is whether the latter can be considered a departure stage leading the Tetrasporales towards the Ulotrichales. On the other hand, there is the question concerning the manner of formation of vegetative cells in the blades of *Monostroma bullosum*. Does it follow the pattern of autospore formation? Careful ultrastructural examination will certainly provide an answer. Thus, at present it seems that the taxonomic boundaries between the Tetrasporales and Chlorococcales on the one hand and the Ulotrichales on the other are not so distinct as often suggested. The separation of some branched genera (*Pseudendoclonium, Gongrosira, Leptosira, Entocladia, Endophyton*) from the Chaetophorales and their inclusion in the present order Ulotrichales may seem to be artificial. However, an explanation will follow after the mechanism of mitosis and cell division is discussed.

The shape of the basal and apical cells is of little diagnostic value for the characterisation of the Ulotrichales. This feature can only be used at the species level (e.g. *Ulothrix*, Lokhorst, 1974; *Ulva*, Koeman and Hoek, 1981).

Bold and Wynne (1978) and Silva (1982) claimed the absence of rhizoids as a diagnostic trait for the Monostromataceae, although *Monostroma bullosum* (G. M. Lokhorst and E. J. P. Delbecque, unpublished) may show a tendency to produce rhizoidal outgrowths protruding from the periphery of the prostrate attaching disc. Tupa (1974) considered the presence and the relative size of the prostrate and erect system of major importance in the classification of the heterotrichous algae that are treated as Ulotrichales in the present paper (*Pseudendoclonium, Gongrosira,* etc.).

3. Reproduction and Life History

In their comprehensive reviews, both Hoek (1981) and Tanner (1981) are of the opinion that the kind of alternation of generations in the life history deserves to be appreciated at the rank of order: The Ulotrichales (including *Monostroma*) are primarily heteromorphic and the Ulvales are primarily isomorphic. The argument for assigning the ulvalean algae to the present Ulotrichales is based on the fact that in the course of evolution various (partly overlapping) life history types were apparently realised in fundamentally similar thalloid growth habits. Therefore, preference is given to common conservative features like the basic architecture of the individual

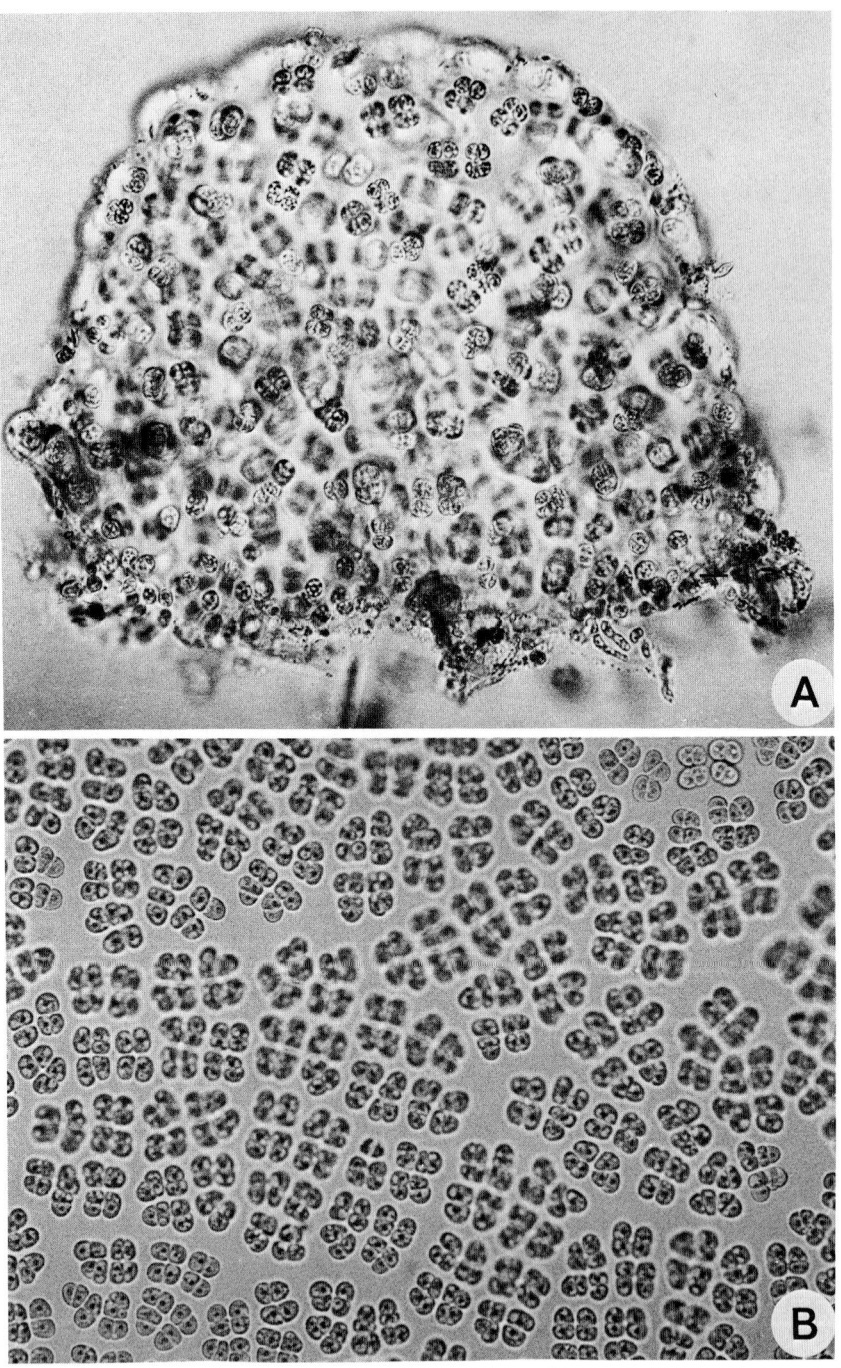

Fig. 1. (A) *Pseudotetraspora marina*; juvenile sac-like thallus with cells arranged in three-dimensional packets (× 300). (B) *Monostroma bullosum*, part of a blade with cells arranged in a monostromatic layer (× 300).

vegetative cell, the mechanism of mitosis–cytokinesis and the morphological organization of flagellate motile cells.

It is postulated here that different combinations of features, including the type of life history, the ontogenetic development of the thallus and its shape at maturity, evolved from the original thalloid gene pool. For example, in *Ulvaria oxysperma*, belonging to the isomorphic diplohaplontic category (Ulvales in the sense of Tanner, 1981), Kapraun and Flynn (1973) also observed a haplobiontic life history reproducing exclusively by bi- and quadriflagellate swarmers. Moreover, when this genus is kept in the traditional Ulvales, it takes up an artificial position caused by its monostromatic growth habit. I agree with Tatewaki (1972) in retaining this entity in *Monostroma*, if desired as a subgenus. Tanner (1981) reviewed the overwhelming number of detailed studies of *Ulva mutabilis*, and also referred to strains showing the ability of parthenogenetic gametes to develop into parthenosporophytes as well as into gametophytes of the same mating type. In addition, the findings of Hoxmark (1975) on the developmental behaviour of bi- and quadriflagellate swarmers arising from parthenosporophytes also point to the possible existence of another variation of the basic isomorphic diplobiontic life history in *Ulva mutabilis*. In *Enteromorpha*, both sexual and asexual life histories have been observed (Kapraun, 1970).

Corresponding variations in types of life histories have been shown in *Ulothrix*. Perrot (1968, 1970, 1972) reported an alternation of isomorphic filamentous generations in strains collected from the French coast. On the contrary, our studies have demonstrated the existence of heteromorphic life histories (Lokhorst, 1974, 1978). In the predominantly estuarine *Ulothrix implexa* the rate of production of asexual spores differed. In clones originated from purely freshwater habitats asexual reproduction was slow and zoospores could hardly be induced. If sexual reproduction occurred, it was always laborious and was only observed in clones originating from plants collected in estuarine or marine habitats.

In the heterotrichous category, asexual reproduction by quadriflagellate zoospores is recorded for *Pseudendoclonium* (Yarish, 1975) and *Gongrosira* (Tupa, 1974), respectively. O'Kelly and Yarish (1981) and O'Kelly (1982) provided evidence for the occurrence of an isomorphic diplohaplontic life history in *Entocladia* and *Endophyton*. Although a complete set of ultrastructural data for these two genera is not yet available, the development and structure of gametangia and sporangia (O'Kelly, 1982) along with the ultrastructure of vegetative cells (Floyd and Yarish, 1978) favour their close affinity with the thalloid genera, here included in the Ulotrichales.

Returning to the whole spectrum of light microscopical features covering the thalloid genera (*Ulva*, *Enteromorpha*, *Ulvaria*, *Chloropelta*, *Gayralia*, *Kornmannia*, *Protomonostroma*, *Capsosiphon* and *Blidingia*) it seems that whatever taxonomic lines will be drawn in order to recognise clear-cut groups, they are almost always artificial. For example, the uniseriate growth habit of juvenile plants producing rhizoids may favour the recognition of a taxonomic category encompassing *Ulva, Protomonostroma, Gayralia, Ulvaria, Enteromorpha* and *Chloropelta,* and the prostrate disc in juvenile plants supports the grouping of *Monostroma, Kornmannia* and *Blidingia* as a second assemblage. By advocating such a classification, one disregards the type of organizational level of the mature thallus, since *Protomonostroma, Gayralia* and *Ulvaria* are essentially monostromatic at maturity and hence resemble *Monostroma sensu stricto* and *Kornmannia* in this respect. The pattern of ontogenetic development of the zoospore, leading to variously-shaped juvenile stages in thalloid algae, is rejected here as the main diagnostic feature for making separations at the level of the family as advocated by Silva (1982). Our findings on the germination pattern of the zoospore of the marine *Ulothrix flacca* (Lokhorst, 1978) support this view. The germination of its zoospore can be bipolar, whereby the attached hyaline part develops into the basal cell and the upper part into a uniseriate row of cells (Fig. 2A) or, after one cell division, the two daughter cells independently give rise to uniseriate filaments which often coalesce as twin filaments (Fig. 2B) or, the attached *U. flacca* zoospore develops into a pseudoparenchymous plate which, after repeated cell divisions, produces erect uniseriate filaments of indeterminate length (Fig. 2C).

In the case of thalloid genera, each incipient germination of the zoospore results in a few-celled uniseriate filament. In the subsequent ontogenetic pattern, the longitudinal cell division may commence early, leading to a prostrate disc (*Monostroma,* Fig. 4 in Kornmann, 1965) or, when this longitudinal cell division is delayed, a more advanced growth stage of an uniseriate filament arises (*Ulva,* Fig. 9 in Koeman and Hoek, 1981). Thus, in my opinion there is virtually no fundamental difference in thallus development in the thalloid forms. In addition, the findings of small monostromatic plants in cultures of the normally distromatic *Ulva lactuca* (Bonneau, 1978) and *Ulva fasciata* (Tanner, 1979) once more illustrates the enormous morphological variability which exists in the thalloid algae.

Despite the absence of one strict type of life history in the Ulotrichales as conceived here, there seems to exist a striking similarity in the developmental pattern of reproductive cells. Firstly, it occurs in ordinary cells (Fig. 2H); however, the shape of these cells may change when utilised as reproductive cells (e.g. *Entocladia;* O'Kelly and Yarish, 1980). Secondly, gametes and zoospores supposedly originate as a result of sequential cleavage of the chloroplast (e.g. Lind, 1932, Nordby, 1974; O'Kelly and Yarish, 1980). Another criterion for the characterisation of the Ulotrichales is provided by the architecture of reproductive cells and the type of sexual fusion. Motile

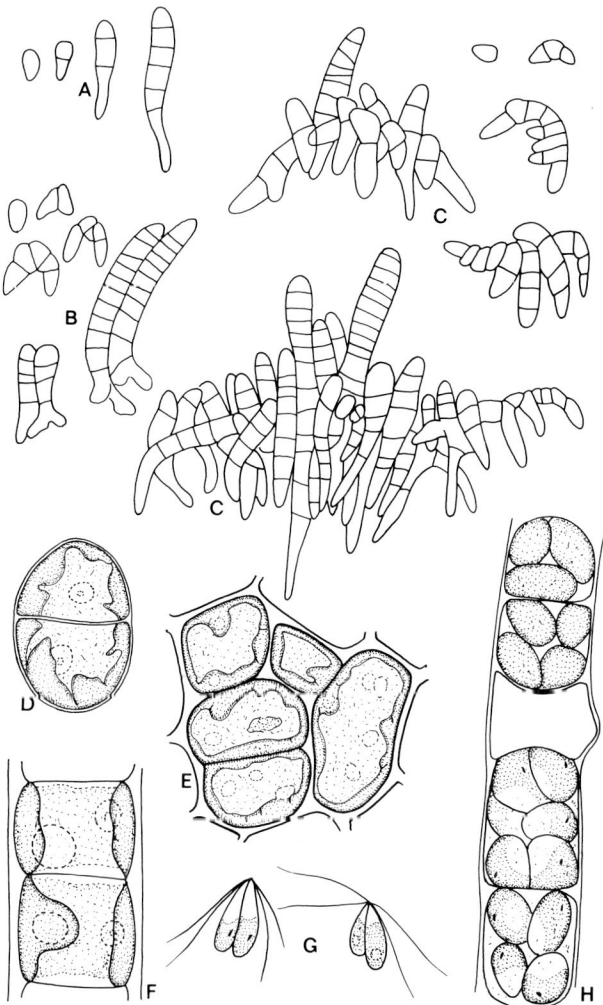

Fig. 2. (A–C) *Ulothrix flacca*. (A) bipolar germination of the zoospore; (B) germination of the zoospore, giving rise to twin filaments; (C) germination of the zoospore, giving rise to a pseudoparenchymous prostrate portion and an erect portion of indeterminate length. (D–F) Examples of the typical ulotrichalean cellular organization: (D) *Friedmannia* (after Chantanachat and Bold, 1962); (F) *Ulothrix*, and (E) *Ulva* (after Bliding, 1968). (G) Iso- and anisogamous fusion of gametes. (H) Zoosporogenesis occurring in ordinary cells in *Ulothrix*.

cells always have apically inserted flagella. Asexual cells are normally quadriflagellate, although often also biflagellate, but in that case it is not always clear whether they have to be regarded as true asexual spores or as gametes which behave as parthenospores in the absence of the dioecious counterpart. Sexual cells are biflagellate and fusion is iso- to anisogamous (Fig. 2G).

4. *The Mechanism of Mitosis and Cytokinesis*

Hoek (1981) reviewed four different types of mechanism for mitosis–cytokinesis that might occur among the green algae. On the basis of our findings on *Urospora* (Lokhorst and Star, 1983), we support Hoek's contention that recognition of only four division types is an oversimplification because of the existence of intermediate types.

Examples quoted by Hoek (1981) demonstrate the heterogeneous character of the Ulotrichales as conceived in the past. *Ulothrix zonata* (Stewart *et al.*, 1973) conforms to his type I pattern while *U. fimbriata* (Floyd *et al.*, 1972a), *Uronema confervicolum* and *U. belkae* (Stewart *et al.*, 1973) show a mitosis–cytokinesis pattern similar to his type III.

Our present knowledge of the ultrastructure of the mitosis–cytokinesis mechanism in the Ulotrichales is still fragmentary, despite the availability of data on *Ulva mutabilis* (Løvlie and Bråten, 1968, 1970), *Trichosarcina polymorpha* and *Pseudoclonium basiliense* (Mattox and Stewart, 1974), an ulotrichalean alga named *Cylindrocapsa involuta* (Pickett-Heaps and McDonald, 1975; to judge from the micrographs, an *Ulothrix* species?), *Enteromorpha intestinalis* (McArthur and Moss, 1978), *Ulothrix zonata* and *U. tenuissima* (Sluiman *et al.*, 1983), and *U. implexa* and *U. verrucosa* (unpublished). Nevertheless, despite these limited number of studies, the following typical mitosis–cytokinesis model can be described for the Ulotrichales (see also Sluiman *et al.*, 1983).

1. Late *interphase* cell (just prior to nuclear division); two centrioles lie in association with a network of dictyosomes in between the centrally-located nucleus and the cell wall from which the chloroplast has been withdrawn (Fig. 9 in Mattox and Stewart, 1974).

2. *Pre-prophase;* the cleavage furrow appears as an incipient ingrowth of the plasmalemma. It usually encircles the whole circumference of the cell, hence it is also involved with the now-commencing constriction of the chloroplast.

3. *Prophase;* the centrioles duplicate and both sets move away from each other to take a distinct lateral position near the spindle poles.

4. *Metaphase;* the spindle is closed except for polar fenestrations and the chloroplast has not yet divided. The poles of the nucleus are flattened (Fig. 8 in Pickett-Heaps and McDonald, 1975).

5. *Anaphase* (Fig. 4A); the spindle elongates considerably by expansion of the interzonal spindle region (Løvlie and Bråten, 1970), but a simultaneous shortening of the chromosome-to-pole spindle may occur as well (Mattox and Stewart, 1974).

6. Early *telophase* cell (Fig. 4B); the daughter nuclei are kept apart widely by the persistent interzonal spindle (not so described for *Trichosarcina* and *Pseudendoclonium* by Mattox and Stewart, 1974, but my interpretation of their published micrographs). Later on, the ingrowing furrow bisects the spindle. As a result the daughter nuclei may come back together. Before cytokinesis is completed the centrioles migrate back to their lateral position. The chloroplast cleaves completely and the formation of the new cross wall is by the coalescence of vesicles (Sluiman et al., 1983), or the precocious cleavage apparently pushes its way between the daughter nuclei (Fig. 3A).

All cited authors are unanimous in their conclusion that no phycoplast microtubules are involved in cytokinesis, despite Pickett-Heaps and McDonald (1975) observing some microtubules near the ingrowing furrow, oriented in the plane of cleavage. The latter authors assumed that they are of little consequence for the formation of a new cross wall. Moreover, they stated that a phycoplast system must be formed before cytokinesis is underway. However, this prerequisite is not always possible in relation to the internal organization of the cell.

Recently, Lokhorst and Star (1983) have described a microtubular system involved in cytokinesis in the multinucleate *Urospora*, showing a mitosis–cytokinesis mechanism basically identical to that described here for the Ulotrichales. The term phycoplast has been emended, because in *Urospora* the presence of a huge vacuole prevents the manufacture of a phycoplast, preceding cytokinesis. Yet, I doubt whether microtubules are entirely excluded from any involvement in cytokinesis in ulotrichalean algae. Certainly, their presence is not as manifest as in cells showing cytokinesis by cell plate formation (e.g. *Draparnaldia*, G. M. Lokhorst and M. E. Bakker, unpublished). On the other hand, careful examination of several *Ulothrix* species (e.g. *U. implexa* in Fig. 3B, *U. verrucosa* in Fig. 3E) revealed an inconspicuous microtubular system presumably associated with cytokinesis. Because of the internal organization of the cell and the persistent character of the interzonal spindle, there is apparently no need for the cell to produce a well-developed phycoplast system. One might speculate as to whether the Ulotrichales produce a phycoplast system that has gradually been reduced after the departure of the ulotrichalean line from a possible chlorococcalean ancestor. This would explain why in the sarcinoid (primitive ulotrichalean alga) *Friedmannia israelensis* a rather peculiar (intermediate?) phycoplast system has been detected (Deason et al., 1979). Plasmodesmata do not occur in the cell walls of the Ulotrichales. On the other hand, they are invariably

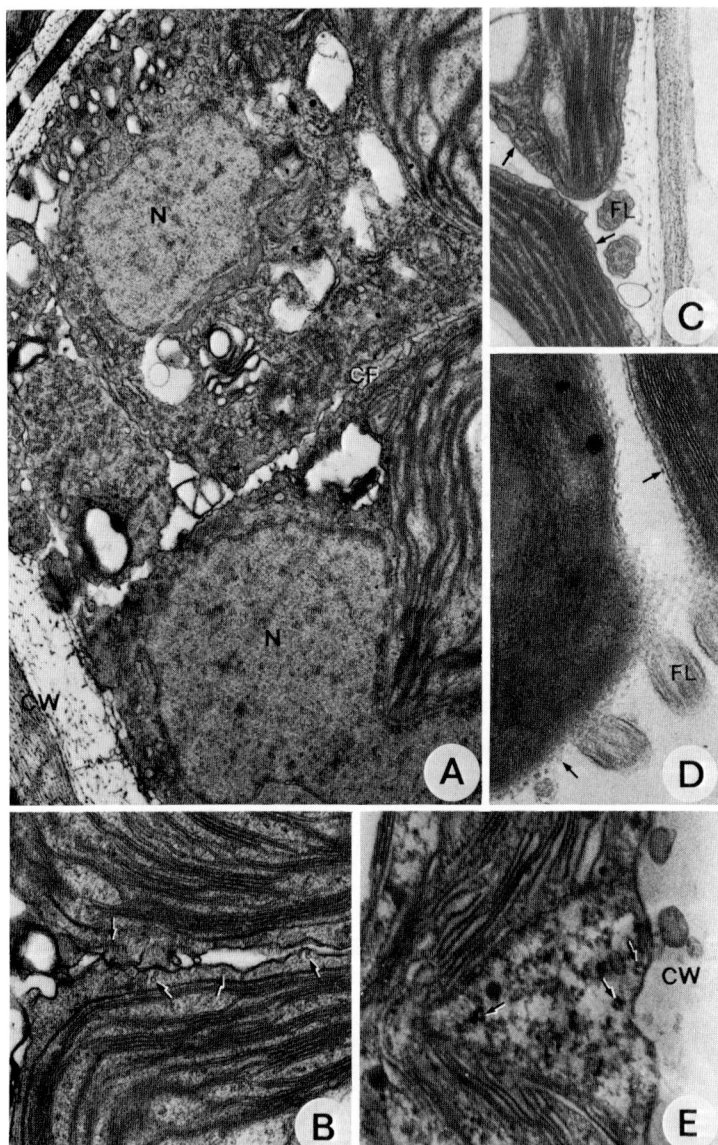

Fig. 3. (A–C) *Ulothrix implexa:* (A) late telophase cell with ingrowing furrow and separated nuclei (× 15 100); (B) detail of an ingrowing furrow with surrounding microtubules (*arrows*) (× 35 300); (C) detail of zoospores in an almost mature zoosporangium, showing (*arrows*) their naked cell body (× 25 200). (D) *Monostroma bullosum;* detail of the peripheral region of the zoospore showing its covering of scales (*arrows*) (× 28 500). (E) *Ulothrix verrucosa;* incipient cleavage of the cytoplasm through the chloroplast showing (*arrows*) microtubules which precede the furrow (× 40 300). Abbreviations: CF, cleavage furrow; CW, cell wall; FL, flagellum; N, nucleus.

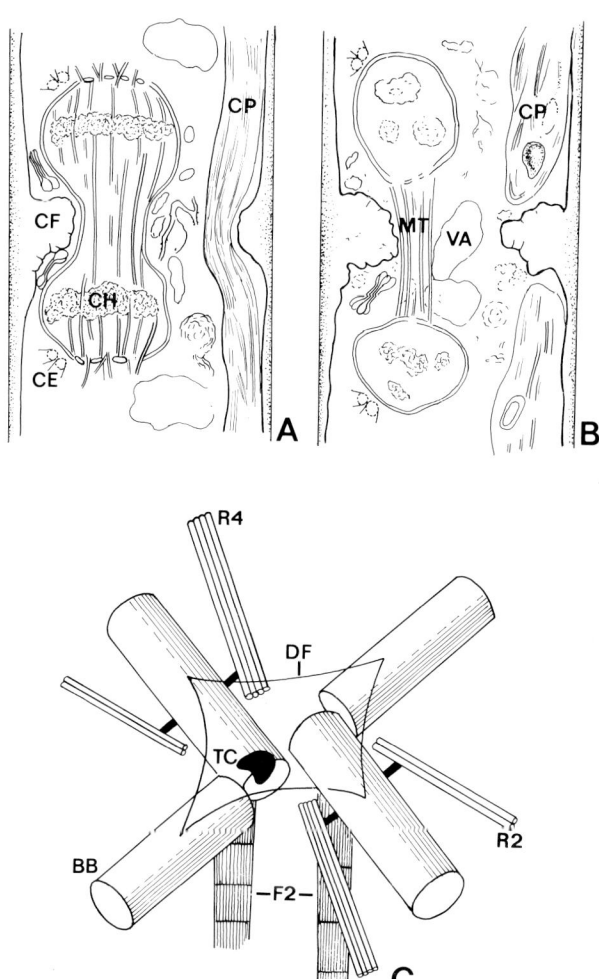

Fig. 4. (A–C) Schematic representation of the typical ulotrichalean anaphase cell (A), telophase cell (B), and absolute configuration of the flagellar apparatus of quadriflagellate swarmers showing most of its components (C). Abbreviations: BB, basal body; CE, centriole; CF, cleavage furrow; CH, chromosome; CP, chloroplast; DF, distal connecting fibre; F2, rhizoplast; MT, microtubule; R2, R4, two- and four-membered microtubular root, respectively; TC, terminal cap; VA, vacuole.

present in the filamentous algae showing cytokinesis by cell plate formation. Apart from a number of other diagnostic features (furrow versus cell plate, persistent telophase spindle versus non-persistent, etc.), the absence of plasmodesmata in the Ulotrichales is primarily used to separate the Ulotrichales from the Chaetophorales. I agree with the supposition of Stewart *et al.* (1973) that due to the absence of plasmodesmata the organizational level of the thallus remained relatively simple in the Ulotrichales, despite the presence of relatively large thalli in *Ulva*. The heterotrichous *Ctenocladus* has been kept apart from the present Ulotrichales, because of the presence of coarse pit-like intercellular connections, hence showing a close relationship to the Trentepohliales (Blinn and Morrison, 1974).

The position of the Klebsormidiales remains controversial. In recent classification systems, this group has been separated from the traditional ulotrichalean algae and assigned to the Charophyceae *sensu* Mattox and Stewart. Silva (1982) stated that the Klebsormidiales and the charophycean phyletic line are similar in having a persistent interzonal spindle, and a mode of cytokinesis that is effectuated by the centripetal growth of a furrow without a phycoplast. Thus, this is virtually the same mechanism as is suggested here as being typical for the Ulotrichales. Unpublished observations on the events of mitosis–cytokinesis in *Ulothrix verrucosa* strengthen its earlier supposed (Lokhorst and Star, 1980) affinity to *Klebsormidium*. From the above-mentioned features, *U. verrucosa* has in common with *Klebsormidium* the polar position of the centrioles at prophase, the early division of the chloroplast, and the complete disintegration of the nuclear envelope at metaphase. The active role attributed to vacuoles in the separation of daughter nuclei at prophase (Floyd *et al.*, 1972b) is not restricted to *Klebsormidium*. In both *Ulothrix verrucosa* and *Trichosarcina polymorpha* (Mattox and Stewart, 1974) an identical involvement has been observed. Based on its reduced phycoplast system and its typical ulotrichalean cellular organization, *U. verrucosa* is for now retained in the Ulotrichales. A detailed electron microscopical inspection of the whole *Klebsormidium/Ulothrix* complex is needed before a final decision can be made on the taxonomic position of species. In this special case it is very unfortunate that *U. verrucosa* has apparently lost its ability to produce swarmers. The structure of its flagellar apparatus would certainly help to clarify the taxonomic boundary between the Ulotrichales and Klebsormidiales. On the other hand, it is hard to imagine that the cells of *U. verrucosa* (with a maximum diameter of *ca.* 20 μm) would be able to produce one giant *Klebsormidium*-like zoospore per cell.

5. The Structure of the Flagellated Reproductive Cells

Within the present Ulotrichales, the type of flagellar apparatus of the motile reproductive cells is exclusively of the cruciate type. As a consequence, the traditionally ulotrichalean *Klebsormidium,* producing swarmers possessing a unilateral flagellar apparatus, is removed from the Ulotrichales despite its rather similar mechanism of mitosis–cytokinesis (see previous section).

In the ulotrichalean swarmers, the flagella are limited in number (2 or 4), apically inserted, naked or only covered by a thin surface coat (e.g. *Monostroma grevillei,* Moestrup, 1982). In most species a so-called hair-point is reported (e.g. *Enteromorpha intestinalis,* McArthur and Moss, 1979).

Complete critical ultrastructural studies on the flagellar apparatus of swarmers (mostly pertaining to zoospores) have been made on relatively few members of the Ulotrichales: *Enteromorpha linza* (Melkonian, 1980a), *Ulothrix zonata* (Sluiman *et al.,* 1980b), *Ulva lactuca* [Melkonian, 1979; (gamete) 1980b], *Ulvaria oxysperma* (Hoops *et al.,* 1982), and *Friedmannia israelensis* (Melkonian and Berns, 1983). Incomplete reports have come from studies on swarmers of *Blidingia minima* (Swanson and Floyd, 1978), *Enteromorpha intestinalis* [Evans and Christie, 1970; (gamete) McArthur and Moss, 1979], *Pseudendoclonium basiliense* and *Trichosarcina polymorpha* (Mattox and Stewart, 1973), and *Monostroma grevillei* (Chesnoy and Jónsson, 1973; Jónsson and Chesnoy, 1974; Moestrup, 1978). The salient characteristics of most of the above-mentioned motile cells are summarised in Table I in Hoops *et al.* (1982). Despite this fragmentary knowledge, it is possible to describe the basic structure of the flagellar apparatus typical of ulotrichalean swarmers (Fig. 4C).

In both gametes and zoospores, the arrangement of the basal bodies and associated components displays a 180° rotational symmetry (terminology of Floyd *et al.,* 1980). There is evidence that the basal bodies are arranged in a so-called 11/5 o'clock configuration (see discussion in Melkonian and Berns, 1983). In zoospores, one basal body pair is directly opposite to one another. Both in gametes and zoospores, the remaining pair of basal bodies are not exactly opposite to one another, especially in gametes where they overlap (slightly) at their bases (McArthur and Moss, 1979; Hoops *et al.,* 1982). The proximal ends of the basal bodies are (partially) covered by an electron-dense terminal cap (Sluiman *et al.,* 1980b; Melkonian, 1979). The basal bodies are distally connected by the distal connecting fibre that consists of two arching halves which are linked midway between opposite basal bodies (Melkonian, 1979, 1980b; Sluiman *et al.,* 1980b). This distal fibre has no striations at all. This non-striated characteristic would apply to the whole ulvophycean class (Sluiman *et al.,* 1980a). As argued earlier (Bakker and Lokhorst, 1984), however, the taxonomic significance of the shape and striation pattern of the distal connecting fibre is overestimated at class level. Recent observations on the gametes of the ulvophycean *Pseudobryopsis* (Roberts *et al.,* 1982) support this postulation, in my opinion.

As in the chaetophoralean algae (e.g. *Draparnaldia,* Bakker and Lokhorst,

1984), proximal fibres seem to link adjacent basal bodies though information is scarce (Hoops *et al.*, 1982). However, Melkonian (1980a) reported that the distal connecting fibre is the only basal body-linking fibre in *Enteromorpha* and *Ulva*.

In the ulotrichalean swarmers, the flagellar root system consists of two structural components, the root microtubules and an associated striated component. The microtubular roots show an alternating pattern which can be described by the typical ulotrichalean root formula x-2-x-2 ($x = 4$), despite the reported occurrence of slight variations in *Friedmannia* ($x = 4$ or 5; Melkonian and Berns, 1983) and in *Ulva lactuca* ($x = 3$, Micalef and Gayral, 1972; $x = 4$, Melkonian, 1979). The two-membered root (R2) is closely associated with a striated component (SMAC after Floyd *et al.*, 1980, or system I fibre after Melkonian, 1980a). There does not seem to exist a uniform striation pattern in this associated fibre; compare, for example, *Ulothrix zonata* (Sluiman *et al.*, 1980b) and *Ulvaria oxysperma* (Hoops *et al.*, 1982).

In the Chlorophyta, a striated component, associated with the four-membered root (R4), called the rhizoplast or system II fibre, seems to occur only in the Ulvophyceae (Sluiman *et al.*, 1980a). Hence, this feature can be used to separate the ulvophycean Ulotrichales from the Chaetophorales belonging to the Chlorophyceae *sensu* Stewart and Mattox (1975). The periodicity of the rhizoplast is very different from that of the system I fibre. Melkonian (1980a) reported that a different fixation method may result in variously spaced rhizoplast striations. The position of the rhizoplast and the number per swarmer may vary somewhat from species to species (see Hoops *et al.*, 1982).

Scales on the body of motile swarmers in ulotrichalean algae may be present (e.g. *Monostroma bullosum*, Fig. 3D; *Ulothrix zonata*, Sluiman *et al.*, 1980b) or absent (e.g. *Ulothrix implexa*, Fig. 3C), and hence this criterion cannot be used for the delimitation of the order.

The isolated position of the (transitional?) genus *Friedmannia* in the Ulotrichales as conceived here is evidenced by its lack of a system I fibre, whilst its rhizoplast has to be considered as strongly reduced (Melkonian and Berns, 1983). In view of the proposed typical architecture of the ulotrichalean flagellar apparatus, *Chlorosarcinopsis*, like *Friedmannia* a member of the Chlorosarcinales *sensu* Groover and Bold (1969), has to be referred to the Chlorophyceae since Melkonian (1978) found *Chlorosarcinopsis* zoospores to be very similar to *Chlamydomonas*. In addition, Rogers *et al.* (1980) demonstrated that the genus *Chlorokybus*, also previously placed in the Chlorosarcinales, has to be assigned to the Charophyceae. At present, it is obvious that the Chlorosarcinales *sensu* Groover and Bold comprise an assemblage containing genera that depart from the chlorococcalean ancestors towards the main phyletic lines *sensu* Stewart and Mattox (1978) in the Chlorophyta.

ACKNOWLEDGEMENTS

The author is much indebted to Professor Dr. C. Kalkman and Mr. H. J. Sluiman for critically reading the manuscript, to Mr. W. Star for his great enthusiasm and expert technical assistance in electron microscopy, and to Mr. J. H. van Os and Mr. B. N. Kieft for preparing the figures and photographs for publication.

REFERENCES

Abbas, A. and Godward, M. B. E. (1964). Cytology in relation to taxonomy in Chaetophorales. *J. Linn. Soc. London* **58**, 499–507.

Bakker, M. E. and Lokhorst, G. M. (1984). Ultrastructure of *Draparnaldia glomerata* (Vauch.) Agardh (Chaetophorales; Chlorophyceae) I. The flagellar apparatus of the zoospore. *Nord. J. Bot.* **4**, 261–273.

Berger-Perrot, Y. (1980). Trois nouvelles espèces d'*Urospora* à cellules uninuclées sur les côtes de Bretagne. *Cryptogam.: Algol.* **1**, 141–160.

Bliding, C. (1963). A critical survey of European taxa in Ulvales. Part I. *Capsosiphon, Percursaria, Blidingia, Enteromorpha. Opera bot.* **8**, 1–160.

Bliding, C. (1968). A critical survey of European taxa in Ulvales. II. *Ulva, Ulvaria, Monostroma, Kornmannia. Bot. Not.* **121**, 535–629.

Blinn, D. W. and Morrison, E. (1974). Intercellular cytoplasmic connections in *Ctenocladus circinnatus* Borzi (Chlorophyceae) with possible ecological significance. *Phycologia* **13**, 95–97.

Bold, H. C. and Wynne, M. J. (1978). "Introduction to the Algae, Structure and Reproduction." Prentice-Hall, Englewood Cliffs, New Jersey.

Bonneau, E. R. (1978). Asexual reproduction capabilities in *Ulva lactuca* L. (Chlorophyceae). *Botanica mar.* **21**, 117–121.

Borzi, A. (1895). "Studi Algologici," Fasc. II. Reber, Palermo.

Bourrelly, P. (1966). "Les Algues d'eau douce. Initiation à la Systématique. Tome I. Les Algues Vertes." Boubée, Paris.

Bråten, T. (1971). The ultrastructure of fertilization and zygote formation in the green alga *Ulva mutabilis* Føyn. *J. Cell Sci.* **9**, 621–635.

Bråten, T. and Løvlie, A. (1968). On the ultrastructure of vegetative and sporulating cells of the multicellular green alga *Ulva mutabilis* Føyn. *Nytt Mag. Bot.* **15**, 209–219.

Cáceres, E. J. and Robinson, D. G. (1980). Ultrastructural studies on *Sphaeroplea annulina* (Chlorophyceae). Vegetative structure and mitosis. *J. Phycol.* **16**, 313–320.

Cáceres, E. J. and Robinson, D. G. (1981). Ultrastructural studies on *Sphaeroplea annulina* (Chlorophyceae). II. Spermatogenesis and male gamete structure. *J. Phycol.* **17**, 173–180.

Chantanachat, S. and Bold, H. C. (1962). Phycological studies. II. Some algae from arid soils. *Univ. Tex. Publ.* **6218**.

Chesnoy, L. and Jónsson, S. (1973). Étude ultrastructurale du développement du zygote calcicole d'une Chlorophycée marine, le *Monostroma grevillei* (Thuret) Wittrock. *C. r. hebd. Seanc. Acad. Sci., Ser. D* **276**, 299–302.

Christensen, T. (1966). Alger. In "Botanik" (T. W. Böcher, M. Lange, and T. Sørensen, eds), Vol. 2. No. 2, pp. 1–178. Munksgaard, Copenhagen.

Deason, T. R., Ryals, P. E., O'Kelley, J. C. and Bullock, K. W. (1979). Fine structure of mitosis and cleavage in *Friedmannia israelensis* (Chlorophyceae, Chlorosarcinaceae). *J. Phycol.* **15**, 452–457.

De Toni, G. B. (1889). "Sylloge algarum omnium hucusque cognitarum. I. Sylloge Chlorophycearum . . ." Sumptibus auctorius, Padua.

Evans, L. V. and Christie, A. O. (1970). Studies on the ship-fouling alga *Enteromorpha*. I. Aspects of fine structure and biochemistry of swimming and newly settled zoospores. *Ann. Bot. (London)* [N.S.], **34**, 451–466.

Floyd, G. L. and Yarish, C. (1978). Comparative ultrastructure of vegetative cells of selected marine Chaetophoraceae (Chlorophyta). *J. Phycol.* **14**, Suppl., 32.

Floyd, G. L., Stewart, K. D. and Mattox, K. R. (1972a). Comparative cytology of *Ulothrix* and *Stigeoclonium*. *J. Phycol.* **8**, 68–81.

Floyd, G. L., Stewart, K. D. and Mattox, K. R. (1972b). Cellular organization, mitosis and cytokinesis in the ulotrichalean alga, *Klebsormidium*. *J. Phycol.* **8**, 176–184.

Floyd, G. L., Hoops, H. J. and Swanson, J. A. (1980). Fine structure of the zoospore of *Ulothrix belkae* with emphasis on the flagellar apparatus. *Protoplasma* **104**, 17–31.

Fott, B. (1959). "Algenkunde." Fischer, Jena.

Fott, B. (1971) "Algenkunde," 2nd ed. Fischer, Stuttgart.

Fritsch, F. E. (1935). "The Structure and Reproduction of the Algae," Vol. I. Cambridge Univ. Press, London and New York.

Gayral, P. (1971). Mise au point sur la systématique de l'ordre des *Ulvales*. *Bull. Soc. phycol. Fr.* **16**, 63–67.

Godward, M. B. E. (1966). "The Chromosomes of the Algae." Edward Arnold, London.

Greville, R. K. (1824). "Flora Edinensis." Blackwood, Edinburgh.

Groover, R. D., and Bold, H. C. (1969). Phycological studies. VIII. The taxonomy and comparative physiology of the *Chlorosarcinales* and certain other edaphic algae. *Univ. Tex. Publ.* **6907**, 1–165.

Hansgirg, A. (1886). Prodromus der Algenflora von Böhmen. I. *Arch. naturw. LandesDurchforsch. Boehmen* **5**(6), 1–288.

Harvey, W. H. (1860). "Index Generum Algarum." Van Voorst, London.

Hauck, F. (1885). Die Meeresalgen Deutschlands und Oesterreichs. In "Kryptogamen-Flora von Deutschland, Oesterreich und der Schweiz," (L. Rabenhorst, ed.), Vol. II, pp. 1–575. Akad. Verlagsges., Leipzig.

Heering, W. (1914). Chlorophyceae. III. Ulothrichales, Microsporales, Oedogoniales. In "Die Süsswasser-Flora Deutschlands, Österreichs und der Schweiz" (A. Pascher, ed.), Part 6, pp. 2–250. Fischer, Jena.

Hindak, F. (1962). Beitrag zur Phylogenese und Systematik der Ulothrichales. *Biologia (Bratislava)* **17**, 641–649.

Hoek, C. van den (1981). Chlorophyta: Morphology and classification. In "The Biology of Seaweeds" (C. S. Lobban and M. J. Wynne, eds), pp. 86–132. Blackwell, Oxford.

Hoek, C. van den, and Jahns, H. M. (1978). "Algen, Einführung in die Phykologie." Thieme, Stuttgart.

Hoops, H. J., Floyd, G. L. and Swanson, J. A. (1982). Ultrastructure of the biflagellate motile cells of *Ulvaria oxysperma* (Kütz.). Bliding and phylogenetic relationships among ulvaphycean algae. *Am. J. Bot.* **69**, 150–159.

Hori, T. (1973). Comparative studies of pyrenoid ultrastructure in algae of the *Monostroma*-complex. *J. Phycol.* **9**, 190–199.

Hoxmark, R. C. (1975). Experimental analysis of the life cycle of *Ulva mutabilis*. *Botanica mar.* **18**, 123–129.
Jónsson, S. and Chesnoy, L. (1974). Étude ultrastructurale de l'incorporation des axonèmes flagellaires dans les zygotes du *Monostroma grevillei* (Thuret) Wittr., Chlorophycée marine. *C. r. hebd. Seanc. Acad. Sci., Ser. D* **278**, 1557–1560.
Kapraun, D. F. (1970). Field and cultural studies of *Ulva* and *Enteromorpha* in the vicinity of Port Aransas, Texas. *Contrib. mar. Sci.* **15**, 205–285.
Kapraun, D. F. and Flynn, E. H. (1973). Culture studies of *Enteromorpha linza* (L.) J. Ag. and *Ulvaria oxysperma* (Kützing) Bliding (Chlorophyceae, Ulvales) from Central America. *Phycologia* **12**, 145–152.
Kirchner, O. (1878). Algen. In "Kryptogamen-Flora von Schlesien" (F. Cohn, ed.), Vol. 2, pp. 1–284. Kern, Breslau.
Koeman, R. P. T. and Hoek, C. van den (1981). The taxonomy of *Ulva* (Chlorophyceae) in the Netherlands. *Br. phycol. J.* **16**, 9–53.
Koeman, R. P. T. and Hoek, C. van den (1982). The taxonomy of *Enteromorpha* Link, 1820, (Chlorophyceae) in the Netherlands. II. The section Proliferae. *Cryptogam.: Algol.* **3**, 37–70.
Kornmann, P. (1965). Ontogenie und Lebenszyklus der Ulotrichales in phylogenetischer Sicht. *Phycologia* **4**, 163–172.
Kornmann, P. (1973). Codiolophyceae, a new class of Chlorophyta. *Helgol. Wiss. Meeresunters.* **25**, 1–13.
Kützing, F. T. (1843). "Phycologia generalis . . ." Brockhaus, Leipzig.
Kützing, F. T. (1849). "Species algarum." Brockhaus, Leipzig.
Lind, E. M. (1932). A contribution to the life-history and cytology of two species of *Ulothrix*. *Ann. Bot. (London)* **46**, 711–725.
Lokhorst, G. M. (1974). Taxonomic studies on the freshwater species of *Ulothrix* in The Netherlands. Thesis, Free University, Amsterdam.
Lokhorst, G. M. (1978). Taxonomic studies on the marine and brackish-water species of *Ulothrix* (Ulotrichales, Chlorophyceae) in Western Europe. *Blumea* **24**, 191–299.
Lokhorst, G. M. and Star, W. (1980). Pyrenoid ultrastructure in *Ulothrix* (Chlorophyceae). *Acta bot. neerl.* **29**, 1–15.
Lokhorst, G. M. and Star, W. (1983). Fine structure of mitosis and cytokinesis in *Urospora* (Acrosiphoniales, Chlorophyta). *Protoplasma* **117**, 142–153.
Løvlie, A. and Bråten, T. (1968). On the division of cytoplasm and chloroplast in the multicellular green alga *Ulva mutabilis* Føyn. *Exp. Cell Res.* **51**, 211–220.
Løvlie, A. and Bråten, T. (1970). On mitosis in the multicellular alga *Ulva mutabilis* Føyn. *J. Cell Sci.* **6**, 109–129.
McArthur, D. M. and Moss, B. L. (1978). Ultrastructural studies of vegetative cells, mitosis and cell division in *Enteromorpha intestinalis* (L.) Link. *Br. phycol. J.* **13**, 255–267.
McArthur, D. M. and Moss, B. L. (1979). Gametogenesis and gamete structure of *Enteromorpha intestinalis* (L.) Link. *Br. phycol. J.* **14**, 43–57.
Mattox, K. R. and Stewart, K. D. (1973). Observations on the zoospores of *Pseudendoclonium basiliense* and *Trichosarcina polymorpha* (Chlorophyceae). *Can. J. Bot.* **51**, 1425–1430.
Mattox, K. R. and Stewart, K. D. (1974). A comparative study of cell division in *Trichosarcina polymorpha* and *Pseudendoclonium basiliense* (Chlorophyceae). *J. Phycol.* **10**, 447–456.
Melkonian, M. (1978). Structure and significance of cruciate flagellar root systems in green algae: Comparative investigations in species of *Chlorosarcinopsis* (Chlorosarcinales). *Plant Syst. Evol.* **130**, 265–292.
Melkonian, M. (1979). Structure and significance of cruciate flagellar root systems in green algae: Zoospores of *Ulva lactuca* (Ulvales, Chlorophyceae). *Helgol. wiss. Meeresunters.* **32**, 425–435.

Melkonian, M. (1980a). Ultrastructural aspects of basal body associated fibrous structures in green algae: A critical review. *BioSystems* **12**, 85–104.
Melkonian, M. (1980b). Flagellar roots, mating structure and gamete fusion in the green alga *Ulva lactuca* (Ulvales). *J. Cell Sci.* **46**, 149–169.
Melkonian, M. and Berns, B. (1983). Zoospore ultrastructure in the green alga *Friedmannia israelensis:* An absolute configuration analysis. *Protoplasma* **114**, 67–84.
Micalef, H. and Gayral, P. (1972). Quelques aspects de l'infrastructure des cellules végétatives et des cellules reproductrices d'*Ulva lactuca* L. (Chlorophycées). *J. Microsc. (Paris)* **13**, 417–428.
Moestrup, Ø. (1978). On the phylogenetic validity of the flagellar apparatus in green algae and other chlorophyll a and b containing plants. *BioSystems* **10**, 117–144.
Moestrup, Ø. (1982). Flagellar structure in algae: A review, with new observations particularly on the Chrysophyceae, Phaeophyceae (Fucophyceae), Euglenophyceae, and *Reckertia. Phycologia* **21**, 427–528.
Nichols, H. W. and Bold, H. C. (1965). *Trichosarcina polymorpha* gen. et sp. nov. *J. Phycol.* **1**, 34–38.
Nordby, Ø. (1974). Light microscopy of meiotic zoosporogenesis and mitotic gametogenesis in *Ulva mutabilis* Føyn. *J. Cell Sci.* **15**, 443–445.
O'Kelly, C. J. (1982). Observations on marine Chaetophoraceae (Chlorophyta). III. The structure, reproduction and life history of *Endophyton ramosum. Phycologia* **21**, 247–257.
O'Kelly, C. J. and Yarish, C. (1980). Observations on marine Chaetophoraceae (Chlorophyta). I. Sporangial ontogeny in the type species *Entocladia* and *Phaeophila. J. Phycol.* **17**, 549–558.
O'Kelly, C. J. and Yarish, C. (1981). Observations on marine Chaetophoraceae (Chlorophyta). II. On the circumscription of the genus *Entocladia* Reinke. *Phycologia* **20**, 32–45.
Papenfuss, G. F. (1955). Classification of the algae. *In* "A Century of Progress in the Natural Sciences—1853–1953," pp. 115–224. Calif. Acad. Sci., San Francisco, California.
Pascher, A. (1914). Ueber Flagellaten und Algen. *Ber. Dtsch. bot. Ges.* **32**, 136–160.
Pascher, A. (1931). Systematische Übersicht über die mit Flagellaten in Zusammenhang stehenden Algenreihen und Versuch einer Einreihung dieser Algenstämme in die Stämme des Pflanzenreiches. *Beih. bot. Zentralbl., Abt.* 2, **48**, 317–332.
Perrot, Y. (1968). Sur le cycle de deux formes d'*Ulothrix flacca* (Dillw.) Thuret de la région de Roscoff. *C. r. hebd. Seanc. Acad. Sci., Ser. D,* **266**, 1953–1955.
Perrot, Y. (1970). Sur la spécificité et le cycle de l'*Ulothrix subflaccida* Wille des côtes françaises. *C. r. hebd. Seanc. Acad. Sci., Ser. D* **270**, 932–933.
Perrot, Y. (1972). Les *Ulothrix* marins de Roscoff et le problème de leur cycle de reproduction. *Mem. Soc. bot. Fr.* pp. 67–74.
Pickett-Heaps, J. D. (1973). Cell division and wall structure in *Microspora. New Phytol.* **72**, 347–355.
Pickett-Heaps, J. D. (1975). "Green Algae, Structure, Reproduction and Evolution in Selected Genera." Sinauer Assoc., Sunderland, Massachusetts.
Pickett-Heaps, J. D. and McDonald, K. L. (1975). *Cylindrocapsa:* Cell division and phylogenetic affinities. *New Phytol.* **74**, 235–241.
Printz, H. (1964). Die Chaetophoralen der Binnengewässer... *Hydrobiologia* **24**, 1–376.
Rabenhorst, L. (1863). "Kryptogamen-Flora von Sachsen, der Ober-Lausitz, Thüringen und Nordböhmen mit Berücksichtiging der benachbarten Länder." Vol. I. Kummer, Leipzig.
Rabenhorst, L. (1868). "Flora europaea algarum aquae dulcis et submarinae." Vol. III. Kummer, Leipzig.
Ramanathan, K. R. (1964). "Ulotrichales." Singh, New Delhi.

Reinbold, T. (1891). Die Chlorophyceen (Grüntange) der Kieler Föhrde. *Schr. naturwiss. Ver. Schleswig-Holstein* **8,** 109–144.
Roberts, K. R., Stewart, K. D. and Mattox, K. R. (1982). Structure of the anisogametes of the green siphon *Pseudobryopsis* sp. (Chlorophyta). *J. Phycol.* **18,** 498–508.
Rogers, C. E., Mattox, K. R. and Stewart, K. D. (1980). The zoospore of *Chlorokybus atmophyticus,* a charophyte with sarcinoid growth habit. *Am. J. Bot.* **67,** 774–783.
Round, F. E. (1963). The taxonomy of the Chlorophyta. *Br. Phycol. Bull.* **2,** 224–235.
Round, F. E. (1971). The taxonomy of the Chlorophyta. II. *Br. Phycol. J.* **6,** 235–264.
Sarma, Y. S. R. K. (1958). Chromosome number in Ulotrichales and allied groups. *Br. phycol. Bull.* **1,** 22–24.
Sarma, Y. S. R. K. (1963). Contributions to the karyology of the Ulotrichales. I. *Ulothrix. Phycologia* **2,** 173–183.
Silva, P. C. (1982). Chlorophycota. In "Synopsis and Classification of Living Organisms" (S. P. Parker, ed.), Vol. 1, pp. 133–162. McGraw-Hill, New York.
Singh, R. N. (1945). Nuclear phases and alternation of generations in *Draparnaldiopsis indica* Bharadwaja. *New Phytol.* **44,** 118–129.
Sluiman, H. J., Stewart, K. D. and Mattox, K. R. (1980a). Moderne opvattingen over de fylogenie van groenwieren en landplanten. *Vakbl. Biol.* **60,** 204–212.
Sluiman, H. J., Roberts, K. R., Stewart, K. D. and Mattox, K. R. (1980b). Comparative cytology and taxonomy of the Ulvaphyceae. I. The zoospore of *Ulothrix zonata* (Chlorophyta). *J. Phycol.* **16,** 537–545.
Sluiman, H. J., Roberts, K. R., Stewart, K. D. and Mattox, K. R. (1983). Comparative cytology and taxonomy of the Ulvophyceae. IV. Mitosis and cytokinesis in *Ulothrix* (Chlorophyta). *Acta bot. neerl.* **32,** 257–269.
Stewart, K. D., and Mattox, K. R. (1975). Comparative cytology, evolution and classification of the green algae with some consideration of the origin of other organisms with chlorophyll a and b. *Bot. Rev.* **41,** 104–135.
Stewart, K. D. and Mattox, K. R. (1978). Structural evolution in the flagellated cells of green algae and land plants. *BioSystems* **10,** 145–152.
Stewart, K. D., Mattox, K. R. and Floyd, G. L. (1973). Mitosis, cytokinesis, the distribution of plasmodesmata, and other cytological characteristics in the Ulotrichales, Ulvales, and Chaetophorales: Phylogenetic and taxonomic considerations. *J. Phycol.* **9,** 128–141.
Swanson, J. A. and Floyd, G. L. (1978). Fine structure of the zoospore and thallus of *Blidingia minima. Trans. Am. microsc. Soc.* **97,** 549–558.
Tanner, C. E. (1979). The taxonomy and morphological variation of distromatic ulvaceous algae (Chlorophyta) from the Northeast Pacific. Ph.D. Dissertation, University of North Columbia, Vancouver.
Tanner, C. E. (1980). *Chloropelta* gen. nov., an ulvaceous green alga with a different type of development. *J. Phycol.* **16,** 128–137.
Tanner, C. E. (1981). Chlorophyta: Life histories. In "The Biology of Seaweeds" (C. S. Lobban and M. J. Wynne, eds), pp. 218–247. Blackwell, Oxford.
Tatewaki, M. (1972). Life history and systematics in *Monostroma. In* "Contributions to the Systematics of Benthic Marine Algae of the North Pacific" (I. A. Abbott and M. Kurogi, eds), pp. 1–15. Jpn. Soc. Phycol., Kobe.
Tschermak-Woess, E. (1982). Über die Abgrenzung der Chlorosarcinales von den Chlorococcales. *Plant Syst. Evol.* **139,** 295–301.
Tupa, D. (1974). An investigation of certain chaetophoralean algae. *Beih. Nova Hedw.* **46,** 1–155.
Vinogradova, K. L. (1969). K sistematike poriadka Ulvales (Chlorophyta). *Bot. Zh. (Leningrad)* **54,** 1347–1355.

Vinogradova, K. L. (1974). "Ul'vovye vodorosli (Chlorophyta) mor'ej SSSR." Nauka, Leningrad.
West, G. S. (1904). "A Treatise on the British Freshwater Algae." Cambridge Univ. Press, London and New York.
West, G. S. and Fritsch, F. E. (1927). "A Treatise on the British Freshwater Algae" (rev. ed.). Cambridge Univ. Press, London and New York.
Wille, N. (1890). Conjugatae und Chlorophyceae. In "Die natürlichen Pflanzenfamilien" (A. Engler and K. Prantl, eds.), Vol. I, pp. 1–175. Engelmann, Leipzig.
Yarish, C. (1975). A cultural assessment of the taxonomic criteria of selected marine Chaetophoraceae (Chlorophyta). Nova Hedw. **26,** 385–430.

7 | On the Systematics of the Chaetophorales

D. M. JOHN

*Department of Botany, British Museum (Natural History),
London, England*

Abstract: The name Chaetophorales was first used at ordinal level by Wille in 1901. This was originally a broadly conceived order consisting of a diverse assemblage of filamentous and parenchymatous chlorophytes, branched and unbranched, and generally containing a parietal chloroplast in each often uninucleate cell. Differences in opinion concerning its membership and limits are discussed, including the development of a much more restrictive concept of the order. The circumscription of the Chaetophorales is now based largely on ultrastructural rather than the traditionally used morphological criteria. In the most recent reviews of the classification of the green algae, those members of the order lacking at cytokinesis a phycoplast or a phragmoplast have been transferred to the Ulvales or, alternatively, to a tentatively proposed new order known as the Ctenocladales. The Chaetophorales has become much reduced, as many of the families traditionally included in it have been given ordinal status or else transferred to other orders. The order as currently recognised contains just three families (Chaetophoraceae, Aphanochaetaceae and Schizomeridaceae) and a total of about 26 genera (116 species). Culture studies have shown that morphological features traditionally used for defining families, genera, and subgeneric taxa are often very variable and hence of questionable taxonomic validity. Those criteria considered to be of taxonomic significance are evaluated and the current status of the genera is discussed. Additional information from ultrastructural, biochemical and cultural studies is required before a fundamental re-organisation can be attempted at all taxonomic levels of this still little-known and unnatural grouping.

INTRODUCTION

Opinions continue to differ and few taxonomic treatments are in agreement concerning the delimitation and classification of the diverse assemblage of largely filamentous algae containing a parietal chloroplast in each cell. This

---

Systematics Association Special Volume No. 27, "Systematics of the Green Algae", edited by
D. E. G. Irvine and D. M. John, 1984, pp. 207–232. Academic Press, London and Orlando.
ISBN 0 12 374040 1   *Copyright © by the Systematics Association*
All rights of reproduction in any form reserved

group has in the past been assigned either to the order Ulotrichales Borzi or to the Chaetophorales, which Wille (1901) first used at ordinal level. West (1904) preferred Wille's name to the Ulotrichales as used in some earlier treatments, and removed from the Chaetophorales the oogamous Oedogoniaceae and the parenchymatous Ulvaceae. Some authors have recognised the Chaetophorales and the Ulotrichales as separate orders while in other treatments they are combined under one or the other name.

The heterotrichous condition was considered by Fritsch (in West and Fritsch, 1927) to be a feature of fundamental taxonomic importance. He considered heterotrichy to be an outstanding characteristic of the Chaetophorales though it was by no means evident in all representatives of the order. The order Ulotrichales was conceived by Fritsch as including all unbranched genera, though in it he placed the family Cladophoraceae. Subsequent authors such as Smith (1933, 1938, 1950, 1955) and Printz (1964) have questioned the importance of branching for the delimitation of subdivisions in the green algae, though some support for Fritsch's restrictive concept of the order has come from life history studies. Unlike the ulotrichalean algae the representatives of the Chaetophorales do not exhibit a distinct alternation of heteromorphic generations, and for this reason the two orders are placed by Kornmann (1973) and Kornmann and Sahling (1977) in separate chlorophyte classes. Little support has come for the suggestion to combine the Ulotrichales and Chaetophorales based on similarities in chromosome number and chromosome morphology as advocated by Abbas and Godward (1964) and Sarma (1964). For an historical review of the classification of the Ulotrichales see Lokhorst (this volume, Chapter 6).

It is widely accepted that several families (e.g. Chlorosarcinaceae, Coleochaetaceae, Trentepohliaceae) traditionally placed under one or the other of the two orders (Chaetophorales or Ulotrichales) are distinct enough morphologically, reproductively and, to a lesser extent, ultrastructurally to be given ordinal status. In the most recent classificatory schemes the orders of green algae are circumscribed for the first time on cytological features observed only with the electron microscope (see Stewart and Mattox, 1975; Silva, 1982; Mattox and Stewart, this volume, Chapter 2).

CIRCUMSCRIPTION OF THE TAXA

1. Orders and Families

The Chaetophorales *sensu stricto* is one of the least specialised green algal orders and is classified on morphological and anatomical criteria, which has resulted in what have long been recognised as unnatural families, genera, and

Table I. The order Chaetophorales together with included families as recognised and circumscribed by Silva (1982).

ORDER CHAETOPHORALES

Interzonal spindle collapses at telophase resulting in the nuclei remaining in close juxtaposition; cytokinesis in vegetative cell division effected by a cell plate in the presence of a phycoplast; centrioles remain on side of nuclei opposite plane of cytokinesis; plasmodesmata present; motile cells with basal bodies associated with one or more relatively narrow bands of microtubules, two to several apically-inserted flagella. Thallus filamentous, uniseriate or, more rarely, parenchymatous, occasionally of small cells forming sarcoidal packets; cells uninucleate, with single laminate parietal chloroplast, and pyrenoids bounded or perforated/traversed by thylakoids; hairs present or absent, unicellular or multicellular. Zoospores with 2 or 4 anteriorly-inserted flagella. Sexual reproduction by bi- or quadriflagellate iso- or anisogametes, or by heterogametes.

FAMILY CHAETOPHORACEAE

Thallus generally differentiated into a prostate and upright system of branching filaments, with few forms wholly erect and these either lacking branches or with rudimentary branches; hairs sometimes present, multicellular; pyrenoids single or several in each cell, bounded but not traversed by thylakoids. Zoospores formed by metamorphosis or cleavage of contents of ordinary vegetative cells to form 2 to 16 zoospores, each with 2 or 4 flagella; zoospores with four cruciately-arranged microtubular roots (2-microtubule roots alternating with 4-microtubule roots). Sexual reproduction by bi- or quadriflagellate gametes, isogamy more common than anisogamy. Life histories are known to be either haplobiontic or diplobiontic (isomorphic phases).

FAMILY APHANOCHAETACEAE

Thallus of a dominant prostrate system, with upright filaments absent or rudimentary; hairs often present, usually unicellular, unsheathed, and sometimes bulbose at base; pyrenoids one to several, shallowly penetrated but not bounded by thylakoids. Asexual reproduction by quadriflagellate zoospores formed in ordinary vegetative cells, surrounded by a delicate vesicle on first release. Sexual reproduction by quadriflagellate anisogametes. Motile cells with four cruciately-arranged microtubular roots (number of microtubules unknown).

FAMILY SCHIZOMERIDACEAE

Thallus parenchymatous, simple and cylindrical, and develops from a uniseriate filamentous stage; chloroplast perforate in mature thallus, and as an incomplete ring in juvenile stage; pyrenoids several in each cell, and traversed by several undulating thylakoids. Asexual reproduction by zoospores formed in ordinary vegetative cells, with 2, 3, 4, 5, 6, or more flagella; zoospores with four cruciately-arranged flagellar roots and all with equal number of microtubules. Life history uncertain, possibly an alternation of heteromorphic generations.

---

subgeneric taxa. Traditionally the most important criteria used for defining the order include the organisation and elaboration of the vegetative thallus together with certain cytological features as observed under the light microscope alone. Silva (1982) circumscribed the Chaetophorales mainly on features associated with mitosis–cytokinesis (Table I) which were originally

Table II. The morphological criteria presently used in the classification of chaetophoralean algae. For characters considered to be of major taxonomic importance by Tupa (1974) and Yarish (1975) for discriminating genera and species, see Table III.

1. Thallus form:
   —uniseriate or, more rarely, pluriseriate filaments; openly branched or pseudoparenchymatous.
   —largely pseudoparenchymatous and cuboidal (packets of cells), discoidal, spherical, subspherical, lenticular, cushion-like, or crustaceous.
   —filamentous, simple or openly branched, with a wholly erect, wholly prostrate, or an erect and prostrate system of branches (heterotrichous habit), and rhizoids present or absent.
2. Thallus organisation:
   —erect branch systems undifferentiated or main branches clearly distinguishable from lower orders of limited growth.
   —prostrate branch system pseudoparenchymatous, unlayered or distinctly layered.
   —relative predominance of prostrate or erect system.
3. Branching systems:
   —extent of development (sparse to abundant), form (loose or densely clustered), and arrangement (verticillate, unilateral, multilateral).
   —type of branching (dichotomous, pseudodichotomous, alternate, opposite).
   —differentiation of cells within and/or between the different orders of branching.
4. Hairs or setae:
   —presence or absence, unicellular or multicellular, position (intercalary, terminal, subterminal), size and shape of unicellular structures (bulbiform, or not swollen basally).
5. Cell wall and extracellular material:
   —presence or absence of calcareous encrustations, or a gelatinous sheath/envelope.
   —thickness of wall, sheath or envelope.
6. Cell form:
   —cell shape (cylindrical, barrel-shaped, clavate, spherical, elliptical, cuneate), size, and length-to-breadth ratio.
7. Cytology:
   —number, form (entire, band-shaped, plicate, lobed, net-like, perforate), and distribution (throughout thallus, confined to apical portions) of the parietal chloroplast.
   —presence or absence of pyrenoids.
8. Ecological:
   —nature of habitat (subaerial, neustonic, aquatic-freshwater, brackish water, marine).
   —habit (epiphytic, epilithic, epizoic, endolithic, endozoic, planktonic, symbiotic) and substratum including host plant or animal.
9. Asexual reproduction:
   —presence or absence or zoospores, gametes, akinetes, and aplanospores.
   (a) Zoospores:
       —the flagella number
       —shape, size, symmetry, and presence or absence of a pyrenoid or stigma
       —fate of the flagella at attachment of zoospore
       —mode of germination

Table II. (cont.)

(b) Zoosporangia:
—swollen or similar to ordinary vegetative cells, position (intercalary, apical, confined or generally distributed throughout thallus), nature of wall (smooth, ornamented), and presence or absence of tubular outgrowths
—number of zoospores or aplanospores per sporangium
—zoospores released with or without an enclosed membrane
(c) Akinetes:
—shape, size, colour and thickness of wall, position
10. Sexual reproductive features are rarely used.
11. Culture characteristics:
—development of secondary carotenoids pigments upon aging of cultures.
—plant mass (or configuration) on agar.
—comparative growth in selected media.

proposed by Stewart and Mattox (1975). The re-defined order included the unbranched genus *Uronema* (formerly in the Ulotrichales), though a number of branched algae traditionally placed in the Chaetophorales were now excluded. Those algae which lack a system of microtubules associated with the cleavage furrow formed at cytokinesis are accommodated by Silva (1982) in a newly created order, the Ctenocladales. Most of the members of this new order are included in the Ulvales by O'Kelly and Floyd (this volume, Chapter 4), while the genus *Microthamnion* (Microthamniaceae) is placed in the Pleurastrales by Mattox and Stewart (this volume, Chapter 2). Though ultrastructural criteria have become increasingly important for the ordinal subdivision of the green algae over the last decade, they are still little used for defining taxa at lower levels.

The families included by Silva (1982) in the order Chaetophorales are delimited on traditional morphological traits except for some mention of the ultrastructure of the pyrenoids and flagellated cells (see Table I). O'Kelly and Floyd (1983) have re-defined one of the families (Ulvellaceae) included in the order Ctenocladales and proposed its transfer to the Ulvales. They took into consideration recently acquired information on the cytoskeleton of the motile stages and found these to be correlated with features of life history, structure and ontogeny of sporangia and gametangia, chloroplast pigments, and ultrastructural details of vegetative cells. In the Ulvellaceae are now several genera which had been previously included in the marine "Chaetophoraceae" (see Yarish, 1975, 1976; O'Kelly and Yarish, 1980, 1981; O'Kelly, 1982). Doubt attaches to the systematic position of those mostly freshwater and ultrastructurally little-known genera which were often included in this family (e.g. *Chamaetrichon, Jaoa*).

2. Genera and Species

Much of the information from the earlier literature is summarised and evaluated in Printz's (1964) global revision of the freshwater representatives of the order Chaetophorales *sensu lato*. This was shortly followed by the equally valuable taxonomic treatise *Les Algues d'eau douce . . . Les Algues Vertes* by Bourrelly (1966, 1973) in which are provided keys and descriptions of chaetophoralean algae along with about 600 genera in other orders. In these comprehensive taxonomic treatments the genera (and species) of chaetophoralean algae were, and still continue to be, distinguished almost solely on morphological grounds (see Table II). Some of the more obvious morphological features commonly used for defining taxa within this group have proven to be extremely variable in recent collaborative field and culture studies. Tupa (1974) and Yarish (1975, 1976) have attempted to evaluate critically those morphological criteria used to delimit certain algae traditionally placed in the order Chaetophorales. The findings of these two studies are in broad agreement (Table III) even though O'Kelly and Floyd (1983) have now transferred the marine genera of the "Chaetophoraceae"

Table III. The morphological characters considered by Tupa (1974; 5, 6a, 12–16), Yarish (1975; 6b, 11), and both Tupa and Yarish (1–4, 7–10), to be of major importance in the classification of certain genera and subgeneric taxa traditionally considered to be chaetophoralean algae.

GENERA

1. Flagella number of zoospores
2. Presence or absence of hairs/setae
3. Presence or absence of a prostrate and erect system
4. Presence or absence of pyrenoids
5. Extent of prostrate and erect system
6a. Shape and symmetry of zoospores

SUBGENERA

6b. Shape, size and symmetry of zoospores
7. Fate of the flagella at zoospore attachment*
8. Presence or absence of a gelatinous sheath
9. Size and/or degree of branching
10. Cellular shapes and dimensions
11. Mode of zoospore germination
12. Development of secondary carotenoid pigments upon aging
13. Production of aplanospores
14. Development of akinetes
15. Comparative growth in selected media
16. Number of zoospores or aplanospores produced per sporangium

* According to Tupa (1974, p. 128) this character may also be significant at the generic level.

Table IV. A critical assessment of the current status of the genera Bourrelly (1966, 1973) recognised in two families (Chaetophoraceae, Aphanochaetaceae) under the Chaetophorales and still retained in this order. Included are genera described subsequent to Bourrelly's treatment of the freshwater chlorophytes and others requiring mention but not considered by him. The designated type is mentioned where known and an estimate given of the number of species now recognised in each genus. For the scheme of classification followed see Table II.

Order Chaetophorales	No. of species	Remarks
Family Chaetophoraceae*		
†*Caespitella* Vischer *pascheri* Vischer	1	Transferred by Cox and Bold (1966) to *Stigeoclonium*; see Shyam and Sarma (1981) for further comments on transfer.
Cedercreutziella Vischer *savoniensis* Vischer	1	Motile swarmers unknown, only non-motile spores (aplanospores, akinetes).
Chaetophora Schrank *globosa* Schrank (nom. illeg.)	12	Species distinguished on macroscopic appearance of thallus and branching of filaments; requires re-investigation.
Cloniophora Tiffany *willei* Tiffany	3	See Islam (1961) for a discussion and keys to known species.
Crenacantha Kützing *orientalis* Kützing	1	According to Bourrelly (1966, 1973) this little-known taxon is possibly a form of *Draparnaldia* with reduced verticills of branchlets, lacking hairs, and embedded in mucilage.
Draparnaldia Bory *mutabilis* (Roth) Bory	11	For review of the genus, see Forest (1956). Implication of phenotypic response of isolates grown in culture to the taxonomy of the genus is discussed by Johnstone (1978a,b).
Draparnaldiella C. Meyer & Skabitschevsky *baicalensis* (C. Meyer) C. Meyer & Skabitschevsky	9	Differs from the closely-related *Draparnaldia* by its reticulate chloroplast and generally larger dimensions. It is necessary to rename this genus as there is an earlier homonym (*Draparnaldiella* Gaillon) which is a synonym of *Draparnaldia* (see entry in Farr *et al*., 1979). Forest (1957) considered the nine species in this genus (all apparently endemic to Lake Baikal) to represent just three species.
Draparnaldiopsis G. M. Smith & Klyver *alpinis* G. M. Smith & Klyver	4	See Forest (1956) for review of genus.
Endoclonium Szymanski *chroolepiforme* Szymanski	3	Little-known genus with type description fragmentary and not accompanied by illustration. One species (*E. rivulare* Hansgirg) possibly referable to *Thamniochaete*, see Bourrelly (1966, p. 320). Printz (1964, p. 164) followed Heering (1914) in considering the type species to be a *Stigeoclonium* (*S. chroolepiforme* (Szym.) Heering).

(*cont.*)

Table IV. (cont.)

Order Chaetophorales	No. of species	Remarks
Fritschiella Iyengar tuberosa Iyengar	1	The fine structure of the vegetative cells and the zoospores have been studied by McBride (1970), Stewart et al. (1973) and Melkonian (1975).
Ireksokonia K. J. Meyer formosa K. J. Meyer	1	Zoosporogenesis and sexuality unknown. Close to *Stigeoclonium* but generally taller and with a reticulate rather than an entire chloroplast.
Iwanoffia Pascher terrestris (Iwanoff) Pascher	1	Separated from the closely-related genus *Stigeoclonium* by always having biflagellate zoospores, lacking hairs, and absence of sexuality (see Bourrelly, 1966, p. 272).
Klebahniella Lemmermann elegans Lemmermann	1	Incompletely described (no information on pyrenoid or flagella number of zoospores). Bourrelly (1966, p. 274) believed it to be probably no more than a form of *Stigeoclonium* deformed due to growth within the gelatinous colony of *Nostoc verrucosa*.
Myxonemopsis C. Meyer crassimembranacea C. Meyer	1	Zoosporogenesis and sexuality unknown. Possess characters of both *Ireksokonia* and *Draparnaldia* (see Bourrelly, 1966, 1973, p. 278). Known only from Lake Baikal.
Pseudochaete W. & G. S. West crassiseta (W. & G. S. West) W. & G. S. West	2	Genus of questionable validity, with zoosporogenesis and sexuality unknown. Tiffany (1937) considered the type to be a species of *Stigeoclonium*.
†*Saprochaete* Coker & Shanor saccharophila Coker & Shanor	1	Wagner and Dawes (1970) have shown this genus to belong to the Fungi Imperfect based on fungal characteristics revealed by electron microscopy, x-ray diffraction, and bioassay studies.
Skvortzoviothrix Bourrelly terrestris (Skvortzov) Bourrelly	1	Substitute name for an earlier homonym *Chlorodendron* Skvortzov, a marine prasinophyte.
Stigeoclonium Kützing tenue (C. Agardh) Kützing	29?	Islam (1963) monographed this extremely polymorphic genus, while Cox and Bold (1966) developed a culture-based taxonomy.
†*Thamniolum* Woronichin elegans Woronichin	1	Bourrelly (1966, 1973, p. 284) considered this genus to be probably a mere growth form of *Saprochaete* (q.v.) and there is no entry for it in the Index Nominum Genericorum (1979).
Trichodiscus Welsford elegans Welsford	1	Epiphytic on the water fern *Azolla*.
Uronema Lagerheim confervicolum Lagerheim	14	Included in the order Ulotrichales by Bourrelly (1966, 1973) and the authors of most other taxonomic treatments, but its ultrastructure (see Stewart et al., 1973) aligns it with the Chaetophorales (Silva, 1982, p. 146). Two *Ulothrix* species (*U. belkae*, *U. fimbriata*) require to be transferred to this genus as they possess a phycoplast (see Mattox and Stewart, this volume, Chapter 2).

Taxon		Notes
Family Aphanochaetaceae		
Aphanochaete A. Braun	7	Tupa (1974) has re-evaluated the genus from descriptions in the earlier literature and from growth in culture, and believes only four species to be valid, two others doubtful, and another (*A. polychaete*) to be a stage of *Stigeoclonium*.
repens A. Braun		
Chaetobolus Rosenvinge	1–2	Little-known genus with no information on reproduction or nature of chloroplast. Type species is probably conspecific with *Ochlochaete hystrix* Thwaites ex Harvey (order Ctenocladales/Ulvales), see Yarish (1975, p. 405).
gibbus Rosenvinge		
Chaetomnion Skuja	2	Zoospores unknown; characterised by the large or pyriform aplanospores.
pyriferum Skuja		
Chaetonema Nowakowski	2	Bourrelly (1966, 1973, p. 320) suggested that *Chaetonema ornatum* Transeau belongs to the genus *Chaetonemopsis* (q.v.).
irregulare Nowakowski		
†*Chaetonemopsis* Gauthier-Lièvre	1	According to Thompson (1972) this genus is a mere artifact and is actually a *Bulbochaete* parasitised by an amoeba forming stalked cysts resembling zygotes.
pseudobulbochaete Gauthier-Lièvre		
Friedaea W. Schmidle	1	Sporangia bottle-shaped, but zoospores unknown; akinetes uncertain.
torrenticola W. Schmidle		
†*Gonatoblaste* Huber	1	Tupa (1974, p. 25) considered this monotypic genus to be a synonym of *Aphanochaete confervicola* (Näg. ex Kütz.) Rabenh.
rostrata Huber		
Micropoa Moewus	1	Sexual reproduction unknown.
leptochaete Moewus		
†*Nordstedtia* Borzi	1	Poorly known and most doubtful genus (no information on reproduction). Type is according to Tupa (1974, p. 81) the basionym for *Chaetosphaeridium globosum* (Nordst.) Klebahn, and included in the Chaetosphaeridiaceae (order Chaetosphaeridiales) in Farr et al. (1979).
globosa (Nordstedt) Borz		
Thamniochaete Gay	1–3?	Bourrelly (1966, 1973, p. 320) expressed doubt concerning the placement of the British species (*T. aculeata* W. & G. S. West) in this genus due to differences in the nature of the setae compared to the type; see also note accompanying *Endoclonium* entry.
huberi Gay		
Family Schizomeridaceae‡		
Schizomeris Kützing	2	There has been no general acceptance of the view of Campbell and Sarafis (1972) that this genus is simply a growth form of *Stigeoclonium*.
leibleinii Kützing		

---

* Equivalent to the subfamily Chaetophoroideae (as Chaetophoroidées) in Bourrelly (1966, 1973) and Chaetophoreae in Printz (1964) of the family Chaetophoraceae.

† Rejected or doubtful genus.

‡ This family not recognised in Bourrelly (1966, 1973) or Printz (1964).

considered by Yarish to the order Ulvales. Some attempt has been made to find diagnostic features less variable and less susceptible to environmental modification, while culture characteristics are now considered as useful taxonomic attributes (see Cox and Bold, 1966; Tupa, 1974).

The genera and species of chaetophoralean algae should ideally be delimited by characters exhibiting the least amount of variability, though extremely variable attributes can be of taxonomic significance providing they show little or no overlap. Confusion exists in the circumscription of many members of this group with the information entirely lacking or else inadequate for a number of those characters found by Tupa (1974) and Yarish (1975, 1976) to be of particular taxonomic importance. The original description of a number of taxa is vague, meagre or fragmentary, and often with little or no account taken of either the constancy and range of expression of the distinguishing features or the potential of the alga to produce reproductive stages. Some are only known from culture and so evaluation of the morphological and perhaps physiological attributes on which they have been defined is only possible by a study of living material. The current status of the genera in the order Chaetophorales is critically assessed in Table IV and the approximate number of species in each is estimated.

CRITERIA USED IN CLASSIFICATION

The chaetophoralean algae are unlike many other groups of green algae in having specific, generic and family distinctions based largely on vegetative rather than reproductive features. This reliance on vegetative features occurs because the reproductive process in members of the order Chaetophorales is relatively unspecialised and exhibits little variation in those taxa in which it has been investigated.

1. Vegetative Features

(a) Thallus structure. The majority of the genera in the order Chaetophorales are filamentous and branched, with only a few unbranched (*Uronema, Schizomeris*). Those algae forming sarcinoid packets of small cells which only produce filaments under special conditions (*Apatococcus, Coccobotrys, Desmococcus*) are transferred to the Ctenocladales by Silva (1982) and to the Ulvales by O'Kelly and Floyd (this volume, Chapter 4). In most of the genera the filaments are uniseriate throughout, though in just a few they become pluriseriate in part by successive vertical divisions (Table V). The heterotrichous condition is common in this group (cf. Fritsch, in West and Fritsch, 1927; Fritsch, 1939), that is, the differentiation of the thallus

into an upright (or projecting portion) and a basal system of branched filaments. The absence of the upright or the prostrate system and the relative predominance of one or another of the branching systems have long been used for classifying this group. Features associated with these branching systems are often very variable, and are known in some genera (e.g. *Draparnaldia, Stigeoclonium*) to be greatly influenced by environmental conditions.

The prostrate system provides a number of characters of taxonomic importance, including its shape, size, branching habit, presence or absence of rhizoids, and the form and the dimensions of the cells. Four types of prostrate system are recognised in the group: (1) little-branched filaments of only limited growth; (2) openly branched filaments of more or less unlimited growth; (3) closely branched filaments lying in juxtaposition to one another to form a one- to several-layered crust or stratum, and (4) branched filaments coalescing laterally to form a generally one-layered disc. Sometimes the filamentous construction is only evident from the radial arrangement of the cells, though in some algae having discoid thalli the filaments become free along the margin. In some genera the prostrate system is represented by a few attaching rhizoids (e.g. *Draparnaldia, Draparnaldiopsis*), while in others only an upright system is present which is attached by the basal cell (e.g. *Lochmium, Schizomeris*). The genus *Microthamnion* consists solely of an erect system and has been transferred to a new order, the Pleurastrales (see Mattox and Stewart, this volume, Chapter 2). Many of those algae traditionally placed in the Chaetophorales which lack an upright system, or in which it is represented by just a few short and rudimentary branches, are placed by O'Kelly and Floyd (1983) in the order Ulvales. They include the "marine Chaetophoraceae" in which the form of the prostrate thallus (compactness, degree of branching) has been found to undergo variation in response to changes in light and temperature when grown in culture (Yarish, 1976).

The heterotrichous condition is most evident in the family Chaetophoraceae, whose members show differing degrees of development and elaboration of the prostrate and upright systems of branches. In the very polymorphic genus *Stigeoclonium* the prostrate system is variable and yet provides more reliable taxonomic characters for distinguishing species than those associated with the uprights (Cox and Bold, 1966; see also Francke and Simons, this volume, Chapter 16). Within this genus the prostrate system ranges from a few attaching rhizoids (*S. helveticum*) and well-developed uprights through to a pseudo-parenchymatous basal system bearing only the occasional upright branch (*S. farctum*). Several genera closely resemble *Stigeoclonium* in showing no significant differentiation between the various orders of branching, but instead are delimited on features including the presence of an

Table V. The principal taxonomic features used for distinguishing the genera currently recognised in the order Chaetophorales (see Table IV).

	Thallus		Hairs/ setae	Number of pyrenoids	Asexual spores‡	Gametes§	Habitat
	Type*	Construction†					
Aphanochaete	h	ob	+	Several	z(4), ap	h(4)	Aquatic (freshwater), epiphytic
Cedercreutziella	h	ob	0	0	ap, ak	—	Subaerial
Chaetobolus	h	ps/ob	+	1	—	—	Aquatic (marine/freshwater)
Chaetomnion	h	ob	+	1–2	ap	—	Aquatic (freshwater); epiphytic
Chaetonema	h	ob	+	1–2	z(4)	h(2)	Aquatic (freshwater); in gelatinous envelope of other algae
Chaetophora	h	ob	+	1+	z(4)	iso(2)	Aquatic (freshwater)
Cloniophora	h	ob	0	1–2	z(2)	—	Aquatic (freshwater)
Crenacantha	h	ob	0	?	—	—	Aquatic (freshwater); Palestine
Drapamaldia§§	h	ob	+	Several	z(4), ap, ak	iso(4)	Aquatic (freshwater)
Drapamaldiella§§	h	ob	+	Several	—	—	Aquatic (freshwater); endemic to Lake Baikal
Drapamaldiopsis§§	h	ob	+	1	z(2/4), ap, ak	iso(4)	Aquatic (freshwater)
Endoclonium	h	ob/ps	+	1	z(2/4)	iso(2)	Aquatic (freshwater); epi- or endophytic
Friedaea#	h	ob	+	1–4	z(?), ak?	—	Aquatic (freshwater)
*Fritschiella***	h	ob	0	Several	z(2/4)	iso(2)	Subaerial
Ireksokonia‡‡	h	ob	+	Several	—	—	Aquatic (freshwater); endemic to Lake Baikal

Iwanoffia	h	ob	+	1+	z(2)	—	Subaerial (on soil)
Klebahniella	h	ob	0	?	z(?)	—	In gelatinous envelope of *Nostoc*
Micropoa	h/p	ob	+	1	z(2)	—	Aquatic (freshwater); epiphytic on *Lemna*
Myxonemopsis‡‡	h	ob	+	2–4	—	—	Aquatic (freshwater)
Nordstedtia ∥,††	h/p	ob	+	1	—	—	Aquatic (freshwater); neustonic
Pseudochaete	p	ob	+	1	—	—	Aquatic (freshwater)
Schizomeris**,‡‡	e	ub	0	Several	z(4)	—	Aquatic (freshwater)
Skvortzoviothrix	h	ob	0	Several	z(2)	—	Subaerial
Stigeoclonium	h	ob/ps	+	Several	z(4), ap	iso(2/4)	Aquatic (freshwater)
Thamniochaete	e	ob?	+	1	z(2), ak	—	Aquatic (freshwater); epiphytic
Trichodiscus	p	ps	+	1	—	iso(2)	Aquatic (freshwater); epiphytic on *Azolla*
Uronema	e	ub	0	1–4	z(4)	—	Aquatic (freshwater/marine)

* e, erect; p, prostrate; h, heterotrichous
† ps, pseudoparenchymatous; ob, openly branched; ub, unbranched.
‡ z, zoospores; ak, akinetes; ap, aplanospores; ( ), flagella number
§ iso, isogametes; h, heterogametes; ( ), flagella number of male gametes/isogametes.
∥ Enclosed in mucilage sheath or envelope.
Lime encrusted.
** Pluriseriate (at least in part).
†† Star-like chloroplast.
‡‡ Perforate/reticulate chloroplast.
§§ Band-like chloroplast.

enclosing envelope of firm mucilage (*Chaetophora*), the nature of the habitat or substratum (e.g. *Klebahniella, Skvortzoviothrix*), or the form of the chloroplast (*Ireksokonia*).

Some genera have the upright branches strikingly differentiated into main axes and smaller lateral branches of limited growth (e.g. *Draparnaldia, Draparnaldiella, Draparnaldiopsis*), while the prostrate axis is represented by a system of attaching rhizoids. Further differentiation is observed in *Draparnaldiopsis* where the main axis consists of long cells alternating with short cells from which arise the branches of limited growth. Other closely-related genera show no such cellular differentiation, though *Draparnaldiella* is distinguished (perhaps doubtfully) by having a reticulate rather than a band-like chloroplast in the cells of the main axes. Another genus close to *Draparnaldia* is *Myxonemopsis*, but this differs in the pinnate form of its branching. Often the distinction between the different orders of branching is lost in cultures of *Draparnaldia*, and isolates grown in the absence of calcium come to resemble *Stigeoclonium* and at high nitrate levels the genus *Cloniophora* (Johnstone, 1978a). Several features of the upright system are traditionally used to define *Draparnaldia* species (size of axial cells, degree of development of lateral branches, form of division of laterals), though they are now known to exhibit considerable phenotypic variation. Johnstone (1978a,b) concluded from his study of isolates of *Draparnaldia* that many species are probably no more than forms of a single taxon and develop in response to strictly local conditions of the environment. In an earlier study Forest (1956) came to a similar conclusion, and was of the opinion that variability of expression rather than constancy of expression was the expectation in this genus.

(b) Hairs and setae. The hairs possessed by some chaetophoralean algae are either elongated cells or rows of colourless cells arising at the tips of prolonged and attenuated filaments. Hair-like outgrowths or projections of the cell wall that lack a nucleus are commonly termed setae, and are present in many of the marine genera which have been transferred by O'Kelly and Floyd (1983) to the order Ulvales. Such hyaline structures are implicit in the name Chaetophorales and are reported in 70% of the genera currently recognised in the order Chaetophorales (see Table IV). In the past the shape or form, size, number, and position of these structures has been used to delimit genera and species of chaetophoralean algae, but less significance is now placed on these structures, which have been shown to be of questionable taxonomic validity (see Cox and Bold, 1966; Tupa, 1974; marine "Chaetophoraceae", Yarish, 1975, 1976; O'Kelly and Yarish, 1981).

(c) Cytology. The chloroplast is parietal in all members of the group except for the doubtfully included genus *Nordstedtia* in which it is axial, lobed, and star-like. These are fairly conservative structures in the chaetophoralean algae and form a complete or partial ring lying just within the cell

wall, though in a few genera the chloroplast is reticulate (*Ireksokonia, Myxonemopsis*) or band-like (*Draparnaldia, Draparnaldiella*). Though pyrenoids may be present in vegetative cells they are sometimes lacking from motile stages. Future ultrastructural studies may reveal the presence of small, obscure or incipient pyrenoids in those few chaetophoralean genera for which they are unreported or of doubtful occurrence (see Table IV).

Information on chromosome numbers and chromosome morphology is still lacking, and it appears premature to discuss correlations of such karyological features with those morphological criteria traditionally used to delimit taxa of chaetophoralean algae. The principal genera currently recognised in this group which have been studied karylogically are *Stigeoclonium, Chaetophora* and *Draparnaldia* (Sarma, 1964; Sinha and Das, 1965; Godward, 1966; Chowdary, 1967; Sarma and Jayaraman, 1980; Francke and Simons, this volume, Chapter 16, among others). In general, these algae have small chromosomes (2 μm in size), with numbers commonly about 10 to 14 (range = 5–20). Reports in the literature are sometimes conflicting, with what is believed to be the same taxon having different chromosome numbers. Such disparities reflect either confusion concerning the identification of the taxa in question or real differences in chromosome numbers between strains or isolates. The search for a basic chromosome number seems unprofitable in this group, where aneuploidy is of common occurrence and leads to diversification of diploid and polyploid chromosome numbers by additions and subtractions.

(d) Extracellular wall material. The sheath or envelope of mucilage formed by the excretion of polysaccharides is a feature which assists in delimiting some members of this order. In the majority of the species the layer of mucilage is so thin as to be little evident, but in a few it either completely surrounds each filament as a definite sheath or forms an envelope enclosing the entire thallus (*Chaetophora, Gonatoblaste*). The form of the firm mucilage envelope surrounding the *Stigeoclonium*-like filaments of *Chaetophora* is a character used for distinguishing species in this freshwater genus. Little or no information exists on the reliability and constancy of such sheath or envelope characteristics. Some crustose or cushion-like chaetophoralean algae are characteristically lime-encrusted (e.g. *Crenacantha, Friedaea*), though the taxonomic validity of this feature is questionable.

2. Asexual Reproduction.

(a) Zoosporogenesis and zoospore release. The vegetative cells that function as spore mother cells provide few characters of taxonomic importance unless they enlarge to form clearly differentiated sporangia. Zoosporogenesis

is usually confined to the uprights when this sytem is well-developed, and may occur simultaneously in widely scattered cells. The position of the sporangia, their degree of enlargement, and the number of zoospores (or aplanospores) produced per sporangium are all features used for distinguishing between species and, to a lesser extent, genera.

The development pattern or ontogeny of the sporangial mother cells has been shown by Yarish (1975) and O'Kelly and Yarish (1980, 1981) to be a reliable and consistent criterion for distinguishing certain "marine Chaetophoraceae" (now transferred to the order Ulvales; see O'Kelly and Floyd, 1983). In this group of marine genera O'Kelly and Yarish found that the contents of the sporangial mother cells of *Entocladia* undergo free-nuclear divisions and simultaneous cleavage into biflagellate zoospores. This feature separates *Entocladia* from the closely related genus *Phaeophila* in which there is no formation of a multinucleate cell during zoosporogenesis. Nielson (1979) proposed the transfer of *Entocladia viridis* to *Acrochaete*, but O'Kelly and Yarish (1981, p. 43) keep these two genera separate, considering that they differ in several features including Nielson's (Nielson, 1979) reported presence of a mucilage envelope surrounding the newly released spores in *Acrochaete repens*. In a later paper Nielson (1983) questions the identity of the algae Yarish (1975, 1976) and O'Kelly and Yarish (1981) were studying, as they deviated from the original description of the species. For instance, she considers that *Entocladia viridis* invariably has a single pyrenoid whereas the alga referred to this species by O'Kelly and Yarish (1981) had a variable number of pyrenoids. In the freshwater genus *Aphanochaete* (order Chaetophorales) the zoospores are released as in *Acrochaete repens* into what appears to be a surrounding membrane (see Tupa, 1974). Another feature of the sporangial cell is the distinctive membraneous inclusion in *Phaeophila*, which is extruded as a hyaline "plug" through the apex of the sporangial neck before zoospore release (O'Kelly and Yarish, 1980; O'Kelly and Floyd, this volume, Chapter 4). Only future investigations will show if features associated with sporangial ontogeny and zoospore release are of taxonomic significance in the order Chaetophorales.

(b) Zoospores. Some of the most significant characters now used for classifying chaetophoralean algae are provided by these motile cells. Of these attributes one of the most taxonomically important is the flagellar number. The majority of the genera in which the flagellar number is known (see Table V) have quadriflagellate (47%) rather than biflagellate zoospores (33%). The remainder of the genera (19%) have been observed to release differently-sized quadriflagellate and biflagellate swarmers. Often the smaller cells are quadriflagellate and sometimes these are assumed to be gametes, though syngamy has been demonstrated on only a few occasions. *Schizomeris leibleinii* is an unusual member of this order with aberrant zoospores possess-

ing more than one papilla, accounting for the reports of the flagella numbers varying from two to eight (see G. L. Floyd and H. J. Hoops, personal communication, in Mattox and Stewart, this volume, Chapter 2). In most genera the zoospores are regular in shape and symmetry, and possess at least a single pyrenoid and an organelle known as a stigma. Ultrastructural studies of these motile cells are still sadly lacking for many genera (see Table VI), though already they are beginning to provide features of considerable taxonomic importance in this group (see Melkonian, this volume Chapter 3; O'Kelly and Floyd, this volume, Chapter 4).

The flagella of the zoospores may on attachment of the unicell to the substratum become abscissed, withdrawn, or abscissed and withdrawn. Tupa (1974) found the flagella were withdrawn in *Aphanochaete,* while in other genera traditionally included in this group (*Chamaetrichon, Protoderma, Pseudendoclonium*) they were abscissed or discarded. Such a feature might be of some taxonomic importance, but must be treated with caution until it is known whether external factors influence the fate of the flagella at zoospore attachment. In this order three types of zoospore germination are recognised (see Islam, 1963; Cox and Bold, 1966; Francke and Simons, this volume, chapter 16): (1) unipolar germination to form an upright filament; (2) unipolar germination to form an upright filament from the lowermost cell of which develops a prostrate system giving rise to further erect branches; and (3) bipolar germination to form a prostrate system from which the upright filaments arise. In most of the representatives of the order studied so far germination is bipolar and horizontal to the surface of attachment, though unipolar germination has been observed in *Chaetonema.* This has been used as a supplementary character to the nature of the prostrate system for the separation of intraspecific taxa in *Stigeoclonium* (see Cox and Bold, 1966; Francke and Simons, this volume, Chapter 16).

(c) Resting spores. Of some taxonomic significance in this group is the ability to produce non-motile resting spores. About 22% of the recognised genera (see Table V) produce what are known as aplanospores. These are believed to be formed as the result of arrested zoosporogenesis caused by conditions unfavourable for the production of motile cells. The form of the aplanospores is distinctive in just a few genera, such as *Chaetomnion* where they are relatively large and pyriform or ellipsoidal in shape. Other thick-walled, starch-laden and often darkly coloured spores are known as akinetes and are reported in about 18.5% of the genera. These resting spores may be distinctive and are commonly produced in cultures grown in depleted media or at low nitrogen levels (see Tupa, 1974). Cox and Bold (1966) have expressed doubt as to whether a true palmella stage is produced in the chaetophoralean algae and suggested that reports of coccoid cells embedded in mucilage may simply represent the formation of akinetes.

3. Sexual Reproduction

Though features associated with sexual reproduction are considered to be fundamental and conservative in most orders, they assume only minor importance in the Chaetophorales, where sexuality is reported in only 33% of recognised genera. Asexual reproduction predominates under most culture conditions while sexual stages are often observed only in field-collected material. It appears that conditions that induce sexuality are rarely encountered in culture. Some of the earlier reports in the literature of sexual stages are confusing, sometimes contradictory and usually in need of further confirmation.

In the majority of the genera in which sexual reproduction has been reported the gametes are biflagellate (55%) rather than quadriflagellate (see Table V). Only in two genera (*Aphanochaete, Chaetonema*) in the family Aphanochaetaceae has oogamy been observed. The form of the oogonia, antheridia and the oospore has little taxonomic value in discriminating intraspecific taxa in these freshwater genera.

4. Life History

Few life histories are completely known in this group of algae and those that are provide little information of any taxonomic significance. Asexual reproduction predominates and accounts of sexuality are scarce and sometimes contradictory. Fusion of gametes results in a zygote which germinates immediately or after a period of time into a plant normally identical in size and form to the parent. There is an alternation of isomorphic generations in *Stigeoclonium* which probably represents a haplodiplontic life history (see Cox and Bold, 1966). Confirmation is required of the report of an alternation of short uniseriate filaments (gametophyte stage) with the larger and pluriseriate thallus (sporophyte stage) in *Schizomeris* (see Sarma and Chaudhary, 1975). The heterogamous species of *Chaetonema* are monoecious, but the male and female plants exhibit no sexual dimorphism. Direct cytological proof of the stage in the life history where reduction division occurs (gametic, sporic or zygotic meiosis) is lacking in the majority of the chaetophoralean algae.

5. Habitat and Substrate

A great diversity of growth form is shown in this order and the members occupy a very wide range of habitats (aquatic: freshwater to marine; sub-aerial). They are commonly epiphytes on other algae as well as on higher plants and also grow as epiliths, epizoically, endophytically and endozoically. Taxonomic significance was formerly given to the nature of the habitat and the surface (including the host) upon which an alga was growing. Doubt now attaches to several of those chaetophoralean taxa created almost solely on the basis of such ecological considerations. Some of these described only from the surface film are probably no more than growth forms of taxa having a completely aquatic existence. For example, the surface-growing *Rhexinema* is believed by Tupa (1974) to be a form of *Pleurastrum paucicellulare* (now transferred to Pleurastrales; see Mattox and Stewart, this volume, Chapter 2). Some of those growing in plants or on animals (*Gloeoplax* in the moss *Sphagnum*, *Trichophilus* on hairs of sloth; transferred to Ctenocladales by Silva, 1982, and to Ulvales by O'Kelly and Floyd, this volume, Chapter 4), or associated with the mucilaginous envelope of larger algae (*Klebahniella*), are probably no more than growth forms of algae having a free-living existence (e.g. *Stigeoclonium, Entocladia*). Tupa (1974) suggested that some algae (then placed in the order Chaetophorales) described principally from soil enrichment cultures may be similar to, or perhaps identical with, ones growing naturally in aquatic habitats. The influence of the environment on phenotypic expression has been studied in very few genera (e.g. *Draparnaldia, Stigeoclonium*) in this order.

6. Culture Features

A number of characters are evident only in living cultures, and these often supplement the taxonomic criteria which traditionally have been used to distinguish species and infraspecific taxa. Cultural characteristics such as colony form and colour were attributes Cox and Bold (1966) used for defining *Stigeoclonium* species. Some taxonomic importance was attributed by Tupa (1974) to the relative growth of species under axenic culture conditions on different defined media. She also used the ability of aging cultures to produce a preponderance of orange secondary carotenoid as a character. The texture and configuration of an alga in culture is useful for ascertaining that certain isolates are identical or otherwise.

TAXONOMIC SURVEY OF ORDERS AND FAMILIES

1. Order Chaetophorales

Though a restrictive view of the order was advocated by Fritsch (in West and Fritsch, 1927), it was still considered by Bourrelly in the latest edition of his *Les Algues d'eau douce . . . Les Algues Vertes* (Bourrelly, 1973) to be a broadly-conceived and very diverse assemblage of algae. In this taxonomic

Table VI. The genera and species currently recognised in the order Chaetophorales whose motile and/or vegetative cells have been studied ultrastructurally and/or available as living cultures in major Culture Collections.

	Motile cells	Vegetative cells	Cultures available
Aphanochaete			
confervicola (Naeg. ex Kütz.) Rabenh.	—	—	+
elegans Tupa	—	+	+
magna Godward	—	+	+
species	+	+	+
Chaetophora			
elegans (Roth) C. Ag.	—	—	+
incrassata (Huds.) Hazen	—	+	+
species	—	+	+
Draparnaldia			
glomerata (Vauch.) C. Ag.	—	+	—
mutabilis (Roth) Cederg. (= *plumosa*)	—	+	+
species		+	—
Fritschiella			
tuberosa Iyeng.	+	+	+
species	+	—	—
Schizomeris			
leibleinii Kütz.	+	+	+
Stigeoclonium			
aestivale (Hazen) Collins	—	+	—
amoenum Kütz.	—	—	+
farctum Berth.	—	+	+
helveticum Visch.	—	+	+
huberi Heering	—	—	+
nanum (Dillw.) Kütz.	—	—	+
tenue Kütz. (= *pascheri, subsecundum, variabile*)*	—	+	
species	+	—	+
Trichophilus *polymorpha* Nichols & Bold	—	—	+
Ulothrix			
belkae Mattox & Bold†	+	+	+
fimbriata Bold†	—	+	+
gigas Visch.	—	+	+
Uronema			
confervicolum Lagerh. (= *Ulothrix confervicolum*)	—	+	+
marinum ?	—	—	+
schwiakofii Buddenbrook	—	—	+

* Considered to be synonyms of *Stigeoclonium tenue* (see Francke and Simons, this volume, Chapter 16).

† Should be transferred to the genus *Uronema* (see Mattox and Stewart, this volume, Chapter 2).

review he credits the order with five families. The subsequent removal of three of these families from the Chaetophorales is now widely accepted. Two of these families (Chaetosphaeridaceae, Coleochaetaceae) are now included in the Coleochaetales while a third (Dicranochaetaceae) is transferred to the Chlorococcales (see Silva, 1982). In Silva's (1982) recently published review of the classification of the green algae, the order Chaetophorales includes just three families which are still distinguished principally on morphological traits (see Table I). This order contains about 26 currently recognised genera and 116 species (see Table V). Only 34% of the genera (19% species) have been presently studied ultrastructurally and most of these are available for investigation as they are deposited in major Culture Collections (Table VI). Just four of the named chaetophoralean algae in the major Culture Collections have yet to be studied ultrastructurally.

(a) Family Chaetophoraceae. This closely corresponds to the subfamily "Chaetophoroidées" in Bourrelly (1966, 1973) and to the subfamily Chaetophoreae as recognised by Printz (1964). Some element of doubt attaches to several of the genera included in Bourrelly's subfamily, while others are no longer recognised (*Saprochaete, Thamniolum*) and a few have recently been transferred to the Chaetophoraceae from other families (*Fritschiella, Uronema*). About 18 genera and 96 species are currently recognised in this family of largely subaerial and freshwater aquatic algae. Of the genera included in this family nine are monotypic, only five have been investigated ultrastructurally at present and seven are deposited in major Culture Collections.

(b) Family Aphanochaetaceae. The generic name *Aphanochaete* Braun has been conserved against the earlier taxonomic synonym *Herposteiron naegeli*, and so the family name Herposteiraceae G. S. West is unavailable (see Silva, 1980). The family as conceived by Printz (1964) contained only oogamous algae bearing unicellular hairs (*Aphanochaete, Chaetonema*), while similar forms for which sexuality had not been reported were assigned to the Chaetophoraceae. In most other taxonomic treatments no such distinction is made on sexuality, and a number of prostrate forms with hyaline structures are placed in the Aphanochaetaceae. Bourrelly (1973) included 11 genera and about 20 species in this family of freshwater algae. One of the genera assigned to this family by Bourrelly (*Ectochaete*, now *Entocladia*) has subsequently been transferred by O'Kelly and Floyd (1983) to the family Ulvellaceae (order Ulvales), the genus *Chaetonemopsis* is rejected as an artifact and *Gonatoblaste* and *Nordstedtia* are of doubtful validity (see Table V). Of the seven genera (about 18 species) currently recognised in this family only *Aphanochaete* has more than three species. This is the only genus which is deposited in major Culture Collections and has been studied ultrastructurally. Further cytological studies will no doubt reveal the incorrect placement of some of the genera presently included in this very artificial family.

(c) Family Schizomeridaceae. The genus *Schizomeris* is included in the Ulotrichaceae (order Chaetophorales/Ulotrichales) in some taxonomic treatments (Heering, 1914; Fritsch, in West and Fritsch, 1927; Printz, 1964), while in others it has been assigned to the Ulvaceae (see Bourrelly, 1966, 1973) or to its own family (Schizomeridaceae) under the order Ulvales (see Table I). On ultrastructural grounds the affinities of the Schizomeridaceae are with the Chaetophorales rather than with the Ulvales as a phycoplast of microtubules is formed at cytokinensis and plasmodesmata are present (Mattox *et al.,* 1974). Differences in the number of flagellar roots (Birkbeck *et al.,* 1974), and the distinctive features of the vegetative cells, all support its placement in a separate family within the order Chaetophorales (see Silva, 1982). The describing authors of the genus *Trichosarcina* (Nichols and Bold, 1965) assigned it to this family because of its morphological similarity to *Schizomeris,* but now it is included in a different class (the Ulvophyceae) as ultrastructurally the two genera are very different (see Mattox *et al.,* 1974).

DISCUSSION

The green algal cell has been intensively studied ultrastructurally over the last decade or so and the newly acquired data have profoundly influenced our ideas on the classification of the chlorophytes. Several tentative proposals have been put forward to re-organise the traditional classification with a view to replacing it with a more natural scheme based on cytological features as observed with the electron microscope. In one of the more recent reviews of the classification, Silva (1982) mainly used the ultrastructural features relating to mitosis–cytokinesis, as proposed by Stewart and Mattox (1973, 1975), to circumscribe the subdivision of the green algae. Further supporting evidence for the separation of the class Chlorophyceae (includes the order Chaetophorales) and the class Ulvophyceae has come from studies of the motile cell ultrastructure (see O'Kelly and Floyd, this volume, Chapter 4). Silva (1982) tentatively proposed the order Ctenocladales and this is not recognised in the two most recent reviews of the classification (Mattox and Stewart, this volume, Chapter 2; O'Kelly and Floyd, 1983, this volume, Chapter 4). O'Kelly and Floyd place all algae without a phycoplast in the Ulvales (class Ulvophyceae), except for those having a Chlorophyceae-like motile cell ultrastructure, which Mattox and Stewart (this volume, Chapter 2) include in the class Pleurastrophyceae. Further changes in the classification will occur from time to time and stability will only be reached after the acquisition of considerably more ultrastructural information for a greater range of taxa than have currently been investigated.

A priority for future research is a re-appraisal of the morphological traits currently used for defining chaetophoralean algae involving the correlation of careful observations of the same taxon in laboratory culture and in nature. Only in a few relatively recent collaborative field and culture studies (e.g. Tupa, 1974; Johnstone, 1978a,b) has the validity of some of the traditionally used morphological criteria been seriously questioned, and some attempt made to discover new characters. Often such studies have revealed that the phenotypic expression of several traditionally used taxonomic attributes vary widely, and several taxa have been found to be identical and considerable doubt now attaches to the validity of others. One solution to the problem of distinguishing taxa where a high degree of polymorphism is established is to develop a culture-based taxonomy as pioneered by Starr and others (see Silva and Starr, 1953; Starr, 1955). They recognised the impossibility of identifying living chlorococcalean algae from the descriptions of such algae in the literature and from an examination of herbarium specimens, and so distinguished species on characters they expressed when grown under standard conditions in the laboratory. Cox and Bold (1966) have successfully adopted this approach for the well-known chaetophoralean genus *Stigeoclonium*. I believe this culture approach along with careful observations of morphological variation in the field is necessary if we are ever to obtain a better understanding of the many inadequately defined and very polymorphic taxa in this order.

It is conceivable that in the future ultrastructural features (especially of the motile cells) may be used for the recognition of genera and species, or both, though every attempt will need to be made to ascertain that the new taxonomic criteria are no less variable than the traditional ones they are to replace. Other approaches including the study of isoenzyme patterns, immunological reactions, nutritional and metabolic products and pigments, and intercrossing experiments might have potential significance in assisting us to obtain a better understanding of this one of the least-known and taxonomically-confused orders of green algae. Finally, it is to be hoped that every attempt will be made to correlate obscure or esoteric characters, or those only observed in living cultures, with more readily observable attributes which are evident in field-collected material so as to serve the needs of those who simply wish to identify algae.

REFERENCES

Abbas, A. and Godward, M. B. E. (1964). Cytology in relation to taxonomy in Chaetophorales. *J. Linn. Soc. London* **58,** 499–597.

Birkbeck, T. E., Stewart, K. D. and Mattox, K. R. (1974). The cytology and classification of *Schizomeris leibleinii* (Chlorophyceae). II. The structure of quadriflagellate zoospores. *Phycologia* **13,** 71–79.

Bourrelly, P. (1966). "Les Algues d'eau douce. Initiation à la Systématique. Tome I. Les Algues Vertes." Boubée, Paris.
Bourrelly, P. (1973). "Les Algues d'eau douce. Initiation à la Systématique. Tome I. Les Algues Vertes" (rev. ed.). Boubée, Paris.
Campbell, E. O. and Sarafis, V. (1972). *Schizomeris*—a growth form of *Stigeoclonium tenue* (Chlorophyta: Chaetophoraceae). *J. Phycol.* **8**, 276–282.
Chowdary, Y. B. K. (1967). The chromosome numbers of some species of the genus *Stigeoclonium* Kütz. *Cytologia* **32**, 174–179.
Cox, E. R. and Bold, H. C. (1966). Phycological studies. VII. Taxonomic investigations of *Stigeoclonium*. *Univ. Tex. Publ.* **6618**, 1–167.
Farr, E. R., Leussink, J. A., and Stafleu, F. A., eds. (1979). "Index Nominum Genericorum (Plantarum)," Regnum Veg., Vols. 100, 101, and 102. Bohn, Scheltema & Holkema, Utrecht and Antwerp.
Forest, H. S. (1956). A study of the genera *Draparnaldia* Bory and *Draparnaldiopsis* Smith and Klyver. *Castanea* **21**, 1–29.
Forest, H. S. (1957). The remarkable *Draparnaldia* species of Lake Baikal, Siberia. *Castanea* **22**, 126–134.
Fritsch, F. E. (1939). The hetcrotrichous habit. *Bot. Not.*, 1939, pp. 125–133.
Godward, M. B. E. (1966). "The Chromosomes of the Algae." Edward Arnold, London.
Heering, W. (1914). Chlorophyceae, III. Ulothrichales, Microsporales, Oedogoniales. *In* "Die Süsswasser-Flora Deutschlands, Österreichs und der Schweiz" (A. Pascher, ed.), Part 6, pp. 1–250. Fischer, Jena.
Islam, A. K. M. N. (1961). The genus *Cloniophora* Tiffany. *Rev. Algol.* [N.S.] **6**, 7–32.
Islam, A. K. M. N. (1963). A revision of the genus *Stigeoclonium*. *Beih. Nova Hedw.* **10**, 1–164.
Johnstone, I. (1978a). Phenotypic plasticity in *Draparnaldia* (Chlorophyta: Chaetophoraceae). I. Effects of the chemical environment. *J. Phycol.* **14**, 302–308.
Johnstone, I. (1978b). Phenotypic plasticity in *Draparnaldia* (Chaetophoraceae). II. The physical environment and conclusions. *Am. J. Bot.* **65**, 608–614.
Kornmann, P. (1973). Codiolophyceae, a new class of Chlorophyta. *Helgol. Wiss. Meeresunters.* **25**, 1–13.
Kornmann, P. and Sahling, P.-H. (1977). Meeresalgen von Helgoland. Benthische Grün-, Braun- und Rotalgen. *Helgol. wiss. Meeresunters.* **29**, 1–289.
McBride, G. E. (1970). Cytokinesis and ultrastructure in *Fritschiella tuberosa* Iyengar. *Arch. Protistenkd.* **112**, 365–375.
Mattox, K. R., Stewart, K. D. and Floyd, G. L. (1974). The cytology and classification of *Schizomeris leibleinii* (Chlorophyceae). I. The vegetative thallus. *Phycologia* **13**, 63–69.
Melkonian, M. (1975). The fine structure of the zoospores of *Fritschiella tuberosa* Iyeng. (Chaetophorineae, Chlorophyceae) with special reference to the flagellar apparatus. *Protoplasma* **86**, 391–404.
Nichols, H. W. and Bold, H. C. (1965). *Trichosarcina polymorpha* gen. et sp. nov. *J. Phycol.* **1**, 34–38.
Nielsen, R. (1979). Culture studies on the type species of *Acrochaete, Bolbocoleon* and *Entocladia* (Chaetophoraceae, Chlorophyceae). *Bot. Not.* **132**, 441–449.
Nielsen, R. (1983). Culture studies of *Acrochaete leptochaete* comb. nov. and *A. wittrockii* comb. nov. (Chaetophoraceae, Chlorophyceae). *Nord. J. Bot.* **3**, 689–694.
O'Kelly, C. J. (1982). Observations on marine Chaetophoraceae (Chlorophyta). III. The structure, reproduction and life history of *Endophyton ramosum*. *Phycologia* **21**, 247–257.
O'Kelly, C. J. and Floyd, G. L. (1983). The flagellar apparatus of *Entocladia viridis* motile cells and the taxonomic position of the resurrected family Ulvellaceae (Ulvales, Chlorophyta). *J. Phycol.* **19**, 153–164.

O'Kelly, C. J. and Yarish, C. (1980). Observations on marine Chaetophoraceae (Chlorophyta). I. Sporangial ontogeny in the type species of *Entocladia* and *Phaeophila*. *J. Phycol.* **16**, 549–558.

O'Kelly, C. J. and Yarish, C. (1981). Observations on marine Chaetophoraceae (Chlorophyta). II. On the circumscription of the genus *Entocladia* Reinke. *Phycologia* **20**, 32–45.

Printz, H. (1964). Die Chaetophoralen der Binnengewässer (eine systematische Übersicht). *Hydrobiologia* **24**, 1–376.

Sarma, Y. S. R. K. (1964). Cytology in relation to systematics of algae with particular reference to Chlorophyceae. *Nucleus* **7**, 127–136.

Sarma, Y. S. R. K. and Chaudhary, B. R. (1975). On a new cytological race of *Schizomeris leibleinii* Kütz. *Hydrobiologia* **47**, 171–181.

Sarma, Y. S. R. K. and Jayaraman, S. (1980). Karyological studies on certain taxa of *Stigeoclonium* and *Chaetophora* (Chaetophorales, Chlorophyceae). *Phycologia* **19**, 253–259.

Shyam, R. and Sarma, Y. S. R. K. (1980). Observations on the morphology, reproduction and cytology of *Stigeoclonium pascheri* (Vischer) Cox et Bold (Chaetophorales - Chlorophyceae) and their bearing on the validity of the genus *Caespitella* Vischer. *Hydrobiologia* **70**, 83–93.

Silva, P. C. (1980). Names of classes and families of living algae. *Regnum Veg.* **103**, 1–156.

Silva, P. C. (1982). Chlorophycota. *In* "Synopsis and Classification of Living Organisms" (S. P. Parker, ed.), Vol. 1, pp. 133–161. McGraw-Hill, New York.

Silva, P. C. and Starr, R. C. (1953). Difficulties in applying the International Code of Botanical Nomenclature to certain unicellular algae, with special reference to *Chlorococcum*. *Sven. bot. Tidskr.* **47**, 235–247.

Sinha, J. P. and Das, R. N. (1965). Cytological study of two species of *Chaetophora* Shrank. *Phykos* **4**, 74–75.

Smith, G. M. (1933). "The Freshwater Algae of the United States." McGraw-Hill, New York.

Smith, G. M. (1938). "Cryptogamic Botany," 1st ed., Vol. I. McGraw-Hill, New York.

Smith, G. M. (1950). "The Freshwater Algae of the United States," 2nd ed. McGraw-Hill, New York.

Smith, G. M. (1955). "Cryptogamic Botany," 2nd ed., Vol. I. McGraw-Hill, New York.

Starr, R. C. (1955). A comparative study of *Chlorococcum* Meneghini and other spherical, zoospore-producing genera of the Chlorococcales. *Indiana Univ. Sci. Ser.* **10**, 1–111.

Stewart, K. D. and Mattox, K. R. (1975). Comparative cytology, evolution and classification of the green algae with some consideration of the origin of other organisms with chlorophyll a and b. *Bot. Rev.* **41**, 104–135.

Stewart, K. D., Mattox, K. R. and Floyd, G. L. (1973). Mitosis, cytokinesis, the distribution of plasmodesmata, and other cytological characteristics in the Ulotrichales, Ulvales and Chaetophorales: phylogenetic and taxonomic considerations. *J. Phycol.* **9**, 128–141.

Thompson, R. H. (1972). On the genus *Chaetonemopsis* Gauthier-Lièvre. *J. Phycol.* **8**, Suppl., 9.

Tiffany, L. H. (1937). The filamentous algae of the west end of Lake Erie. *Am. Midl. Nat.* **18**, 911–951.

Tupa, D. D. (1974). An investigation of certain Chaetophoralean algae. *Beih. Nova Hedw.* **46**, 1–155.

Wagner, D. T. S. and Dawes, C. J. (1970). Revision of the systematic position of *Saprochaete saccharophila*. *Mycologia* **62**, 791–796.

West, G. S. (1904). "A Treatise on the British Freshwater Algae." Cambridge Univ. Press, London and New York.

West, G. S. and Fritsch, F. E. (1927). "A Treatise on the British Freshwater Algae" (rev. ed.). Cambridge Univ. Press, London and New York.

Wille, N. (1901). Algologische notizen. VII and VIII. *Nytt Mag. Naturvid.* **39,** 1–22.
Yarish, C. (1975). A cultural assessment of the taxonomic criteria of selected marine Chaetophoraceae (Chlorophyta). *Nova Hedw.* **26,** 385–430.
Yarish, C. (1976). Polymorphism of selected marine Chaetophoraceae (Chlorophyta). *Br. phycol. J.* **11,** 29–38.

8 | An Assessment of the Current State of Our Knowledge of the Trentepohliaceae*

R. L. CHAPMAN

Department of Botany, Louisiana State University, Baton Rouge, Louisiana, USA

Abstract: The Trentepohliaceae are subaerial green algae some of which are obligate epiphytes (and perhaps parasites) of vascular plants. The family name Chroolepidaceae is incorrect, although it has priority over Trentepohliaceae. The major published study on the Trentepohliaceae is that of Printz (1939). The unpublished studies of R. H. Thompson have provided important albeit incomplete observations and taxonomic treatments, including a proposed genus, *Printzia*. Although neither Printz nor Thompson included aquatic taxa in the family, other authors have. Ultrastructural studies of motile cells of Trentepohliaceae have revealed important, consistent features including: bilaterally keeled flagella, a cruciate arrangement of parallel basal bodies, and microtubular splines which originate adjacent to the basal bodies in a multilayered structure similar (but not identical) to those present in some other green algae and lower land plants. The presence of simple plasmodesmata in the central area of cross walls provides an additional feature for comparison with aquatic taxa like *Ctenocladus* and *Smithsoniella*. Many features of the Trentepohliaceae indicate close affinity to the class Ulvophyceae *sensu* Stewart and Mattox. Additional ultrastructural studies (especially on karyokinesis and cytokinesis in vegetative cells), biochemical studies, and even verification of the life histories are needed to clarify the phylogenetic position of this intriguing family.

INTRODUCTION

The first objective of this assessment is to introduce the Trentepohliaceae to phycologists and others. Although current research in several laboratories

---

*This report is dedicated to the late Dr. Rufus H. Thompson, whose keen powers of observation and dedication to the study of algae have provided numerous, important contributions to our understanding of the Trentepohliaceae and have provided a basis for discussion and further scientific inquiry.

Systematics Association Special Volume No. 27, "Systematics of the Green Algae", edited by D. E. G. Irvine and D. M. John, 1984, pp. 233–250. Academic Press, London and Orlando.

ISBN 0 12 374040 1   *Copyright © by the Systematics Association All rights of reproduction in any form reserved*

has brought some aspects of the Trentepohliaceae to the attention of many phycologists, such an introduction *per se* is fitting because to many more the family remains virtually unknown. In fact, some mycologists may have erred recently in their descriptions of new fungal taxa because they did not know of this family! A second objective is a brief review of systematic research on the Trentepohliaceae in the twentieth century. This review raises the basic question "Which genera belong in the Trentepohliaceae?"

The circumscription of the family and the question of diagnostic features leads into a discussion of recent ultrastructural and biochemical studies relating current research on the Trentepohliaceae to research dealt with in various chapters in this volume. The final objective of this assessment of our knowledge of the Trentepohliaceae is to answer the question "Where does this family belong in recently proposed classifications of green algae?"

FAMILY NAME AND TAXONOMIC STUDIES

At the start of a discussion of this intriguing family of green algae, it is logical to provide the family name. Although "Trentepohliaceae" is the better known and more widely used name, Papenfuss (1962) cited the precedence of "Chroolepidaceae" Rabenhorst (1868) over "Trentepohliaceae" Hansgirg (1886). Recently, P. C. Silva (personal communication) indicated that Rabenhorst had incorrectly constructed "Chroolepidaceae" and that the family name should have been "Chroolepaceae." Thus, "Chroolepidaceae" has been eliminated, and it is reasonable to recommend the continued use of Trentepohliaceae because "Chroolepaceae" is virtually unknown and *Chroolepus,* although available as a valid name, is not currently applied to any genus. Despite the fact that *Byssus* Linnaeus may be an earlier taxonomic synonym of *Trentepohlia* Martius, there is no need to alter current nomenclatural usage (Ross and Irvine, 1967) and, again, use of Trentepohliaceae is appropriate.

The major published taxonomic work on the Trentepohliaceae is the "Vorarbeiten" of Printz (1939). A brief synopsis of the number of taxa recognized by Printz is presented in Table I. It should be noted that Printz included only five genera (*Cephaleuros, Phycopeltis, Physolinum, Stomatochroon,* and *Trentepohlia*) all of which are subaerial. Flint (1959) eliminated the genus *Physolinum* by returning this monotypic genus to *Trentepohlia.* On the basis of morphological features, Printz (1964) included the Trentepohliaceae among the Chaetophorales; this treatment should be reviewed in the light of current thoughts as presented by Mattox and Stewart (this volume, Chapter 2) and by O'Kelly and Floyd (this volume, Chapter 4). Unfortunately, Printz did not publish a complete monograph on the Tren-

Table I. Numbers of species recognized by Printz (1939).

Genus		No. of species
Trentepohlia		36
Chroolepus	22	
Heterothallus	8	
Nylandera	6	
Cephaleuros		13
Phycopeltis		12
Eucophycopeltis	4	
Hansgirgia	8	
Stomatochroon lagerheimii		1
Physolinum		1
	Total	63 species

tepohliaceae and, to this author's knowledge, no one else has attempted or is attempting to meet the challenge of doing so.

The late Rufus H. Thompson prepared a detailed study of some of the Trentepohliaceae, but did not undertake a revision of the species of *Trentepohlia*. His work has not been published, but is cited frequently in this text because of the wealth of important observations it contains. Table II presents the number of taxa recognized by Thompson, and several points should be noted. First, although the total number of species of *Phycopeltis* and *Cephaleuros* in Tables I and II are the same, Thompson did not recognize all of the species recognized by Printz, but did describe some new species. Second, the new species described by Thompson have not been published and, in a sense, do not exist officially! Third, Thompson described a new genus, *Printzia,* which is an "inconspicuous" foliicolous epiphyte "usually

Table II. Numbers of species recognized by R. H. Thompson (unpublished).

Genus	No. of species
Trentepohlia	N/A
Cephaleuros	13
Phycopeltis	12
Stomatochroon	3

Table III. Genera included in the Trentepohliaceae by Fritsch (1935).

Gongrosireae
    Chloroclonium   *Lochmium*
    Chlorotylium   *Pleurastrum*
    Endophyton   *Pseudendoclonium*
    Gongrosira   *Pseudodictyon*
    Leptosira   *Sporocladus*
Gomontieae
    Gomontia   *Tellamia*
Trentepohliceae
    Trentepohlia   *Stomatochroon*
    Cephaleuros   *Rhizothallus*
    Phycopeltis   *Physolinum*

green in color" in which the prostrate sytem is well developed (unlike that in *Trentepohlia*). This genus has not been published and therefore, like the new species of *Cephaleuros, Phycopeltis,* and *Stomatochroon,* has no official standing. Like Printz, Thompson had a special interest in the Trentepohliaceae and like Printz he did not include any aquatic taxa in the family.

Aquatic taxa have been included in the Trentepohliaceae by some authors such as Fritsch (1935) and Smith (1950). Tables III and IV present the taxa included in the family by these authors. At the risk of oversimplification and obvious omission of other phycologists who have circumscribed the Trentepohliaceae, it can be stated that the principal character upon which aquatic taxa have been assigned to the Trentepohliaceae is the production of differentiated reproductive cells. There is clearly no agreement among phycologists on which, if any, aquatic taxa belong in the family; however, it is equally clear that *Cephaleuros, Phycopeltis, Stomatochroon,* and *Trentepohlia*

Table IV. Genera included in the Trentepohliaceae by Smith (1950).

Trentepohlia (Martius, 1817)
Cephaleuros (Kunze, 1829)
Leptosira (Borzi, 1883)
Gomontia (Bornet & Flahault, 1888)
Ctenocladus (Borzi, 1883)
Gongrosira (Kützing, 1843)
Fridaea (Schmidle, 1905)
Physolinum (Printz, 1921)

are the core of the family in all taxonomic treatments. Although the following section is limited to the presentation of these four subaerial genera, circumscription of the family will return to consideration of aquatic taxa and the characters that can be used to include or exclude these algae in or from the family. The reader is directed to other sections of this volume wherein several of the genera listed in Tables III and IV are presented, but definitely not as belonging among the Trentepohliaceae!

FOUR PRINCIPAL GENERA

Discussion of the major characters of the Trentepohliaceae will be more meaningful after the reader has been introduced to the characters through a description of the four principal taxa that are the starting point for a circumscription of the family. It would perhaps not be too unfair to deviate from this approach by mentioning that the Trentepohliaceae occur most abundantly in tropical and subtropical regions (*Trentepohlia* and *Phycopeltis* also occur in temperate areas). The onset of the author's interest in the Trentepohliaceae was coincident with his move to the subtropical environs of Baton Rouge, Louisiana where all four genera occur. Also, lichenized forms of these algae (see Santesson, 1952) are found in Louisiana (Chapman, 1976a; Meier and Chapman, 1983).

Trentepohlia Martius is the best known genus among these rather poorly known algae. It often grows on rock, wood, or bark and is often red-orange due to the presence of haematochrome pigments in the cytoplasm. The alga is a branched filament with a weakly developed prostrate system and abundant erect system (Fig. 1a). Vegetative cells give rise to gametangia or zoosporangia that are morphologically distinct from vegetative cells and that release their respective motile cells through single exit pores. The zoosporangia abscise from their subtending cells and can be dispersed by wind, rain, and insects. The abscission is mediated by a specialized cross wall between the zoosporangium and subtending cell. Although *Trentepohlia* can be epiphytic it is not obligately so and is never parasitic.

Phycopeltis Millardet often forms a tight discoidal monostromatic thallus of compacted branched filaments (Fig. 1b). Gametangia differentiate from cells within the thallus, and zoosporangia are borne singly at the end of erect branches (Good and Chapman, 1978b). As in *Trentepohlia* the zoosporangia abscise. R. H. Thompson (unpublished) found *Phycopeltis* on a variety of substrates, but most references (e.g., Millardet, 1870) mention only an epiphytic habitat (wherein the inconspicuous alga is more easily observed!). *Phycopeltis* is always supracuticular on plant hosts and is never parasitic.

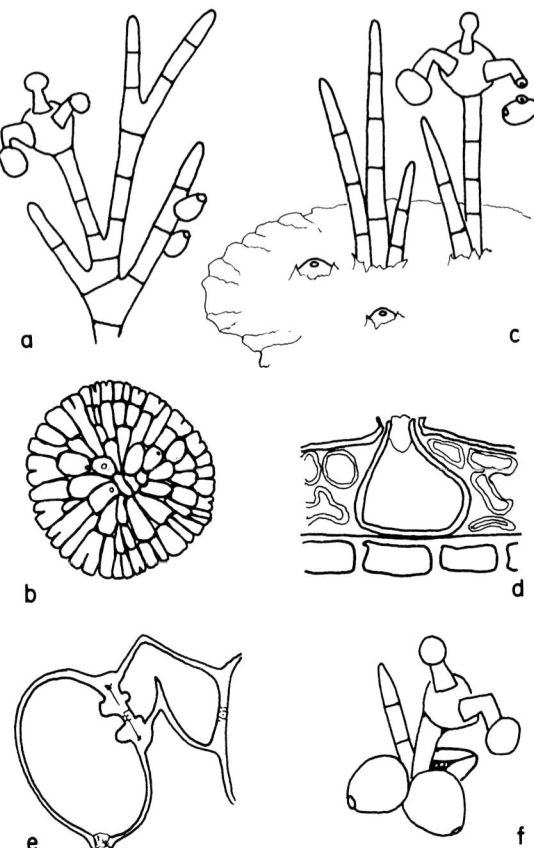

Fig. 1. Diagrammatic illustrations of four genera of Trentepohliaceae. (a) *Trentepohlia*, showing filamentous growth pattern, head cell (with three developing sporangiate laterals, and two gametangia with single exit pores) (× 400). (b) Discoid thallus of *Phycopeltis* (× 500). Four gametangia are present. (c) *Cephaleuros*, growing subcuticularly except for erect filaments and mature gametangia (× 400). One zoosporangium with its single exit pore has abscised from the suffultory cell at the specialized cross wall. Abscission scars are visible on both cells. (d) Section through maturing *Cephaleuros* gametangium, subtended by thickened epidermal cells of host leaf and surrounded by subcuticular vegetative cells of the alga (× 1000). The enlarging gametangium has broken through the leaf cuticle and the exit pore is occluded by a pectinaceous plug. (e) Section through a sporangiate lateral of *Cephaleuros* (× 2000). The specialized thickened rings of wall material between the zoosporangium and the suffultory cell surround a "pit" containing simple plasmodesmata, as does the cross wall between the suffultory cell and the head cells. A pectinaceous plug fills the single exit pore. (f) Composite representation of *Stomatochroon* thallus emergent from leaf stomate (× 400). Typical trentepohliacean gametangia and zoosporangia (which possibly do not occur together on the same thallus in nature) are represented together with a single vegetative filament.

Cephaleuros Kunze is obligately epiphytic and may be parasitic. The alga grows on leaves, fruits, or young stems (Batista and Lima, 1949; Suematu, 1957; Holcomb, 1975; Chapman, 1976b; Marlatt and Alfieri, 1981) and is often orange-red. On the leaves of *Magnolia grandiflora* and *Camellia* species, *Cephaleuros virescens* forms velvet-like circular spots that can be ten to 15 mm in diameter. Thus, *Cephaleuros* can often be easily observed and recognized with the naked eye (unlike *Phycopeltis*). The prostrate thallus is well developed and is comprised of compacted branched filaments organized into branch systems or ramuli (Fig. 1c). Although the general thallus form may be circular, the thallus is not as clearly discoidal as that of *Phycopeltis epiphyton*, which is a very regular circular disk. The *Cephaleuros* thallus produces rhizoidal filaments that create the impression that the thallus is multilayered. The prostrate portion of the thallus is always subcuticular when mature, and gametangia formed in the thallus break through the host cuticle as they enlarge (Fig. 1d). Zoosporangia are produced on specialized branches that are part of a well-developed erect system of filaments that also includes numbers of sterile, multicellular, uniseriate trichomes. The fertile branches, or sporangiophores, terminate in an enlarged head cell that ultimately bears eight or more sporangiate laterals. The sporangiate laterals consist of two cells, a short stalk or pedicel (also called a suffultory cell) and the terminal zoosporangium (Fig. 1e). Abscission of the zoosporangium occurs at the specialized cross wall between the pedicel and the zoosporangium as it does in other Trentepohliaceae (Chapman and Good, 1978, 1983).

The obligate epiphytism and subcuticular habitat of *Cephaleuros* may indicate a physiological dependence on the host and raises the question, "Is *Cephaleuros* a parasite?" Although necrosis of the subtending host tissue can occur in response to the presence of *Cephaleuros* (Chapman and Good, 1976), R. H. Thompson (unpublished) stated that in most instances there was no host response. Furthermore, he indicated that those species of *Cephaleuros* that grow into the leaf tissue do so only because other infections (e.g. fungal) have disrupted the tissue. In the late 1800s and early 1900s, however, *Cephaleuros* was specifically studied because of its destructive effects as a pathogen of coffee and tea plants (Swingle, 1894, Went, 1895; Mann and Hutchinson, 1907), and has more recently been studied because of its pathogenicity (Wellman, 1965, 1972; Golato, 1970; Roth, 1971); thus it is appropriate to expand on Thompson's ideas. He considered the destructive infections by *Cephaleuros* to be merely opportunistic growth facilitated by poor host nutrition or previous infection by other organisms. Very little is known about the exact mode of 'infection' by *Cephaleuros*, but it is likely that a broken or disrupted host cuticle is a prerequisite to establishment of the thallus (Joubert and Rijkenberg, 1971a,b). Despite some physiological studies (Joubert, 1969; Chowdary, 1969, 1970; Vidhyasekaran and Parambaramani, 1971a,b; Joubert et al., 1975; Jose and Chowdary, 1979; Chowdary and Jose, 1979), there is no clear evidence of a nutritional dependence on the host. Also, *Cephaleuros* can be grown autotrophically in many media and is thus certainly not totally dependent on a vascular plant host. It is interesting to note that *Cephaleuros* grown in culture does not normally produce zoosporangia (Chowdary, 1969), but the developmental block is not known.

Whether parasitic or not, *Cephaleuros* is a subaerial alga and an obligate epiphyte. One can compare *Trentepohlia, Phycopeltis,* and *Cephaleuros* and suggest that there is a continuum of specialization and morphological differentiation. *Trentepohlia* is the least specialized in form and substrate requirements, and *Cephaleuros* is the most.

Stomatochroon Palm is endophytic and emerges from the leaf through the stomates (Fig. 1f). Since only a few cells are produced on the abaxial leaf surface, the alga is scarcely visible unless there is a heavy infection, in which case the color of the leaf surface can be noticeably changed. Pigmented prostrate filaments grow into the intercellular spaces of the leaf mesophyll and may produce emergent thalli at stomates; thus several thalli visible on the under-surface of the leaf may be interconnected (Good and Timpano, 1980). The emergent thalli consist of gametangia, short trichomes, and sporangiate laterals that bear zoosporangia typical of the Trentepohliaceae (Chapman and Good, 1983). There is no evidence indicating that infections by *Stomatochroon* are very deleterious; however, a heavy infection would interfere with normal gas exchange. *Stomatochroon* is obligately endophytic, but is pigmented and can be grown autotrophically in culture. Although morphologically simple, *Stomatochroon* occupies a very specific ecological niche and can be considered the most specialized of the Trentepohliaceae.

The four principal genera share several features: (1) all four are branched filaments and form prostrate and erect systems; (2) they produce orange-red haematochrome pigment that is stored in the cytoplasm; (3) all four produce differentiated reproductive cells; (4) both gametangia and sporangia bear single papillate exit pores for the release of motile cells; (5) all four genera are subaerial; (6) all produce zoosporangia that abscise *via* a specialized cross wall; (7) plasmodesmata are present in the central region of cross walls in vegetative cells; and (8) pyrenoids are lacking. There are additional common features that are discussed in the sections that follow; however, the eight already mentioned allow comparison of the Trentepohliaceae with other families of green algae and provide characters for possible use in circumscription of the family. Although both gametangia and zoosporangia have been mentioned, life histories *per se* have not, but are discussed separately below.

LIFE HISTORIES

The Trentepohliaceae have been studied since the mid-1800s, but even now the life histories of the four principal genera are not well understood nor thoroughly documented. As mentioned elsewhere in this volume, life histories can be considered in the circumscription of green algal orders, and whether useful in taxonomy or not, knowledge of the life histories of the Trentepohliaceae is basic to understanding the biology of the family. The most recent and intriguing observations on life histories are those of Thompson (1961 and unpublished). For *Cephaleuros* and *Stomatochroon,* he described (1961) an alternation of heteromorphic generations in which the haploid gametophyte thallus produces quadriflagellate zoospores or biflagellate gametes (Fig. 2). The former develop immediately after release to reproduce the haploid thallus. The isogamous gametes from the same thallus mate and produce a diploid dwarf thallus that bears small zoosporangia, the presumed site of meiosis, and produces quadriflagellate meiozoospores ("microzoospores" in the original published report). The meiozoospores directly develop into haploid gametophyte thalli. Thus, for these two genera there is a heteromorphic alternation of generations with isogamy and homothallic mating. Tentative confirmation of Thompson's observations has been provided by the discovery of synaptonemal complexes in sporangia of *Cephaleuros virescens* (Chapman and Henk, 1981).

Phycopeltis and perhaps *Trentepohlia* are thought to exhibit an alternation of isomorphic generations (Fig. 3). The haploid gametophyte produces isogametes that mate and produce a diploid sporophyte thallus that bears only meiozoosporangia. Quadriflagellate meiozoospores develop immediately after release into haploid gametophytes. In some species, thalli will produce only gametangia or meiozoosporangia (and are thus readily distin-

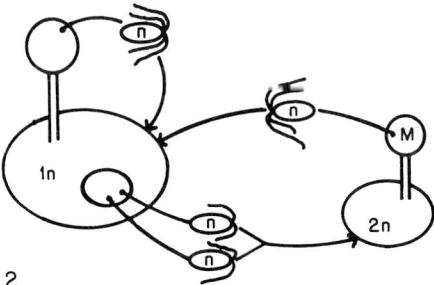

Fig. 2. Diagram of alternation of heteromorphic generations in *Cephaleuros* and *Stomatochroon* as described by Thompson (1961). The diploid dwarf thallus is the presumed site of meiosis. 1n and n = haploid, 2n = diploid, M = meiosis.

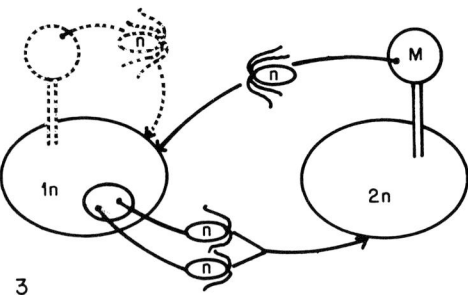

Fig. 3. Diagram of alternation of isomorphic generations in *Phycopeltis* and *Trentepohlia* based on unpublished reports by Thompson. The haploid gametophyte of some species may bear haploid zoosporangia as well as gametangia. 1n and n = haploid, 2n = diploid, M = meiosis.

guishable), but in others (e.g., *P. epiphyton*) the gametophyte can produce haploid zoosporangia from which quadriflagellate zoospores are released. The zoospores develop directly into gametophyte thalli. If an alternation of isomorphic generations is more primitive than an alternation of heteromorphic generations, then it is interesting that the simpler two genera have the more primitive life histories. In no instance do the motile cells or the zygotes form a resting stage. Thompson noted, however, that under dry conditions sporangiophore intitials form akinetes, which eventually produce either sporangiophores or zoospores.

Confirmation and documentation of Thompson's observations are clearly needed, but his observations are probably more accurate than numerous conflicting reports in the literature (see Chapman, 1976b). One difficulty encountered by previous investigators was the fact that the quadriflagellate motile cells of the Trentepohliaceae often appear biflagellate because the paired flagella adhere to each other. The use of culture-grown Trentepohliaceae has not facilitated confirmation of life histories because not all reproductive structures and generations are produced (e.g. fusion of gametes and further development has never been observed in culture-grown material).

ULTRASTRUCTURE OF MOTILE CELLS

The importance of ultrastructural features of motile cells in the systematics of green algae is both well established and well covered in this volume. Three important features of motile cells in the Trentepohliaceae were first reported by Graham and McBride (1975) for the biflagellate motile cells (presumably gametes) of *Trentepohlia*. First, the flagella are bilaterally keeled or winged. Second, the basal bodies are antiparallel, side-by-side, and overlapping. Third, a multilayered structure (MLS) is associated with each basal body and incorporates a microtubular spline that extends beneath the plasmalemma toward the posterior of the cell. These three key features have subsequently been reported for the gametes and zoospores of *Cephaleuros* (Chapman, 1976b, 1980, 1981; Chapman and Henk, 1982, 1983) and of *Phycopeltis* (Good, 1978). *Stomatochroon* gametes appear to share these key features (Good and Timpano, 1980, and personal communications).

In the biflagellate gametes of *Cephaleuros* (Chapman, 1980; Chapman and Henk, 1983) the flagellar apparatus includes four microtubular roots in a cruciate arrangement exhibiting 180° rotational symmetry. Although this arrangement differs from that originally proposed for *Trentepohlia* motile cells, Roberts (this volume, Chapter 13) and L. E. Graham (personal communication) have confirmed that the arrangement in *Trentepohlia* is similar to that in *Cephaleuros*. It should be noted that the MLSs in the Trentepohliaceae are certainly not morphologically identical to the MLSs that have been found in some other green algae and in the bryophytes. The MLSs do consist of three definite layers and the most anterior layer is a microtubular spline. The evolutionary and biochemical homologies among the MLSs in green algae and lower land plants are important in the systematics of green algae and investigation of these homologies has begun (see Melkonian, this volume, Chapter 3).

Although the ultrastructure of both biflagellate gametes and quadriflagellate zoospores has been examined, the quadriflagellate meiozoospores have not. One can predict that they will share the key features mentioned above. Similarly, the motile cells of all species of the four genera may consistently exhibit these features and, with appropriate caveats, one could consider the use of these features in circumscription of the family.

PLASMODESMATA, KARYOKINESIS, AND CYTOKINESIS

Ultrastructural features of vegetative cells can be used in assessing the phylogenetic relationships among green algae, but our knowledge about important features in the Trentepohliaceae is still rather meager. One important feature that has been established clearly is the presence of simple plasmodesmata in the central region of the cross walls in vegetative cells (Chapman and Good, 1978; Chappel et al., 1978). In *Phycopeltis* the central portion of the cross wall that contains the plasmodesmata is bounded by a ring of lightly stained wall material (Chapman and Good, 1978; Good, 1978). Plasmodesmata also occur in the specialized zoosporangium–pedicel cross wall, the site of sporangial abscission. In this cross wall the plasmodesmata occur in the area delineated by a thickened ring of lightly stained wall material (cf. the cross walls in *Phycopeltis*). A second ring of thickened wall material occurs around the perimeter of the cross wall and functions in the abscission process.

Whether bounded by a ring of wall material or not, the plasmodesmata occur only in the central region of the cross walls, a region that has been termed a "pit" (Chappell et al., 1978). In the freshwater genus *Ctenocladus* (Blinn and Morrison, 1974) and the marine genus *Smithsoniella* (Sears and Brawley, 1982), a single opening or pore occurs in the central region of the cross walls and one can cite this similarity (with appropriate caution!) as a possible indication of phylogenetic affinity.

The central localization of plasmodesmata in the cross wall prompts questions about the formation of the plasmodesmata and hence also about the process of cytokinesis. The cross walls in *Phycopeltis* are composed of two distinct sections: the central region, bound by the ring of thickened wall material, and the outer region. One can easily speculate that the outer region is formed by an incomplete infurrowing of the wall and that the central plasmodesmata-containing region is formed by partial cell plate formation. Unfortunately, speculation is all that is available at this time. Despite the interest in and importance of the features of cytokinesis in vegetative cells, there are no ultrastructural studies of this phenomenon. Floyd is working on specimens provided by L. E. Graham (personal communication) that may provide the information needed. Light microscopic observations by various workers (see Chapman, 1976b, for references) do not provide much information, but investigators (see, e.g., Cunningham, 1879; Thomas, 1913; Wolf, 1930) observed that cell division was accomplished by ingrowth of the cell walls. Jose and Chowdary (1977) reported that cross wall formation starts during telophase and "cell-plate formation takes place after the daughter nuclei are well separated" in *Cephaleuros*. R. H. Thompson (unpublished) frequently referred to cell divisions during thallus development, but did not specifically describe or discuss the mode of cross wall formation.

Although nuclear division and cell division in a vegetative cell may be separated temporally, the ultrastructural features of karyokinesis in the Trentepohliaceae have not been well recorded. Some stages of nuclear divisions in gametangia and sporangia have been observed (Chapman and Good, 1977; Graham and McBride, 1978). Mitosis in the "sessile sporangium" of *Trentepohlia aurea* was described as closed and centric at late anaphase, and neither phycoplast nor phragmoplast microtubules were observed during cytokinesis (Graham and McBride, 1978). Similar features were described for the nuclear divisions in zoosporangia of *Cephaleuros virescens* (Chapman and Good, 1977). Based on light microscopic observations, Chowdary

(1960) described nuclear division in *Cephaleuros* and *Trentepohlia* and stated that "no distinct spindle could be observed." As indicated both in this volume and in numerous recent publications on green algae, the ultrastructural details of cytokinesis and karyokinesis are important parameters for assessing phylogenetic relationships among green algae and one can only hope that the current gaps in our knowledge of these phenomena for the Trentepohliaceae will soon be filled.

BIOCHEMISTRY

Just as the advent of ultrastructural studies provided new and important taxonomic characters for use in green algae systematics, so also have biochemical studies. Pigments have been a basic, useful character in algal taxonomy from the earliest days of phycology. Comparison of accessory pigments remains a useful parameter in comparing green algal taxa (see O'Kelly and Floyd, this volume, Chapter 4), but such information does not seem to be particularly pertinent in the taxonomy of the Trentepohliaceae. Comparisons of storage products have begun to indicate interesting similarities among various taxa of algae (see, e.g., Kremer and Kirst, 1982). The Trentepohliaceae have been included in some of these studies (Feige and Kremer, 1980; Kremer and Kirst, 1982) and demonstrate a very wide spectrum of accumulated alditols. The comparative analysis of accumulated photosynthates will very likely provide useful taxonomic information at various levels. Similarly, the variety of biochemical parameters included in the studies of Kessler (this volume, Chapter 18) and the enzyme studies of Schlösser (this volume, Chapter 19) provide clear examples of approaches that can and should be applied to the Trentepohliaceae to elucidate relationships within the family and among related families.

Both the structure and biochemistry of the cell walls in the Trentepohliaceae remain poorly known, although sporopollenin has been found to occur in the walls of *Phycopeltis* and *Cephaleuros*, but not of *Trentepohlia* (Good and Chapman, 1978a, and unpublished). Further biochemical studies of the wall as well as perhaps more sophisticated analyses of proteins (e.g. ribosomal) and nucleic acids are needed for the Trentepohliaceae.

There are two major categories of taxonomic biochemical investigations of the Trentepohliaceae that are needed: first, those studies that will be useful in relating the Trentepohliaceae to other families of green algae; and second, those that will be useful in refining our concepts of species and genera within the family. Of these two, the former are of most interest at the moment. The latter will become increasingly important as the taxonomy of the family is defined, but at present there are many unanswered basic questions about the overall position among green algae and the circumscription of the family for which answers are more urgently needed than detailed analyses of interspecific relationships.

SYSTEMATICS AND PHYLOGENY

The various objectives of this assessment of our knowledge of the Trentepohliaceae mentioned in the Introduction must now be focused on the systematics of the green algae. Thus, the implications of the foregoing information to the systematics of the Trentepohliaceae are summarized.

The basic taxonomy of the Trentepohliaceae must be revised. First, it is important that Thompson's studies be published if at all possible, that his new taxa (several species and one genus) be validly published, and that vouchers be deposited. Second, specimens must be collected extensively, while they still can be, from tropical areas threatened with destruction. Third, there should be a reconsideration of Thompson's work as part of a thorough study that must include *Trentepohlia* and should utilize the latest systematic techniques and approaches.

Circumscription of the Trentepohliaceae will be basic to the needed study and the features common to the four major taxa must be carefully considered. Initially, the Trentepohliaceae can be proposed as a neatly defined, distinct group based, in part, on characters including, but not limited to, the following: subaerial habitat, zoosporangia that abscise, gametangia and zoosporangia that are differentiated and bear a single papillate exit pore, biflagellate gametes, quadriflagellate zoospores (and meiozoospores), isogamy, bilaterally keeled flagella, multilayered structures in motile cells, alternation of isomorphic or heteromorphic generations, sporic meiosis, simple plasmodesmata in the central portion of cross walls, the absence of pyrenoids, and cytoplasmic accumulation of hematochrome pigment. If there is no *a priori* basis for limiting the family to subaerial algae, some of the preceding characters might not be valid for circumscribing the family. Aquatic forms might not produce zoosporangia that abscise, and bilaterally keeled flagella and hematochrome accumulation could be absent. Also, the biochemistry of storage compounds and wall components might not be useful in comparing closely related aquatic and subaerial taxa. Thus, it is quite possible that on the basis of motile cell ultrastructure and details of karyokinesis and cytokinesis an aquatic genus may be assigned to the Trentepohliaceae. Despite some similarities that can be cited (e.g., between *Ctenocladus* or *Smithsoniella* and the Trentepohliaceae), no aquatic taxon has been shown to be very similar to the four principal taxa of the Trentepohliaceae.

A search for green algae similar to the Trentepohliaceae clearly points to the Ulvophyceae (see Mattox and Stewart, this volume, Chapter 2); however, as similar as the architecture of motile cells may be, the Trentepohliaceae lack rhizoplasts, well-developed capping plates, septate B-tubules in basal bodies, and striated microtubule-associated components (SMACs). Although the relative importance of various features of the Ulvophyceae is a topic of discussion in this volume (see O'Kelly and Floyd, Chapter 4), the Trentepohliaceae are certainly different from taxa in the Ulvophyceae. A review of the recent proposals for a new class, the Pleurastracaeae (see Mattox and Stewart, this volume, Chapter 2), shows that the Trentepohliaceae are more similar to the Ulvophyceae. Thus, the Trentepohliaceae should be acknowledged as a distinct group (the order Trentepohliales) with affinities to the Ulvophyceae. Melkonian (1982) has recognized that other subaerial green algae (e.g. *Friedmannia* and *Trebouxia*) have ulvophycean characteristics, but are different enough to warrant separation from the Ulvophyceae proper. The phylogenetic relationship between these algae and the Trentepohliaceae will become clearer as more data become available, and attempts to classify the family further should await the arrival of these data. The presence of multilayered structures should not be construed as evidence for close affinity to the Charophyceae and certainly must not be used as a basis for assigning the Trentepohliaceae to the Charophyceae. Conversely, structures that are clearly multilayered and associated with basal bodies and microtubular spline should not be branded "false multilayered structures" or "multilayered structure-like structures" until and unless definitive biochemical data demonstrate that there is only a single type of "true multilayered structure."

An assessment of the current state of our knowledge of the Trentepohliaceae must conclude that an enormous amount of work is needed from the most basic collection of specimens and confirmation of life histories to the molecular characterization of cellular components. One must hope that the Trentepohliaceae are intriguing enough to attract the necessary attention from phycologists.

ACKNOWLEDGEMENTS

The author wishes to acknowledge the assistance of Dr. Peter Timpano, who has provided some unpublished observations of the late Dr. Rufus H. Thompson as well as helpful comments and ideas about the Trentepohliaceae. The expert and patient assistance of Margaret C. Henk in the preparation of this article and in phycological research at Louisiana State University is acknowledged with sincere thanks.

REFERENCES

Batista, A. C. and Lima, D. A. (1949). Lista de suscetiveis da alga *Cephaleuros mycoidea* Karst em Pernambuco. *Bol. Secr. Agric., Ind. Com. Est. Pernambuco* **16**, 32–46.

Blinn, D. W. and Morrison, E. (1974). Intercellular cytoplasmic connections in *Ctenocladus circinnatus* Borzi (Chlorophyceae) with possible ecological significance. *Phycologia* **13**, 95–97.

Chapman, R. L. (1976a). Ultrastructural investigation on the foliicolous pyrenocarpous lichen *Strigula elegans*. (Fee) Mull. Arg. *Phycologia* **15**, 191–196.

Chapman, R. L. (1976b). Ultrastructure of *Cephaleuros virescens* (Chroolepidaceae; Chlorophyta). I. Scanning electron microscopy of zoosporangia. *Am. J. Bot.* **63**, 1060–1070.

Chapman, R. L. (1980). Ultrastructure of *Cephaleuros virescens* (Chroolepidaceae; Chlorophyta). II. Gametes. *Am. J. Bot.* **67**, 10–17.

Chapman, R. L. (1981). Ultrastructure of *Cephaleuros virescens* (Chroolepidaceae; Chlorophyta). III. Zoospores. *Am. J. Bot.* **68**, 544–556.

Chapman, R. L. and Good, B. H. (1976). Observations on the morphology and taxonomy of *Phycopeltis hawaiiensis* King (Chroolepidaceae). *Pac. Sci.* **30**, 187–195.

Chapman, R. L. and Good, B. H. (1977). Some comparisons among *Cephaleuros, Phycopeltis, Trentepohlia* and other green algae. *J. Phycol.* **13**, Suppl., 12.

Chapman, R. L. and Good, B. H. (1978). Ultrastructure of plasmodesmata and cross walls in *Cephaleuros, Phycopeltis* and *Trentepohlia* (Chroolepidaceae; Chlorophyta). *Br. Phycol. J.* **13**, 241–246.

Chapman, R. L. and Good, B. H. (1983). Subaerial symbiotic green algae: Interactions with vascular plant hosts. *In* "Algal Symbiosis: A Continuum of Interaction Strategies" (L. J. Goff, ed.), pp. 173–204. Cambridge Univ. Press, London and New York.

Chapman, R. L. and Henk, M. C. (1981). Observations on the putative dwarf sporophyte of *Cephaleuros* Kunze (Chlorophyta; Chroolepidaceae). *Proc. Int. Bot. Congr., 13th, 1981* p. 162.

Chapman, R. L. and Henk, M. C. (1982). Cruciate flagellar apparatus and multilayered structures in *Cephaleuros virescens* gametes. *Proc. Int. Phycol. Congr., 1st*, 1982, a8.

Chapman, R. L. and Henk, M. C. (1983). Ultrastructure of *Cephaleuros virescens* (Chroolepidaceae; Chlorophyta). IV. Absolute configuration analysis of the cruciate flagella apparatus and multilayered structures in the pre- and post-release gametes. *Am. J. Bot.* **70**, 1340–1355.

Chappell, D. F., Stewart, K. D. and Mattox, K. R. (1978). On pits and plasmodesmata of trentepohlialean algae (Chlorophyta). *Trans. Am. microsc. Soc.* **97**, 88–94.

Chowdary, Y. B. K. (1960). Cytology of *Trentepohlia* and *Cephaleuros*. *Proc. Symp. Algol., 1959*, pp. 65–69.

Chowdary, Y. B. K. (1969). Induction of reproductive organs in *Cephaleuros virescens*. *Indian J. Microbiol.* **8**, 153–158.

Chowdary, Y. B. K. (1970). Cultural and nutritional requirements of *Cephaleuros virescens* Kunze. *Indiana Biol.* **2**, 75–79.

Chowdary, Y. B. K. and Jose, G. (1979). Biology of *Cephaleuros* Kunze in nature. *Phykos* **18**, 1–9.

Cunningham, D. D. (1879). On *Mycoidea parasitica*, a new genus of parasitic algae, and the part which it plays in the formation of certain lichens. *Trans. Linn. Soc. London* [2] **1**, 301–316.

Feige, G. B. and Kremer, B. P. (1980). Unusual carbohydrate pattern in *Trentepohlia* species. *Phytochemistry* **19**, 1844–1845.

Flint, E. A. (1959). The occurrence of zoospores in *Physolinum* Printz. *New Phytol.* **58**, 267–270.

Fritsch, F. E. (1935). "The Structure and Reproduction of the Algae," Vol. 1. Cambridge Univ. Press, London and New York.
Golato, C. (1970). Una grave malattia dell'annarcardio (*Anacardium occidentale* L.) in Tanzania. *Riv. Agric. subtrop. trop.* **64**, 334–340.
Good, B. H. (1978). Ultrastructural and biochemical studies on the epiphytic subaerial green alga *Phycopeltis epiphyton* Millardet. Ph.D. Dissertation, Louisiana State University, Baton Rouge.
Good, B. H. and Chapman, R. L. (1978a). The ultrastructure of *Phycopeltis* (Chroolepidaceae; Chlorophyta). I. Sporopollenin in the cell walls. *Am. J. Bot.* **65**, 27–33.
Good, B. H. and Chapman, R. L. (1978b). Scanning electron microscope observations on zoosporangial abscission in *Phycopeltis epiphyton* (Chlorophyta). *J. Phycol.* **14**, 374–376.
Good, B. H. and Timpano, P. (1980). Preliminary observations on the subaerial green alga *Stomatochroon*. *J. Phycol.* **16**, Suppl., 14.
Graham, L. E., and McBride, G. M. (1975). The ultrastructure of multilayered structures associated with flagellar bases in motile cells of *Trentepohlia aurea*. *J. Phycol.* **11**, 86–96.
Graham, L. E. and McBride, G. E. (1978). Mitosis and cytokinesis in sessile sporangia of *Trentepohlia aurea* (Chlorophyceae). *J. Phycol.* **14**, 132–137.
Hansgirg, A. (1886). Prodromus der Algenflora von Böhmen. Erster Theil. . . *Arch. naturw. Landes Durchf. orsch. Boehmen* 5(6), 1–96.
Holcomb, G. E. (1975). Hosts of the alga *Cephaleuros virescens* in Louisiana. *Proc. Am. Phytopathol. Soc.* **2**, 134.
Jose, G. and Chowdary, Y. B. K. (1977). Karyological studies on *Cephaleuros* Kunze. *Acta bot. Indica* **5**, 114–122.
Jose, G. and Chowdary, Y. B. K. (1979). Effect of three nitrogen sources on the growth of *Cephaleuros* Kunze isolates. *Phykos* **18**, 69–72.
Joubert, J. J. (1969). Cultivation of *Cephaleuros virescens* Kunze on an artificial medium. *Rev. Biol. (Lisbon)* **7**, 1–6.
Joubert, J. J. and Rijkenberg, F. H. J. (1971a). Studies on the host range of *Cephaleuros* spp. in Natal. *Rev. Biol. (Lisbon)* **7**, 185–193.
Joubert, J. J. and Rijkenberg, F. H. J. (1971b). Parasitic green algae. *Annu. Rev. Phytopathol.* **9**, 45–64.
Joubert, J. J., Rijkenberg, F. H. J. and Steyn, P. L. (1975). Studies on the physiology of a parasitic green alga *Cephaleuros* sp. *Phytopathol. Z.* **84**, 147–152.
Kremer, B. P. and Kirst, G. O. (1982). Biosynthesis of photosynthates and taxonomy of algae. *Z. Naturforsch., C: Biosci.* **37**, 761–771.
Mann, H. H. and Hutchinson, C. M. (1907). *Cephaleuros virescens* Kunze, the "red rust" of tea. *Mem. Dep. Agric. India, Bot. Ser.* **1**, 1–35.
Marlatt, R. B. and Alfieri, S. A. (1981). Hosts of *Cephaleuros*, a parasitic alga in Florida. *Proc. Fla. State hortic. Soc.* **94**, 311–317.
Meier, J. L., and Chapman, R. L. (1983). Ultrastructure of the lichen *Coenogonium interplexum* Nyl. *Am. J. Bot.* **70**(3), 400–407.
Melkonian, M. (1982). Structural and evolutionary aspects of the flagellar apparatus in green algae and land plants. *Taxon*, **31**, 255–265.
Millardet, M. A. (1870). De la germination des zygospores dans les genres *Closterium* et *Staurastrum* et sur un genre noveau d'algues chlorosporées. *Mem. Soc. Sci. nat. Strasbourg* **6**, 37–50.
Papenfuss, G. F. (1962). On the circumscription of the green algal genera *Ulvella* and *Pilinia*. *Phykos* **1**, 6–12.
Printz, H. (1939). Vorarbeiten zu einer Monographie der Trentepohliaceen. *Nytt Mag. Naturvid.* **80**, 137–210.

Printz, H. (1964). Die Chaetophoralen der Binnengewasser. Eine systematische Ubersicht. *Hydrobiologia* **24,** 1–76.
Rabenhorst, L. (1868). "Flora europaea algarum aquae dulcis et submarinae. Sectio III. Algas chlorophyllophyceas, melanophyceas et rhodophyceas complectens." Kummer, Leipzig.
Ross, R. and Irvine, L. M. (1967). The typification of the genus *Byssus* L. (1753). *Taxon* **16,** 184–186.
Roth, G. (1971). An algal leafspot disease on avocado pears (*Persea americana* Mill.) in South Africa. *Phytopathol. Z.* **70,** 323–334.
Santesson, R. (1952). Foliicolous lichens. I. A revision of the taxonomy of the obligately foliicolous, lichenized fungi. *Symb. bot. ups.* **12,** 1–590.
Sears, J. R. and Brawley, S. H. (1982). *Smithsoniella* gen. nov., a possible evolutionary link between the multicellular and siphonous habits in the Ulvophyceae, Chlorophyta. *Am. J. Bot.* **69,** 1450–1461.
Smith, G. M. (1950). "The Fresh-water Algae of the United States," 2nd ed. McGraw-Hill, New York.
Suematu, S. (1957). Notes on *Cephaleuros* and *Phycopeltis,* parasitic and epiphytic aerial-algae III. Lists of infected plants. *Bot. Mag.* **70,** 276–281.
Swingle, W. T. (1894). *Cephaleuros mycoidea* and *Phyllosiphon,* two species of parasitic algae new to North America. *Proc. Am. Assoc. Adv. Sci.* **42,** 260.
Thomas, N. (1913). Notes on *Cephaleuros. Ann. Bot. (London)* **27,** 781–797.
Thompson, R. H. (1961). The life cycles of *Cephaleuros* and *Stomatochroon. Proc. Int. Bot. Congr., 9th, 1959* Vol. 2, p. 397.
Vidhyasekaran, P. and Parambaramani, C. (1971a). Carbon metabolism of alga infected plants. *Indian Phytopathol.* **24,** 369–374.
Vidhyasekaran, P. and Parambaramani, C. (1971b). Nitrogen metabolism of alga infected plants. *Indian Phytopathol.* **24,** 500–504.
Wellman, F. L. (1965). Pathogenicity of *Cephaleuros virescens* in the neotropics. *Phytopathology* **55,** 1082.
Wellman, F. L. (1972). "Tropical American Plant Disease (Neotropical Phytopathology Problems), Chapter 26, pp. 639–668. Scarecrow Press, Inc., Metuchen, New Jersey.
Went. F. A. C. (1895). *Cephaleuros coffeae,* eine neue parasitische Chroolepidee. *Zentralbl. Bakteriol., Parasitenkd. Infektionskr.* **1,** 681–687.
Wolf, F. A. (1930). A parasitic alga, *Cephaleuros virescens* Kunze, on *Citrus* and certain other plants. *J. Elisha Mitchell Sci. Soc.* **45,** 187–205.

9 | Comparative Studies in a Polyphyletic Group, The Desmidiaceae—30 Years On

A. J. BROOK

School of Life Sciences, University of Buckingham, Buckingham, England

Abstract: Thirty years ago, the distinguished phycologist Professor F. E. Fritsch, then President of the Linnean Society of London, offered as his Presidential Address a paper called "Comparative studies in a polyphyletic group, the Desmidiaceae". As Fritsch carefully pointed out in introducing his paper, the state of desmid taxonomy at that time left much to be desired because of the very variable nature of these algae, so that innumerable species and varieties had been described which differ from established ones only by what he described as trivial detail. The fathers (or strictly father and son) of desmid taxonomy, the Wests, were in part responsible for this state of affairs in that they "over-exaggerated the importance of minor detail" and as Fritsch also stated "the authoritative lead they gave in this respect was followed by most subsequent workers" . . . including himself! Fritsch went on to predict that he believed that many species and varieties which differ from one another in small ways do not deserve such taxonomic rank and will in time merely be "assigned as forms to the better-known species". The aim of the present paper is to examine some of the trends in desmid taxonomy over the 30 years since Professor Fritsch made these statements. Much of desmid taxonomy is still purely morphological or typological (alpha taxonomy) and even for very many so-called well-established species, descriptions and illustrations are surprisingly inadequate. Hence even with the aid of large and extensive taxonomic monographs published during very recent years, it is far from easy to identify certain species and the Wests' legacy of trivial detail remains. On the other hand, some detailed knowledge has been gained about a few taxa, natural populations of which have been studied over extended periods to explore their normal range of variability and this has led to some useful taxonomic revisions. Laboratory cultures have also led to some recognition not only of the extent of desmid variation but also of the constancy of certain characters. Scanning electron microscopy, the ultimate morphological tool has revealed little or nothing that cannot be seen with good, critical light microscopy. However, especially the examination of the ultrastructure of the desmid cell wall has revealed the false basis of certain aspects of some omega taxonomic criteria which has led to a reclassification of the Desmidiaceae which reflects natural relationships. Other criteria which have been used to elucidate

---

Systematics Association Special Volume No. 27, "Systematics of the Green Algae", edited by D. E. G. Irvine and D. M. John, 1984, pp. 251–269. Academic Press, London and Orlando.
ISBN 0 12 374040 1

Copyright © by the Systematics Association
All rights of reproduction in any form reserved

problems of desmid taxonomy over the past 30 years and which will be discussed, include nuclear cytology and also aspects of their sexual reproduction and genetic compatibility.

INTRODUCTION

It is not altogether clear why Professor Fritsch chose desmids as the subject for his Presidential Address to the Linnean Society in 1952 (Fritsch, 1953) for he published few papers on these remarkable chlorophytes himself (Fritsch, 1907, 1930, 1933; Fritsch and Rich, 1928, 1937). However, I suspect that the most important factor that made him so aware of the inadequacy of much of desmid taxonomy was an inevitable consequence of the compilation of his extensive iconography, the basis of the Fritsch Collection now lodged and still fortunately being maintained by the Freshwater Biological Association at Windermere. This great "labour of love" in which all available illustrations of freshwater algae were pasted onto foolscap sheets, filed and catalogued, brought home to him as he said (Fritsch, 1953, p. 258) that "innumerable species and varieties had been described which differ from established ones only by trivial detail".

The Wests (W. and G. S.), the most prodigious and significant of all contributors to the literature on desmid taxonomy, with publications extending from 1888 to 1914, culminating in their five-volume Ray Society monograph, British Desmidiaceae* (1904, 1905, 1908, 1912, 1923) were unfortunately partly responsible for this state of affairs. They, too, over-exaggerated the importance of minor detail and "the authoritative lead they gave in this respect was followed by most subsequent workers"—including, as Fritsch (1953) generously admitted, himself. As Silva (this volume, Chapter 20) comments, "Current taxonomic practice is often strongly influenced by tradition . . .".

The purpose of this contribution is to review the progress that has been made during the past 30 years in finding answers to some of the problems of desmid taxonomy which were posed in Professor Fritsch's Linnean Society address.

A REVISED CLASSIFICATION OF THE DESMIDS

Thirty years ago it was generally accepted that desmids belonged to the Order Conjugales within the Class Chlorophyceae. With the recognition that most green algae should be grouped in the Division Chlorophyta, desmids along with other conjugate algae were placed in the Class Conjugatophyceae (Fott, 1959; Meyer, 1962; Round, 1963). Bourrelly (1966), however, chose to rename the class the Zygophyceae. Because Round (1971) rightly felt it preferable to have the names of classes based on generic names, he proposed the name Zygnemaphyceae.

---

*Volume V was completed posthumously by Dr. N. Carter.

We know little or nothing of the possible phylogenetic relationships of the Zygnemaphyceae with other classes of the Chlorophyta (see, however, Pickett-Heaps and Marchant, 1972). It is not only their mode of sexual reproduction by conjugation and their complete lack of zoospores or motile gametes that clearly separates them from other green algae, but they are also unique in their modes of cell division (see review by Brook, 1981). Also on the basis of their nuclear cytology (Godward, 1966), it seems more than evident that the Zygnemaphyceae are a discrete class of green algae. In her review Godward focusses attention on the common, though not invariable (Ling and Tyler, 1976), absence from the chromosomes of localised centromeres, and the presence during mitosis of stainable material derived

Table I. Class Zygnemaphyceae.

ORDER I ZYNEMATALES

Family 1. Mesotaeniaceae (Saccoderm desmids)
Family 2. Zygnemataceae
  Cell wall features:
    i. Wall of each cell consisting of one piece
    ii. No shedding of primary wall
    iii. Outer hyaline (mucous) layer smooth
    iv. No (or only weak) incrustations
    v. No pores in walls

ORDER II DESMIDIALES (Placoderm desmids)

Sub-Order I. Archidesmidiinae

Family 3. Gonatozygonaceae
Family 4. Peniaceae
Family 5. Closteriaceae
  Cell wall features:
    i. Wall may be formed in several segments that are divided by only very slight constrictions
    ii. No shedding of primary wall
    iii. Compact, structured outer layer, with warts, spines and ridges originating from outer layer
    iv. Differences in nature of ornamentation in three families
    v. Pores or pore-like gaps present only in outer wall layer

Sub-Order II. Desmidiinae

Family 6. Desmidiaceae
  Cell wall features:
    i. Wall consisting of two segments divided by a constriction (isthmus) where they slightly overlap
    ii. Primary wall shed
    iii. No continuous outer layer but mucilaginous envelope originating from pore organs
    iv. Strongly ornamented in many cases; ornamentation originates from secondary wall
    v. Complex pore organs occurring in secondary wall

from the nucleolus; the latter being a structure of remarkable organisation and persistence.

There was early recognition of the unusual nature of the cell walls of the Zygnemaphyceae (Lütkemueller 1902), and this has been amply confirmed, especially by the electron microscope studies of Mix (1972, 1975). Her investigations have revealed that all the cell walls of members of the Zygnemaphyceae possess three distinct layers, and cell wall characteristics now provide the basis on which the two orders of the class, the Zygnematales and Desmidiales, are distinguished and also the two sub-orders within the latter (Table I).

Although the sub-orders Archidesmidiinae and Desmidiinae seemed to be clearly defined, it has now been discovered that the filamentous genus *Phymatodocis* has pores typical of the first sub-order but its primary wall is shed and the secondary wall overlaps in the isthmus region, so in these respects it is characteristic of the Desmidiinae (Engels and Lorch, 1981).

THE ELECTRON MICROSCOPE AND DESMID GENERA

Not only have electron microscope studies of desmid cell walls provided a new basis for the separation of orders and families, but ultrastructural detail has been of help in the reordering of some misplaced genera and species. Light microscopy revealed long ago that the walls of many *Closterium* species are ridged. The electron microscope has shown that the ridges are formed by an indeterminate electron-impervious, amorphous substance superimposed upon the primary wall. The arrangement, height and width of the ridges are characteristic of different *Closterium* species and hence are of taxonomic significance.

Various types of ridges, some longitudinally arranged, some spirally, can be seen with the light microscope in species of *Penium,* for example *P. polymorphum* and *P. spirostriolatum,* while the wall of *P. margaritaceum* appears to be covered by small wart-like excrescences. Scanning electron microscope studies of the surfaces of desmids of this genus (see especially Couté and Tell, 1981) have shown that the ornamentation of the outer-wall layer is interconnected by a system of meshes, lateral in the longitudinally- or spirally-ridged species, stellate in those with warts. In the past there have been some difficulties in assigning certain small, cylindrical, baton-shaped desmids to the correct genus but electron microscope studies of their walls have helped in this connection. Gerrath (1969) was thus able to demonstrate that the desmid originally named *Pleurotaenium spinulosum* was in fact a *Penium*. Similarly, it has been shown that the ultrastructure of the cell wall of *Gonatozygon chadefaudii* clearly indicates that this too, must be assigned to the genus *Penium*.

THE ALPHA TAXONOMY OF PLACODERM DESMIDS

Although electron microscope studies have made valuable contributions to desmid systematics at the higher taxonomic levels of orders, families and genera, they have done nothing to solve the sorts of taxonomic problems about which Professor Fritsch expressed concern and which are intrinsic to the desmid cell itself. The crucial problems of desmid taxonomy relate to species of placoderms of the Family Desmidiaceae. They are mostly solitary unicells, generally consisting of two almost identical semicells joined in the region of a usually well-defined median constriction, the isthmus. The adjoining semicells, however, are separate entities of different ages. Each time a cell divides it separates at the isthmus and new semicells are regenerated as a consequence of which the parent semicells are progressively pushed apart.

Because each placoderm cell is divided across its middle by an isthmus, adjacent semicells are symmetrical and ideally should be perfect mirror images of one another, so that there is a plane of symmetry across the isthmus. There is also another plane of symmetry at right angles to the isthmus. There is yet a third plane of symmetry perpendicular to the vertical axis so that, except when the cells are cylindrical, they appear when viewed from the cell apex to be elliptical, triangular or to have higher orders of angularity. The corners of the angles may be rounded but, especially in desmids with a basically triangular apical view, the angles are commonly considerably extended to form the ray-like processes typical of the genus *Staurastrum*. Desmids with this type of morphology are termed radiate and possess particular degrees of radiation (triradiate, quadriradiate, etc.). Elliptical cells with only two corners (most forms of *Cosmarium*, *Euastrum* and *Micrasterias*) are deemed to be biradiate.

Each time a desmid cell divides, two new semicells are reconstructed. These, however, may not always be mirror images of the parent semicells and in extreme cases (which are by no means rare) differences, often striking, are apparent in the shape and/or ornamentation of adjacent semicells, so that they may be distinctly asymmetric with respect to one or more planes of symmetry. Moreover, the differences which arise in this way between adjacent semicells may be replicated in succeeding divisions so that as the population of the taxon in question grows, distinctly new forms may appear with increasing frequency and they may even become the dominant morph.

Desmids are commonly found in small numbers in collections of plankton, and more especially of metaphyton or moss squeezings, that usually contain diverse assemblages of other algae. Once a particular specimen has been found it may take considerable searching or be virtually impossible to find other specimens of the same taxon in the collection. Hence it can be very difficult to make a preserved collection of a particular taxon for reference purposes, and equally there are difficulties in the preparation of mounted specimens. Moreover, some characters to which taxonomic importance has been attached can exhibit great variability. However, it would seem that in the past new taxa (even species) have been described following the examination of one or at best a small number of specimens, and these could well have been atypical of the species-population as a whole. Despite these limitations, the iconotype has become accepted as the basis for the typification of taxa. But the illustrations, often poorly executed, are frequently of only one specimen and no indication is given of the range of variability of the taxon being described. In consequence much confusion has arisen from what, with some hesitation, I describe as this "stamp-collecting" alpha taxonomy.

As Mayr (1982) has written in "The Growth of Biological Thought" in relation to the practical importance of taxonomy, "Much confusion in various branches of biology . . . has been caused by the imprecise if not erroneous identification of the species with which the investigator worked", and he goes on to stress the importance of sound taxonomy for the applied biologist. It was as an applied biologist working in a freshwater fisheries laboratory that I became lured and then trapped in the meshes of desmid taxonomy, for in the late 1940s and early 1950s it began to be recognised that the presence of desmids in the plankton not only could be used to indicate the trophic status of lakes, but might also be sensitive indicators of environmental change in freshwaters (for review, see Brook, 1981, pp. 211–236). As a result I made detailed examinations of the phytoplankton, paying particular attention to the desmids, of over 300 British lakes, 20 of them on a long-term basis (Brook, 1965). For this I used the Wests' British Desmidiaceae (1904, 1905, 1908, 1912, 1923) as my taxonomic mentor.

Desmids of the genus *Staurastrum* are frequent in the plankton of a wide range of lake types, and "the published records suggest that *S. paradoxum* Meyen and *S. gracile* Ralfs are the most widely-distributed planktonic desmids in British, as well as in other European freshwaters" (Brook, 1959d). However, following the careful examination of over 250 plankton samples from many regions of the British Isles and from lakes of many types, none of the Staurastra occurring in them could be referred to either of these "species". I had suggested in an earlier paper (Brook, 1959b) that the species *S. paradoxum* could not be maintained because of Meyen's (1828) very imperfect and thus far from adequate drawing (the iconotype) of this species, the first of the genus *Staurastrum* to be described. Also Ralfs (1848), with whom desmid nomenclature begins, had used Meyen's inadequate description and illustration of *S. paradoxum*. The examination of the material in the Jenner Herbarium of the British Museum (Natural History), believed to have been used by Ralfs (1848) in the preparation of his monograph on British Desmidiaceae, was found to be of specimens of *S. micron* (Brook, 1959c), which bears little resemblance to Ralfs' description of *S. paradoxum*. In consequence I suggested that *S. paradoxum* should be shelved as a "nomen inquirendum", since the detailed examination of all the figures (80 in number) published in the 128 years since Meyen established this taxon indicates that they could all be assigned to other, well-described and well-illustrated (that is "good"), *Staurastrum* species (see also Brook, 1960a).

An inevitable consequence of the existence of such an ill-defined species as *S. paradoxum* was the establishment of a considerable number of varieties (14), but as with the species several of these had also become merely convenient names to attach to "difficult" forms. Since the species has no reality, it was clear that none of these varieties could be maintained. In another paper (Brook, 1960b) I attempted to place them in the taxa to which I considered they belonged. Regrettably, more than 20 years later one still finds *S. paradoxum* or some of its varieties mentioned in species lists.

I was a little more fortunate with *S. gracile* Ralfs, type material of which was still extant in the Jenner Herbarium in the BM (NH). Through the illustrations and description in West and Carter (1923) and Smith (1924) *S. gracile* had become accepted, erroneously, as a planktonic desmid with slender and usually long, divergent processes. The Jenner Herbarium material, on the other hand, indicates a small benthic species with short, parallel or convergent processes. Again, scattered throughout the desmid literature are well over 50 illustrations purporting to be of *S. gracile*. However, very few of these correspond with the type material and most can be related to other well defined taxa. On the other hand I suggested that eight taxa were but synonyms of *S. gracile* (Brook, 1959d).

The *S. anatinum* group serves to emphasize how in the past students of desmids, owing to the fact that descriptions of new taxa were based on the examination of only a small amount of material, have placed too much importance on decorations of the cell wall (granules, spines, verrucae) in establishing species and varieties. Sometimes in a single sample, more especially in samples taken at intervals from the same body of water, it is possible to observe a complete range in the development of the apical and lateral ornamentation of the cells. These can range from single granules to large spines, stout verrucae or even doubly bifurcate verrucae. That these are merely developmental differences is shown by the fact that in a single plant one may observe small granules on one semicell and bifurcate verrucae on the adjacent one (Brook, 1959d, Plate VII).

Of the numerous varieties of *S. anatinum* which have been established very few would seem to be sufficiently distinctive and fixed to warrant being maintained. Many of the varieties, such as *curtum, longibrachiatum, denticulatum* (and also the species *S. vestitum* and its variant *subanatinum*), which were based on such variable characters as length of processes and the extent of the development of ornamentation on the cell body, I proposed should be reduced to the rank of formae. On this basis it was possible to distinguish 11 forms (Brook, 1959d).

But there is an added complication, which is that the forms listed above are distinguished only by one particular character. Hence, forms *longibrachiatum* and *curtum* refer only to the length of the semicell processes in contrast to the forms *hirsutum, denticulatum* or *glabrum* which are based on the ornamentation of the semicell body. But these characters are in no way discrete, so that one individual in a population of *S. anatinum* possesses both *longibrachiatum* and *hirsutum* characteristics, and another *longibrachiatum* and *glabrum* characteristics. Hence I suggested for this very variable desmid (and possibly it should be used for others) the adoption of a system of nomenclature which at the infra-specific level involved the use of a combination of two or even three of these "form-epithets", e.g. *S. anatinum* forma *longibrachiatum–denticulatum–paradoxum*.

That there is a reluctance on the part of desmidiologists to recognise the great variability of their organisms was pointed out by Heimans (1969) with reference to another *Staurastrum*, again inadequately described by Ralfs (1948). Heimans had produced detailed drawings and a revised diagnosis of this species in 1926 but in his 1969 paper complains that, despite his detailed study, many authors have during the intervening 40 years either named this species as quite a different species or have described as new species desmids which are merely forms of *S. echinatum*.

It would seem that the major problem still to be solved (and the result agreed upon) by desmid taxonomists is to decide on what are the characters of such morphologically variable entities as desmid cells that show a high degree of constancy and therefore may be said to be "reliable". As I shall explain in more detail later, the main thesis of Fritsch's (1953) Linnean Society Address was that there would seem to be certain basic cell shapes, and that these should form the basis for a new taxonomy erected on species–groups. This eminently reasonable idea has yet to be taken up, though a few small species–groups have been treated to careful measurement and statistical analysis; for example the *Closterium setaceum–kuetzingii* group by Tassigny (1966) and the *C. peracerosum-strigosum–littorale* group by Watanabe and Ichimura (1978). Such lines of investigation could well produce most worthwhile results, especially in relation to the blurred distinctions which exist between genera such as *Euastrum* and *Cosmarium*, and *Cosmarium* and *Staurastrum*. It is envisaged that computer graphics could be most helpful in the comparative analysis of cell morphologies and subsequent treatment of data in such endeavours.

If it is accepted that basic cell shape (other than radiation) is a good alpha-taxonomic criterion, then slight morphological variations on which some species would seem to be based, (such as number of undulations in the cell wall in certain obviously related forms of *Cosmarium*) must be examined critically (Ružička, 1966). So also should the extent of lobe development in *Micrasterias* (see below; also Lenzenweger, 1979), the development of spines in *Xanthidium* (Gerrath, 1982), and the length and angles at which the arms of *Staurastrum* species are disposed (Brook, 1959d, 1982).

It is my strong conviction, supported by the examination of very large numbers of cells in natural populations and in culture, that undue emphasis has been placed on cell wall ornamentation. Not, I hasten to emphasize, on the disposition of such ornament which I am convinced is one of the most reliable of taxonomic characters, but rather on the extent of its development (Brook, 1959d). Thus varietal or in some cases specific rank (e.g. *Staurastrum vestitum*) has been incorrectly conferred on the basis of whether the ornamentation occurs as almost indiscernible single granules or large single or bifurcate spines, stout verrucae or even doubly bifurcate verrucae. All are but varying expressions of a single character. My views about the significance of the consistency of the disposition of cell wall ornamentation would appear to be supported recently by Gerrath (1979) in his study of a natural population of *Cosmarium taxichondrum* var. *taxichondrum*. One of the distinguishing features of this widely distributed desmid is the presence of large granules, one on either side of the isthmus. Gerrath found only one cell in 400 examined that lacked such granules.

Fortunately there has been increasing recognition over the past 30 years that desmid taxonomy must be based not on the examination of a few specimens but on populations, and that descriptions and illustrations must give an account of the species variability. So far, however, for only a small number of taxa have these requirements been fulfilled. In this connection it is most important to study populations on a seasonal basis and also in various habitats and localities, and to look for dichotypic forms linking often remarkably dissimilar morphs. Recent examples of such studies are those by Gerrath (1982) on *Xanthidium tetracentrotum*, Brook (1982) on the *Staurastrum tetracerum* group and Gerrath (1983) on *S. pentacerum*.

Because of the great potential for variation possessed by desmids, largely as a consequence of their mode of cell division and subsequent enlargement, and the fact that the iconotype has become the accepted basis of their taxonomy, I would urge that the editors of phycological or botanical journals use considerable discretion about accepting for publication descriptions of supposedly new taxa. Before new taxa are committed to print, editors should ensure that there is evidence that the author has examined a significant number of cells of the desmid in question to explore the range of its variability. Also it should be a condition for acceptance for publication that this variability is adequately illustrated by a corresponding range of good figures.

RADIATION AND THE TAXONOMY OF PLACODERM DESMIDS

Teiling (1950) directed attention to the study of radiation, "that element of structure which is decisive in the shape of desmids according to their vertical symmetry plane"; this relates particularly to the genus *Staurastrum*. Earlier, taxonomists gave varietal rank to otherwise identical desmids merely on whether they were biradiate or triradiate. Dichotypic links clearly pointed to the fact that such a distinction was unwarranted and so many varieties were dispensed with. As Teiling pointed out, there are no rules of nomenclature to cope with desmids with variable radiation, and because it may be important to record the degree of radiation he proposed the term "facies" for use as a nomenclatural expression for this morphological feature.

It is of interest to note that there is now evidence from laboratory studies of *Staurastrum polymorphum* (Brandham and Godward, 1963), and from field observations of a population of *S. sebaldi* var. *ornatum* (Lind and Croasdale, 1966), that the degree of radiation is related to the temperature at which development takes place. The higher degrees of radiation tend to occur at lower temperatures and therefore would seem to be an environmentally-determined condition related to the rate of cell development. Although there is no evidence that differences in the ploidy of the cell nucleus are involved in the examples quoted above, it has been demonstrated by Kallio (1953) and Starr (1958) that diploid cells commonly show a higher degree of radiation than haploids and that diploid cells usually multiply less rapidly than haploids.

The distinctions drawn between some placoderm genera are arbitrary. This is especially so in relation to the two largest and artificial genera, *Cosmarium* and *Staurastrum*. As Fritsch (1953) pointed out, "a very significant variation in the shape of the *Cosmarium* semicell is the assumption of a triangular outline in cross-section. The majority of such variants have been described as species of *Staurastrum*, because in them the triangular cross-section is the rule". He pointed out that one can also find "triangular varieties (new facies) of species of *Euastrum*, *Xanthidium* and *Micrasterias*" which, except for this particular feature, are "essentially like the typical forms". As he went on to complain, "the position is quite illogical" since for many of the simpler trigonal species of *Staurastrum*, comparable *Cosmarium* species are known (see Brook and Williamson, 1983). Fritsch suggested that it might be sensible to group together all the *Cosmarium*-like forms which in apical view are three- or more sided, and whose angles are not produced as processes, under one new genus, namely *Cosmostaurastrum*. Since, however, it is now fully recognised that changes in radiation can be environmentally (or even cytologically) induced, and that almost any placoderm desmid is potentially capable of producing cells with varying degrees of radiation, this may not be the answer to this problem. Clearly as increasing numbers of tri- or quadriradiate forms of *Cosmarium* are found, so the proposed genus *Cosmostaurastrum* would grow at the expense of *Cosmarium*, which could become a genus without species!

The diagnosis of the genus *Staurastrum* in the authoritative taxonomic work on green algae by Bourrelly (1966) is that "Ces espèces ne peuvant pas être confondues avec les *Cosmarium*, car la vue frontale où la corps de l'hemisomate se prolonge de chaque côte par un ou 2 processus ou bras allongés est très caracteristique". It must seem quite obvious, even to the uninitiated looking at the figures illustrating the range of forms within this artificial genus *Staurastrum*, that the only unifying taxonomic feature is the variable character of radiation. It is a consequence of this illogicality that the genus *Staurastrum* now contains the greatest range of morphologies of all desmid genera. As stated above, the type species *S. paradoxum nomen dubium* certainly was a triradiate desmid with "processus ou bras allongés". However, it would seem that as more and more desmids were described the morphological character that took precedence was radiation, so almost any specimen which was not obviously a *Xanthidium* or *Micrasterias* but which was other than biradiate was assigned to *Staurastrum*.

In an attempt to solve some anomalies of the consequences of this obvious illogicality, Palamar-Mordvintseva (1976) has proposed three new genera. The first is *Cylindriastrum*, whose triangular recto-cylindrical cells are typified by the desmid originally named *S. pileolatum*. The second is *Cosmoastrum*, with *Cosmarium*-shaped cells whose walls are covered with short spines as in *S. polytrichum*, and thus would include the *muricatum–hirsutum* group of *Staurastrum*. Third is the genus *Raphidiastrum* with its smooth walls whose angles bear two or three spines as in *S. brasiliense*. For *Staurastrum*-like cells with smooth cell walls but whose angles bear a single spine and so are identical with species of *Arthrodesmus* (except for the fact that they are more than biradiate) Teiling (1948) erected the genus *Staurodesmus* (see, however, Bicudo 1975). This new genus has been widely accepted by European taxonomists but surprisingly Prescott *et al.* (1982) reject it as "not a real help to desmid taxonomy". Sadly, neither have they made any attempt to unravel the profound taxonomic problems associated with the genus *Staurastrum*.

Bourrelly (1966) has stressed the need for a thorough-going revision of

Staurastrum. In the interim he has followed the lead given long ago by Turner (1894) and much later Hirano (1955–1959), and divides the genus into two sub-genera: *Prostaurastrum*, whose mostly triangular semicells are without processes developed at the angles, and *Staurastrum* in which the processes are clearly developed as arms. Within these sub-genera are recognised 13 sections which in any other group of algae would I feel be surely recognised as separate genera. It is to be hoped that within the next 30 years someone is brave enough to take up the daunting challenge of the confused and confusing taxonomy of *Staurastrum*.

DESMID PHYTOGEOGRAPHY AND TAXONOMY

Because of their poor powers of dispersal and because most desmids very rarely reproduce sexually many populations, especially of phytoplankters, are probably clones which may have propagated themselves by purely vegetative means over periods of possibly thousands of years (Brook, 1959a). Such populations are of interest in evolutionary terms with respect to the "Founder Principle" (Mayr, 1942), i.e. chance factors may be associated with natural selection and be responsible for their differentiation with reference to certain characters which have little apparent adaptive value. One immediately questions in this connection the adaptive value of the elaborate ornamentation of cell walls of many desmid species. Thus local races of a desmid (formenkries) may arise in a lake or group of lakes within a restricted area. Indeed there is good evidence that several planktonic taxa have a fairly circumscribed distribution. Distinctive forms of a particular species have thus been described from different lakes even within such a comparatively small geographic area as the British Isles. In some cases, depending on the predelictions of the investigator, these have been named as distinct varieties or even given specific rank. Examples are *Xanthidium controversum* var. *planctonicum* (probably a local race of *X. antilopeum*) restricted to the north-west of Scotland and lochs in the Hebrides, *Staurastrum cingulum* var. *affine*, confined to Caithness, the Orkneys and Shetlands, and *X. subhastiferum* var. *murrayi*, fac. *triquetra*, found only in the plankton of Loch Lomond. In the plankton of Lough Corrib in western Ireland there are two very distinctive varieties of *S. furcigerum*, the var. *reductum* and var. *simplicissimum*, and a reduced form of *S. pelagicium* and also the unique *S. dorsidentiferum* (Brook, 1958, 1959a).

"That desmids provide an interesting study of geographical distribution within a natural group" (Fritsch, 1953) has been confirmed by the long-term observations of Heimans (1969) on *Micrasterias* populations in the Netherlands. Among several species of the genus, *M. mahabuleshwarensis* var. *wallichi* was found to occur in its triradiate condition in the central fens (vennen) while the biradiate form existed elsewhere. In 1952 he found for the first time in the Netherlands the closely related but distinctive *M. americana*. It is interesting to note that a short time later Teiling informed him that *M. americana* appeared suddenly in a Swedish pool where *M. mahabuleshwarensis* was common. Heimans could not accept that "the one had originated from the other on the spot: the differences between the two species are too great". However, in North America a whole range of intermediate forms have been found by Prescott and Scott (1952) connecting the two taxa, and they express the view that possibly they are extremes of a very variable species–complex; but if only the European forms were known, "no-one would hesitate to recognize them as two separate species" (Heimans, 1969). As Heimans points out there is a parallel to be found in the distribution of phanerogams. In North America many genera contain series of demonstrably closely-related species, whereas in Europe the same genera contain only a few markedly different species. The explanation would seem to be provided by the great north–south orientation of the American continent as a consequence of which during the Pleistocene Ice Age many plants found ice-free refuges. From these they gradually re-extended their range as the glacial era came to an end. In Europe by contrast only a few species could survive. Heimans goes on to point out that conjugation has never been observed in either taxon of *Micrasterias* under discussion, and thus it must be assumed that all the existing, isolated populations are clones (see, however, Ramanathan, 1962).

Blackburn and Tyler (1980) have isolated strains of the varieties *wallichii*, *typica*, *reducta* and *europa* from localities in Tasmania, Victoria and New South Wales but they suggest that these varietal taxonomic distinctions "are of scant value", and took no account of them in their cytological study of this desmid. They did, however, find a difference in the chromosome numbers between *M. americana* ($n = 89$–93; see Kasprik, 1973) and *M. mahabuleshwarensis* ($n = 55$–64).

NUCLEAR CYTOLOGY AND DESMID TAXONOMY

1. Chromosome Complements

Thirty years ago there was little knowledge of the nuclear cytology of desmids (Wisselingh, 1911, 1912). In the intervening period considerable interest in this aspect of desmid biology was developed by Godward (herself one of Fritsch's students) and her students. They provided, for example, valuable accounts of the structure of the remarkable and very variable nucleolus (King, 1959) of polyploidy (Brandham, 1965) and of meiosis in various desmid species (Brandham and Godward, 1963).

The most extensive study of desmid karyotypes to date has been that of Kasprik (1973) who by using a new microdissection technique, examined 23 species of *Micrasterias* from Scandinavia, Germany and the U.S. As a result he was able to distinguish four distinct karyotype groups on the basis of size, shape and stainability, the clarity of the chromosome boundaries, and the "stickiness" of the matrix. He found evidence for the occurrence of polycentric chromosomes, and also for agametoploidy. For example, from different strains of *M. americana* he counted from 89 to 93 chromosomes and also 135, and noted that chromosomes from cells with high counts were smaller than strains that had fewer chromosomes. Although the highest recorded count was about 205, the chromosomes of this clone were not measurably smaller than those of the 135 clone. Kasprik was uncertain as to the extent to which fusion or fragmentation produced the observed differences in chromosome size and number, nor was he able to deduce a basic chromosome complement for the genus *Micrasterias*. However, it seems possible that the predominance of small chromosomes may represent the primitive condition. At this stage it is not possible to reorganize *Micrasterias* species-groups on the basis of their karyotypes. For example, *Micrasterias* species with the largest and morphologically most complex cells (*M. denticulata, M. rotata, M. torreyi*) and which might reasonably be expected to have some affinity, in fact possess three quite distinctive chromosome complements. On the other hand, Kasprik emphasises that in the evolution of karyotypes simple quantitative changes do not appear to lead to marked morphological changes—at least not in desmids grown in culture.

As a consequence of his *Micrasterias* study Kasprik (1973) suggested that desmid taxonomy might benefit by adding karyological data to descriptions of external cell morphology. In support of this view he points out that Krieger (1939) commented that occasionally *M. denticulata, M. rotata* and *M. thomasiana* can be confused morphologically. However, if the chromosomes can be examined in metaphase then the three species can be readily distinguished (one of course needs populations of actively-dividing cells). It is possibly significant that these taxa and *M. papillifera* are all recorded as reproducing sexually (Krieger, 1939; Ružička, 1981), but zygospores are unknown in *M. americana,* the inference drawn being that sexual reproduction has a stabilizing influence on the karyotype since only closely similar chromosome complements will permit unimpeded pairing at meiosis. Further investigations of the relationship between karotypes and the incidence of sexuality, not only in *Micrasterias* but in other desmid genera, would be instructive.

2. Meiosis in Desmids

The first fully detailed account of meiosis in a desmid was that of Brandham and Godward (1963) in *Cosmarium botrytis*. Ling and Tyler (1974) made a similarly detailed study by observing germinating zygospores from intercrosses between the "species" *Pleurotaenium ehrenbergii, P. mamillatum*, and *P. coronatum* and found a remarkable flexibility in chromosome complements between strains without a corresponding failure in meiosis. The basic number for *P. ehrenbergii* and *P. mamillatum* is $n = 53$, but viable zygospores were formed between typical haploids and presumed diploids ($n = 106$) which developed spontaneously in his cultures. In the ensuing meiosis which accompanied zygospore germination, chromosome counts of 79 were made at metaphase II. Hence there must have been an approximately even distribution of chromosomes to each pole following metaphase I. Of considerable interest to the desmid taxonomist is the fact that what was initially presumed to be interspecific compatability was later treated with caution (Ling and Tyler, 1976). They based the distinction between their three species entirely on comparative iconography but were later informed by Bourrelly that their three presumed taxa were but forms of *P. mamillatum*. However, they conclude their paper (Ling and Tyler, 1976) by stating "That is not quite the end of the matter". They point out that if they relied entirely on comparative iconography then they could quite legitimately identify their desmids as distinct species, and go on to quote examples of figures by several "well-respected doyens of desmidology", for example, Irene-Marie, West and West, and Prescott and Scott. They graciously accept Bourrelly's authoritative opinion that all their strains are of *P. mamillatum*—but one can envisage a somewhat frustrated pair of hardworking phycologists saying (in the outback!) "What are we to do?" Writing with some warmth—if not a little heat—at the end of their paper "Other workers in other parts of the world are likely to be faced with the same problems of the iconotype and the bewildering array of taxa in the genus *Pleurotaenium*".

3. Polyploidy and Taxonomy

Kasprik (1973) reported an interesting case of an aneuploid series correlated with morphological differences in *Micrasterias thomasiana*. In cultures of this desmid, clearly-distinguishable variants of "normal" forms were found to be hyperhaploids. The basic chromosome number is $n = 39$ but one variant with over-lapping lateral processes was an $n = 40$ while an even more abnormal form was an $n = 46$. Two further variants which were triradiate had compliments $n = 70$ and $n =$ about 75. Some cells of the latter clone were inclined to develop more or less typical *M. thomasiana* morphologies except that they were significantly larger. Brandham (1965) found that diploids which arose spontaneously in an originally haploid clone of the normally triradiate *Staurastrum denticulatum* became almost exclusively quadriradiate. The haploids were 26.10 μm long while the diploids, in addition to being greater than 40.00 μm long, showed other differences in their morphology. In a non-sexual clone of *S. dilatatum* which contained both 3- and 4-radiate and also three/four janus cells measuring 23.75 μm long, larger cells (34.5 μm long) were found. When these were isolated, the resulting population consisted entirely of large, mostly 5-radiate cells with a few that were 6-radiate. Based on his experience of other cells, Brandham inferred that the larger cells were probably diploids or at least hyperhaploids. A very large cell (49.5 μm long) was seen in one of the "diploid" clones and this possessed two whorls of 8–10 processes on each semicell. Brandham presumed that this cell represented an even higher degree of polyploidy and suggested that it might even be a tetraploid.

This latter observation especially may be of significance in relation to Fritsch's urgings that desmids should be reorganised into species–groups, for he wrote

> In a polyphyletic assemblage like that of the desmids in which evolution appears to have exhausted almost all possibilities and homoplastic development is so widespread, generic distinction is bound to be arbitrary and highly artificial and to fail altogether among borderline species which are transitional in character; there is, however, no way of classifying the huge number of forms involved. I am of the opinion that a true assessment of affinities can only be undertaken when these artificial genera are ignored and the recorded forms of desmids are assembled on the basis of their fundamental cell shape. . . . The species groups are an expression of the multiplicity of evolutionary series among desmids, they transcend present generic limits and . . . their value lies in placing comparable forms side by side and displaying what are probably the true inter-relationships of their members.

If we examine, for example, Fritsch's omniradiate *Cosmarium moniliforme*– group it is quite clear that in terms of basic body shape no distinction can be drawn between species called *Cosmarium* and those assigned to *Staurastrum*. The latter differ only in the possession of processes, mostly apical in position. In view of our recent knowledge of what an extra chromosome or two can do to desmid morphology is it too much to suggest that this group could be part of a polyploid series? In this connection it is also of interest to note that those species with smooth outlines are cosmopolitan in their distribution, the granulate types are mostly restricted to the northern hemisphere while the more elaborate taxa are from the southern hemisphere, largely South America. No-one has taken up Fritsch's radical but eminently sensible suggestion. There is still considerable "overexaggeration of the importance of minor detail" and Mayr's (1982) recent comments therefore surely apply: "The detractors of taxonomy have not without reason ridiculed those taxonomists who seem to have no other research objective than to describe ever more new species as if this activity were the alpha and omega of taxonomic science".

REFERENCES

Bicudo, C. E. M. (1975). Polymorphism in the desmid *Arthrodesmus mucronulatus* and its taxonomic implications. *Phycologia* **14,** 145–148.

Blackburn, S. I. and Tyler, P. A. (1980). Conjugation, germination and meiosis in *Micrasterias mahabuleshwarensis* Hobson (Desmidiaceae). *Br. phycol. J.* **15,** 83–93.

Bourrelly, P. (1966). "Les Algues d'eau douce. Initiation à la Systématique. Tome I. Les Algues Vertes" (rev. ed.). Boubée, Paris.

Brandham, P. E. (1965). Polyploidy in desmids. *Can. J. Bot.* **43,** 405–417.

Brandham, P. E. and Godward, M. B. E. (1963). Mating types and meiosis in desmids. *Br. phycol. Bull.* **2,** 280–281.

Brook, A. J. (1958). Desmids from the plankton of some Irish loughs. *Proc. R. Ir. Acad., Sect. B.* **59B,** 71–91.

Brook, A. J. (1959a). The status of desmids in the plankton and the determination of phytoplankton quotients. *J. Ecol.* **47,** 429–445.

Brook, A. J. (1959b). The published figures of the desmid *Staurastrum paradoxum*. *Rev. Algol.* [N.S.]**4,** 239–255.

Brook, A. J. (1959c). Notes on desmids of the genus *Staurastrum*. III. *S. paradoxum* Meyen in the Jenner Herbarium of the British Museum. *Naturalist, London* July–Sept. 1959, 81–83.

Brook, A. J. (1959d). *Staurastrum paradoxum* Meyen and *S. gracile* Ralfs in the British freshwater plankton, and a revision of the *S. anatinum*-group of radiate desmids. *Trans. R. Soc. Edinburgh* **33,** 589–628.

Brook, A. J. (1960a). Some additional figures of the desmid *Staurastrum paradoxum*. *Rev. algol.* [N.S.]**5,** 208–210.

Brook, A. J. (1960b). The varieties of *Staurastrum paradoxum* Meyen - nomen dubium. *Nova Hedw.* **1,** 431–442.

Brook, A. J. (1965). Planktonic algae as indicators of lake types with special reference to the Desmidiaceae. *Limnol. Oceanogr.* **10,** 403–411.

Brook, A. J. (1981). "The Biology of Desmids," Bot. Monogr. Vol. 16. Blackwell, Oxford.

Brook, A. J. (1982). Desmids of the *Staurastrum tetracerum*-group from a eutrophic lake in mid-Wales. *Br. phycol. J.* **17,** 259–274.

Brook, A. J. and Williamson, D. B. (1983). On *Staurastrum botrophilum* Wolle, a rare and inadequately described desmid. *Br. phycol. J.* **18,** 69–72.

Couté, A. and Tell, G. (1981). Ultrastructure de la paroi cellulaire des Desmidiacées au microscope electronique à balayage. *Beih. Nova Hedw.* **68,** 1–228.

Engels, M. and Lorch, D. W. (1981). Some observations on cell wall structure and taxonomy of *Phymatodocis nordstedtiana*. *Plant Syst. Evol.* **138,** 217–225.

Fott, B. (1959). "Algenkunde." Fischer, Jena.

Fott, B. (1965). Evolutionary trends among algae and their position in the plant kingdom. *Preslia* **37,** 117–26.

Fritsch, F. E. (1907). The desmids of the tropics. The subaerial and freshwater algal flora of the tropics. A phytogeographical and ecological study. *Ann. Bot. (London)*, **21**, 235–275.
Fritsch, F. E. (1930). Uber Entwicklungstendenzen bei Desmidiaceen. *Z. Bot.* **23**, 402–418.
Fritsch, F. E. (1933). Evolutionary sequence among the desmids. *J. Bot., Br. Foreign* **71**, 200–201.
Fritsch, F. E. (1953). Comparative studies in a polyphyletic group, the Desmidiaceae. *Proc. Linn. Soc. London* **164**, 258–280.
Fritsch, F. E. and Rich, F. (1928). Contributions to our knowledge of the freshwater algae of Africa. 7 Freshwater algae (exclusive of diatoms) from Griqualand West. *Trans. R. Soc. S. Afr.* **18**, 1–92.
Fritsch, F. E. and Rich, F. (1937). Contributions to our knowledge of the freshwater algae of Africa. Algae from the Belfast Pan, Transvaal. *Trans. R. Soc. S. Afr.* **25**, 153–228.
Gerrath, J. F. (1969). *Penium spinulosum* (Wolle) nov. comb. a taxonomic correction based on cell wall ultrastructure. *Phycologia* **8**, 109–118.
Gerrath, J. F. (1979). Polymorphism in the desmid *Cosmarium taxichondrum* Lundell. *Br. phycol. J.* **14**, 211–217.
Gerrath, J. F. (1982). Morphological variation in two populations of *Xanthidium tetracentrotum* Wolle var. *hexagonum* G. M. Smith (Desmidiaceae). *Br. phycol. J.* **17**, 411–418.
Gerrath, J. F. (1983). Polymorphism in the desmid *Staurastrum pentacerum* (Wolle) G. M. Smith. *Br. phycol. J.* **18**, 141–150.
Godward, M. B. E. (1966). "The Chromosomes of Algae." Edward Arnold, London.
Heimans, J. (1969). Ecological phytogeographical and taxonomic problems with desmids. *Vegetatio* **17**, 50–82.
Hirano, M. (1955–1959). Flora Desmidiarium Japonicum. I–VI. *Contrib. Biol. lab. Kyoto Univ.* pp. 1–386.
Ichimura, T. (1982). Isolating mechanisms in speciation of *Closterium*. *Jpn. J. Phycol.* **30**, 332–343.
Kallio, P. (1953). On the morphogenetics of the desmids. *Bull. Torrey Bot. Club* **80**, 247–263.
Kasprik, W. (1973). Beiträge zur Karyologie der Desmidiaceen - Gattung *Micrasterias* Ag. *Beih. Nova Hedw.* **42**, 115–137.
King, G. C. (1959). The nucleoli and related structures in the desmids. *New Phytol.* **58**, 20–28.
Krieger, W. (1939). Die Desmidiaceen Europas mit Berücksichtigung der aubereuropäischen Arten. *In* "Kryptogamen-Flora von Deutschland, Oesterreich und der Schweiz" (L. Rabenhorst, ed.), Vol. XIII, Sect. 1, Part 2, pp. 1–117. Akad. Verlagsges., Leipzig.
Lenzenweger, R. (1979). Algologische Notizen. IV. Beobachtungen zur Variabilitat von *Micrasterias americana* (Ehr.) Ralfs. *Linzer biol. Beitr.* **11**, 271–278.
Lind, E. M. and Croasdale, H. (1966). Variation in the desmid *Staurastrum sebaldi* var. *ornatum*. *J. Phycol.* **2**, 111–116.
Ling, H. U. and Tyler, P. A. (1974). Interspecific hybridity in the desmid genus *Pleurotaenium*. *J. Phycol.* **10**, 225–230.
Ling, H. U. and Tyler, P. A. (1976). Meiosis, polyploidy and taxonomy of the *Pleurotaenium mamillatum* complex (Desmidiaceae). *Br. phycol. J.* **11**, 315–330.
Lütkemuller, J. (1902). Die Zellmembran der Desmidiaceen. *Beitr. Biol. Pflanz.* **8**, 347–414.
Mayr, E. (1942). "Systematics and the Origin of Species." Columbia Univ. Press, New York.
Mayr, E. (1982). "The Growth of Biological Thought." Belknap Press, Cambridge, Massachusetts.
Meyen, F. J. F. (1828). Beobachtungen über einige niedere Algenformen. *Nova Acta phys. med.* **14**, 768.
Meyer, K. I. (1962). Über das phylogenetischen System der grünen Algen (Chlorophycophyta). *Preslia* **34**, 147–158.

Mix, M. (1972). Die Feinstruktur der Zellwände bei Mesotaeniaceae und Gonatozygaceae mit einer vergleichenden Betrachtung der verschiedenen Wandtypen der Conjugatophyceae und über deren systematischten Wert. *Arch. Mikrobiol.* **81,** 197–220.

Mix, M. (1975). Die Feinstruktur der Zellwände der Conjugaten und ihre systematische Bedeutung. *Beih. Nova Hedw.* **42,** 179–194.

Palamar-Mordvintseva, G. M. (1976). A taxonomic analysis of the genus *Staurastrum* Meyen. *Ukr. bot. Zh.* **33,** 31–38.

Pickett-Heaps, J. D. and Marchant, H. J. (1972). The phylogeny of the green algae: A new proposal. *Cytobios* **6,** 255–64.

Prescott, G. W. and Scott, A. M. (1952). The genus *Micrasterias*. *Trans. Am. microsc. Soc.* **71,** 229–252.

Prescott, G. W., Croasdale, H. T., Vinyard, W. C. and Bicudo, C. E. de M. (1982). "A Synopsis of North American Desmids. Part II. Desmidiaceae: Placodermae Section 4." Univ. of Nebraska Press, Lincoln.

Ralfs, J. (1848). "The British Desmidieae." Reeve, Benham & Reeve, London.

Ramanathan, K. R. (1962). Zygospore formation in some South Indian desmids. *Phykos* **1,** 38–43.

Round, F. E. (1963). The taxonomy of the Chlorophyta. *Br. phycol. Bull.* **2,** 224–235.

Round, F. E. (1971). The taxonomy of the Chlorophyta. II. *Br. phycol. J.* **6,** 235–264.

Ružička, J. (1966). Zur Variabilität der infraspezifischen Taxa der Desmidiaceen (*Cosmarium laeve* Rab. f. *Majus* Borge). *Arch. Protistenkd.* **109,** 125–138.

Ružička, J. (1981). "Die Desmidiaceen Mitteleuropas," Vol. I, No. 2. Schweizerbart'sche, Stuttgart.

Smith, G. M. (1924). The Phytoplankton of the inland lakes of Wisconsin. *Bull.—Wis. geol. nat. Hist. Surv.* **1270,** 1–227.

Starr, R. C. (1958). The production and inheritance of the triradiate form in *Cosmarium turpinii*. *Am. J. Bot.* **45,** 243–248.

Tassigny, M. (1966). Étude critique de genre *Closterium* (Desmidiales) le group *setaceum - kuetzingii*. *Rev. algol.* [N.S.]**8,** 228–250.

Teiling, E. (1948). *Staurodesmus,* genus novum. *Bot. Not.* **101,** 49–83.

Teiling, E. (1950). Radiation in desmids, its origin and consequences as regards taxonomy and nomenclature. *Bot. Not.* **103,** 299–327.

Turner, W. B. (1894). Algae aquae dulcis Indiaeorientalis. Fresh-water algae (principally Desmidieae) of East India. *K. sven. Vetenskaps Akad. Handl.* **25**(5), 1–187 (1892).

Watanabe, M. M. and Ichimura, T. (1978). Biosystematic studies of the *Closterium peracerosum-strigosum-littorale* complex. II. Reproductive isolation and morphological variation among several populations from Northern Kanto area in Japan. *Bot. Mag.* **91,** 1–10.

West, W. and Carter, N. (1923). "The British Desmidiaceae," Vol. V. Ray Society, London.

West, W. and West, G. S. (1904). "A Monograph of the British Desmidiaceae," Vol. I. Ray Society, London.

West, W., and West, G. S. (1905). "A Monograph of the British Desmidiaceae," Vol. II. Ray Society, London.

West, W., and West, G. S. (1908). "A Monograph of the British Desmidiaceae," Vol. III. Ray Society, London.

West, W., and West, G. S. (1912). "A Monograph of the British Desmidiaceae," Vol. IV. Ray Society, London.

Wisselingh, C. van (1911). On the structure of the nucleus and karyokinesis in *Closterium ehrenbergii*. *Proc. Sect. Sci. ned. Akad. Wet.* **13,** 365–375.

Wisselingh, C. van (1912). Über die Kernstructur und Kernteilung bei *Closterium*. *Beih. bot. Zentralbl.* **29,** 409–432.

10 | Systematics of the Siphonales

L. HILLIS-COLINVAUX

Department of Zoology, Ohio State University, Columbus, Ohio, USA

Abstract: Early concepts of the Siphonales are reviewed and a definition, somewhat broader than that used by G. M. Smith, is selected to provide the framework of the paper. Criteria for identifying genera and elucidating generic relationships are discussed, with emphasis on the 16 genera of the *Caulerpa-Halimeda-Udotea* (CHU) group. These criteria include habit, anatomy, chemistry and structure of filament walls, growth patterns, morphology and physiology of sexual (or specialized) reproductive structures, life history data, cloning mechanisms, allelochemicals and other metabolic compounds, cytology, fossil data, and geographic distribution. Three other categories, the *Bryopsis-Derbesia-Codium* (BDC) group, the *Ostreobium* group, and the *Dichotomosiphon* group, are included in the discussion. The limited data available for many of the characteristics make an analysis that includes all the genera difficult. Nonetheless, there appear to be two definite lines of development that generally follow the lines of the BDC and CHU groups. Characters separating them are not considered to be of ordinal stature; hence they are designated as suborders and given the names Bryopsidineae and Halimedineae in the order Bryopsidales. A full scheme of classification, and a protoscheme of suggested phylogenetic relationships are developed.

DEFINITION OF SIPHONALES: PHASE I

Greville (1830) introduced the descriptive term Siphoneae for plants with a "frond either composed of membranaceous, filiform, continuous, simple or branched tubes, or formed of a combination of similar tubes", and assigned to it *Codium, Bryopsis, Botrydium,* and *Vaucheria. Caulerpa* he placed in a separate group, the Caulerpeae, while *Valonia* and *Anadyomene* were assigned to the Ulvaceae (Table I).

Some years later Harvey (1858) provided an early perceptive synthesis of a much larger group of siphonaceous algae in his treatise on North American algae. He broke from the pattern of several other systematists of the

Table I. Classification schemes important in the definition of siphonaceous algae.

Greville (1830)	Harvey (1858)	Blackman and Tansley (1902)	Fritsch (1935, 1948) (includes fossil genera)	Smith (1938, 1955)
		Ser. Siphonales	Siphonales	Siphonocladales
Ulvaceae	Valoniaceae	Siphonocladaceae		
Anadyomene	Anadyomene, Apjohnia,	Valoniaceae	Valoniaceae	
Valonia	Blodgettia, Chamaedoris, Dic-	Anadyomene, Apjohnia,	Anadyomene, Boodlea,	
Alysium	tyosphaeria, Microdictyon,	Boodlea, Chamaedoris,	Chamaedoris, Cladoprop-	
Enteromorpha	Penicillus (includes Rhi-	Cystodictyon, Dictyosphaeria,	sis, Dictyosphaeria, Ernodes-	
Porphyra	pocephalus), Struvea, Valonia	Halicystis, Microdictyon, Si-	mis, Microdictyon, Siphono-	
Tetraspora		phonocladus, Struvea,	cladus, Struvea, Valonia	
Ulva		Valonia		
		Gomontiaceae		
		Gomontia		
		Cladophoraceae		Cladophorales
		Chaetomorpha, Cladophora,		
		Pithophora, Rhizoclonium,		
		Urospora		
		Sphaeropleaceae		
		Sphaeroplea		
Siphoneae	Siphonaceae	Siphoneae		Siphonales
Botrydium		Protosiphonaceae	Protosiphonaceae	
Bryopsis		Codiolum, Protosiphon	Follicularia, Halicystis, Pro-	
Codium			tosiphon, Sphaerosiphon	
Vaucheria				
		Bryopsidaceae		Bryopsidaceae
		Bryopsis		Bryopsis
		Derbesiaceae	Derbesiaceae	Halicystidaceae
		Derbesia	Derbesia	Derbesia, Halicystis

Caulerpeae	Caulerpeae	Caulerpaceae	Caulerpaceae
Caulerpa	*Caulerpa*	*Bryopsis, Caulerpa, Pseudobryopsis*	*Caulerpa*
Codieae	Codiaceae	Codiaceae	Codiaceae
calcareous *Halimeda, Udotea* noncalcareous *Botrydium Bryopsis, Chlorodesmis, Codium, Vaucheria*	*Avrainvillea, Callipsygma, Codium, Halimeda, Penicillus, Pseudocodium, Rhipocephalus Udotea*	*Avrainvillea, Boodleopsis, Boucina, Codium, Dimorphosiphon, Halimeda, Ovulites, Paleoporella, Penicillus, Pseudocodium, Rhipidodesmis, Udotea* (including *Flabellaria*)	*Codium*
		Vaucheriaceae	Dichotomosiphonaceae
		Dichotomosiphon, Pseudodichotomosiphon, Vaucheria, Vaucheriopsis	*D. pusillus, tuberosum*
Dasycladeae	Verticillatae	Dasycladaceae	Dasycladales
Acetabularia, Cymopolia, Dasycladus, Neomeris, Polyphysa	Subfamily Dasycladeae *Botryophora, Chlorocladus, Dasycladus* Subfamily Cymopolieae *Bornetella, Cymopolia, Neomeris* Subfamily Acetabularieae *Acetabularia, Acicularia, Chalmasia, Halicoryne*	*Acetabularia, Acicularia, Batophora, Bornetella, Coelosphaeridium, Cymopolia, Dactylopora, Dasycladus, Digitella, Diplopora Gyroporella, Halicoryne, Mizzia, Neomeris, Palaeodasycladus, Primicorallina, Rhabdoporella, Thyrsoporella, Triploporella, Vermiporella*	
		Chaetosiphonaceae	
		Blastophysa, Chaetosiphon,	
		Phyllosiphonaceae	
		Ostreobium, Phyllosiphon, Phytophysa	

time by not placing *Caulerpa* in its own "order" (= family) on the basis of the many filamentous extensions (trabeculae) of the wall into the central part of the filament, but included it in the Siphonaceae. He writes that "an unwillingness needlessly to multiply families, and a belief that *synthesis,* much more than *analysis,* ought to be the study of a system framer, has prevented my adopting these views".

By 1902, in a revision of the classification of green algae, Blackman and Tansley clearly treated the Siphonales as an order, and subdivided it into two, the Siphonocladeae, with septate thalli, and the Siphoneae, with thalli lacking cross walls except at the bases of reproductive structures. The broadness of their definition is shown by a glance at the genera included (see Table I).

Almost immediately some portions were split off into new orders [the Cladophorales of West (1904) and the Siphonocladales of Oltmanns (1905) are examples], but a broad definition, modified by Fritsch (see Table I), continued to the midpoint of the twentieth century. Concurrent with the Fritschian concept were classifications that interpreted the Siphonales more strictly. They included in the order only multinucleate algae that were without septa except at the bases of reproductive units. The widely accepted system of Smith (1938, 1955) is a good example.

The time period embraced by these taxonomic developments, 1830 through part of the 1940s, I interpret as phase I in the definition of the Siphonales. At the end of it most of the genera had been described, good morphological studies existed for many, and progress had been made toward achieving taxonomic order. Additional aspects of the changes that occurred are provided by Egerod (1952), Hillis (1959), Round (1963, 1971, 1973), Ducker (1965), and Hillis-Colinvaux (1980).

SIPHONALES: PHASE II

From the mid-1940s research has tended to be concentrated on genera, and has provided an opportunity for the development and some assessment of an increasing array of systematic characters. Generally the taxonomic framework chosen embodies the stricter definition of the Siphonales, one that is close to Smith (1955) but that is extended in a number of classifications to include families from Fritsch such as the Phyllosiphonaceae.

A broadened Smith definition is the one I will use in this paper. It includes 24 genera that, for convenience of analysis, are subdivided into the following categories: *Bryopsis-Derbesia-Codium,* or BDC, group (6 genera); *Caulerpa-Halimeda-Udotea,* or CHU, group (16 genera); *Ostreobium* group (1 genus); and *Dichotomosiphon* group (1 genus).

Since the second category is the largest, and much less well known than the first, emphasis will be on systematic characters now used or of potential importance in this category. The criteria are not peculiar to the CHU group, however, as will be apparent in the discussions of the characters, and of interrelationships among the four groups.

BRYOPSIS-DERBESIA-CODIUM GROUP (BDC)

The genera of this category are *Bryopsis, Bryopsidella, Codium, Derbesia (Halicystis), Pedobesia,* and *Pseudobryopsis*. Life history and morphological characters are important systematic criteria, and recently *Derbesia* and *Bryopsis* have been brought together into the same family (Rietema, 1975; Hillis-Colinvaux, 1980) on the basis of life history interrelationships as elucidated by a number of workers including Neumann (1969) and Rietema (1975).

Life history studies on *Bryopsidella* (Rietema, 1975) and *Pseudobryopsis* (Mayhoub, 1974; Tatewaki, 1979) indicated close relationships to the preceding genera. *Pedobesia* also shares a number of traits with these genera, including stephanokont zoospores, but is the only member of the group in which sexual reproduction is unknown, and is unusual in possessing an internal calcareous skeleton (MacRaild and Womersley, 1974).

Codium, with its multiaxial construction, stands apart from the preceding taxa, which are constructed of simple filaments or vesicles (morphological data are provided by Silva, 1951, 1960, 1962). It nonetheless shares many criteria with the preceding taxa, while all of them differ from the CHU group in a number of important characters. Discussion of these aspects will be incorporated into the next sections.

CAULERPA-HALIMEDA-UDOTEA GROUP (CHU)

For convenience of analysis I have placed in this group the taxa I assigned to the Caulerpales in 1980 and the genus *Johnson sea linkia* (Eiseman and Earle, 1983). The complete list is *Halimeda, Penicillus, Rhipocephalus, Tydemania, Udotea* (which are all calcareous), *Avrainvillea, Boodleopsis, Callipsygma, Caulerpa, Chlorodesmis* (including *Rhipidodesmis;* see Ducker, 1967), *Cladocephalus, Johnson-sea-linkia, Pseudochlorodesmis, Pseudocodium, Rhipilia,* and *Rhipiliopsis*, which includes *Geppella* in part or entirely (cf. Farghaly and Denizot, 1979) and *Siphonoclathrus* (for description and comment on *Siphonoclathrus* see Earle and Young, 1972, and Farghaly and Denizot, 1979).

The largest green algae belong to this group, and the fact that it is so poorly known is somewhat of a paradox. The collections for several of the genera are meagre, and since we know little more than their superficial morphology, their assignment to the CHU group must remain tentative. Nonetheless, for morphological and other reasons that will be discussed below, all the taxa except *Pseudocodium* and possibly *Boodleopsis* appear more suitably placed in the CHU category than in any other. The misfit of *Pseudocodium* will be discussed separately.

1. *Criteria Useful for Generic Separation*

The principal characters are habit, anatomy, and calcification.

(*a*) *Habit*. General appearance ranges from a loose aggregation of typical coenocytic filaments as in *Chlorodesmis, Boodleopsis, Pseudochlorodesmis,* and *Penicillus sibogae*, to a variety of distinctive phylloid, flabellate, cyathiform, capitate (*Penicillus*), catenulate (*Halimeda*), and globuliferous (*Tydemania*) shapes that are characteristic for many of the genera. Stipes, rhizomes, and much enlarged holdfasts are associated with some.

(*b*) *Anatomy*. (i) Axial filaments. The complexities of both branching and organization of main (medullary) filaments as well as their diameters are important characteristics in the separation of genera and species within the CHU group. New main filaments are produced by branching and may be dichotomous, trichotomous, or occasionally whorled as in *Tydemania expeditionis, Boodleopsis,* and in the capitulum of *Rhipocephalus*. Variations include the lateral displacement of one of the dichotomies, or the absence of wall thickening at one of the junctures. In some genera the portions between successive forkings may be variously swollen and constricted, or one of the dichotomies may support a series of short lateral branches as in *Cladocephalus* or *Rhipilia*.

Dichotomous branches, if contiguous and in one plane, form a thin flat blade as in *Callipsygma, Udotea javensis, U. glaucescens, Tydemania,* and *Rhipocephalus*. Where the branches are interwoven a thicker blade results, as in *Avrainvillea*. If branches diverge and are in different planes a glomerulus (*Tydemania expeditionis*) or capitulum (*Penicillus*) develops. Gepp and Gepp (1911) provided illustrations and additional details.

(ii) Papillas, outgrowths, and lateral branches. The main filaments of some genera and species also bear papillae, short outgrowths, or lateral branches of varying shapes and complexity. All are of some taxonomic significance. In genera such as *Rhipiliopsis* the structures are to the sides and some of the adjacent short branches characteristically adhere (called pseudoconjugation by Gepp and Gepp, 1911). In other genera (e.g. *Udotea* sp.) outgrowths occur mostly on the two surfaces of the filament that face the exterior of the frond.

(iii) Cortex. In some genera lateral branches (either true or displaced dichotomous ones) proliferate into a series of short, closely placed branches that touch or adhere. This lateral branch system or cortex represents the most complex morphological development of the group and occurs in the fronds of many species of *Udotea*, all *Halimeda* and *Pseudocodium* species, as well as the stipes of *Penicillus, Rhipocephalus,* and all *Udotea* species except *javensis* and *papillosa*. In the BDC group only *Codium* is corticated.

(c) *Calcification.* The genera *Halimeda, Penicillus, Rhipocephalus, Tydemania* and *Udotea* [with the exception of the species *U. petiolata* (= *U. minima*)], consistently deposit calcium carbonate in the conspicuous phase in their life history. In all five genera the calcium carbonate species is aragonite (McConnell and Colinvaux, 1967). For *Halimeda,* calcium carbonate deposition begins in the newly formed segment when it is about 36 h old (Wilbur *et al.,* 1969). The remaining genera of the group are uncalcified.

2. Criteria for Elucidating Generic Relationships

Characteristics that are currently or potentially useful are discussed below. Only some aspects of these can be considered here.

(a) *Habit and anatomy.* These two characters, described in the preceding section, are the best known and most traditional. In the CHU group the variety in plant architecture is extensive, and ranges from basic filamentous forms as in *Chlorodesmis* to the corticated complexity of *Pseudocodium* and *Halimeda.* These complex forms are also multiaxial except for the species *H. cryptica* (Colinvaux and Graham, 1964; Hillis-Colinvaux, 1980). The remaining genera, those with capitate, globuliferous, and a variety of phylloid habits, are of intermediate complexity. Dimorphy is exhibited by the uniaxial *Tydemania expeditionis,* which sometimes has monostromatic flabella near the base of its glomeruliferous thalli.

Although a number of genera appear very similar, close relationships cannot be assumed, nor can one determine the direction of evolution on the basis of morphological complexity alone.

(b) *Chemistry and structure of filament walls.* (i) Chemistry. For the siphonous algae the predominant wall compound is not cellulose. In the CHU group, as far as is known, the main component is β-1, 3 xylan, which is reported for *Caulerpa, Chlorodesmis, Halimeda, Udotea* (Miwa *et al.,* 1961; Parker, 1970), and *Penicillus* (Frei and Preston, 1964). Absence of cellulose in filament walls was one of two characters used by Feldmann (1946) to divide the Siphonales into the orders Caulerpales (heteroplastic genera possessing filament walls without cellulose) and Eusiphonales (homoplastic genera with walls containing cellulose). Feldmann's source of data on wall chemistry was Mirande (1913), who reported the prominence of callose, a polymer of β-1, 3-linked glucose residues (Aspinall and Kessler, 1957) in 17 genera (22 species) of siphonaceous algae.

Wall chemistry data for other genera of Siphonales demonstrate the complexity of this criterion and the difficulties of applying it broadly. The principal wall polysaccharide of *Pseudocodium floridanum* (Dawes and Mathieson, 1972) and the sporophytic stages of *Bryopsis plumosa* and *Derbesia tenuissima* is a mannan (Huizing and Rietema, 1975), whereas xylan and cellulose together are important components of the filament walls of gametophytes of the latter two taxa. Mannan is the main polysaccharide in the conspicuous form of *Codium*, but undifferentiated *Codium* filaments contain much glucan as well (Hoek, 1981).

Although xylan may be a principal wall component in the CHU series (with the known exception of *Pseudocodium*) this character, as previously pointed out (Hillis-Colinvaux, 1980), can have but a supporting role until data are available on more taxa as well as on holdfasts and life history stages.

(ii) Structure. Filament walls of some genera have an elaborate series of inward extensions or trabeculae. They are well known for *Caulerpa*, an early description being provided by Montagne (1838), and in many classifications are the sole justification given for its assignment to a separate family. It is noteworthy, therefore, that trabeculae or trabecula-like structures occur in species of *Codium* (Silva, 1951) and *Halimeda* (Borowitzka and Larkum, 1977).

(c) *Growth patterns.* This character is subdivided into the type of growth, method of initiation of new thalli, and nature of growth axis.

(i) Type of growth. A distinction can be made between thalli with a capacity for extensive and essentially indeterminate growth such as *Caulerpa*, *Halimeda*, and *Tydemania*, and those with determinate growth, such as *Penicillus* (Colinvaux et al., 1965), in which there is essentially no growth after a typical "brush" is formed. *Rhipocephalus* and some species of *Udotea* (e.g., *flabellum*) show a capacity for at least limited extended growth by the addition of new zones, lobes, or blades (Colinvaux et al., 1965). Indeterminate growth of the photosynthetic thallus is not to be confused with growth that leads to cloning.

(ii) Initiation of new thalli. Many, and in fact probably all, of the genera can develop new thalli asexually (Colinvaux et al., 1965, Colinvaux, 1968b, Hillis-Colinvaux, 1973, 1980). Friedmann and Roth (1977), in their detailed description of the initiation of new thalli of *Penicillus capitatus*, indicated that a new plant develops from a single initial. They report a single initial for *Rhipocephalus* also, but a group initial for *Avrainvillea* (Roth and Friedmann, 1981).

(iii) Growth axis. In *Caulerpa* the principal growth axis is a rhizome. The main axis of *Tydemania* also may be rhizomatous, at least in part. However, among morphologically complex CHU genera this condition is unusual. The principal expression of this character is as a single (uniaxial) erect filament as in the stipe of *Udotea javensis* or a multiple (multiaxial) group of erect filaments as in *Udotea flabellum* stipes.

(d) *Morphology and physiology of sexual (or specialized) reproductive structures.*
(i) Background. Sexual characteristics are likely to be especially important in elucidating evolutionary trends among genera of the CHU series. At present, however, specialized structures have been discovered only for approximately half the genera, and for a number of these we do not know yet whether the units are sexual or asexual. We have some information for *Avrainvillea* (Howe, 1907; Gepp and Gepp, 1911; Roth, 1977; Young, 1977), *Boodleopsis* (Taylor *et al.,* 1953; Cribb, 1954; Trono, 1971), *Caulerpa* (Montagne, 1838; Dostal, 1928; Iyengar, 1933, 1940; Miyake and Kunieda, 1937; Schussnig, 1929, 1939; Kajimura, 1968a, 1968b, 1977; Goldstein and Morrall, 1970; Price, 1972; Lohr and Dawes, 1974; Meinesz, 1979), *Chlorodesmis* (Ducker, 1965), *Halimeda* (Hillis, 1959; Hillis-Colinvaux, 1980), *Penicillus* (Colinvaux, 1969, Hillis-Colinvaux, 1973; Meinesz, 1975), *Pseudocodium* (Dawes and Mathieson, 1972), and *Udotea* (Yamada, 1934; Nasr, 1939; Nizamuddin, 1963; Meinesz, 1969, 1972c). Much of the information is available also in the review of reproduction in the Udoteaceae by Meinesz (1980b).

Events and structures are best known for *Halimeda,* and the following baseline information (Hillis-Colinvaux, 1980) pertains to the genus. Sexual reproduction is signalled by: development of stalked gametangia on the outside of segments (they look like minature bunches of grapes); transfer of the free organic matter of thallus, including reserves, to gametangia (holocarpy), a process that turns parent thallus white overnight; and release of biflagellated gametes approximately 36 h after gametangial stalks are formed, leaving a white disintegrating thallus that collapses within a day or two in the field, or in 2–3 weeks in culture.

It is this transient nature of the structures coupled with death of the plants that makes sexual reproduction expecially difficult to study in holocarpic genera, and appear rarer than it is. Nonetheless, for *Halimeda* gametangia now have been found for approximately 60% of the species.

(ii) Criteria. Analysis of the data available for specialized reproductive structures indicates that the characters discussed below are likely to provide useful information about relationships within the CHU series and to close relatives. A full evaluation of these is premature, and at best information exists for only about half the genera. Most is known concerning the morphology and location of reproductive structures, characters that are readily evaluated in preserved material.

(iii) Morphology of reproductive structures. These structures range from the essentially unspecialized gametangia of *Caulerpa* to the conspicuous, branched, and swollen external structures of *Halimeda*.

Relatively unspecialized: In *Caulerpa* species the photosynthetic blade, the
   equivalent of a filament, functions as a gametangial structure A number

of simple or branched white discharge papillas develop on it (Goldstein and Morrall, 1970). In *Penicillus capitatus* (Colinvaux, 1969; Hillis-Colinvaux, 1973, 1980; Meinesz, 1975, 1980b) and *Udotea petiolata* (Meinesz, 1969, 1972c, 1980b) filaments of capitulum and blade, respectively, produce or develop into reproductive structures that may be somewhat swollen and have narrowed uncalcified apices that function as discharge tubes. These soft portions extend beyond the capitulum or frond, and in *Penicillus* produce a distinctive fuzz or halo.

Moderately specialized: This category includes genera in which a portion or all of the reproductive filament becomes considerably swollen. In *Udotea orientalis* (Yamada, 1934), *U. indica* (Nizamuddin, 1963), and *U. javensis* (Nasr, 1939; Meinesz, 1980a) the elongate reproductive filament terminates in an uncalcified discharge region. In *Avrainvillea nigricans* (Howe, 1907), *A. erecta* (Gepp and Gepp, 1911), and *Pseudocodium floridanum* (Dawes and Mathieson, 1972) the clavate to spherical swellings occur at the tips of branches and are without an obvious discharge region.

Specialized: Halimeda is assigned to this category. *Chlorodesmis* appears to belong also, but there is information only for the species *baculifera* (= *bulbosa*) (Ducker, 1965). Reproductive structures for both taxa are the greatly swollen tips of a special branch system. There is one (possibly two, shared) central discharge tube. In the calcified genus *Halimeda* these structures, like the holdfast region, are uncalcified.

(iv) Location of reproductive structures. In those genera with thalli more complex than the free filament, two categories can be delimited according to whether reproductive structures exclusive of the discharge region are internal or external. They are internal in *Pseudocodium* and probably so in *Penicillus capitatus, U. petiolata, U. orientalis,* and *U. indica,* but are external in *Avrainvillea, Cladocephalus, Halimeda,* and *U. javensis.* These categories are not directly equivalent to the terms "intrinsic" and "extrinsic" of Meinesz (1980b). The latter are broader and to some extent combine my criteria of morphology and position.

(v) Presence of distinct wall or a plug separating vegetative and reproductive portions. True walls at the base of reproductive structures are absent in most members of the CHU series for which reproductive features have been reported. Exceptions as noted in the literature are plugs at the bases of some gametangia of *Halimeda scabra* and *H. tuna* (Howe, 1905), and walls at the base of the reproductive structures of *Boodleopsis carolinensis* (Trono, 1971). For *Halimeda* subsequent examination has not confirmed their presence (Feldmann, 1951; Hillis, 1959; Colinvaux *et al.*, 1965). Howe's observation may be an artifact of preservation, but healthy live material of different ages should be critically studied.

For *Boodleopsis* the situation is different. In the species *pusilla* walls at the base of reproductive structures were not noted, but were reported in lower or basal filaments by Taylor et al. (1953). Although this genus still needs much careful study, the presence of walls in any part of the filament may indicate that the taxon does not belong to the main group.

The general absence of septa at the base of reproductive structures in the CHU series presents a direct contrast to *Codium* and *Bryopsis,* in which such walls or plugs are an established feature.

(vi) Gametogenesis. The presence of a membrane or of other substances including mucilage around the "gametes" during maturation, as well as basic data on cleavage, may eventually contribute to the systematics of the group. At present, however, there are almost no data for comparative purposes.

Membranes encasing gametes have been reported for *Caulerpa* (Goldstein and Morrall, 1970), *Penicillus capitatus* (Meinesz, 1975), and *Udotea javensis* (Meinesz, 1980a), while a thick layer of unidentified material was observed beneath gametangial walls in micrographs of late stages of development of microgametes of *Halimeda incrassata* (Hillis-Colinvaux, 1980). Progressive cleavage has been reported for *Caulerpa* (Goldstein and Morrall, 1970).

(vii) Gamete characteristics. Because live reproductive material of CHU genera is difficult to acquire, considerably more so than for BDC genera, there are few data on what could be a critical character. Biflagellated zooids have been observed for *Caulerpa* (Goldstein and Morrall, 1970), *Chlorodesmis bulbosa* (Ducker, 1965), *Halimeda* (Hillis, 1959; Hillis-Colinvaux, 1980), *Penicillus* (Hillis-Colinvaux, 1973; Meinesz, 1975), *Udotea petiolata* (Meinesz, 1969, 1972c), and *U. indica* (Nizamuddin, 1963).

Anisogamy, to differing degrees, is reported for *Caulerpa* (seven species), *Halimeda* (five species), and *Udotea petiolata,* but probably occurs for all the above taxa with the possible exception of *Chlorodesmis*. These data are tabulated by Hillis-Colinvaux (1980) and Meinesz (1980b). Among taxa where anisogamy has been observed, only in *Caulerpa* have monoecious species been reported. In *C. serrulata* and *C. cupressoides* macro- and microgametes form in separate portions of the same frond (Goldstein and Morrall, 1970).

Electron microscope studies have been made of the gametes of *Caulerpa* (Lohr and Dawes, 1974) and *Halimeda* (Gori, 1979a, 1979b, released micro- and macrogametes of *H. tuna;* Hillis-Colinvaux, 1980, two stages of unreleased microgametes of *H. incrassata*).

(viii) Region of "gametic" discharge. Two major categories are delimited: reproductive structures with many specialized release sites and those with one to few specialized release sites. Further subdivision is inappropriate until considerably more information is available.

Multiple specialized sites: In *Caulerpa* simple or branched whitish papillas develop over the erect frond and sometimes on the rhizome approximately 24 h before gametes are released (Kajimura, 1969, 1977). Gametes are discharged through these papillas and, at least in laboratory environments, through ruptures in filament walls (Goldstein and Morrall, 1970). The pores reported by Kajimura (1968a, 1968b) are perhaps best interpreted as reflecting the appearance of the frond 24 h after gamete discharge when the papillas would have been shed (Kajimura, 1969), rather than a morphologically different structure.

Few specialized sites: In *Chlorodesmis* (Ducker, 1965) the reproductive units are forcefully discharged through the apex of a central siphon. Ducker describes the following sequence of events: "movement among the mature zooids starts in the most basal fertile branch. Its plug is pushed into the central siphon and up the long corridor until it forces open a discharge pore. Then plugs of other fertile branches are pushed into the central siphon by the increasing pressure of the zooids. The mucilage of some of the plugs appears to line the apical part of the siphon, thus forming a canal through which the zooids are channelled at great speed." We lack such specific details for other CHU genera but the process seems similar. In *Halimeda* discharge appears to be essentially through 1 (–2) central siphons, whereas in *Penicillus* and the *Udotea* species in which release has been observed discharge is through a single (i.e. tapered end of gametangium as in *U. javensis*) or branched (*U. petiolata*) system of tubes formed at the apex of the gametangium (Meinesz, 1980b).

(ix) Holocarpy. Feldmann (1954) suggested that holocarpy be considered a trait for separating Caulerpales from Codiales and Derbesiales, but listed only the genera *Caulerpa* and *Halimeda* as examples. Other genera now known to be holocarpic are *Chlorodesmis* (Ducker, 1965), *Penicillus* (Hillis-Colinvaux, 1973, 1980; Meinesz, 1975), and *Udotea* (Meinesz, 1969, for *U. petiolata;* Meinesz, 1980a, for *U. javensis*). For *Avrainvillea* there have been conflicting data; holocarpy is implied in the brief account by Young (1977), but the genus is stated to be non-holocarpic by Roth and Friedmann (1981). Recently, the genus has been studied by J. Stojkovich, who confirms its holocarpic behavior (personal communication).

(e) *Life history data.* Development of the zygote has been followed only for *Halimeda tuna* (Meinesz, 1972b, 1973, 1980b), *Udotea petiolata* (Meinesz, 1972c, 1980b), and *Caulerpa serrulata* (Price, 1972). In all three the zygotes initially enlarge to form a large sphere with many chloroplasts (protosphere), In the first two taxa this event took 7 and 5 months, respectively, and amyloplasts were not observed. They were reported, however, for *Caulerpa*. In *Halimeda* and *Udotea* germination and growth of the protosphere over a several-month period involved division of a primary nucleus, development of amyloplasts, and an eventual production of a mass of creeping and erect filaments. A typical *Halimeda* thallus has not been produced by subsequent growth, but for *Udotea* a thallus similar to that of *Udotea minima* was observed. In *Caulerpa* the protosphere developed into a filamentous phase differentiated into rhizoids and broader green axes. Eventually, after more than 5 months, a branch that was pinnate and hence unlike the parental type developed from the axis.

Available life history data can be summarized as shown in Fig. 1. A diplontic life history is strongly implied, but data on meiosis other than that supplied by Schussnig (1939) for *Caulerpa* are required before life histories can be compared meaningfully. Some of the variations to be encountered as life histories are examined may follow the pattern of the "*espera*" variant of *Penicillus capitatus* (Woronin, 1862; Bornet, 1892; Gepp and Gepp, 1911; Huvé and Huvé, 1961; Meinesz, 1972a; Roth and Friedmann, 1976; Hillis-Colinvaux, 1980). It seems likely too that *Pseudochlorodesmis* and possibly *Boodleopsis* may be but filamentous stages of other CHU genera, as suggested by Papenfuss (1962) and Taylor et al. (1953), respectively. Recently Meinesz (1980b) has linked *Pseudochlorodesmis furcellata* with *H. tuna*.

(f) Cloning mechanisms. Sexual reproduction in holocarpic taxa is a single event in the life span of an individual. To be abundant or ecologically successful such genera must have well-developed strategies for asexual reproduction. One such mechanism for *Halimeda, Penicillus, Rhipocephalus,* and *Udotea* is the development of clones by the horizontal growth of rhizoidal filaments (Colinvaux et al., 1965; Colinvaux, 1968; Hillis-Colinvaux, 1973, 1980). Variations occur in the construction of these extensions (the pattern may be of a few filaments, or a tangled mass with some differentia-

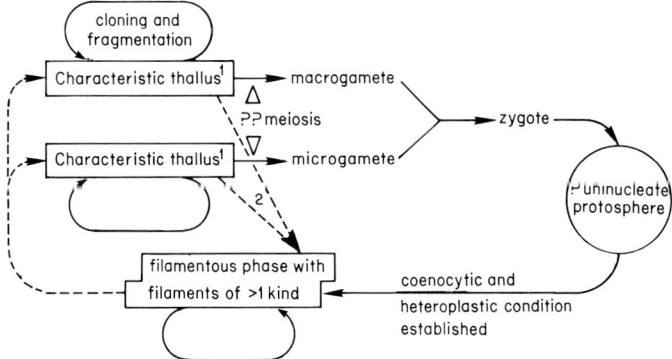

Fig. 1. Life history of the *Caulerpa-Halimeda-Udotea* group as currently known. 1. Some taxa are monoecious. 2. In *Penicillus capitatus* an "*espera*" filamentous phase may be produced asexually from a typical thallus (Hillis-Colinvaux, 1980). Such cycling may occur in other genera as well.

tion in diameter of the filaments) and the distance of separation of new thallus from parent. In *Avrainvillea*, for example, new plants commonly grow out of or near the base of the parent. Fragmentation, another cloning mechanism, occurs in *Caulerpa* and some species of *Halimeda*.

(g) Allelochemicals and other metabolic compounds. Unusual and often toxic compounds are produced by many genera of the CHU series (Norris and Fenical, 1982). The list includes compounds such as norcycloartene triterpenoids in *Tydemania expeditionis* (Paul et al., 1982b), sesquiterpenoids in *Caulerpa bikensis* (Paul and Fenical, 1982), udoteal (a linear diterpenoid) in *Udotea flabellum* (Paul et al., 1982a), avrainvilleal (a brominated diphenylmethane derivative) in *Avrainvillea longicaulis* (Sun et al., 1983), and halimedrial in *Halimeda* (W. Fenical, personal communication). In addition, some taxa contain inclusions in their filaments, such as the calcium oxalate crystals reported for all species of *Chlorodesmis* (Ducker et al., 1965) and for species of *Penicillus*, *Rhipocephalus*, and *Udotea* (Friedmann et al., 1972). It is expected that this character will contribute additional understanding of evolutionary relationships within the group. Our knowledge at this stage, however, permits hypotheses rather than conclusive statements.

(h) Cytology. Of the many cytological characters that might be anticipated as useful the only ones that have been used are plastid, and to a lesser extent nuclear, characteristics.

(i) Heteroplasty. Feldmann (1946) building on the microscopical observations of Ernst (1904), Czurda (1928), and Chadefaud (1941), subdivided the Siphonales into heteroplastic and homoplastic orders. The former was composed of taxa with both chloroplasts and starch-storing amyloplasts, although it has been observed subsequently that the chloroplasts of such taxa also may contain starch (Hillis-Colinvaux, 1980). Members of the second group contained only chloroplasts.

Heteroplasty now has been reported for *Avrainvillea, Caulerpa, Chlorodesmis, Halimeda,* and *Udotea* (Hori and Ueda, 1967, 1975), *Penicillus* (Turner and Friedmann, 1974), *Cladocephalus* (Roth and Friedmann, 1980), *Tydemania* (personal observation), as well as *Rhipocephalus, Pseudochlorodesmis, Pseudocodium, Dichotomosiphon pusillus,* and *D. tuberosus* (Feldmann, 1946). Homoplasty has been noted for *Bryopsis, Codium,* and *Derbesia* (Hori and Ueda, 1967, 1975), *Pseudocodium* (Silva, 1982), *Johnson-sea-linkia* (Eiseman and Earle, 1983), and the protospheres of *H. tuna* and *U. petiolata* (Meinesz, 1980b). As can be seen by the taxa listed we have more data on this characteristic than on most. There are obvious inconsistencies, and more observations are needed. Nonetheless, heteroplasty appears to be a useful and valid characteristic for separating CHU and BDC groups of siphonaceous algae. It seems likely that these comments apply to the next character as well.

(ii) Concentric lamellar systems (CLS). Chloroplasts and amyloplasts of at least a number of CHU taxa possess a system of concentric lamellae, sometimes referred to as a "thylakoid organizing body" (Borowitzka and Larkum, 1974a, 1974b), although such a function has yet to be demonstrated, or "dome-shaped body" (Turner and Friedmann, 1974) at one end of the plastid. This system is known for *Avrainvillea, Caulerpa, Chlorodesmis, Halimeda,* and *Penicillus.*

(iii) Pyrenoids. These structures are present in some members of the CHU series (e.g. some species of *Caulerpa*) and absent from others (e.g. all species of *Halimeda*). More data are needed.

(iv) Nucleus. In *Avrainvillea* and *Cladocephalus* the nucleolus appears to be zoned into granular and fibrillar components, and in this way differs from ten other genera of the CHU series and three genera of the BDC series included in the study (Roth and Friedmann, 1980). These workers also reported a nucleus-microbody association in the same ten CHU genera but not in *Avrainvillea, Cladocephalus* or the three BDC genera.

(i) *Fossil data.* A substantial fossil record can be expected and has been partially worked out for calcareous green siphons (Elliott, 1960, 1965, 1978, 1981). The best known, albeit limited, history is for the genus *Halimeda,* which, through its ancestors, can be traced to Ordovician time (Elliott, 1982). Morphological development is exceedingly conservative, yet a variety of changes in cortex (Elliott, 1982) and external form (Hillis-Colinvaux, 1980) provide some guidance to the direction of evolution.

(j) *Geographic distribution.* This criterion, if used in conjunction with others, can contribute usefully to the understanding of evolutionary relationships among genera and species. Categories of distribution include pan-oceanic, Indo-Pacific, and Atlantic.

Genera of the CHU series are essentially tropical or subtropical, although a few CHU taxa such as *Chlorodesmis baculifera* (Ducker, 1967), *Avrainvillea longicaulis, Caulerpa prolifera, C. racemosa* var. *peltata, Udotea cyathiformis, U. conglutinata,* and *U. flabellum* (Schneider, 1976) extend into temperate latitudes. However, water temperatures at the specific sites involved for these examples, the rock pools of Point Lonsdale, Victoria, southern Australia, and Onslow Bay, North Carolina, are in both cases subtropical. In contrast members of the BDC series have an extensive distribution in temperate waters.

The only freshwater representative in the list of the green siphons is *Dichotomosiphon tuberosus.*

OSTREOBIUM GROUP

Ostreobium now is the only genus in the Phyllosiphonales, and the order should be renamed *Ostreobiales* (cf. Hillis-Colinvaux, 1980; Silva, 1982).

The six species live endophytically within shells and corals in tropical and temperate oceans. *Ostreobium* possesses the pigments siphonein and siphonaxanthin (Jeffrey, 1968, for *O. reineckei*), which are characteristic of the Siphonales, and an *O. queketti*-like stage is associated with the life history of *Pseudobryopsis* (Mayhoub, 1974). Silva (1982) pointed out as difficulties of such an alignment the facts that *Pseudobryopsis* normally does not grow out of *O. queketti*-infested substrates, and that geographic distributions do not coincide. These problems must be considered in interpreting *Ostreobium* but do not exclude the possibility of *Ostreobium* species being a phase in some life history.

Ostreobium at times is septate, and has a temperate as well as tropical-subtropical distribution. The species *quekettii* produces quadriflagellate zoospores in a sporangium that is not separated from the vegetative filament by a cross wall (Kornmann and Sahling, 1980). It departs in some characteristics from both the CHU and BDC groups, but seems more closely aligned to the latter, where I have tentatively placed it (Fig. 2). Alternatively, its separate status could be maintained until relationships are clearer.

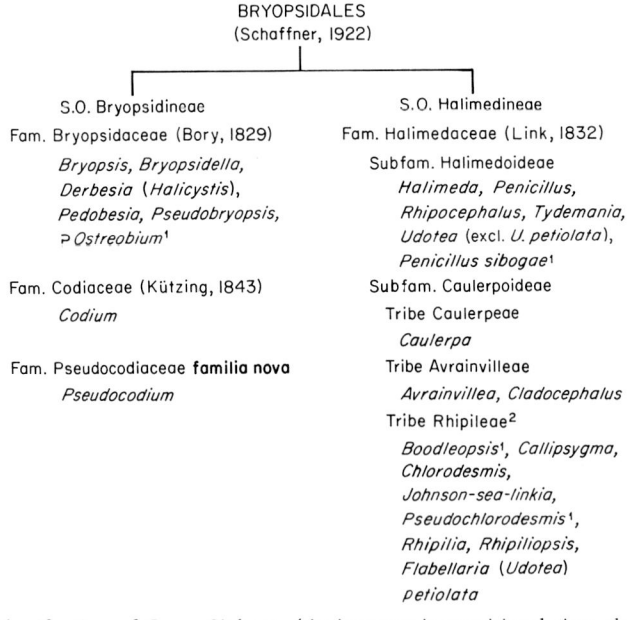

Fig. 2. Classification of Green Siphons. [1]Assignment is provisional since these taxa may represent stages in the life histories of other taxa. [2]If *Flabellaria* is valid, then Flabellarieae (Gepp and Gepp, 1911) has priority over Rhipileae.

DICHOTOMOSIPHON GROUP

The species *D. tuberosus* is the sole representative of this group. It is a freshwater, oogamous plant containing the pigment siphonein but not siphonaxanthin (Kleinig, 1969). In these three characters it differs from the BCD and CHU groups, which are marine, anisogamous, and contain both siphonein and siphonaxanthin. Although *Dichotomosiphon* is heteroplastic (Feldmann, 1946), contains xylan in its filament walls, and is essentially lacking in cross walls, it seems misplaced among the green siphons. This concept is reflected in my proposed classification (Fig. 2), and is indicated also by Hoek (1978, 1981) in his phylogenetic diagrams.

CLASSIFICATION OF THE GREEN SIPHONS

Two definite lines of development appear to exist for the green siphons: the BDC and CHU series. Characters that have been useful in separating them are listed in Table II. In a classification of the green algae these two groups might be given ordinal status, or alternatively, recognition of the two lines could be provided by giving them subordinal status within the same order. I have chosen the latter (see Fig. 2). A strong argument for doing so is the consistency provided at the ordinal level in the currently evolving system of classification of green algae. The scheme of Fig. 2 therefore reflects the viewpoint that characters separating the CHU and BDC series are not of ordinal stature. Subordinal status, although somewhat cumbersome, signals the distinct evolutionary trends within the group, trends that are not apparent by a simple listing of families.

Table II. Characters separating suborders of Bryopsidales.

Bryopsidineae (BDC series)	*Halimedineae (CHU series)*
Homoplasty	Heteroplasty
Concentric lamellar system absent	Concentric lamellar system present
Non-holocarpic reproduction	Holocarpic reproduction
Broad geographical range (tropical to temperate)	Restricted geographic range (principally tropical-subtropical)
Allelochemicals generally lacking	Allelochemicals often present
Principal wall polysaccharides: mannan, or xylan and cellulose	Principal wall component: xylan, with cellulose absent or essentially so
Septa commonly present at base of reproductive structure	Septa absent from base of reproductive structures

1. Suborder Halimedineae

The choice of epithet for the suborder is based on the family name Halimedaceae having priority over both Caulerpaceae and Udoteaceae (Silva, 1980).

To reflect perceived evolutionary development I have subdivided the single family into two subfamilies based on ability to calcify (see also Fig. 2). In doing so I have removed the uncalcified taxon *Udotea petiolata*, thereby eliminating a major objection to such a separation (Hillis-Colinvaux, 1980). Gepp and Gepp (1911) also delimited calcareous (Udoteae) and non-calcareous (Codieae and Flabellarieae) groups and included *U. petiolata* as *Flabellaria petiola* in the Flabellarieae. It would seem logical to assign *U. petiolata* (incl. *U. minima*) to *Flabellaria,* but the epithet may not be available (Farghaly, 1980) so any renaming must be tentative.

In the subfamily Caulerpoideae a series of tribes are delimited that reflect different trends as expressed in the distinctive rhizomatous habit of *Caulerpa*, or the morphology, different nuclear condition, and allelochemistry of *Avrainvillea* and *Cladocephalus*. Assignment to the third tribe is based, in part, on morphological similarities, with some of the taxa only tentatively placed therein.

2. Suborder Bryopsidineae

Subdivision of this group into the families Bryopsidaceae and Codiaceae reflects some of the relationships and differences discussed earlier in the paper. The epithet Bryopsidaceae has priority over Derbesiaceae (Silva, 1980). The genus *Ostreobium* for reasons discussed earlier is tentatively placed in the Bryopsidaceae.

3. Genus Pseudocodium

Pseudocodium is placed by some workers in the CHU group on the basis of general resemblance to *Halimeda* (Levring, 1938) and supposed heteroplasty (Feldman, 1946), while others align it with *Codium* (Dawes and Mathieson, 1972), with which it shares the characters of a mannan component in the wall, sympodial origin of utricles (Silva, 1982), and possible homoplasty, which is implied by Dawes and Mathieson (1972) and indicated by Silva (1982).

Reproductive structures, which are known only for one of the three species (*P. floridanum*), are similar to those of *Codium* in being internal to the peripheral boundary of the thallus and in apparently lacking obvious central siphons for discharge. They differ in lacking a septum between reproductive and vegetative portions.

On the basis of the above criteria, including homoplasty, the overall alliance of the genus is to the suborder Bryopsidineae. It fits neither family, however, and consequently I have placed it in the new family Pseudocodiaceae. Should heteroplasty be demonstrated, then the appropriate assignment in this scheme would be to separate and new suborder, the Pseudocodiineae.

4. Ordinal Epithet

Rules of nomenclature currently permit the use of a descriptive epithet such as "Siphonales" (P. C. Silva, personal communication). Nonetheless, it is more appropriate to base the name on that of a genus within this restructured group. Both Bryopsidales and Caulerpales are available, with the former having priority (P. C. Silva, personal communication).

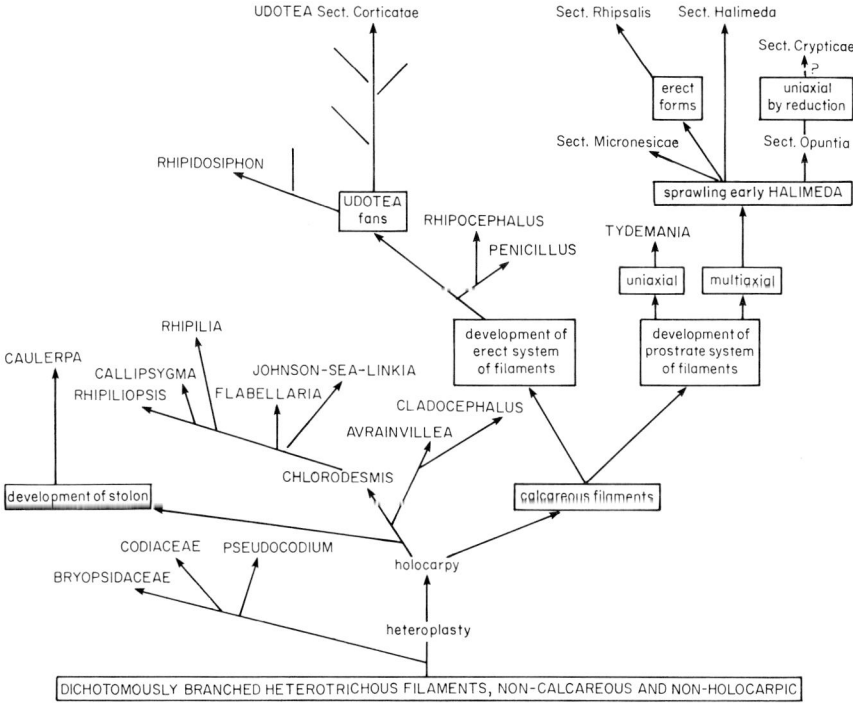

Fig. 3. Protoscheme of phylogenetic relationships in the Bryopsidales. *Boodleopsis* and *Pseudochlorodesmis* are not included since they are probably filamentous stages in the life cycles of other taxa.

5. Intergeneric Relationships

It is evident from the preceding discussion that data that permit an evaluation of interrelationships among the Bryopsidales, and particularly of the Halimedineae, are limited. But if alert to the many limitations, some expression of conceived relationships is important in providing a basis for subsequent research and discussion. I therefore have developed a protoscheme of relationships (Fig. 3), which for *Halimeda* includes the sections of the genus, and is an extension of my 1980 scheme. Morphology is a strong although not exclusive component, since available data are predominantly of this type.

In developing the scheme I generally assumed that complex taxa are more recent, although there is little empirical evidence of the direction of evolution. An exception to such a pattern is the proposed development in *Halimeda* of a uniaxial group (section Crypticae) from section Opuntia, a multiaxial group.

CONCLUDING REMARKS

In recent classifications the "green siphons" generally are assigned, as an order or small group of orders, to class Chlorophyceae (Silva, 1982) or class Bryopsidophyceae (Round, 1971; Hillis-Colinvaux, 1980), but there are a few notable variations. One is the designation of a phylum Siphonophyta in a recent scheme of Margulis (1974). Another is the suggestion that these algae belong to the class Ulvophyceae in the Pickett-Heaps (1975) and Stewart and Mattox (1978) subdivision of green algae (Roberts *et al.*, 1980), which in the papers cited contains also the classes Charophyceae and Chlorophyceae. Roberts *et al.* (1982) provided the following characterization of the Ulvophyceae: absence of phycoplast; closed mitotic spindle; occasional presence of scales on flagellated reproductive cells; distinctly overlapping basal bodies in biflagellate swarmers; cruciate flagellar root system; presence of rhizoplasts ("system II striated roots"); and presence of a non-striated distal fiber (capping plate) connecting basal bodies.

For the Bryopsidales we currently have data pertaining to motile cell characters for *Bryopsis* (Hori, 1977; Melkonian, 1980, 1981), *Derbesia* (Roberts *et al.*, 1981), *Pseudobryopsis* (Roberts *et al.*, 1982), and *Codium* (C. J. O'Kelly, personal communication) of the Bryopsidineae, and *Halimeda* (Gori, 1979a, 1979b; Hillis-Colinvaux, 1980 and unpublished) of the Halimedineae. The sample is limited, especially for the second suborder, but the characters as known fit the Ulvophyceae, with variation on details being greater among female gametes (Roberts *et al.*, 1982; C. J. O'Kelly, personal communication). With additional data on motile cells of the Halimedineae assignment to the Ulvophyceae may become increasingly comfortable.

ACKNOWLEDGEMENTS

This paper, the outgrowth of a thoughtful and timely invitation, was skilfully nurtured throughout by Dr. D. E. G. Irvine. During the weeks of concentrated library research and writing, my requirements of assistance, discussion, and understanding were most willingly provided by Dr. D. M. John, my laboratory and home associates, and Professor E. S. Barghoorn, while taxonomic advice was freely given by Professor P. C. Silva. I gratefully acknowledge this support. I thank also the Royal Society for its important contribution to this endeavor.

REFERENCES

Aspinall, G. O. and Kessler, G. (1957). The structure of callose from the grape vine. *Chemy Ind.* 1296.
Blackman, F. F. and Tansley, A. G. (1902). A revision of the classification of the green algae. *New Phytol.* **1,** 17–24, 114–120, 133–144, 238–244.
Bornet, É. (1892). Les algues de P.K.A. Schousboe récoltées au Maroc et dans la Méditerranée de 1815–1829. *Mem. Soc. natn. Sci. nat. Math. Cherbourg* **28,** 165–376.
Bory de Saint-Vincent, J. B. G. M. (1829). Cryptogamie. *In* "Voyage autour du monde . . . 'La Coquille' pendant . . . 1822, 1823, 1824 et 1825" (L. I. Duperrey, ed.), pp. 201–301. Paris, France.
Borowitzka, M. A. and Larkum, A. W. D. (1974a). The caulerpalean thylakoid organizing body. *Eighth Int. Congr. Electron Microsc.* **2,** 588–589.
Borowitzka, M. A. and Larkum, A. W. D. (1974b). Chloroplast development in the caulerpalean alga. *Halimeda. Protoplasma* **81,** 131–144.
Borowitzka, M. A. and Larkum, A. W. D. (1977). Calcification in the green alga *Halimeda*. I. An ultrastructure study of the thallus development. *J. Phycol.* **13,** 6–16.
Chadefaud, M. (1941). Les pyrénoides des algues. *Annls Sci. Nat.* (Botanique) sér. 2, **11,** 1–44.
Colinvaux, L. Hillis (1968). Vegetative reproduction of *Halimeda* and related Siphonales in laboratory culture. *J. Phycol.* **4** (suppl.), 4.
Colinvaux, L. Hillis (1969). Research on and culture of calcareous green algae. *In* "Oceanic Biology," pp. 53–54. Office of Naval Research Report Dr-151.
Colinvaux, L. Hillis and Graham, E. A. (1964). A new species of *Halimeda*. *Nova Hedwigia* **7,** 5–10.
Colinvaux, L. Hillis, Wilbur, K. M. and Watabe, N. (1965). Tropical marine algae: growth in laboratory culture. *J. Phycol.* **1,** 69–78.
Cribb, A. B. (1954). Records of marine algae from Southeastern Queensland I. *Pap. Dep. Bot. Univ. Qd.* **3,** 15–37.
Czurda, V. (1928). Morphologie und Physiologie der Algen-Starkenkornes. *Beih. bot. Zbl.* **45,** 97–270.
Dawes, C. J. and Mathieson, A. C. (1972). A new species of *Pseudocodium* (Chlorophyta, Siphonales) from the west coast of Florida. *Phycologia* **11,** 273–277.
Dostal, R. 1928. Zur Frage der Fortpflanzungsorgane der Caulerpaceen. *Planta* **5,** 622–634.

Ducker, S. C. (1965). The structure and reproduction of the green alga *Chlorodesmis bulbosa*. *Phycologia* **4**, 149–162.

Ducker, S. C. (1967). The genus *Chlorodesmis* (Chlorophyta) in the Indo-Pacific region. *Nova Hedwigia* **13**, 145–182.

Earle, S. A. and Young, J. R. (1972). *Siphonoclathrus*, a new genus of Chlorophyta (Siphonales: Codiaceae) from Panama. *Occ. Pap. Farlow Herbm cryptogam. Bot.* **3**, 1–4.

Egerod, L. E. (1952). An analysis of the siphonous Chlorophycophyta with special reference to the Siphonocladales, Siphonales and Dasycladales of Hawaii. *Univ. Calif. Publs. Bot.* **25**, 325–454.

Eiseman, N. J. and Earle, S. A. (1983). *Johnson-sea-linkia profunda*, a new genus and species of deep-water Chlorophyta from the Bahama Islands. *Phycologia* **22**, 1–6.

Elliott, G. F. (1960). Fossil calcareous algal floras of the Middle East with a note on a Cretaceous problematicum, *Hensonella cylindrica* gen. et sp. nov. *Q. J. geol. Soc. Lond.* **115**, 217–232.

Elliott, G. F. (1965). The interrelationships of some Cretaceous Codiaceae (calcareous algae). *Palaeontology* **8**, 199–203.

Elliott, G. F. (1978). Ecologic significance of post-Palaeozoic green calcareous algae. *Geol. Mag.* **115**, 437–442.

Elliott, G. F. (1981). The tethyan dispersal of some chlorophyte algae subsequent to the Palaeozoic. *Paleogeogr. Palaeoclimat. Palaeoecol.* **32** (1980/1981), 341–358.

Elliott, G. F. (1982). A new calcareous green alga from the Middle Jurassic of England: its relationships and evolutionary position. *Palaeontology* **25**, 431–437.

Ernst, A. (1904). *Siphoneen*-Studien. II. Beiträge zur Kenntnis der Codiaceen. *Beih. bot. Zbl.* **16**, 199–236.

Farghaly, M. S. (1980). Algues benthiques de la mer Rouge et du bassin occidental de l'Ocean Indian. Ph.D. Thesis, Academy of Montpellier, Languedoc University of Science and Techniques.

Farghaly, M. S. and Denizot, M. (1979). Le genre *Rhipilopsis*. Définition et place dans les Caulerpales (Chlorophycées). *Revue algol.* **14**, 169–184.

Feldmann, J. (1946). Sur l'Hétéroplastie de certaines Siphonales et leur classification. *C. r. hebd. Séanc. Acad. Sci., Paris* **222**, 752–753.

Feldmann, J. (1951). Sur la reproduction sexuée de l'*Halimeda tuna* (Ell. et Sol.) Lamour. f. *platydisca* (Decaisne) Barton. *C. r. hebd. Séanc. Acad. Sci., Paris* **233**, 1309–1310.

Feldmann, J. (1954). Sur la classification des Chlorophycées Siphonées. VIII. Congrès International de Botanique, Rapports et Communications, Section 17, pp. 96–98.

Frie, E. and Preston, R. D. (1964). Non-cellulosic structural polysaccharides in algal cell walls. I. Xylan in siphoneous green algae. *Proc. R. Soc.* **1960B**, 293–313.

Friedmann, E. I. and Roth, W. C. (1977). Development of the siphonous green alga *Penicillus* and the *espera* state. *Bot. J. Linn. Soc.* **74**, 189–214.

Friedmann, E. I., Roth, W. C., Turner, J. B. and McEwen, R. S. (1972). Calcium oxalate crystals in the aragonite-producing green alga *Penicillus* and related genera. *Science, N.Y.* **177**, 891–893.

Fritsch, F. E. (1935). "The Structure and Reproduction of the Algae," Vol. I. Cambridge University Press, Cambridge.

Fritsch, F. E. (1948). "The Structure and Reproduction of the Algae," Vol. I. Cambridge University Press, Cambridge.

Gepp, A. and Gepp, E. S. (1911). "The Codiaceae of the Siboga Expedition Including a Monograph of Flabellarieae and Udoteae." E. J. Brill, Leiden.

Goldstein, M. and Morrall, S. (1970). Gametogenesis and fertilization in *Caulerpa*. *Ann. N.Y. Acad. Sci.* **175**, 660–672.

Gori, P. (1979b). Ultrastructure of the spermatozoid in *Halimeda tuna* (Chlorophyceae). *Gamete Res.* **2**, 345–355.
Gori, P. (1979b). The Ultrastructure of the female gamete in *Halimeda tuna* (Chlorophyceae). *Caryologia* **32**(4), 499–504.
Greville, R. K. (1830). "Algae Britannicae." MacClachlan and Stewart, Edinburgh.
Harvey, W. H. (1858). Nereis boreali-americana; Part III. Chlorospermae, including supplements. *Smithson. Contr. Knowl.* **10**, 1–140.
Hillis, L. (1959). A revision of the genus Halimeda (order Siphonales). *Publs Inst. mar. Sci. Univ. Tex.* **6**, 321–403.
Hillis-Colinvaux, L. (1973). Reproduction in the calcareous green algae of coral reefs. *J. mar. biol. Ass. India* **14**, 328–334.
Hillis-Colinvaux, L. (1980). Ecology and taxonomy of *Halimeda:* Primary Producer of coral reefs. *Adv. mar. Biol.* **17**, 1–327.
Hoek, C. van den (1978). "Algen. Einführung in die Phykologie." G. Thieme, Stuttgart.
Hoek, C. van den (1981). Chlorophyta: morphology and classification. *In* "The Biology of Seaweeds" (C. S. Lobban and M. J. Wynne, eds), pp. 86–132. Blackwell, Oxford, London, Edinburgh, Boston and Melbourne.
Hori, T. (1977). Electron microscope observations on the flagellar apparatus of *Bryopsis maxima* (Chlorophyceae). *J. Phycol.* **13**, 238–243.
Hori, T. and Ueda, R. (1967). Electron microscope studies on the fine structure of plastids in siphonous green algae with special reference to their phylogenetic relationships. *Sci. Rep. Tokyo Kyoiku Daig.* Section B **12**, 225–244.
Hori, T. and Ueda, R. (1975). The fine structure of algal chloroplasts and algal phylogeny. *In* "Advance of Phycology in Japan" (J. Tokida and H. Hirose, eds), pp. 11–42. Fischer, Jena.
Howe, M. A. (1905). Phycological studies I. New Chlorophyceae from Florida and the Bahamas. *Bull. Torrey bot. Club* **32**, 241–252.
Howe, M. A. (1907). Phycological studies. III. Further notes on *Halimeda* and *Avrainvillea*. *Bull. Torrey bot. Club* **34**, 491–516.
Huizing, H. J. and Rietema, H. (1975). Xylan and mannan as cell wall constituents of different stages in the life-histories of some siphoneous green algae. *Br. phycol. J.* **10**, 13–16.
Huvé, P. and Huvé, H. (1961). A propos de *Penicillus capitatus* Lamarck, forma *mediterranea* (Decaisne) comb. nov. (Caulerpale, Udotéacée). *Proc. Int. Seaweed Symp.* **4**, 99–111.
Iyengar, M. O. P. (1933). On the formation of gametes in a *Caulerpa*. *J. Indian bot. Soc.* **12**, 325.
Iyengar, M. O. P. (1940). On the formation of gametes in *Caulerpa*. *J. Indian bot. Soc.* **18**, 191–194.
Jeffrey, S. E. (1968). Pigment composition of siphonales algae in the brain coral *Favia*. *Biol. Bull mar. biol. Lab. Woods Hole* **135**, 141–148.
Kajimura, M. (1968a). On fruiting season of *Caulerpa scapelliformis* (R. Br.) Ag. var. *denticulata* (Descn.) Weber van Bosse in the Oki Islands, Shimane Prefecture. *Bull. Jap. Soc. Phycol.* **14**, 38–43.
Kajimura, M. (1968b). On fruiting season of *Caulerpa okamurai* Web. v. Bosse in Shimane Prefecture. *Bull. Jap. Soc. Phycol.* **16**, 14–18.
Kajimura, M. (1969). On swarmer production and discharge in *Caulerpa okamurai* Weber van Bosse from Shimane prefecture. *Bull. Jap. Soc. Phycol.* **17**, 98–103.
Kajimura, M. (1977). On dioecious and isogamous reproduction of *Caulerpa scapelliformis* (R.Br.) Ag. var *denticulata* (Descn.) Weber van Bosse from the Oki Islands, Shimane prefecture. *Bull. Jap. Soc. Phycol.* **25**, 27–33.
Kleinig, H. (1969). Carotenoids of siphonous green algae. a chemotaxonomical study. *J. Phycol.* **5**, 281–285.

Kornman, P. and Sahling, P.-H. (1980). *Ostreobium queketti* (Codiales, Chlorophyta). *Helgoländer. Meeresünters.* **34**, 115–122.

Kützing, F. T. (1843). "Phycologia generalis." Brockhaus, Leipzig.

Levring, T. (1938). Verzeichnis einiger Chlorophyceen und Phaeophyceen von Südafrika. *Acta Univ. lund* **34**, 1–25.

Link, H. F. (1832). Über die Pflanzenthiere überhaupt und die dazu gerechneten Gewächse besonders. *Abh. Königl. Acad. Wiss. Berlin, Phys. Kl.* **1830**, 109–123.

Lohr, C. A. and Dawes, C. J. (1974). Light and electron microscope studies on the gametes of the green alga, *Caulerpa* (Chlorophyta, Siphonales). *Flo. Sci.* **37**, 45–49.

MacRaild, G. N. and Womersley, H. B. S. (1974). The morphology and reproduction of *Derbesia clavaeformis* (J. Agardh) de Toni (Chlorophyta). *Phycologia* **13**, 83–95.

McConnell, D. and Colinvaux, L. Hillis (1967). Aragonite in *Halimeda* and *Tydemania* (order Siphonales). *J. Phycol.* **3**, 198–200.

Margulis, L. (1974). Classification and evolution of prokaryotes and eukaryotes. In "Handbook of Genetics" (I. R. King, ed.), Vol. 1, pp. 1–41.

Mayhoub, H. (1974). Reproduction sexuée et cycle de développement de *Pseudobryopsis myura* (Ag.) Berthold (Chlorophycée, Codiale). *C. r. hebd. Séanc. Acad. Sci., Paris Sér.* D. **278**, 86/–70.

Meinesz, A. (1969). Sur la reproduction sexué de l'*Udotea petiolata* (Turra) Boergesen. *C. r hebd. Séanc. Acad. Sci., Paris Sér. D* **299**, 1063–1065.

Meinesz, A. (1972a). Sur la croissance et le developpement du *Penicillus capitatus* Lamarck forma *mediterranea* (Dec.) P. and H. Huvé (Caulerpale, Udotéacée). *C. r. hebd. Séanc. Acad. Sci., Paris Sér. D* **275**, 667–669.

Meinesz, A. (1972b). Sur le cycle de l'*Halimeda tuna* (Ellis et Solander) Lamouroux (Udotéacée, Caulerpale). *C. r. hebd. Séanc. Acad. Sci., Paris Sér. D* **275**, 1365–1365.

Meinesz, A. (1972c). Sur le cycle de l'*Udotea petiolata* (Turra) Boergesen (Caulerpale, Udotéacée). *C. r. hebd. Séanc. Acad. Sci. Paris Sér. D* **275**, 1975–1977.

Meinesz, A. (1973). Les caulerpales des côtes françaises de la Méditerranée. Biologie et écologie. Thesis, University of Paris.

Meinesz, A. (1975). Premières observations sur la reproduction du *Penicillus capitatus* Lamarck forma *mediterranea* (Decaisne) P. and H. Huvé (Caulerpale, Udotéacée). *Annls Mus. Hist. nat. Nice* **3**, 1920.

Meinesz, A. (1979). Contribution à l'étude de *Caulerpa prolifera* (Forsskål) Lamouroux (Chlorophycée, Caulerpale). II. La reproduction sexuée sur les côtes occidentales de la Méditerranée. *Botanica mar.* **22**, 117–121.

Meinesz, A. (1980a). Sur la reproduction de l'*Udotea javensis* A. et E.S. Gepp (Udotéacée, Caulerpale). *Phycologia* **19**, 82–84.

Meinesz, A. (1980b). Connaissances actuelles et contribution à l'étude de la reproduction et du cycle des Udotéacées (Caulerpales, Chlorophytes). *Phycologia* **19**, 110–138.

Melkonian, M. (1980). Ultrastructural aspects of basal body associated fibrous structures in green algae: a critical review. *BioSystems* **12**, 85–104.

Melkonian, M. (1981). Structure and significance of cruciate flagellar root systems in green algae: female gametes of *Bryopsis lyngbyei* (Bryopsidales). *Helgoländer wiss. Meersünters.* **34**, 355–370.

Mirande, R. (1913). Recherches sur la composition chimique de la membrane et le morcellement du thalle chez les Siphonales. *Annls. Sci. nat.* (Botanique) sér. 2 **18**, 147–264.

Miwa, T., Iriki, Y. and Suzuki, T. (1961). Mannan and xylan as essential cell wall constituents of some siphonous green algae. *Collogues int. Cent. Res. scient.* **103**, 135–144.

Miyake, K. and Kunieda, H. (1937). On sexual reproduction of *Caulerpa. Cytologia* **8**, 205–207.

Montagne, C. (1838). De l'organisation et du mode de reproduction des Caulerpées, et en particulier du *Caulerpa webbiana,* espèce nouvelle des îles Canaries. *Annls Sci. nat. (Botanique) sér.* 2 **9,** 129–150 + pl. 6.
Nasr, A. H. (1939). Reports on the preliminary expedition for the exploration of the Red Sea in the R. R. S. *Mabahith.* Algae. *Publs mar. biol. Stn Ghardaqa* **1,** 47–76.
Neumann, K. (1969). Beitrag zur Cytologie und Entwicklung der siphonalen Grünalge *Derbesia marina. Helgoländer wiss. Meeresünters.* **19,** 355–75.
Nizamuddin, M. 1963. Studies on the green alga *Udotea indica* A. & E. S. Gepp, 1911. *Pacif. Sci.* **17,** 243–5.
Norris, J. J. and Fenical, W. (1982). Chemical defense in tropical marine algae. *In* "The Atlantic Barrier Reef Ecosystem at Carrie Bow Cay, Belize I: Structure and Communities" (K. Rutzler and I. G. Macintyre, eds), pp. 417–431. Smithsonian Contributions to the Marine Sciences.
Oltmanns, F. (1905). "Morphologie und Biologie der Algen." Vol. 2. Fischer, Jena.
Papenfuss, G. F. (1962). Clearing old trails in systematic phycology. *Proc. Ninth Pacif. Sci. Congr.* **1957, 4,** 229–233.
Parker, B. C. (1970). Significance of cell wall chemistry to phylogeny in the algae. *Ann. N. Y. Acad. Sci.* **175,** 417–428.
Paul, V. J. and Fenical, W. (1982). Toxic feeding deterrents from the tropical marine alga *Caulerpa bikinensis* (Chlorophyta). *Tetrahedron Letters* **23,** 5017–5020.
Paul, V. J., Su, H. H. and Fenical, W. (1982a). Udoteal, a linear diterpenoid feeding deterrent from the tropical green alga *Udotea flabellum. Phytochemistry* **21,** 468–469.
Paul, V. J., Fenical, W., Raffii, S. and Clardy, J. (1982b). The isolation of new norcycloartene triterpenoids from the tropical marine alga *Tydemania expeditionitis* (Chlorophyta). *Tetrahedron Letters* **23,** 3459–3462.
Pickett-Heaps, J. D. (1975). "Green Algae. Structure, Reproduction and Evolution in Selected Genera." Sinauer, Sunderland, Massachusetts.
Price, I. R. (1972). Zygote development in *Caulerpa* (Chlorophyta, Caulerpales). *Phycologia* **11,** 217.
Rietema, H. (1975). Comparative investigations on the life-histories and reproduction of some species in the siphoneous green algal genera *Bryopsis and Derbesia.* Doctoral Dissertation, Rijksuniversity of Groningen.
Roberts, K. R., Sluiman, H. J., Stewart, K. D. and Mattox, K. R. (1980). Comparative cytology and taxonomy of the Ulvaphyceae. II. Ulvalean characteristics of the stephanokont flagellar apparatus of *Derbesia tenuissima. Protoplasma* **104,** 223–38.
Roberts, K. R., Sluiman, H. J. Stewart, K. D. and Mattox, K. R. (1981). Comparative cytology and taxonomy of the Ulvaphyceae. III. The flagellar apparatuses of the anisogametes of *Derbesia tenuissima* (Chlorophyta). *J. Phycol.* **17,** 330–340.
Roberts, K. R., Stewart, K. D., Mattox, R. R. (1982). Structure of the anisogametes of the green siphon *Pseudobryopsis* species (Chlorophyta). *J. Phycol.* **18,** 498–508.
Roth, W. C. (1977). Observations on the reproduction of some siphonous green algae. *J. Phycol.* **13** (suppl.), 59.
Roth, W. C. and Friedmann, E. I. (1976). Occurrence of the *espera* stage of *Penicillus* in North American. *J. Phycol.* **12** (suppl.), 27.
Roth, W. C. and Friedmann, E. I. (1980). Taxonomic significance of nucleus-microbody associations, segregated nucleoli and other nuclear features in siphonous green algae. *J. Phycol.* **16,** 449–464.
Roth, W. C. and Friedmann, E. I. (1981). Persistent macrosegregated nucleoli in the siphonous green algae *Avrainvillea, Cladocephalus* and *Blastophysa* and possible implications concerning ribosomal RNA synthesis, nuclear cycles and life histories. *Phycologia* **20,** 193–198.

Round, F. E. (1963). The taxonomy of the Chlorophyta. *Br. phycol. Bull.* **2**, 224–235.
Round, F. E. (1971). The taxonomy of the Chlorophyta. II. *Br. phycol. J.* **6**, 235–264.
Round, F. E. (1973). "The Biology of Algae" (2nd ed.). Edward Arnold, London.
Schaffner, J. H. (1922). The classification of plants. XII. *Ohio J. Sci.* **22**, 129–139.
Schneider, C. W. (1976). Spatial and temporal distributions of benthic marine algae on the continental shelf of the Carolinas. *Bull. mar. Sci.* **26**, 133–151.
Schussnig, B. (1929). Die Fortpflanzung von *Caulerpa prolifera*. *Ost. bot. Z.* **78**, 1–8.
Schussnig, B. (1939). Ein Beitrag zur Entwicklungsgeschichte von *Caulerpa prolifera*. *Bot. Notiser* **92**, 75–96.
Silva, P. C. (1951). The genus *Codium* in California with observations on the structure of the walls of the utricles. *Univ. Calif. Publ. Bot.* **25**, 79–114.
Silva, P. C. (1960). *Codium* (Chlorophyta) of the tropical western Atlantic *Nova Hedwigia* **1**, 497–536.
Silva, P. C. (1962). Comparison of algal floristic patterns in the Pacific with those in the Atlantic and Indian Oceans, with special reference to *Codium*. *Proc. Ninth Pacif. Sci. Congr.* **4**, 201–216.
Silva, P. C. (1980). Names of classes and families of living algae. *Regnum veg.* **103**, 1–156.
Silva, P. C. (1982). Chlorophycota. In "Synopsis and Classification of Living Organisms" (S. P. Parker, ed.), Vol. l, pp. 133–161. McGraw-Hill, New York.
Smith, G. M. (1938). "Cryptogamic Botany, Vol. 1. Algae and Fungi" (1st ed.). McGraw-Hill, New York, London and Toronto.
Smith, G. M. (1955). "Cryptogamic Botany, Vol. I. Algae and Fungi" (2nd ed.). McGraw-Hill, New York, London and Toronto.
Stewart, K. D. and Mattox, K. R. (1978). Structural evolution in the flagellated cells of green algae and land plants. *BioSystems* **10**, 145–52.
Sun, H. H., Paul, V. J. and Fenical, W. (1983). Avrainvilleol, a brominated diphenylmethane derivative with feeding deterrent properties from the tropical green alga *Avrainvillea longicaulis*. *Phytochemistry* **22**, 743–745.
Tatewaki, M. (1979). Life history of *Pseudobryopsis* sp. (Codiales, Chlorophyta) *Jap. J. Phycol. (Sôrui)* **27**, 7–16.
Taylor, W. R., Joly, A. B. and Bernatowicz, A. J. (1953). The relation of *Dichotomosiphon pusillus* to the algal genus Boodleopsis. *Pap. Mich. Acad. Sci.* **38**, 97–108.
Trono, G., Jr. (1971). Some new species of marine benthic algae from the Caroline Islands, western-central Pacific. *Micronesica* **7**, 45–77.
Turner, J. B. and Friedmann, E. I. (1974). Fine structure of capitular filaments in the coenocytic green alga *Penicillus*. *J. Phycol.* **10**, 125–134.
West, G. S. (1904). "A Treatise on the British Freshwater Algae." Cambridge University Press, Cambridge.
Wilbur, K. M., Colinvaux, L. Hillis and Watabe, N. (1969). Electron microscope study of calcification in the alga *Halimeda* (order Siphonales). *Phycologia* **8**, 27–35.
Woronin, M. (1862). Recherches sur les algues marines *Acetabularia* Lamx. et *Espera* Dcne. *Anns. Sci. nat. sér.* 4 **16**, 200–214.
Yamada, Y. (1934). The marine Chlorophyceae from Ryukyu, especially from the vicinity of Nawa. *J. Fac. Sci. Hokkaido Univ.* **3**, 33–88.
Young, J. R. (1977). Ecological observations on the reproduction of the tropical green marine alga *Avrainvillea* from Panama (Siphonales, Codiaceae). *J. Phycol.* **13** (suppl.), 76.

11 | Modern Developments in the Classification of Some Fossil Green Algae

G. F. ELLIOTT

Department of Palaeontology, British Museum (Natural History), London, England

Abstract: The limitations of the fossil green algae available for classification, both the incidence of fossilization and the preservation of the fossils themselves, are summarised. Three groups of macroscopic green algae with reasonably good fossil records are discussed. The history of the classification of the Dasycladales is summarised, and the classification is shown to be in an active phase of development. The fossil Udoteaceae are shown from their nature to be unlikely to support a comparable classification. The fossil Charophyta are in an earlier phase of classification, with numerous and partly synonymous species described and not yet revised into a coherent scheme. It is concluded that, imperfect though it is, the classification of past algae should be borne in mind by taxonomists dealing with the living algae.

THE LIMITATIONS OF FOSSIL ALGAE

We all know what we mean by classification, even if we do not all mean quite the same thing. It can be asked, what is special about the classification of fossil green algae, as opposed to living ones? A fossil can be defined as the remains of a once-living organism, naturally buried and more or less mineralized. There are two principal factors, or sets of factors, which affect the classification of such material. One is the implied content of part of the definition, "more or less mineralized". What happens to your biological characters when they are more or less mineralized? The other, equally important, lies in the incidence of fossilization. Organisms vary enormously in their chances of fossilization, as well as how the mineralization will then affect them if and when it does.

Systematics Association Special Volume No. 27, "Systematics of the Green Algae", edited by D. E. G. Irvine and D. M. John, 1984, pp. 297–302. Academic Press, London and Orlando.
ISBN 0 12 374040 1

Copyright © by the Systematics Association
All rights of reproduction in any form reserved

One of the common sights on some British beaches is the bright green of the drifted sea-lettuce *Ulva*. Yet the chance of this or any other alga without hard parts becoming a fossil is negligible. Our coasts can be gardened with them, but they will leave no trace. There are rocks resulting from the sudden fall in the past of volcanic ash into the sea, smothering the algal flora and preserving it as impressions. But such rocks form a very small part of the total volume of rocks which may be expected to yield fossil algae. It is for this reason that in the brown algae, the kelps, so important now in so many ways, leave a very poor, indeed a negligible, fossil record.

Fossil green algae, then, are almost always records of the hard parts of those which calcify. I am not considering the unicellular, planktonic forms which are sometimes petrified, because I have never made a special study of them. Of those which have consistently been fossilized, the Charophyta, the Dasycladales and the Udoteaceae form the great majority. When we look at their modern representatives, and see how many non-calcified taxa there are, especially in the latter two, we see how incomplete our records, already known to be incomplete, are likely to be, and odd occasional preservations of non-calcified taxa confirm this. When Shelley wrote that he saw "The Deep's untrampled floor, with green and purple seaweeds strown" he had little idea how transient this was.

It is these hard parts then, usually if not invariably separated on the death of the algae, and then often rolled around, mixed and worn by the currents on the sea-floor before burial, which form most of our fossil record. And we do find them "more or less mineralized". The sum total of the changes which take place between an organism becoming available for burial and its survival as a fossil in rock is known as diagenesis. A calcified sea-weed, dead and buried, will be altered in various ways by combinations of pressure, heat and circulating water in the rock which will dissolve some minerals and may deposit others in its place. The original calcification of living green algae is sometimes calcitic, sometimes aragonitic, depending on the taxon. The mineral aragonite is relatively unstable, and in the fossil state normally becomes converted to calcite. Where this is a "fine" replacement biological detail survives recognisably, but in a coarse "window-pane" calcite-crystal replacement the crystals grow across the biological structures of the alga, either obliterating them, to give an "outline-replacement" fossil only, or with the former biological boundaries indicated by faint colour-changes in the transparent calcite. Much worse, the calcite may be converted to dolomite or magnesium carbonate, in which case the alga survives only, if at all, as a fuzzy "granulated-sugar" outline replacement.

This sounds like a catalogue of disaster. But it is what we have to work with, and careful collection and study has built up a surprising picture of green algae, or at any rate, of some of them, in the past. The preservation-

disasters of one rock can be surprisingly well-preserved in another. Thus I once discovered tantalisingly recrystallized outlines of a large new dasycladalean alga in a white limestone from the Middle East. This limestone was the remains of an old organic reef, heavily recrystallized. A few miles away, the limestone changed to a different rock altogether, a relic of the sediments around and outside the reef. In this the same alga was beautifully preserved, enabling its branch-system and its reproductive bodies to be studied in some detail, and its relationships elucidated.

With this picture in mind, I turn to those macroscopic green algae which have left sufficient remains fossil over the last 500 million years or so for a coherent evolutionary story to emerge. I begin with the order Dasycladales, the best known, and the most useful fossil group.

THE DASYCLADALES

Living dasyclads are small shallow-marine warm-water algae, with a central stem-cell, successive whorls of lateral branches of varying complexity, and a nucleate holdfast. A tendency to aragonitic calcification is irregularly distributed through the group, though consistent in different taxa, and can vary from absent to very heavy. The living group is a small one of about seven genera.

When a dasyclad is heavily calcified, it may survive fossilization as a tubular cylinder, the walls perforated by successive lateral whorls or verticils of branching pores and associated cavities of varying complexity, representing the former plant matter. Segmented spherical and club-shaped forms are known, and some come apart after death as separate beads, discs or other pieces. They occur as fossils from the Lower Palaeozoic onwards, and in the right limestone facies can be extremely abundant. In the tropical lagoons of the Triassic reefs of some 230 million years ago, now represented by the mountains of central and southeastern Europe, they were so abundant as to function as sediment-formers. They seem at that time to have had an ecologic function rather like that of the present-day udoteacean *Halimeda*. As fossils in early nineteenth-century studies they were regarded as somewhat problematic, but their algal nature was recognised by Munier-Chalmas (1877). Although many new forms were described, their classification was disorganised until the early twentieth century, when the enthusiastic Julius Pia made them his life-study.

Pia, who worked necessarily in part with thin-sections of hard alpine limestones, and who displayed a genius for reconstructions fuelled by a vivid imagination, handled an immense variety of fossil algae. He referred all his extinct forms believed dasyclad to the family Dasycladaceae, and then divided them into groups known as tribes. In 500 million years or so the incomplete remnants, as preserved, of many successive dasyclads gave a telescoped broken view of evolutionary progress, not especially amenable to the same treatment as a living flora. Pia did indeed ascribe evolutionary trends to some of the phenomena seen, notably the various positions of the reproductive bodies, but his tribes were very much an empirical marshalling of the materials available.

Pia's classification (1920, 1927) survived very much intact until and into a new wave of interest in these algae. This arose after the second World War, as an exploration of their utility as stratigraphic and palaeoecologic microfossils in the search for oil and gas. My own initial involvement came about in this way, and in dealing with the rich floras of the Middle East, identification with what was known and description of what was new took priority for many years. My subsequent interests have been largely in dasyclad palaeogeography and palaeoclimatology, rather than in higher classification. However, in France particularly, team-work by algal workers has put forward a reasoned enquiry into the value of the different characters of these fossils for their classification (Groupe français d'étude des algues fossiles, 1975), with recommendations for their use. The early non-verticillate dasyclads are accommodated in the emended family of Seletonellaceae Korde 1971, within the Order Dasycladales. A later paper (Bassoulet *et al.*, 1979) attempts to classify all described genera into tribes within the three families Seletonellaceae, Dasycladaceae and Acetabulariaceae.

It will be seen from this summary account that very marked progress in the classification of these algae has been made in the last few years. This will continue, as new genera are described and new information comes to light on those incompletely known. It is interesting to compare this with the classification of the living survivors by Valet (1969). Here the present-day genera available to the marine botanist yield information on reproduction, biochemistry, cytology—all manner of things unknown or only to be inferred in the fossils. At the same time the genera are seen to be a relict group from an enormous number of predecessors in the geological past. The two are complementary: palaeontologists appreciate biological studies for the help they can afford to their own problems, and marine botanists should consider the palaeontological record in making their own classifications.

THE UDOTEACEAE

The fossil Udoteaceae, long known as Codiaceae, contrast in many ways with the Dasycladales. Like them, they have a sporadically abundant record from Lower Palaeozoic to Recent, calcify heavily on the whole, and are recognisable by comparison with living survivors. But, unlike the Dasycladales, they are not a relict group; *Halimeda* in particular is now a most successful warm-water marine plant, and indeed ecologically has replaced the Dasycladaceae of the Triassic as a rock-former. But the respective evolution of the two groups has been different. Dasycladales are curiously bizarre plants, diversifying endlessly on a few basic structures, and affording an infinity of new combinations of variants for detailed classification. The Udoteaceae found as fossils are not so much variants on a common plan as new successive populations with a different average of the same characters. *Dimorphosiphon* from the Lower Palaeozoic is very recognisably a plant of the *Halimeda*-kind, and the problem with some of the genera described from the rocks in between, from Palaeozoic and Mesozoic, is not to accommodate them in an evolutionary position within the family; it is difficult to be sure if they are indeed green algae, and not for example red algae, as with the extinct Gymnocodiaceae believed to have been somewhat similar to the living red Chaetangiaceae. These problems are not all solved; compare the views of Vachard and myself (see Elliott, 1982) on certain Permian genera.

Overall, then, the classification of the fossil Udoteaceae has been the recognition of different stages in a continuous process leading up to the modern taxa. They have added much less to knowledge than have the Dasycladales. A recent French publication (Bassoulet *et al.*, 1983) gives an admirable and useful summary of the fossil genera.

THE CHAROPHYTA

The curious group of brackish and freshwater green algae, often known as charophytes, are not uncommon as fossils in the right facies. Although remains of stems and branches are not unknown, it is overwhelmingly the little spiral-surfaced reproductive bodies known as gyrogonites which have received study. These range from the Lower Palaeozoic onwards, and a recent study suggests that some of the early forms (e.g. *Karpinskya* of the family Trochiliscaceae) were definitely marine (Racki, 1982), unlike the modern taxa and, by reasonable inference on palaeoecologic evidence, most of those also in between.

Described genera are relatively few, but an immense profusion of species have been named, based on very small differences in the abundant materials available. Small differences in palaeontological species are not necessarily to be dismissed as biological variation of the kind familiar in living species, since if they are associated with change through geological time they can be genetic in origin, even if not directly adaptive. But the numbers of charophyte species described (recent Chinese work adds noticeably to them) is excessive. It is to be expected that revisionary work will reduce them noticeably, and give a better generic framework for their assignation. Studies of this kind are actively in progress at Montpellier in France, where the work and traditions of the late Louis Grambast are carried on. Charophyte classificäion is backward compared with that of the Dasycladales, but probably has more potential than that of the Udoteaceae, so far as can be foreseen.

CONCLUSIONS

This review of classification in the fossil macroscopic green algae is necessarily brief. The emphasis laid upon the imperfections of the evidence is intentional, and explains the limited variety of common fossils available for classificatory review. It is safe to say that some green algal phylogeny (and hence probably classification) will always be a matter of surmise from living taxa. The evidence from the known green algae of the past, and the nature of that evidence, is what I have tried to summarise here, and I hope that some memory of its existence will sometimes come and remind algal taxonomists that the present is the result of the past.

REFERENCES

Bassoulet, J. P., Bernier, P., Deloffre, R., Génot, P., Jaffrezo, M. and Vachard, D. (1979). Essai de classification des Dasycladales en tribus. *Bull. Cent. Rech. Explor.-Prod. Elf-Aquitaine* **3**, 429–442.

Bassoulet, J. P., Bernier, P., Deloffre, R., Génot, P., Poncet, J. and Roux, A. (1983). Les algues udoteacées du Paléozoique au Cénozoique. *Bull. Cent Rech. Explor. Prod. Elf-Aquitaine* **7**, 449–621.

Elliott, G. F. (1982). A new calcareous green alga from the Middle Jurassic of England: Its relationships and evolutionary position. *Palaeontology* **25**, 431–437.

Groupe français d'étude des algues fossiles (1975). Réflexions sur la systématique des dasycladales fossiles. Étude critique de la terminologie et importance relative des critères de classification. *Geobios* **8**, 259–290.

Munier-Chalmas, E. P. (1877). Observations sur les algues calcaires appartenant au groupe des siphonées verticillées (Dasycladées Harv.) et confondues avec les foraminifères. *C.r. hebd. Seanc. Acad. Sci., Paris* **85**, 814–817.

Pia, J. (1920). Die Siphoneae Verticillatae vom Karbon bis zur Kreide. *Abh. Zool.-Bot. Ges. Wien* **11**(2), 1–263.

Pia, J. (1927). Thallophyta. *In* "Handbuch der Paläobotanik" (M. Hirmer, ed), pp. 31–136. Oldenbourg, Munich and Berlin.

Racki, G. (1982). Ecology of the primitive charophyte algae: A critical review. *Neues Jahrb. Geol. Paläeontol., Abh.* **162**, 388–399.

Valet, G. (1969). Contribution à l'étude des Dasycladales. III. Révision systématique. *Nova Hedwigia* **17**, 574–630.

12 | Cytogeography and Cytosystematics of Charophyta

M. KHAN

Department of Botany, Kamla Nehru Institute of Science and Technology, Sultanpur, India

Y. S. R. K. SARMA

Centre of Advanced Study in Botany, Banaras Hindu University, Varanasi, India

Abstract: This communication deals with the pattern of geographical distribution and speciation in the polyploid complex of haploid Charophyta. The composition of the charophyte flora in each of the zones of the world is considered. The extent of the similarities in the flora between each zone and the rest has been indicated. Depending on the extent of the world distribution, the charophyte taxa have been categorised into four groups: cosmopolitan, subcosmopolitan, restricted and endemic. Cytogeographical and cytosystematic aspects of the charophyte flora of each of the zones have been critically assessed. A detailed survey of chromosome numbers has been undertaken. Relative roles of euploidy, aneuploidy, chromosomal rearrangements, hybridization, allopatry, sympatry, etc. in the speciation of Charophyta have been discussed. The primary and secondary centres of origin for Charophyta have been formulated and phylogenetic relationships between various sections/subsections related to the various genera have been examined. Monoecism and dioecism are reiterated to be valid taxonomic criteria. Further, the pattern of cortication of axis and branches as well as the number of stipulode rows are considered as taxonomic criteria in the genus *Chara*.

INTRODUCTION

The Charophyta (Characeae) are commonly referred to as charophytes and constitute a group of macroscopic chlorophyllous algae placed in two

---

Systematics Association Special Volume No. 27, "Systematics of the Green Algae", edited by D. E. G. Irvine and D. M. John, 1984, pp. 303–330. Academic Press, London and Orlando.
ISBN 0 12 374040 1   Copyright © by the Systematics Association
                                       All rights of reproduction in any form reserved

tribes. In the Tribe Chareae the main axes support whorls of simple branchlets at the nodes while in the Tribe Nitelleae the whorls are made up of rays which may be simple or divided. The Chareae (*Chara, Lamprothamnium, Lychnothamnus, Nitellopsis*) and the Nitelleae (*Nitella, Tolypella*) are distinguished by the number and arrangement of the cells in the coronula at the apex of the oogonium. The coronula in the Chareae is formed of five cells arranged in one tier while two tiers, each with five cells, are found in the Nitelleae. The members of the Tribe Nitelleae are completely ecorticated but in the Chareae the members may be ecorticated or corticated in a distinctive manner.

The characteristic life history involves the germination of the oospore after meiotic division to produce hyaline filaments which develop into non-chlorophyllous, single-celled rhizoids and short horizontal protonema from which the erect shoot develops. Though meiosis is zygotic, the interpolation of a protonemal stage results in a characteristic type of monobiontic, digenic, heteromorphic, haplontic life history (see Chapman and Chapman, 1961) not known to occur in any other division of the algae.

The charophyte species are known to reproduce vegetatively with the help of star-shaped aggregates of cells (amylum stars) developed from the lower nodes, bulbils developed on the rhizoids, protonema-like outgrowths from the nodes, and through agamospermy (e.g. *Chara canescens*).

The organisation of the plant body and the life history pattern in the Charophyta are very characteristic, though the group also shares many features with the Bryophyta, i.e. nodes and internodes, a cortical covering produced from nodal cells, laterals of limited growth (branchlets), scales on the body and flagella of the antherozoids (see Pickett-Heaps, 1968; Moestrup, 1970), differentiation of the plant body into root-like, shoot-like and leaf-like structures, a sterile wall around the reproductive organs, a subapical flagellum in the antherozoid, and protonemal germination of the oospore. These features have led many charologists to suggest the separation of the charophytes from the Chlorophyta to constitute an independent division, the Charophyta (see Pascher, 1931; Imahori, 1954; Papenfuss, 1955; Desikachary and Sundaralingam, 1962; Guerlesquin, 1967; Khan and Sarma, 1967a,b,c; Sarma *et al.*, 1970; Bold and Wynne, 1978; amongst others).

The charophytes are commonly known to occur in freshwater lakes, smaller static water bodies, rivers, streams, swamps and in several man-made habitats. Various ecological and nutrient-impoverished conditions produce a variety of abnormalities (see Forsberg, 1965; Vaidya, 1967; Langangen, 1974) particularly in the cortication of *Chara* species and the formation of "heads" often associated with mucus (Imahori, 1954) in *Nitella*. These features have led many charologists to split and/or combine taxa without any clear understanding of whether the phenotypes are environmentally induced or under genetic control (Wood and Imahori, 1965).

PATTERN OF GEOGRAPHICAL DISTRIBUTION

Only a few attempts have been made in the past to understand the pattern of geographical distribution of the charophytes either globally (Braun, 1867; Braun and Nordstedt, 1882; Wood and Imahori, 1959; Proctor, 1980) or regionally, e.g. Zaneveld (1940) for Malaysia, Imahori (1951, 1954, 1955) for Formosa, Japan and the Ryuku Islands, Corillion (1955, 1957) for Europe, Corillion (1972, 1978) for Africa, and Khan and Sarma (1979) for India.

The present paper is an attempt to reassess the information concerning the general distribution pattern of the individual taxa in order to arrive at an understanding of charophyte migration and to discern the primary and secondary centres of origin which are used for the final assessment of the phylogeny of any particular group. During the last 30 years several distinct geographical regions have been explored extensively for charophytes, and about a dozen new taxa have been recently published.

For assessing geographical distribution of charophytes on a global basis, eight broad zones are recognised in the present work: (1) North American, (2) South American, (3) African, (4) European, (5) Asian (including Japan but excluding India), (6) Indian subcontinent, (7) Pacific Island region, and (8) Australian. These zones are further regrouped into hemispheres. The North American and South American zones constitute the Western Hemisphere while the rest comprise the Eastern Hemisphere. Furthermore, the North American, European and Asian zones constitute the Northern Hemisphere and the South American, African, Indian subcontinent, Pacific Islands and Australian regions are grouped into the Southern Hemisphere.

On the basis of the extent of distribution of the charophytes four arbitrary categories are recognised: (1) Cosmopolitan—in six zones (Indian subcontinent and the Pacific region are not taken into consideration); (2) Sub-cosmopolitan—in at least five zones; (3) Restricted—in at least two to four zones, mostly in one or more hemispheres but quite a few in more than one hemisphere exhibiting discontinuous distribution; and (4) Endemic—in only one zone.

In this paper the classification of Wood and Imahori (1965) has been followed. However, all those monoecious and dioecious taxa regarded by them as synonyms and merged into other taxa are considered as separate entities since many charophyte cytologists do not agree with their views.

Table I. The geographical distribution of charophyte taxa in different zones or regions with details given for East Hemisphere flora.

Categories	Number of genera	Total	Percentage
Cosmopolitan	N, 2 + T, 1 + C, 4	7	(1.50)
Subcosmopolitan	N, 7 + C, 1	8	(1.81)
Restricted Distribution			
East hemisphere	N, 70 + T, 1 + L, 4 + Ly, 1 + C, 58	134	(30.45)
West hemisphere	N, 3 + C, 4 + Nit, 1	8	(1.81)
North hemisphere	N, 4–2 (also in EH)* + T, 1 + C, 5–? (also in EH)*	5 (+5)*	(1.13)
South hemisphere	N, 19–19 (also in EH)* + C, 20–16 (also in EH)* + Ly, 1–1 (also in EH)*	4 (+36)*	(0.94)
Endemic	N, 140 + T, 11 + C, 115 + L, 4 + Ly, 3 + Nit, 1	274	(62.27)
	TOTAL	440	

Comparative distribution

	EH vs. WH	SH vs. NH
Number	134 8	40 10
Ratio	16.75 : 1	4 : 1

Constituents of East Hemisphere flora

Region	Chareae	Nitelleae	Ratio	Total	Percentage
Afro-European	15	2	1:0.08	18	13.23
Eurasian	4	3	1:0.75	7	5.57
Indian	30	40	1.33:1	70	52.10
Afro &/or SE Asian	21	19	1:1.05	39	29.10
TOTALS	70	64		134	100.00

Dioecism vs. Monoecism

I.

	ME vs. M		DE vs. D		II.	ME vs. DE		M vs. D	
number	212	127	62	39		212	62	127	39
ratio	5.43 :	3.25	1.58 :	1		3.41 :	1	3.25 :	1
ratio	1.66 :	1	1.58 :	1		ca. 3 :	1	3 :	1

According-y: ME vs. M   DE vs. D
             381 127   42   (= 677)
              9 : 3   3 : 1

*Abbreviations C, *Chara*; D, dioecious; DE, dioecious endemic; EH, Eastern Hemisphere; L, *Lamprothamnium*; Ly, *Lychnothamnus*; M, monoecious; ME, monoecious endemic; N, *Nitella*; NH, North Hemisphere; Nit, *Nitellopsis*; SH, South Hemispher; T, *Tolypella*; WH, West Hemisphere.

Likewise are considered all those taxa Wood and Imahori merged with other taxa but which have been subsequently found to be distinct and valid on the basis of cytological findings. Only a couple of species described after the publication of Wood and Imahori's monograph show some distinct morphological features depicting the development of a particular phylogenetic line, and these are considered here as independent species and, therefore, retained as such.

1. General Pattern (Table I)

Charophytes occur in all the continents except Antarctica, from 80° North in Spitzbergen to about 49° South in the Kerguelen Islands, and are represented by about 440 taxa in six genera. The contribution to the total number of taxa by the tribes Nitelleae (*Nitella*, 216; *Tolypella*, 15) and Chareae (*Chara*, 194; *Lamprothamnium*, 8; *Lychnothamnus*, 4; *Nitellopsis*, 3) is almost equal.

On an analysis of available data, it has been observed that, out of a total 440 charophyte taxa, 274 (62.27%) are endemic (see Table I), while the rest have either a cosmopolitan, sub-cosmopolitan, or restricted distribution. Thus the ratio of endemic to non-endemic comes to about 2:1. Again, out of 440 taxa, 339 are monoecious while 101 are dioecious, the ratio of monoecious to dioecious taxa coming to 3:1. The same ratio is reflected in the genus *Nitella*, while in *Chara* it comes to 4:1. Amongst the monoecious taxa the ratio of endemics to non-endemics is 1.6:1 while in dioecious taxa it is 1.5:1.

The number of endemic taxa confined to the various zones recognised here are as follows: North America (50, 19.3%), South America (25, 9.1%), Asia (48, 18.5%), Africa (42, 16.15%), Europe (41, 15.8%), Pacific region including Australia (44, 16.8%), and the Indian subcontinent (23, 8.8%). Endemic taxa in the South American zone are likely to increase more as the area has not been explored extensively.

A general grouping of the various other taxa, which are non-endemic (about 166, 37.8%), of the world charophyte flora reveals that seven taxa are cosmopolitan, eight are sub-cosmopolitan and the rest 151 (34.3%) have a restricted distribution. The seven cosmopolitan taxa are *C. braunii*, *C. globularis*, *C. vulgaris*, *C. contraria*, *N. gracilis*, *N. hyalina* and *T. glomerata* (not recorded from South America but expected to be present). *C. zeylanica*, *N. flexilis*, *N. opaca*, *N. acuminata*, *N. mucronata*, *N. oligospira*, *N. tenuissima* and *N. batrachosperma* are cosmopolitan. It is interesting to note that the group *C. zeylanica* var. and f. *zeylanica* is mysteriously absent from the central land mass of the African and European zones except for two records from Africa.

The taxa with restricted distributions exhibit variation in the pattern of distribution, which can be better explained on the basis of their distribution in the various hemispheres (see Table I). Ten taxa are restricted to the North Hemisphere and 40 to the South Hemisphere. The taxa restricted to the West Hemisphere are only eight while the number of taxa is 134 in the East Hemisphere, where they also exhibit a gradual increase in their number eastward: seven in Eurasia, 18 Afro-European, 39 for South East Asian regions, and 70 for the Indian subcontinent.

There is an equal number of taxa in the two tribes (5 in each) in the North Hemisphere as well as in the East (5) and West (5). The numbers of taxa in the South Hemipshere are variable: Nitelleae, 16; Chareae, 10 (23 Eastern, 3 Western). Likewise, in the West Hemisphere taxa representatives of the North and South Hemispheres are equal in number (8) while the Nitelleae are three and the Chareae are five. In the East Hemisphere, members of the Nitelleae (71) and Chareae (63) are almost equal in number and have only five representatives from the North Hemisphere. The remaining 95 taxa in the East Hemisphere are also found in the South Hemisphere and exhibit an interesting increase in the ratio between Nitelleae and Chareae from the Afro-European (1:0.08), Eurasian (1:0.75), Afro- and/or South-East Asian (1:1.05) to the Indian region, where the ratio is greatest (1.33:1).

Apart from the above grouping, some taxa of *Nitella* (13), *Chara* (16), *Nitellopsis* (1), and *Tolypella* (1) are restricted to the North, South, West and East Hemispheres though may be present in distantly-situated zones within these hemispheres (without any existing land connection in between) and are interesting examples of discontinuous distribution or geographical isolation.

2. Patterns in Various Zones (Table II)

In the flora of the various zones (Table II) all the genera are not distributed uniformly: all six genera are present in the European zone and the Indian subcontinent, five in South America, six in Asia and Australia, and four in North America, Africa and the Pacific region. *Nitella, Tolypella* and *Chara* are cosmopolitan, while *Lychnothamnus* is confined to the East Hemisphere only. *Lamprothamnium* and *Nitellopsis* are in six zones, and *Lychnothamnus* is in three zones (see Table II). The number of endemic and non-endemic taxa more or less corresponds with the area of the zone, but the Indian subcontinent has relatively high values for endemic (23) and non-endemic (102) taxa for its area.

The relative proportion of taxa in the two largest genera (*Nitella, Chara*) also varies for the zones. The number of taxa in each genus is equal in Africa (55/55) while there is a gradual increase in the number of *Nitella* taxa as

Table II. The number of taxa in each of the six genera in the eight recognised zones.

Zones*		Zonal flora	Chara	Lamprothamnium	Lychnothamnus	Nitellopsis	Nitella	Tolypella
North America (4)	Total	114	55			2	49	8
	Endemic	50	24			2	21	5
	Dioecious	18	8				7	1
	DE†	9	4				4	1
South America (5)	Total	89	50	1		1	33	4
	Endemic	25	16				8	1
	Dioecious	13	6			1	6	
	DE	6	4				2	
Africa (4)	Total	116	55	4			55	2
	Endemic	42	14				28	
	Dioecious	32	14				18	1
	DE	19	7				12	
Europe (6)	Total	91	53	6	2	1	20	9
	Endemic	41	25	4	1		6	5
	Dioecious	17	10			1	5	1
	DE	4	1				2	1

Asia (5)	Total	122	46		2			71	2
	Endemic	48	15					33	
	Dioecious	19	9				1	9	
	DE	3	1					2	
India (6)	Total	125	53		1	3	1	65	2
	Endemic	23	8			2		13	
	Dioecious	13	4				1	8	
	DE	1						1	
Pacific Regions (4)	Total	72	22		2		1	47	
	Endemic	19	4				1	14	
	Dioecious	15	8				1	6	
	DE	4	2				1	1	
Australia (5)	Total	62	24		1	1		35	1
	Endemic	25	7					18	
	Dioecious	28	10					18	
	DE	17	3					14	
World Charo-phyta	Total	440	195		8	4	3	215	15
	Endemic	274	115		4	3	1	140	11
	Dioecious	101	43				3	53	2
	DE	62	22				1	38	1

* Figures in parentheses indicate the number of genera in each zone.
† DE, dioecious endemic.

compared to *Chara* towards the North East (Asia, 71/46) and South East (Australia, 35/24) via the Indian subcontinent (16/9). On the other hand, a decrease in the number of taxa of *Nitella* is evident towards the North (Europe, 20/53) and West of Africa (South America, 33/50; North America, 49/55).

The flora of any particular region or zone always has some constituents which are also present elsewhere; thereby, critical analysis of these constituents is considered very important in order to understand explicitly the pattern of geographical distribution. Unfortunately it is not possible to consider here the composition of the charophyte flora in the zones.

CYTOSYSTEMATIC AND CYTOGEOGRAPHICAL CONSIDERATIONS

Cytotaxonomic data contributed by various workers on the Charophyta have proved to be of immense value in the cytogeographic assessment of the group. In no other group of algae is such a wealth of cytological information available (see Sarma, 1968, 1982), and this has led to a cytogeographic study of the charophytes of the world.

General Pattern

A perusal of the cytological literature has revealed that globally chromosome numbers have been determined so far for about 383 different individuals belonging to 168 (38.8%) taxa in all six genera. The range of reported chromosome numbers for 76 (40%) taxa of *Chara* are $n = 7-70$; three (37.5%) taxa of *Lamprothamnium* are $n = 14-70$; two (66.6%) taxa of *Nitellopsis* are $n = 14$ and 28; a taxon (100%) of *Lychnothamnus* is $n = 14$ and 28; 76 (35.7%) taxa of *Nitella* are $n = 6-48$; and nine (60%) taxa of *Tolypella* are $n = 9-$ about 42 and $n = 10-50$. These chromosome numbers prove the occurrence of ploidy in the Charophyta with basic numbers of $x = 7$ in the Tribe Chareae, $x = 3$ in *Nitella* and $x = 3$ and 5 in *Tolypella*.

Generally, most of the monoecious taxa have chromosome numbers twice that of the morphologically similar and closely related dioecious taxa, and based on this difference Wood and Imahori (1965) considered that the dioecious taxa are genetic variants of the monoecious ones. The opinion of these workers is untenable, as demonstrated by Corillion and Guerlesquin (1959), Sarma and Khan (1964, 1965a,b), Sawa (1965) and Proctor (1971). Karyological analysis of the dioecious taxa (D) and the monoecious taxa (M) into which they were merged by Wood and Imahori has revealed that (1) even if the chromosome number in dioecious taxa is half that of monoecious counterparts, their karyotypes are entirely different (e.g. *C. canescens*,

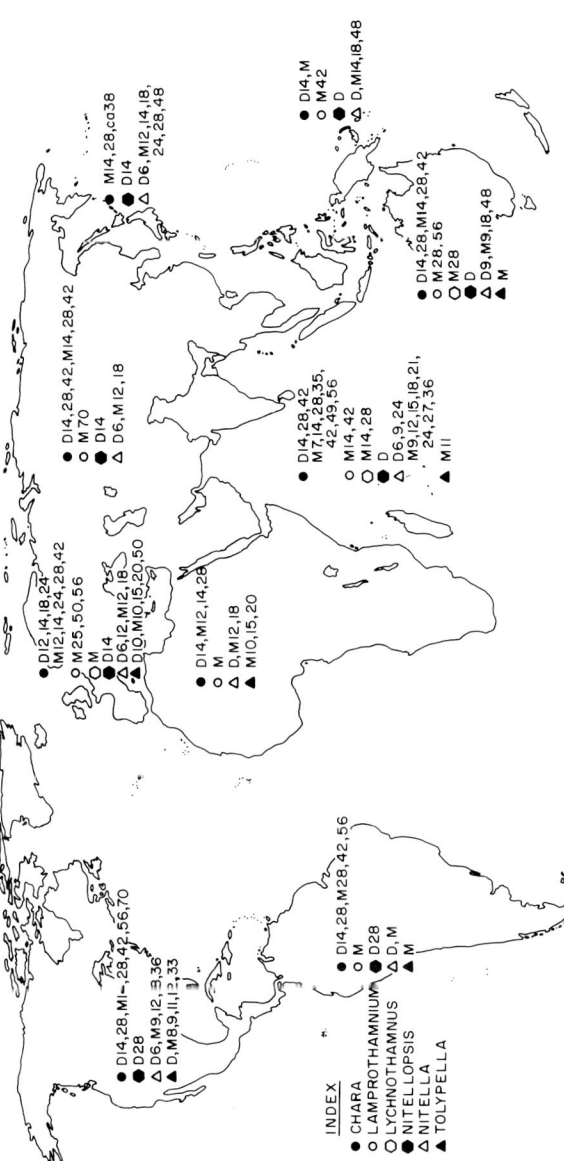

Fig. 1. The global distribution of chromosome numbers of the six genera of Charophyta (D, dioecious taxa; M, monoecious taxa).

$n = 28D$; *C. evoluta, $n = 56M$*), and (2) in some cases, the chromosome numbers of monoecious counterparts are more than twice the number of dioecious taxa (e.g. *C. connivens, $n = 14D$; C. globularis, $n = 14, 28$ and $70M$*). The aneuploid chromosome numbers in various taxa of *Chara, Lamprothamnium, Nitellopsis, Nitella* and *Tolypella* have been recorded mostly from Europe (Fig. 1).

All the known chromosome numbers of charophytes are not distributed evenly in every zone. From the frequency of distribution of chromosome numbers in the six genera (Table III) in the eight geographical zones, it is evident that the chromosome numbers $n = 14, 28$ and 42 in *Chara* and $n = 12$ and 18 in *Nitella* have the highest distribution frequencies (above 60%). Perhaps they represent the more successfully adapted genomes and have been produced through ploidy; however, their origin by hybridisation cannot be ruled out. The chromosome numbers with low frequencies might represent sensitive genomes, with the lower ones being primary or secondary basic numbers and the higher numbers might be due to hybridisation or polyploidy.

The genome and chromosome numbers probably underwent change as populations expanded into new environments. The gradients in chromosome numbers which resulted from this expansion are likely to indicate not only the probable routes of migration from their centre of origin and thereby their relatedness, but also the degree of successful adaptation for the new environments (see Stebbins, 1950, 1971, 1974). Such chromosomal gradients have been observed in the European charophytes (Guerlesquin, 1980).

Out of the seven cosmopolitan taxa (see p. 308), most of them exhibit

Table III. The frequency of distribution of the chromosome numbers for the six charophyte genera in the eight geographical zones or regions.

Distribution frequency	Chara	Lampro-thamnium	Lychno-thamnus	Nitel-lopsis	Nitella	Tolypella
12.50	8,9,20, 26,ca30, 32,35,36, 37,49,70	14,ca24, 25,28 ca50,70	14	—	15,16,17, 21,27,28, 29,30,	8,11,12,25 33,ca42,50
25.00	7,12,16, 18,24,+40	42,56	28	14,28	24,36	9,10,ca11, 15,20
37.50	—	—	—	—	9,14,48	—
50.00	56	—	—	—	6	—
62.50	—	—	—	—	12	—
87.50	28,42	—	—	—	18	—
100.00	14	—	—	—	—	—

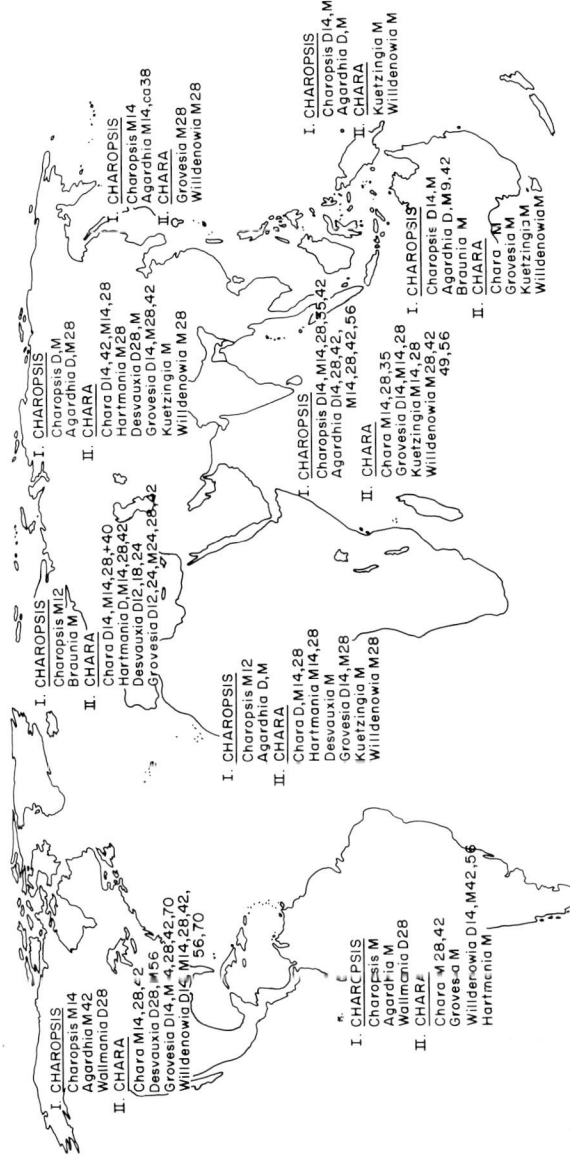

Fig. 2. The global distribution of *Chara* subsections with the chromosome number(s) for dioecious (D) and/or monoecious (M) taxa in the subsections indicated where known.

variations in the level of ploidy in the various subsections (see Figs 2, 3), e.g. *C. vulgaris*, $n = 7$, 14, 28 and 42 (subsection *Chara*), and *C. globularis*, $n = 14$, 28, 42 and 70 (subsection *Grovesia*). Some of the cosmopolitan taxa of the subsection *Chara* show a westward chromosome gradient as highest chromosome numbers are recorded for *C. vulgaris* ($n = 42$), *C. contraria* ($n = 42$), *C. globularis* ($n = 70$) and *T. glomerata* ($n = 33$) in the North American zone, while *C. braunii* ($n = 28$) and *N. hyalina* ($n = 15$ and 18) show an eastward gradient. Out of eight sub-cosmopolitan taxa, some exhibit variation in the level of ploidy in subsections *Willdenowia* (*C. zeylanica*, $n = 28$, 42, 49, 56 and 70), *Nitella* (*N. flexilis*, $n = 12$; *N. opaca*, $n = 6$ and 12D), *Rajia* (*N. acuminata*, $n = 9$, 12 and 18) and *Tieffallenia* (*N. confervacea*, $n = 18$). The sub-cosmopolitan taxa of the subgenus *Chara* also exhibit a westward chromosome gradient with the highest chromosome numbers for *C. zeylanica* ($n = 56$ and 70) confined to North America. However, *N. mucronata*, *N. oligospira* and *N. tenuissima* show an eastward gradient as their highest chromosome number ($n = 36$) is confined to the East Hemisphere. Variation in the level of ploidy is also evident in the various 117 taxa confined to distinct geographical regions (East, West, North, and South Hemispheres) as well as to those 30 taxa in geographical isolation.

The taxa belonging to the Tribe Chareae, and present in the northern zones of the East Hemisphere, show an eastward as well as westward gradient in chromosome number. Others present in both the northern and southern hemisphere exhibit northward and north-westward gradients in chromosome numbers. Such a gradient is not evident for those confined to the southern continents of the East Hemisphere (except the dioecious *C. preissii*). Other examples also exist for the East and West Hemispheres.

The endemic taxa are believed to be mostly of recent origin produced either through hybridisation or polyploidy. In spite of the large number of known endemic taxa no inference can be drawn at the moment due to the paucity of adequate cytological information. However, the prevalence of the chromosome numbers $n = 14$, 28 and 56 in *Chara*, $n = 14$ in *Lynchnothamnus*, $n = 6$, 12, 15, 18 and 36 in *Nitella* and $n = 8$ and about 11 in *Tolypella* indicates their origin through hybridisation and ploidy. Occurrence of large numbers of endemic taxa also indicates that the process of speciation is still actively continuing within the division.

Geographically isolated taxa are known to exist in various zones but about 30 (6.9%) of charophyte taxa have such a unique and discontinuous pattern of distribution that they cannot be classified with any of the above-mentioned groups. For example, several forms of *C. zeylanica* var. *zeylanica* (but not the cosmopolitan f. *zeylanica*) are found in South and North America and the Indian subcontinent, but are mysteriously absent from the central land mass (Africa and Europe) of the North and South hemispheres and

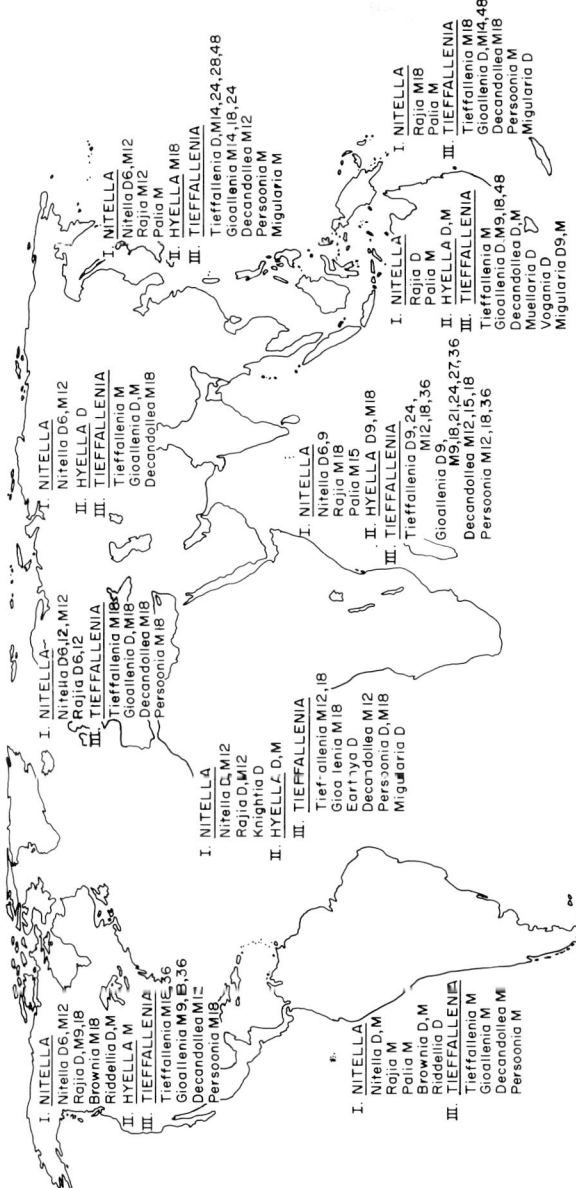

Fig. 3. The global distribution of *Nitella* subsections with the chromosome numbers for dioecious (D) and/or monoecious (M) taxa in the subsections indicated where known.

their chromosome numbers (f. *michauxii*, $n = 28$ India, $n = 42$ and 56 North America) show a north-westward gradient. Other taxa known to occupy different geographical regions have uniform chromosome numbers all over the world, but sometimes at different levels of ploidy. Also there appears to exist some correlation between a variety of chromosome numbers and a large area of distribution of the related taxa (see Guerlesquin, 1967; Kellmann, 1980).

POLYPLOID COMPLEX

A survey of chromosome numbers reported from various regions reveals the occurrence of distinct euploidy in different charophycean taxa.

1. General Pattern

The polyploid complex of the Charophyta contains even and odd ploidy levels with chromosome numbers ranging from $3x$ to $10x$ in the Chareae and up to $16x$ in the Nitelleae. The tetraploids in *Chara* ($n = 28$) and hexaploids in *Nitella* ($n = 18$) have a very widespread and almost continuous distribution as evident from the high values for their frequency of distribution (see Table III). It is worth noting that taxa possessing odd polyploid chromosome numbers in the Chareae ($n = 35$ and 49) and the Nitelleae ($n = 9, 15, 21, 27$ and 33) are all fertile in nature.

The polyploidization has succeeded only to a very limited extent in the dioecious taxa (e.g. *C. tomentosa*, $n = 42$; *C. canescens*, $n = 28$) when compared to monoecious taxa, where much higher orders of polyploids (e.g. *Chara globularis*, *C. zeylanica*, *Lamprothamnium papulosum*, all $n = 70$) are recorded.

On the basis of the euploid series of chromosome numbers recorded in charophytes, basic numbers of $x = 7$ for *Chara* (Moutschen et al., 1956), $x = 5$ for *Tolypella* (Guerlesquin, 1967), and $x = 3$ for *Nitella* (Sarma and Khan, 1964) have been suggested. The numbers $x = 7$ and $x = 3$ also appear to be the basic numbers for the tribes Chareae and Nitelleae, respectively. The lowest recorded chromosome numbers in the living Chareae ($n = 14$) and Nitelleae ($n = 6$) may be considered as secondary basic numbers which might have evolved through ploidy from monoploid ancestors expected to have primary basic numbers of $x = 7$ and $x = 3$.

The occurrence of two sets of morphologically similar chromosomes in the karyotype of the dioecious *N. mirabilis* ($n = 6$), indicates its autoploid nature (Sarma and Khan, 1964; Khan and Sarma, 1967b). Sawa (1965) demonstrated the similarity of the karyotypes of the monoecious *N. flexilis* ($n =$

12) and the dioecious *N. opaca* ($n = 6$) and *N. mirabilis* ($n = 6$; cf. Tindall, 1967), and hypothesized that former monoecious taxa may represent an allopolyploid of the latter two dioecious taxa. However, the exact nature of the prevalent ploidy, whether representing autoploidy or alloploidy, in charophytes must remain unconfirmed till conclusive evidence concerning the behaviour of the meiotic chromosomes is available.

The frequency of polyploids with reference to primary and secondary basic numbers is 84.1% and 49.7% in the Charophyta (70.7% and 26.0% in Chareae; 96.4% and 71.6% in Nitelleae). The value for polyploid frequency varies according to taxa. The highest frequency is recorded for such monoecious taxa as *C. brachypus*, *C. indica*, *N. subglomerata*, *N. pulchella*, *T. salina*, etc. and the dioecious *C. fragifera*; and lowest for the monoecious *C. braunii* and the dioecious *C. tomentosa*.

The occurrence of aneuploid numbers in various taxa may be accounted for by either technical or counting errors, as chromosomes in related taxa are found to be very sticky and fairly large (Khan and Sarma, 1967a,b,c), or may be the result of non-disjunctions which lead to increased or decreased chromosome number before gametogenesis in the antheridial filaments (cf. Moutschen *et al.*, 1956).

2. Polyploidy and Gene Duplication

The electrophoretic analysis of genetic variation in natural populations of charophytes (Grant and Proctor, 1982) indicates that all the extant forms have undergone at least one polyploidization or, more likely, are derived from ancestors which did so. The polyploidization led to the duplication of functional genes which directly resulted in the generation of substantial levels of genic variation both within and between various taxa. The duplication of functional genes in the individual taxa with higher chromosome numbers probably helped in maintaining a much greater degree of enzymatic versatility, and this was of evolutionary significance by allowing a greater ecological tolerance to develop, which was especially necessary for surviving the temperature changes brought about by repeated past glaciations.

DISCUSSION

1. Origin and Migration

The production of resistant wall material in the charophyte oospore has increased the possibility of its preservation in sediments. Fossilized oogonia (gyrogonites) are rare in the Silurian-Devonian and more abundant in the middle Cretaceous deposits (Grambast, 1974), indicating that this group might have originated in the Late Silurian. The existing fossil evidence provides only certain clues about the appearance of the ancestral forms.

It appears that the greatest number of monoecious and dioecious taxa showing primitive features, the comparatively small number of endemic taxa, the large number of species in common with other zones, the low frequency of the highest polyploids (but relatively large number of polyploid series, though at the lower levels), in the Indian subcontinent indicate that it is probably the primary centre of origin for almost all six of the charophyte genera. The occurrence of introgressive hybridization coupled with changes in the chromosome number to form a polyploid series along migratory routes radiating from the Indian subcontinent (see Figs 4, 5) provides further support for this region being the centre of origin where many of the basic diploids are still relatively restricted whereas most of the widespread taxa are polyploids. This is supported by palaeobotanical evidence which indicates the occurrence of charophyte taxa in the Indian subcontinent during the Silurian-Devonian period (cf. Khan, 1973).

The chromosome gradients and the very restricted distribution of the higher polyploid levels in the Indian subcontinent (see Khan and Sarma, 1979) clearly indicate the migration of flora from one area to another.

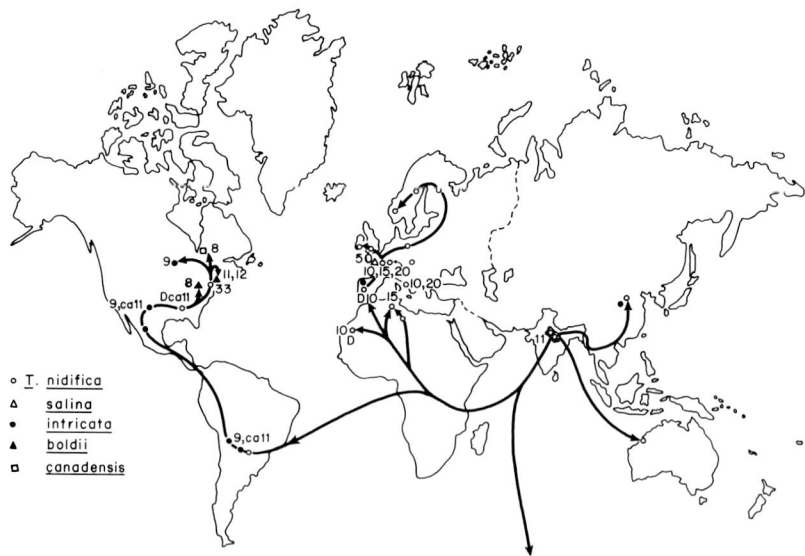

Fig. 4. The suggested migration routes for the five species of *Tolypella* with chromosome numbers indicated where known.

Fig. 5. The suggested migration routes of the Charophyta.

Conversely, occurrence of primitive taxa with lowest recorded chromosome numbers in the Indian subcontinent clearly indicates the migration of charophytes from this region to various other zones, especially to South America, Europe, Asia and Australia in the remote past, and to the North and South-East regions more recently.

The prevalence of highest ploidy levels for various taxa of the Tribe Chareae growing in North-West Asia, North and South America, and for the Nitelleae in North-East Asia and Australia (see Fig. 1), indicate the active role of these zones as secondary centres of origin and for the dispersal of charophytes. The presence of large numbers of endemic taxa in these zones also provides further support to this conclusion.

An understanding of the geological history of the various zones will prove to be very rewarding in understanding the present pattern of the geographical distribution of the Charophyta. The former Gondwanaland constituted by the South American, African and Australian continents together with the Indian subcontinent and Antarctica, moved northward about 270–230 million years ago (mya) and brought the European continent to its present position. It is probable that during this period the charophytes originated near the periphery of the ice-sheets that then existed in that part of Gondwanaland which was to become the Indian subcontinent. The break up of Gondwanaland led to the separation of the Indian subcontinent from Africa and Australia in the Jurassic period (195 mya) and from Antarctica in the late Cretaceous (75 mya). Gondwanaland completely fragmented by the mid-Tertiary Period and all the separated continents came to occupy their present positions (Windley, 1977). The splitting of this unified land mass must have resulted in the genetic isolation and morphological divergence among the separated segments of a formerly homogeneous charophyte flora.

The distribution pattern of the charophytes (Figs 4, 5) reveals that related charophytes are either distributed continuously in adjacent regions, or are widely separated with a disjunct distribution. For instance, the Subsection *Braunia,* restricted in the Australian and European zones as vicariants (see Fig. 2), exhibits a disjunct distribution. It may either represent the scattered survivors of a once widespread group related to Subsection *Agardhia* or have originated from the latter independently in both the zones and shows an evolutionary parallelism/convergence. From the current distribution patterns it is clearly evident that there is a complete lack of migration, and thereby exchange of dioecious taxa between the Western Hemisphere and Australia, as those dioecious taxa growing in these regions are quite distinct and belong to different subsections.

It is also apparent that islands mostly harbour monoecious taxa (Figs 6, 7). Species with dioecious counterparts are more common than those lacking such dioecious counterparts. Prevalence of the dioecious Nitelleae and the absence of dioecious Chareae in Japan may be due to Japan's early

Fig. 6. The global distribution of monoecious and dioecious *Chara* taxa.

Fig. 7. The global distribution of monoecious and dioecious *Nitella* taxa.

separation from the Asian mainland, by which time the ancient dioecious members of the Chareae might not have reached there.

The existence of more taxa in the tropical regions than in temperate ones (see Tables I, II) indicates that tropical environments have more ecological niches which allow for gradual speciation.

2. Speciation

Major ancient sections and subsections of the Charophyta appear to have originated in ecotones where habitat diversity and consequently varying selective pressure led to diversification through adaptive radiation which culminated in distinct evolutionary lines. These lines, through extinctions, have become progressively more distinct from one another.

Speciation in the charophytes appears to have occurred through gradual divergence of geographically separate populations, geographic allopatric speciation, and/or sympatric speciation. It is apparent from the breeding pattern of allopatric and sympatric populations (Grant and Proctor, 1972; Proctor, 1975) that a group like the charophytes generates new diversity within itself by the combined processes of hybridization and uniparental reproduction, which leads to the occurrence of large numbers of localised microspecies together with a few geographically widespread microspecies or species of hybrid constitution.

The inbreeding within a series of small populations promotes local differentiation. Some distinct species may be genetically very close and not only cross but also give fertile and vigorous hybrids. In contrast, some species identical in outward appearance are found to be reproductively isolated, e.g. *C. vulgaris* and *C. contraria* (Grant and Proctor, 1972). Reproductive isolation permits two populations to occupy the same habitat without exchanging genes, and this is an important step towards divergent evolution which is seen as inviability or sterility of hybrids between the species that belong to different subsections within the Charophyta. The pattern of sterility is reflected in the chromosome morphology of the concerned population exhibiting a regular ecological distribution, e.g. *C. zeylanica* (Griffin and Proctor, 1964; Sarma and Khan, 1965c).

The chromosomal evolution in Nitelleae involved polyploidization as well as ascending aneuploidy. The ascending aneuploidy might have altered the primary basic number ($x = 3$) of the Nitelleae and provided a basic number $x = 5$ for the genus *Tolypella*, and ultimately given rise to a basic number $x = 7$ for Chareae. The high secondary basic numbers $x = 6$ (Nitelleae) and $x = 14$ (Chareae) in extant charophytes might have been established by ancient events of polyploidization during an early period of evolution. Additional cycles of ploidy produced further diversification within the individual genera as well as higher ranks of polyploids within the group.

The distribution pattern of diploids and their polyploid derivatives in the Charophyta (*C. zeylanica*, *C. globularis*, $n = 14, 28, 42, 56, 70$, etc.) is favoured by extensive new habitats such as areas vacated by the retreating ice-sheets (North America and Europe) and recently formed islands (Pacific regions and Japan).

The occurrence of the fertile odd-numbered polyploid in charophytes (*Chara*, $n = 35$ and 49; *Nitella*, $n = 9, 15, 21$ and 27; *Tolypella*, $n = 15$ and 33) indicates their amphiploid condition. These are adaptively superior to their diploid taxa as is evident from their comparatively widespread geographical distribution and abundance, e.g. *C. vulgaris*, *C. zeylanica*, *N. acuminata* and *T. glomerata*. The amphiploids in the charophytes appear to have originated from diploid hybrids as a product of hybridization between distinct autoploids. The occurrence of aneuploid chromosome numbers and their probable role in the evolution of the Charophyta has been often emphasised (Guerlesquin, 1967, 1980; Noor and Mukherjee, 1975).

Dioecism in the Charophyta is a primitive character (Sarma and Khan, 1967). The presence of the ratio (3:1) in natural populations which are widespread (127M:39D) and also in endemic (212M:62D) populations probably indicates the Mendelian segregation of characters associated with dioecism and monoecism. The specific segregation pattern may also account for the general non-existence of the higher ploidy level in the dioecious taxa, because higher ploidy is likely to break down the sex-determination mechanism in tetraploids (Muller, 1925).

3. Classification

Wood and Imahori (1965) and Proctor (1980) have proposed models explaining the phylogenetic relations between various subsections of the genus *Chara*. Contrary to the prevalent views concerning the division of the genus *Chara* into haplostephanous and diplostephanous groups, Proctor (1980) preferred to divide the same into Gondwanaland and Lauresia groups on the basis of their geographical distribution. The validity of the stipulode number as one of the major evolutionary trends as well as an important taxonomic criterion gets support from cytotaxonomical studies (Khan and Sarma, 1967a,b; Ramjee and Sarma, 1971; Guerlesquin, 1980).

The pattern of geographic and cytogeographic distribution has also proved invaluable in reassessing the taxonomic position of some taxa in view of their phylogenetic relationships. The taxa related to the subsections *Agardhia, Braunia* and *Wallmania* have been rearranged into the subsections *Agardhia* (axis diplostichous branchlet ecorticated), *Richardwoodia* (axis haplostichous, branchlet ecorticated), *Braunia* and *Wallmania* (axis triplostichous, branchlet with corona), *Corillionia* (axis diplo-triplostichous, branchlet ecorticated), *Imahoria* (axis diplostichous, branchlet corticated except basal-segment), *Desikachariya* (axis diplo-triplostichous, branchlet corticated except basal segment) and *Proctoria* (axis diplostichous, branchlet corticated completely), and two taxa of the section *Grovesia* with verticillate bract cells into a new section *Guerlesquinia*. Similar re-arrangement of the Nitelleae is in progress.

4. Phylogeny

The evidence suggests that genera with a chromosome number in the Chareae of $n = 14$ and in the Nitelleae of $n = 6$ or a higher multiple, have probably been derived by an ancient event of ploidy. Accordingly, the origin of Chareae (Fig. 8) through the genus *Nitellopsis* (intermediate between the Chareae and Nitelleae) by a process of polyploidization and polyploid drop, appears to be more logical (Guerlesquin, 1967; Sarma *et al.*, 1970). The same sequence of events may also give rise to *Nitellopsis* from *Tolypella*, a genus which also shares some features common to both the Chareae and Nitelleae.

The study on the distributional pattern of world charophytes (Khan, 1982) led us to propose an alternative to the earlier proposed model (see Fig.

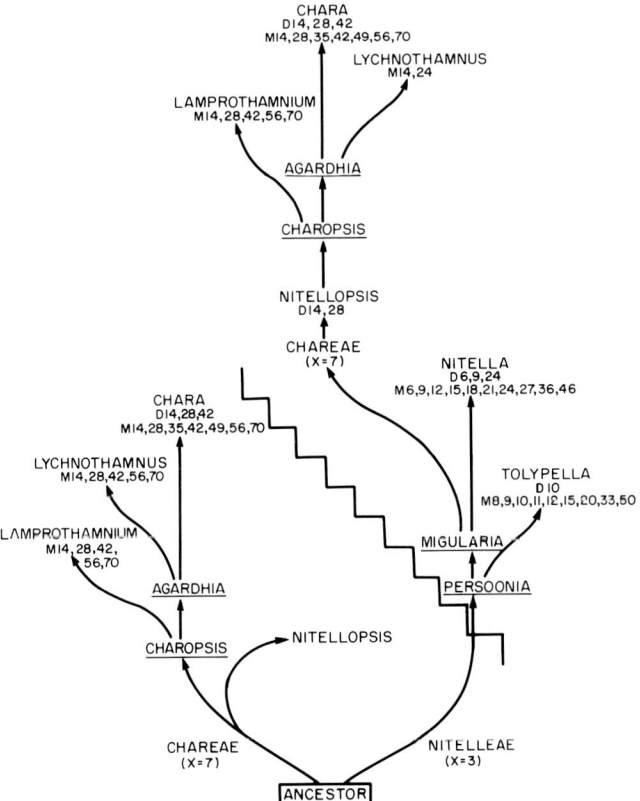

Fig. 8. Interrelationships, phylogeny and speciation in Charophyta.

8) in which it is now envisaged that the Subsection *Migularia* of the Tribe Nitelleae might have given rise to the genus *Nitellopsis* of the Tribe Chareae, from where evolution of various subsections of the genus *Chara* started from Subsection *Charopsis* (ecorticated, haplostichous) and proceeded towards a gradual complication of vegetative structures through diplostichous *Agardhia* (branchlet ecorticated), *Imahoria* (axis diplostichous, branchlet corticated except basal branchlet segment) and *Chara* (diplostichous, spine cells solitary), leading to the establishment of the triplostichous Subsection *Grovesia* (axis triplostichous, branchlet basal segment normal). The Subsection *Charopsis* appears to have given rise to the genus *Lamprothamnium* through *C. corallina*-like forms.

Simultaneously, Subsection *Agardhia* gave rise to subsections *Wallmania* (without terminal corona), *Braunia* (with terminal corona), *Corillionia* (axis 2–3 corticated) and *Richardwoodia* (axis haplostichous), and from the latter appears to have originated the genus *Lynchnothamnus* through a form like *C. myriophylla*. Similarly, subsections *Imahoria* (axis diplostichous, branchlet basal segment ecorticated), *Desikachariya* (axis 2–3 corticated) and *Proctoria* (axis diplostichous, branchlet basal segment corticated) originated from the Subsection *Agardhia*; Subsection *Chara* gave rise to subsections *Hartmania* (spine cells mostly fasciculate) and *Desvauxia* (axis haplostichous); and Subsection *Grovesia* to subsections *Kuetzingia* (basal branchlet segment discoloured), *Wildenowia* (basal branchlet segment ecorticated) and *Guerlesquinia* (bract cells verticillate).

The monoecious taxa of haplostephanous and diplostephanous groups of the genus *Chara* are generally found to have $n = 14$ and $n = 28$ chromosome numbers, respectively (cf. Sarma, 1982). Thus the statement of Proctor (1980) that "no studies are expected to support the contention that stipulode number alone reflect major evolutionary trends" becomes untenable. The non-existence of some dioecious taxa in various sections of *Chara* as shown by Proctor (1980) in his model is also unacceptable as these are known to exist in the geographic regions concerned, e.g. *Chara* (*C. tomentosa,* Africa), *Hartmania* (*C. hispida* f. *crassicaulis,* Europe), *Desvauxia* (*C. canescens,* North America), *Charopsis* (*C. ecklonia,* Africa; *C. fibrosa* f. *hookeri,* Asia; *C. fibrosa* f. *fibrosa* = *preissii,* India), and *Wildenowia* (*C. kenoyeri,* North America). These observations are contrary to the views of Proctor (1980) and support the belief of earlier charologists that haplostephanous and diplostephanous features appear to be valid taxonomic criteria. Similarly, the pattern of cortication of the axes as well as of the branchlets seems also to be a valid taxonomic criterion on cytotaxonomic grounds (see Khan and Sarma, 1967a,b,c; Sarma, 1968; Ramjee and Sarma, 1971; Guerlesquin, 1980). This is contrary to the views of Langangen (1974), who questioned the significance of cortication as a criterion for major subdivision of the genus, and what Proctor (1980) considered only as "a useful descriptive term but not a particularly faithful reflection of evolutionary trends".

ACKNOWLEDGEMENTS

The authors are grateful to the Department of Science and Technology, University Grants Commission, New Delhi, the Chief Minister, Uttar Pradesh, and the Vice-Chancellor, Avadhu University, Faizabad for providing financial assistance. The authors are also indebted to Dr. (Mrs.) M. Guerlesquin, Maître de Recherches, and Professor R. Corillion, I.R.F.A., C.N.R.S., Angers, France for their invaluable help, suggestions and criticism. Thanks are also due to Drs. D. E. G. Irvine and D. M. John for critically revising the manuscript.

REFERENCES

Bold, H. C. and Wynne, M. J. (1978). "Introduction to the Algae." Prentice-Hall, Englewood Cliffs, New Jersey.
Braun, A. (1867). Die Characeen Afrika's. *Monatsber. K. preuss. Akad. Wiss.* pp. 782–800, 873–944.
Braun, A. and Nordstedt, C. F. O. (1882). "Fragmente einer Monographie der Characeen." Abh. K. Akad. Wiss., Berlin.
Chapman, D. J. and Chapman, V. J. (1961). Life histories in the algae. *Ann. Bot. (London)* [N.S.] **25,** 547–561.
Corillion, R. (1955). Les Charophycées de France et d'Europe Occidentale. Thèse, Toulouse. (Also published in 1957 in *Bull. Soc. Sci., Bretagne* **32,** 1–499.)
Corillion, R. (1972). Aspects généraux de la distribution géographique des Characées Africano-Malgaches. *C. r. Somm. Seanc. Soc. Biogeogr.* **49,** 64–81.
Corillion, R. (1978). Les Characées du Nord de l'Afrique: Eléments floristiques et distribution. *Bull. Soc. Etud. Sci. Anjou* **10,** 27–34.
Corillion, R. and Guerlesquin, M. (1959). Premières observations cytotaxonomiques sur le genre *Tolypella* (Charophycées). *Bull. Soc. Etud. Sci. Angers* **2,** 167–179.
Desikachary, T. V. and Sundaralingam, V. S. (1962). Affinities and interrelationships of the Characeae. *Phycologia* **2,** 9–16.
Forsberg, C. (1965). Environmental conditions of Swedish charophytes. *Symb. bot. ups.* **18,** 1–67.
Grambast, L. J. (1974). Phylogeny of the Charophyta. *Taxon* **23,** 463–481.
Grant, M. C. and Proctor, V. W. (1972). *Chara vulgaris* and *C. contraria:* Patterns of reproductive isolation for two cosmopolitan species complexes. *Evolution* **26,** 267–281.
Grant, M. C. and Proctor, V. W.(1982). Electrophoretic analysis of genetic variation in the Charophyta I. Gene duplication via polyploidy. *J. Phycol.* **16,** 109–115.
Griffin, D. G., III and Proctor, V. W. (1964). A population study of *Chara zeylanica* in Texas, Oklahoma and New Mexico. *Am. J. Bot.* **51,** 120–124.
Guerlesquin, M. (1967). Recherches caryotypiques et cytotaxonomiques chez les Charophycées d'Europe occidentale et d'Afrique du Nord. *Bull. Soc. Sci. Bretagne* **41,** 1–265.
Guerlesquin, M. (1980). Recherches récentes sur les Charophycées (Morphologie et systématique cytotaxonomie). *Bull. Soc. Etud. Sci. Anjou, Mem.* **4,** 143–155.
Imahori, K. (1951). Studies on Charophyta in Formosa. I. *Sci. Rep. Kanazawa Univ.* **1,** 201–221.
Imahori, K. (1954). "Ecology, Phytogeography and Taxonomy of Japanese Charophyta." Kanazawa University, Kanazawa.
Imahori, K. (1955). Phytogeographical survey on Charophyta flora in Ryuku Islands. *Sci. Rep. Kanazawa Univ.* **3,** 93–99.
Kellmann, M. C. (1980). "Plant Geography," 2nd ed. Methuen, London.
Khan, M. (1973). Algae through the ages. *Acta bot. Indica* **1,** 55–67.
Khan, M. (1982). Pattern of distribution and speciation in haploid Charophyta. *Int. Phycol. Congr., 1st,* 1982, a25.
Khan, M. and Sarma, Y. S. R. K. (1967a). Studies on cytotaxonomy of Indian Charophyta. I. Chara. *Phykos* **6,** 36–47.
Khan, M. and Sarma, Y. S. R. K. (1967b). Studies on cytotaxonomy of Indian Charophyta. II. Nitella. *Phykos* **6,** 48–61.
Khan, M. and Sarma, Y. S. R. K. (1967c). Some observations on the cytology of Indian Charophyta. *Phykos* **6,** 62–74.

Khan, M. and Sarma, Y. S. R. K. (1979). Cytogeographic study of Charophyta with particular reference to India. *Proc. Recent Adv. Crypto. Bot. Geophytol.* **11**.

Langangen, A. (1974). Ecology and distribution of Norwegian charophytes. *Norw. J. Bot.* **21**, 31–52.

Moestrup, O. (1970). The structure of mature spermatozoids of *Chara corallina* with special reference to microtubules and scales. *Planta* **93**, 295–308.

Moutschen, J., Dahmen, M. and Gillet, C. (1956). Sur les modifications de la spermogénèse de *Chara vulgaris* L. induites par les rayons. X. *Rev. Cytol. Biol. veg.* **17**, 433–450.

Muller, H. J. (1925). Why polyploidy is rarer in animals than plants. *Am. Nat.* **59**, 346–353.

Noor, M. N. and Mukherjee, S. (1975). On the aneuploid chromosome numbers in *Chara hydropitys* Reich. from India. *Cytologia* **40**, 803–807.

Papenfuss, G. F. (1955). Classification of the algae. *In* "A Century of Progress in the Natural Sciences—1853–1953," pp. 115–224. Calif. Acad. Sci., San Francisco, California.

Pickett-Heaps, J. D. (1968). Ultrastructure and differentiation in *Chara*. IV. Spermatogenesis. *Aust. J. Bot.* **21**, 655–690.

Proctor, V. W. (1971). Taxonomic significance of monoecism and dioecism in the genus *Chara*. *Phycologia* **10**, 299–307.

Proctor, V. W. (1975). The nature of charophyte species. *Phycologia* **14**, 97–113.

Proctor, V. W. (1980). Historical biogeography of *Chara* (Charophyta): An appraisal of the Braun–Wood classification plus a falsifiable alternative for future consideration. *J. Phycol.* **16**, 218–233.

Ramjee and Sarma, Y. S. R. K. (1971). Some observations on the morphology and cytology of Indian Charophyta. *Hydrobiologia* **37**, 367–382.

Sarma, Y. S. R. K. (1968). Cytology and cytotaxonomy of Indian Charophyta: A resumé. *Nucleus (Calcutta), Suppl.* pp. 128–137.

Sarma, Y. S. R. K. (1982). Chromosome numbers in algae. *Nucleus (Calcutta)* **25**, 66–108.

Sarma, Y. S. R. K. (1983). Algal karyology and evolutionary trends. *In* "Chromosomes in Evolution of Eucaryotic Groups" (A. K. Sharma and A. Sharma, eds.), pp. 177–227. CRC Press, Florida.

Sarma, Y. S. R. K. and Khan, M. (1964). Chromosome numbers in some Indian species of *Nitella*. *Chromosoma* **15**, 246–247.

Sarma, Y. S. R. K. and Khan, M. (1965a). Chromosome numbers in some Indian species of *Chara*. *Phycologia* **4**, 173–178.

Sarma, Y. S. R. K. and Khan, M. (1965b). A preliminary survey on the chromosome numbers of Indian Charophyta. *Nucleus (Calcutta)* **8**, 33–38.

Sarma, Y. S. R. K. and Khan, M. (1965c). Some new observations on the karyology of *Chara zeylanica* Klein ex Willd. *Curr. Sci.* **34**, 293–294.

Sarma, Y. S. R. K. and Khan, M. (1967). Dioecism and monoecism as taxonomic criteria in Charophyta. *Curr. Sci.* **36**, 245–247.

Sarma, Y. S. R. K., Khan, M. and Ramjee (1970). A cytological approach to phylogeny, inter-relationships and evolution in Charophyta. *Indian Biol.* **2**, 11–19.

Sawa, T. (1965). Cytotaxonomy of the Characeae: Karyotype analysis of *Nitella opaca* and *Nitella flexilis*. *Am. J. Bot.* **52**, 962–970.

Stebbins, G. L. (1950). "Variation and Evolution in Plants." Columbia Univ. Press, New York.

Stebbins, G. L. (1971). "Chromosomal Evolution in Higher Plants." Edward Arnold, London.

Stebbins, G. L. (1974). Adaptive radiation and the origin of form in the earliest multicellular organisms. *Syst. Zool.* **22**, 478–485.

Tindall, D. R. (1967). A new species of *Nitella* (Characeae) belonging to the *Nitella flexilis* species group in North America. *J. Phycol.* **3**, 229–232.

Vaidya, B. S. (1967). Study of some environmental factors affecting the occurrence of charophytes in Western India. *Hydrobiologia* **29**, 256–262.

Windley, B. F. (1977). "The Evolving Continents." Wiley, London.

Wood, R. D. and Imahori, K. (1959). Geographical distribution of Characeae. *Bull. Torrey bot. Club* **86**, 172–183.

Wood, R. D. and Imahori, K. (1965). "A Revision of the Characeae." Cramer, Weinheim.

Zaneveld, J. S. (1940). The charophytes of Malaysia and adjacent countries. *Blumea* **4**, 1–223.

13 | The Flagellar Apparatus in *Batophora* and *Trentepohlia* and Its Phylogenetic Significance

K. R. ROBERTS

Department of Biology, The University of Southwestern Louisiana, Lafayette, Louisiana, USA

Abstract: The flagellar apparatus in the gametes of *Batophora* and *Trentepohlia* has been investigated and the absolute configuration of components determined in each. In *Batophora*, microtubular roots are arranged in a cruciate 5:2:5:2 pattern with the 5:2 root pairs emerging nearly parallel to one another. Basal bodies overlap and are connected to one another anteriorly by a large capping plate that bears a broad, singly striated fibrous mid-region. Two system II fibrous roots (rhizoplasts) emerge from the posterior side of the basal bodies and extend toward the nucleus. In *Trentepohlia,* the basal bodies also overlap and the microtubular roots are arranged in a cruciate 6:4:6:4 pattern, with the 6:4 root pairs also emerging in a near parallel fashion. The six-stranded roots emerge in a 3-over-3 configuration from within an electron-dense terminal cap that covers the entire proximal region of each basal body. The four-stranded roots are associated with a fibrous region along their lower surface at the level of the basal bodies. The fibers of this region appear as short individual extensions of three of the four root microtubules in both longitudinal section and cross section. The basal bodies of *Trentepohlia* are connected by three narrow extensions of opposing basal body triplets. Based upon these flagellar apparatus characteristics, it is proposed that both *Batophora* and *Trentepohlia* be assigned to the Ulvophyceae.

INTRODUCTION

During the past five years, comparative studies of the flagellar apparatus components in several green algae have resulted in a more positive characterization of the Ulvophyceae (see Mattox and Stewart, this volume, Chapter 2). As a result, the ulvophycean flagellar apparatus can be generally characterized by the following features; overlapping basal bodies in biflagellate cells that are arranged in an 11/5 configuration, the presence of a non-striated (for the most part) connective between the basal bodies (the capping plate), the presence of an electron-dense plate that partially or fully covers the proximal end of each basal body (the terminal cap), a cruciate microtubular root system with the larger roots attached to the inner (i.e. adjacent) side of the basal bodies while the smaller roots attach at the outer (i.e. non-adjacent) side of each basal body, and the presence of a striated fibrous root (rhizoplast, system II fibrous root). With an increase in three-dimensional reconstruction studies of the flagellar apparatus of ulvophycean algae, more precise descriptions of the distinguishing characteristics are becoming available (see O'Kelly and Floyd, 1983, this volume, Chapter 4; Melkonian, this volume, Chapter 3).

Several groups of green algae that were previously aligned with the Chlorophyceae are now considered to be most closely aligned with the Ulvophyceae. Some of the algae that have been determined to possess ulvophycean flagellar apparatus characteristics include *Bryopsis* (Hori, 1977; Melkonian, 1980a), *Cladophora* (Floyd, 1981), *Derbesia* (Roberts et al., 1980, 1981), *Entocladia* (O'Kelly and Floyd, 1983), *Friedmannia* (Melkonian and Berns, 1983), *Pseudobryopsis* (Roberts et al., 1982), *Ulothrix zonata* (Sluiman et al., 1980), *Ulva* (Melkonian, 1979, 1980b), *Ulvaria* (Hoops et al., 1982) and *Urospora* (Sluiman et al., 1982).

Batophora was chosen for this study because the closely related genus *Acetabularia* was reported to have a flagellar apparatus with characteristics intermediate between those found in the Chlorophyceae and the Ulvophyceae (Herth et al., 1981). Since all other siphonous green algae examined to date have been assigned to the Ulvophyceae, it was hoped that enough variation would exist between the flagellar apparatus components of *Batophora* and *Acetabularia* to aid in determining their phylogenetic affinities. Similarly, *Trentepohlia* was chosen because it and the closely related *Cephaleuros* have both been reported to have a multilayered structure (MLS) (Graham and McBride, 1974, 1975 for *Trentepohlia;* Chapman, 1980, 1981; Chapman and Henk, 1982, for *Cephaleuros*), which is a characteristic feature of the motile cells of charophycean algae and bryophytes. Since it is clear from the published micrographs of both *Trentepohlia* and *Cephaleuros* that basal bodies overlap like they do in ulvophycean genera, it was decided that a re-examination of *Trentepohlia* might add insight into its affinities with other green algae.

MATERIALS AND METHODS

Swarmers of *Batophora* and *Trentepohlia* were fixed in a 1% glutaraldehyde solution made with Provasoli's Enriched Seawater (PES) medium for 1 h.

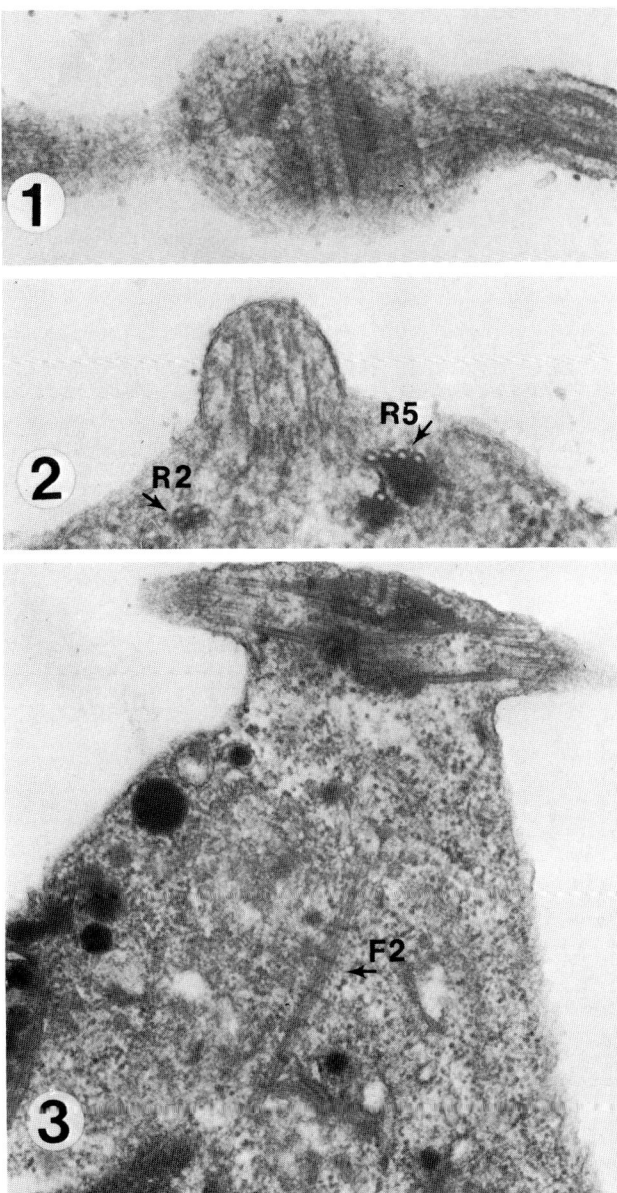

Fig. 1. *Batophora:* Longitudinal section through the compound capping plate showing the broad, singly striated fibrous connective between the two nonstriated capping plate halves.

Fig. 2. *Batophora:* Cross section of the 2-stranded (R2) and 5-stranded (R5) microtubular roots that emerge nearly parallel to one another. Note the electron-dense material and fibrillar connection between the single lower microtubule and the four outer microtubules of the R5 root.

Fig. 3. *Batophora:* Longitudinal section through the anterior region showing one of the two striated system II fibrous roots (F2).

Cells were gently centrifuged and embedded in agar where they were then rinsed in PES and then postfixed in 1% OsO_4 in PES for 30 min. Cells were then stained en bloc in 2% aqueous uranyl acetate for 1 h, dehydrated in a graduated series of acetone, and embedded in Spurr's Resin (Spurr, 1969).

RESULTS

1. Batophora

The flagellar apparatus of *Batophora* consists of overlapping basal bodies that are connected to one another along their anterior surface by a large compound capping plate (Fig. 1). The capping plate consists of two non-striated halves that are directly attached to the anterior surface of each basal body. Lying between and connecting the non-striated halves is a broad fibrous connective region that bears a single broad striation. While faint striations are also observable at the edges of the fibrous zone, these are interpreted as the boundary region between the striated and non-striated region and therefore not true striations like the central striation.

The microtubular roots of *Batophora* are of a modified cruciate type with the roots arranged in a 5:2:5:2 pattern (Fig. 2). The microtubular roots emerge from beneath and lie along two sides of the cell with both the 5- and 2-stranded roots emerging nearly parallel to one another. The 5-stranded roots emerge in a 4-over-1 configuration with the single microtubule attached to the four exterior microtubules by an electron-dense fibrillar extension (Fig. 2, *arrow*). Both the 5- and the 2-stranded microtubular roots are subtended by an electron-dense material that appears to be narrowly striated, having the same periodicity as system I fibrous roots.

A narrow system II fibrous root or rhizoplast emanates from the proximal region of each basal body (Fig. 3). The two striated roots descend posteriorly and appear to terminate near the anterior lobe of the nucleus.

A diagrammatic reconstruction of the flagellar apparatus (Fig. 9a,b) illustrates the three dimensional relationships among the major flagellar apparatus components.

Fig. 4. *Trentepohlia:* Cross section of the papilla showing the cruciate microtubular root system (R4, R6 = 4-stranded, 6-stranded). Note the fine columnar regions beneath the R4 root. View is from posterior toward anterior of cell.

Fig. 5. *Trentepohlia:* Oblique longitudinal section of the papilla showing both the R4 root with its associated columnar portion and two of the microtubules of the R6 root emerging from the electron-dense region at the proximal end of the basal body.

Fig. 6. *Trentepohlia:* Longitudinal section of the electron-dense material that surrounds the R6 root and covers the proximal end of the basal body. Note the fine striations present in the distal portion of this region (R6, *arrow*) and the R4 root.

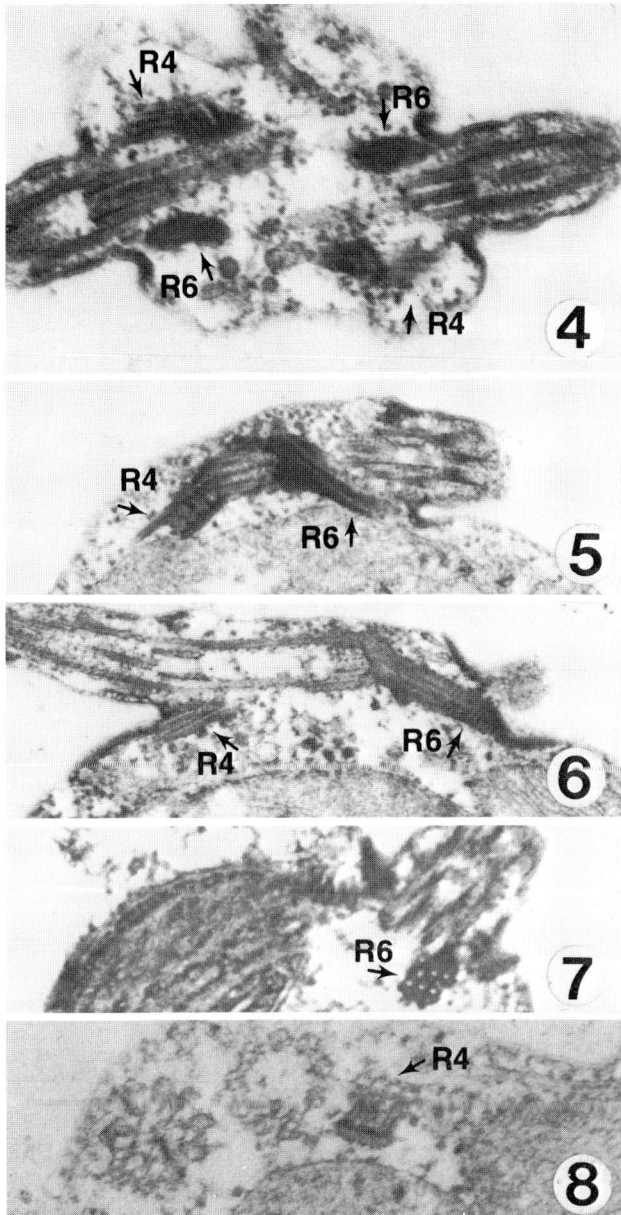

Fig. 7. *Trentepohlia:* Cross section of the R6 root. Note that the microtubules of the R6 root lie in a 3-over-3 configuration while within the electron-dense material.

Fig. 8. *Trentepohlia:* Cross section of the 4-stranded microtubular root with its associated columnar region. The columnar region consists of three columns that emerge from between the base of the microtubules. The columns appear as narrow light regions surrounded by narrow electron-dense material.

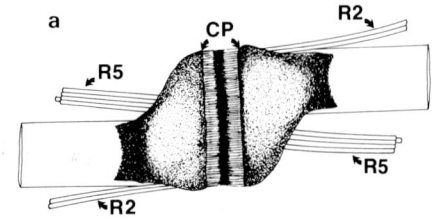

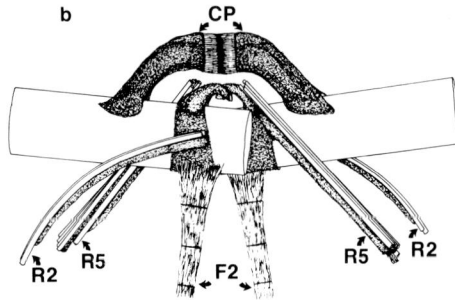

Fig. 9a,b. Diagrammatic reconstructions of the flagellar apparatus of *Batophora*. (a) Anterior view. (b) Lateral view. Abbreviations: CP, capping plate; F2, system II striated fibrous root; R2, R5 = 2-, and 5-stranded microtubular roots, respectively.

2. *Trentepohlia*

Basal bodies overlap with one another in the biflagellate cell of *Trentepohlia*; however, the connection between the basal bodies is small, consisting of a simple fibrillar link between adjacent basal body triplets. The microtubular root system is cruciate (Fig. 4) and as with *Batophora* the microtubular roots descend down two sides of the cell. The microtubular roots are arranged in a 6:4:6:4: configuration. The 6-stranded microtubular roots emerge from the proximal end of each basal body (Fig. 5) from within electron-dense material (Fig. 6). The electron-dense material completely covers the proximal end of each basal body and appears finely striated with a periodicity similar to system I fibrous roots (Fig. 7). The 4-stranded roots emanate from the outer side of each basal body and emerge nearly parallel to the 6-stranded roots. At the level where the 4-stranded roots emanate, the microtubules of the root are proximally associated with narrow fibrous material (Figs 5, 8). This material has the same appearance in both cross and longitudinal section.

In cross section the substructure of the fibrous material subtending the 4-stranded root consists of two regions: a columnar region consisting of three

13. Phylogenetic Significance of Flagellar Apparatus

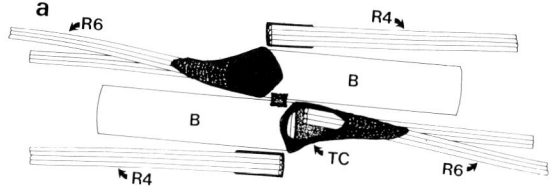

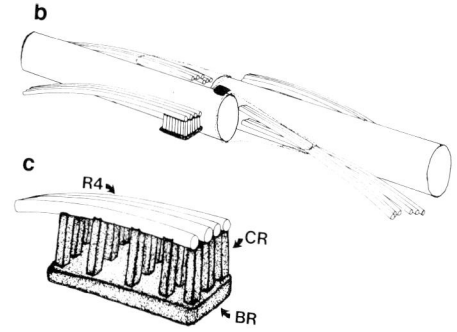

Fig. 10a,b,c. Diagrammatic reconstruction of the flagellar apparatus of *Trentepohlia*. (a) Anterior view. (b) Lateral view. (c) Enlarged columnar region beneath the R4 microtubular root. Abbreviations: B, basal body; BR, basal region, CR, columnar region; R4, R6 = 4- and 6-stranded microtubular roots, respectively; TC, terminal cap.

columns each of which attaches to the root microtubule, and a basal region that subtends and attaches to each column (Fig. 8). Each column appears to consist of a central granular portion flanked by electron-dense material. The columns attach to the root in the region between adjacent microtubules. The basal region has a similar light/dark/light appearance and is directly attached to the base of the three columns.

A diagrammatic reconstruction of the flagellar apparatus of *Trentepohlia* illustrates the three-dimensional relationships between the flagellar apparatus components (Fig. 10a,b).

DISCUSSION

The absolute configuration of the flagellar apparatus of *Batophora* is similar to that of many other siphonous green algae previously described (Roberts et al., 1981, 1982) in that the basal bodies overlap and lie in a 11 o'clock and 5 o'clock (11/5) position as do all basal bodies of ulvophycean cells (see

Roberts et al., 1982; Melkonian and Berns, 1983). As Melkonian and Berns (1983) pointed out, the configuration of the microtubular roots in the 11/5 flagellar apparatus should have the variable-numbered x roots (i.e., $x:2:x:2$) lying in the same plane as one another. This clearly is the case in *Batophora* and *Trentepohlia*, which indicates their affinities with the Ulvophyceae. In *Trentepohlia*, however, the 4-stranded root is subtended by columnar material and a basal region that appear remarkably similar to the multilayered structures (MLSs) of the Charophyceae, bryophytes, pteridophytes, and cycads. Indeed, this structure has been referred to as an MLS in *Trentepohlia* (Graham and McBride, 1975) and in the related *Cephaleuros* (Chapman, 1980, 1981; Chapman and Henk, 1982; Chapman, this volume, Chapter 8).

One of the most important results of this study is the illustration that the columnar regions are discernable in both longitudinal section and cross section. This is certainly not the case for true multilayered structures, in which the columnar portion of the multilayered structure is composed of lamellar sheets with each sheet lying below the microtubular root and traversing its width at an angle. When the lamellar sheets of the MLS are viewed in cross section (oblique to the root microtubules) the sheets appear as thin vertical lamellae. When the lamellar sheets are individually viewed in longitudinal section, however, they appear as horizontally intact structures. It is therefore suggested that the columnar region of the 4-stranded roots in *Trentepohlia* does not represent a multilayered structure similar to that of the Charophyceae because it consists of individual columns each of which appear electron-dark/light/dark in structure and are arranged three columns wide and at least four columns deep (see Fig. 8; see also Figs 7, 17, and 18 in Graham and McBride, 1975).

The exact nature of the columnar material is unknown but it corresponds in location with the crescent bodies described in the female gametes of *Bryopsis* (Melkonian, 1981) and *Pseudobryopsis* (Roberts et al., 1982). It seems clear that the crescent body in these algae is part of the material that surrounds the proximal regions of the basal bodies, which includes the nonstriated terminal caps regularly reported in ulvophycean algae. Although the crescent bodies have been reported to be narrowly striated and the terminal caps nonstriated, the recent report (Melkonian and Berns, 1983) of narrow striations in the terminal caps of *Friedmannia* (sometimes appearing "tubular" in form) indicates that the electron-dense material that is associated with the proximal region of the basal bodies in ulvophycean swarmers may be of similar nature throughout. In this light, the columnar material and basal region of *Trentepohlia* (which also appears tubular in longitudinal section) may more accurately represent a modification of the proximal electron-dense material found throughout the Ulvophyceae rather than a multilayered structure of the Charophyceae.

In *Batophora* and *Trentepohlia* the microtubular roots descend into two opposite regions of the cell as they do in *Friedmannia* (Melkonian and Berns, 1983) and *Microthamnion* (Watson and Arnott, 1973). This feature is doubtless related to the morphology of the mature swarmer of *Microthamnion* and *Friedmannia* (see Melkonian and Berns, 1983). The motile cells of *Trentepohlia* (see Fig. 3a–c in Graham and McBride, 1975) and *Batophora* are not flattened at maturity. The parallel and two-sided nature of the microtubular roots in *Trentepohlia* may, however, be related to the movement of the cell in nature. Sluiman *et al.* (1982) noted that the winged doublets in the stiff flagella of *Urospora* may account for its crawling movement along the substrate. In *Trentepohlia,* which also bears wing-like keels within the flagella, a similar crawling movement may occur due to its epiphytic nature. Upon the wetting and release of swarmers, the habit of *Trentepohlia* may emulate the temporarily emergent habitat of the intertidal region where *Urospora* occurs (see Sluiman *et al.*, 1982). For *Trentepohlia*, this hypothesized habitat and cellular movement could conceivably be responsible for the polarity of the microtubular roots. As yet, the polarity of the microtubular roots in *Batophora* and *Acetabularia* (Herth *et al.*, 1981) is not explainable with regard to function.

Another important feature of the *Batophora* flagellar apparatus is the broad striated connective between capping plate halves. A similar structure has been reported in *Acetabularia* (Herth *et al.*, 1981) and is considered to be most like the multistriated fibrous connective between basal bodies of chlorophycean cells. Herth *et al.* (1981) also noted that the basal bodies overlapped in *Acetabularia* (a feature common to ulvophycean swarmers) and therefore suggested that the flagellar apparatus of *Acetabularia* is intermediate between the ulvophycean and chlorophycean type. In *Batophora*, however, the flagellar apparatus is essentially identical to that of *Acetabularia* except for the presence of two system II fibrous roots, which have been described in several ulvophycean cells (see Melkonian, 1980b; Melkonian and Berns, 1983). Among siphonous green algae, the system II fibrous roots are quite small and are mostly restricted to the male gametes of anisogamous genera, where they are associated directly with two of the microtubular roots. Among cellular ulvophycean genera the system II fibrous roots are larger and directly or indirectly associated with two or more microtubular roots. This is the first report of a comparatively large system II fibrous root in a siphonous green alga. It is also important to note that the only other ulvophycean genera that have system II fibrous roots that descend posteriorly to the level of the nucleus are the quadriflagellate zoospore of *Urospora* (Sluiman *et al.*, 1982) and the biflagellate cells of *Friedmannia* (Melkonian and Berns, 1983) and *Microthamnion* (Watson and Arnott, 1973). The phylogenetic significance of vertical nucleus-associated versus the microtubular root-associated system II fibrous roots is not clear, but if the vertical nucleus-associated system II fibrous root (rhizoplast) is a primitive feature of the flagellar apparatus (see Melkonian and Berns, 1983) then its presence in the Dasycladales and Siphonocladales (T. Hori, personal communication) and near absence in the Caulerpales is additional evidence for the ordinal separation of the Caulerpales from the Siphonocladales and Dasycladales.

REFERENCES

Chapman, R. L. (1980). Ultrastructure of *Cephaleuros virescens* (Chroolepidaceae; Chlorophyta). II. Gametes. *Am. J. Bot.* **67,** 10–17.

Chapman, R. L. (1981). Ultrastructure of *Cephaleuros virescens* (Chroolepidaceae; Chlorophyta). III. Zoospores. *Am. J. Bot.* **68,** 544–556.

Chapman, R. L. and Henk, M. C. (1982). Cruciate flagellar apparatus and multilayered structures in *Cephaleuros virescens* gametes. *J. Phycol.* **18,** Suppl., a8.

Floyd, G. L. (1981). Ultrastructure of motile cells from two species of *Cladophora*. *J. Phycol.* **17,** Suppl., 11.

Graham, L. E. and McBride, G. E. (1974). Multilayered structures in motile cells of *Trentepohlia aurea* and the evolutionary implications. *J. Phycol.* **10,** Suppl., 8.

Graham, L. E. and McBride, G. E. (1975). The ultrastructure of multilayered structures associated with flagellar bases in motile cells of *Trentepohlia aurea*. *J. Phycol.* **11,** 86–97.

Herth, W., Heck, B. and Koop, H. U. (1981). The flagellar root system in the gamete of *Acetabularia mediterranea*. *Protoplasma* **109,** 257–269.

Hoops, H. J., Floyd, G. L. and Swanson, J. A. (1982). Ultrastructure of the biflagellate motile cells of *Ulvaria oxysperma* (Kütz.) Bliding and phylogenetic relationships among ulvaphycean algae. *Am. J. Bot.* **69,** 150–159.

Hori, T. (1977). Electron microscope observations on the flagellar apparatus of *Bryopsis maxima* (Chlorophyceae). *J. Phycol.* **13,** 238–243.

Melkonian, M. (1979). Structure and significance of cruciate flagellar root systems in green algae: Zoospores of *Ulva lactuca* L. (Ulvales, Chlorophyceae). *Helgol. Wiss. Meeresunters.* **32,** 425–435.

Melkonian, M. (1980a). Ultrastructural aspects of basal body associated fibrous structures in green algae: A critical review. *BioSystems* **12,** 85–104.

Melkonian, M. (1980b). Flagellar roots, mating structure and gametic fusion in the green alga *Ulva lactuca* (Ulvales). *J. Cell Sci.* **46,** 149–169.

Melkonian, M. (1981). Structure and significance of cruciate flagellar root systems in green algae: Female gametes of *Bryopsis lyngbyei* (Bryopsidales). *Helgol. wiss. Meeresunters.* **34,** 355–369.

Melkonian, M. and Berns, B. (1983). Zoospore ultrastructure in the green alga *Friedmannia israelensis:* An absolute configuration analysis. *Protoplasma* **114,** 67–84.

O'Kelly, C. J. and Floyd, G. L. (1983). The flagellar apparatus of *Entocladia viridis* motile cells and the taxonomic position of the resurrected family Ulvellaceae (Ulvales, Chlorophyta). *J. Phycol.* **19,** 153–164.

Roberts, K. R., Sluiman, H. J., Stewart, K. D. and Mattox, K. R. (1980). Comparative cytology and taxonomy of the Ulvaphyceae. II. Ulvalean characteristics of the stephanokont flagellar apparatus of *Derbesia tenuissima*. *Protoplasma* **104,** 223–238.

Roberts, K. R., Sluiman, H. J., Stewart, K. D., and Mattox, K. R. (1981). Comparative cytology and taxonomy of the Ulvaphyceae. III. The flagellar apparatus of the anisogametes of *Derbesia tenuissima* (Chlorophyta). *J. Phycol.* **17,** 330–340.

Roberts, K. R., Stewart, K. D. and Mattox, K. R. (1982). Structure of the anisogametes of the green siphon *Pseudobryopsis* sp. (Chlorophyta). *J. Phycol.* **18,** 498–508.

Sluiman, H. J., Roberts, K. R., Stewart, K. D. and Mattox, K. R. (1980). Comparative cytology and taxonomy of the Ulvaphyceae. I. The zoospore of *Ulothrix zonata* (Chlorophyta). *J. Phycol.* **16,** 537–545.

Sluiman, H. J., Roberts, K. R., Stewart, K. D., Mattox, K. R. and Lokhorst, G. M. (1982). The flagellar apparatus of the zoospore of *Urospora penicilliformis* (Chlorophyta). *J. Phycol.* **18,** 1–12.

Spurr, A. R. (1969). A low-viscosity epoxy resin embedding medium for electron microscopy. *J. Ultrastruct. Res.* **26,** 31–43.

Watson, M. W. and Arnott, H. J.(1973). Ultrastructural morphology of *Microthamnion* zoospores. *J. Phycol.* **9,** 15–29.

14 | Ultrastructural Characterization of Taxa in the Genus *Enteromorpha*

A. J. YOUNG, J. C. COLLINS and G. RUSSELL

Department of Botany, University of Liverpool, Liverpool, England

Abstract: The well-documented taxonomic problems of the genus *Enteromorpha* have three principal sources: the relatively small number of morphological features that can be deployed for making taxonomic discrimination, the plasticity of certain of these, and the widespread occurrence of intraspecific genetic variation. The present study concerns two common single-pyrenoid euryhaline species, *E. intestinalis* (L.) Link and *E. prolifera* (O. F. Müll.) J. Ag. In the case of the former the existence of three ecotypes on closely adjacent shores on the Isle of Man (U.K.) has been demonstrated. These have evidently evolved in response to differences in their saline environments. A single population of *E. prolifera*, from Hilbre Island (River Dee), has been examined. The fine structure of these plants has been studied, both in their natural saline environments and in response to hypo- and hypersaline shock, using transmission electron microscopy. These populations provide a means of assessing inter- and intraspecific differences in response to saline stress. The following ultrastructural features have been investigated: thylakoids, pyrenoids, cell walls and membrane structure. The production of vacuoles and the incidence of osmiophilic bodies and vesicles within the cell have also been noted. The taxonomic and physiological implications of these will be discussed.

INTRODUCTION

Recent taxonomic revisions of the genus *Enteromorpha* by Bliding (1963) and Koeman and van den Hock (1982) have revealed a complex of difficulties relating to the definition of taxa. First, the simple structure of the thallus, and of its component cells, generates very few phenetic characters. Most of the available characters are of a quantitative nature and their ranges of expression overlap from one species to another. This can, in part, be

---

Systematics Association Special Volume No. 27, "Systematics of the Green Algae", edited by D. E. G. Irvine and D. M. John, 1984, pp. 343–351. Academic Press, London and Orlando.
ISBN 0 12 374040 1

Copyright © by the Systematics Association
All rights of reproduction in any form reserved

overcome by adopting combinations of attributes for the delimitation of taxa, as Koeman and van den Hoek (1982) have done in the manner recommended by Mathieson et al. (1981). Combination of individually-unreliable criteria may serve simply to compound the problem, however.

Additional taxonomic difficulties arise from the fact that certain of the attributes are highly plastic in nature, either because they vary with the state of the plant development (size, shape) or because they are readily modified by environmental factors. Branch production in *Enteromorpha*, can be influenced by temperature (Moss and Marsland, 1976), by light (de Silva, 1970) and by salinity (Dangeard, 1957; Burrows, 1959; Reed and Russell, 1978). However, it is also evident that branching *per se* or the predisposition to branch in a particular manner under certain circumstances, is a heritable character (de Silva and Burrows, 1973).

It may be construed from these observations that a basis for taxonomic discrimination does not exist. However, the reported sexual incompatibility between species, complete or incomplete, according to Larsen (1981) and de Silva and Burrows (1973), respectively, is indicative of a degree of genetic isolation consistent with species differentiation. A degree of genetic isolation, expressed by the evolution of ecotypes, has also been demonstrated by Reed and Russell (1979) and Goodman et al. (1976), although the status of these biosystematic categories with respect to those of orthodox taxonomy remains unsolved.

Delimitation of taxa within the genus *Enteromorpha* will probably continue to be based on phenetic characters. Plasticity amongst macroscopic features is well documented though plasticity of microscopic features has received little attention. The purpose of the present study was to determine whether any microscopic characters are responsive to external stimuli and hence to assess their taxonomic importance.

MATERIALS AND METHODS

Two euryhaline species of the genus *Enteromorpha* have been investigated, *E. prolifera* (O. F. Müll.) J. Ag. and *E. intestinalis* (L.) Link. *E. prolifera* was sampled from a single eulittoral population at Hilbre Island, River Dee (Grid Ref.: SJ 205860). Three different populations (ecotypes) of *E. intestinalis* were collected from sites at Castletown, Isle of Man (Grid Ref.: SC 257662). These were located in a littoral fringe rock pool, in the lower eulittoral zone of an exposed shore (Scarlett Point) and from the upper reaches of the Silverburn River (see also Reed and Russell, 1978, 1979).

Plants were collected, together with ambient water samples, and transferred to the laboratory in a vacuum flask where they were kept under controlled conditions (15°C, 16:8 h light regime of 30–40 μ Em^{-2} s^{-1}) in filtered aerated ambient water. Whole adult plants were transferred to an experimental range of seawaters (0.25× to 3.00×, relative to natural Irish Seawater of 35‰) for 48 h.

Plants were prepared for transmission electron microscopy (TEM) by first cutting a small piece of tissue from the central region of the thallus. This was fixed overnight at 4°C in 2.5% w/v glutaraldehyde solution; all fixatives used seawater of the relevant experimental concentration to avoid osmotic shock. This treatment was found to buffer the system adequately and additional buffers were not necessary. Postfixation using 4% w/v osmium tetroxide was carried out for 1 h at 4°C, followed by dehydration through an ethanol series. The material was embedded in Spurr's Resin (Spurr, 1969). Sections were poststained with uranyl acetate and lead citrate and examined on a Corinth 300 electron microscope.

RESULTS

E. intestinalis and *E. prolifera* were characterised on the basis of certain morphological criteria (Table I). Cell and outer cell wall sizes were in agreement with those found by de Silva (1970) and by Koeman and van den Hoek (1982), though cell sizes were found to be smaller than those stated by Bliding (1963).

Differences in cell sizes, under ambient conditions, were not significant between species or ecotypes. However, on exposure to hyposaline and hypersaline conditions cell size (volume) was found to vary greatly between these isolates. Inner cell wall thickness was found to be significantly greater in the littoral fringe isolate, resulting in a total wall thickness 2 to 4 times greater than the other plants. Such a feature has probably developed in response to the saline environment.

When plants were examined under TEM a series of bands were found in the cell walls of all plants. These were more prominent in the outer cell wall than the inner cell wall. The bands were noticeably more numerous and thicker in the walls of the estuarine and rock pool ecotypes of *E. intestinalis* than the eulittoral populations (Table I; see Figs 3, 5).

Cells of *E. intestinalis* are highly vacuolate (see Table I) and possess a parietal chloroplast, which takes on a hood-shaped appearance in surface view, while cells of *E. prolifera* are less vacuolate and also possess a parietal chloroplast. Chloroplast thylakoids were arranged in stacks of two to six, though in *E. prolifera* stacks of three thylakoids were frequently observed. Pyriform starch grains (Figs 1, 2) were common throughout the chloroplast, often distorting thylakoid arrangement. Other storage bodies present included large lipid bodies, typically 250–850 nm in diameter, which were found to a greater extent in the eulittoral populations of *E. intestinalis*

Table I. Morphological characters in *Enteromorpha* under ambient conditions.

Character	E. prolifera (eulittoral)	E. intestinalis		
		Eulittoral	Estuarine	Littoral fringe
Mean cell size (μm), middle region of thallus (± S.E.)	8.7 × 11.4 (±1.4) (±0.9)	6.8 × 11.2 (±0.4) (±0.6)	8.4 × 12.6 (±0.3) (±0.6)	8.6 × 15.1 (±0.6) (±0.7)
Outer cell wall thickness (μm)	1.5–3.0	1.5–2.5	1.0–1.5	1.5–2.0
Cell wall banding (frequency)	++	++	+++	++++
Vacuolar volume (% of total cellular volume)	up to 40%	up to 50%	up to 80%	up to 60%

and *E. prolifera*. Small, electron-dense, osmiophilic bodies (100 nm in diameter) were commonly found in groups of six or more within the chloroplast (Figs 1, 5). The estuarine plants of *E. intestinalis* were almost devoid of

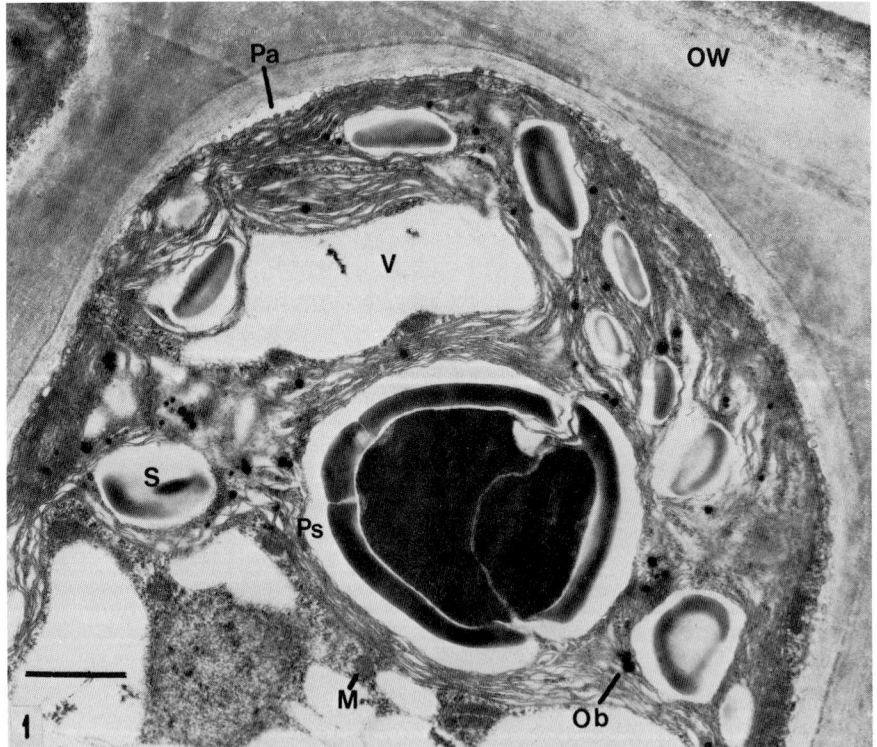

Fig. 1. *E. intestinalis* (littoral fringe ecotype), ambient water. Scale bar, 1.5 μm. Abbreviations: M, mitochondria; Ob, osmiophilic body; OW, outer wall; Pa, paramural space; Ps, pyrenoidal starch sheath; S, starch grain; V, vacuole.

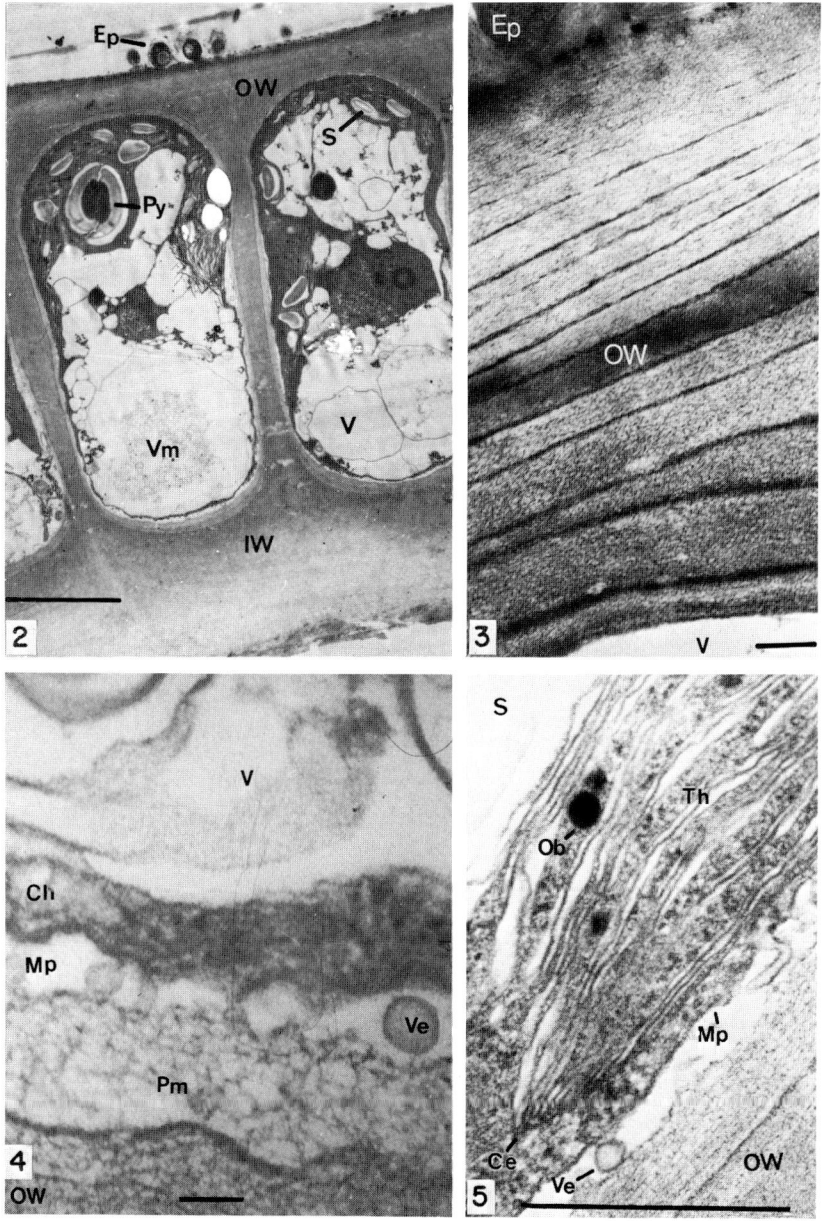

Figs 2–5. *Enteromorpha intestinalis*. Fig. 2. Eulittoral ecotype (0.50× seawater). Scale bar, 5 μm. Fig. 3. Estuarine ecotype (ambient water): outer cell wall. Scale bar, 250 nm. Fig. 4. Eulittoral ecotype (2.00× seawater). Scale bar, 100 nm. Fig. 5. Eulittoral ecotype (ambient water). Scale bar, 1.0 μm. Abbreviations: Ce, chloroplast envelope; Ch, chloroplast; Ep, epiphyte layer; IW, inner wall; Mp, plasmamembrane; Ob, osmiophilic body; OW, outer wall; Pm, paramural matrix; Py, pyrenoid; S, starch grain; Th, thylakoid; V, vacuole; Ve, vesicle; Vm, vacuolar matrix.

storage products, and the starch sheath surrounding the single pyrenoid was often much reduced.

Cells of *E. intestinalis* typically contained a single pyrenoid, agreeing with the taxonomic criteria of Bliding (1963) and de Silva (1970). Two pyrenoid cells of *E. intestinalis* were uncommon. Cells of *E. prolifera* were found to contain two pyrenoids in over 50% of the cells; this conflicts with the findings of Bliding (1963) and de Silva (1970), who both described the species as generally possessing a single pyrenoid. Pyrenoids of both species fit into categories I and II as defined by Hori (1973) for species of *Monostroma*. All pyrenoids possessed a starch sheath of two to four plates. The pyrenoidal matrix was traversed by either a single or, less frequently, a double band of thylakoids (Fig. 1).

The cells of both species are highly vacuolate; typically the volume of the cell occupied by the vacuole decreases in the order; *E. intestinalis*, estuarine > *E. intestinalis*, littoral fringe > *E. intestinalis*, eulittoral > *E. prolifera*, eulittoral. Under ambient conditions the littoral fringe and estuarine plants, possessed one large vacuole, whereas the other plants had a number of smaller vacuoles present (Table II). These differences may be related to the saline environments of the two eulittoral zone samples.

Other cellular components including the nucleus, golgi complex and mitochondria are typical of those found in members of the Chlorophyta, as reviewed by Brawley and Wetherbee (1981). Mitochondria were more numerous in *E. prolifera* than in any of the ecotypes of *E. intestinalis*. The

Table II. Cellular responses of two *Enteromorpha* species to saline shock.

Character*	*E. prolifera* (eulittoral)	*E. intestinalis*		
		Eulittoral	Estuarine	Littoral fringe
No. vacuoles per cell				
ambient	>3	up to 10	1	1
0.25× SW	1	up to 10	1	1
3.00× SW	1	>10	>6	1
Granulosity of vacuole				
ambient	0	+	0	++
0.25× SW	++++†	+++	++	+++
3.00× SW	0	++	0	++
No. vesicles per cell (80 nm diam.)				
ambient	1–3	>20	>20	up to 10
0.25× SW	>10‡	1–2	1–2	1–2
3.00× SW	0	>20	0	0

* The results in 0.25×, 3.00× and ambient seawater (SW) conditions only are presented.
† On a scale 0 to +++++.
‡ Vesicles also found within vacuole.

occurrence of microbodies (in the range of 250–500 nm; see Silverberg, 1975) was noted though the incidence was too low, typically one or two per cell, to provide information on the status of plants studied.

Under ambient conditions all cells contained a number of regularly shaped, single-membrane–bound vesicles (80–100 nm diameter) located in the region between the plasmalemma and the cell wall (the paramural space). These are similar to those described by Burr and West (1971) for *Bryopsis hypnoides* and by Dreher *et al.* (1978) for *Caulerpa simpliciuscula* in response to a physical wounding of the cell. Associated with this response is the formation of a large protein body which degrades to form a granulo-fibrillar matrix (Burr and West, 1971) within the vacuole. The latter was clearly visible within most cells (Fig. 2, the vacuole; Fig. 4, the paramural space).

Two major responses to saline shock were noted for *E. intestinalis* and *E. prolifera*. First, the estuarine and eulittoral ecotypes of *E. intestinalis* showed an increased number of vacuoles under hypersaline conditions (see Table II). Both estuarine and littoral fringe ecotypes of *E. intestinalis* maintained their single vacuole state under the saline conditions they are likely to experience in nature. The single vacuole of the littoral fringe cells was constant under all treatments. Second, the response of the cells to saline shock was similar to the physical wounding response described earlier, i.e. vesicle production (Figs 4, 5) and the formation of the granulo-fibrillar matrix within the vacuole (Fig. 2). Dreher *et al.* (1978) have suggested that vesicle production may be stimulated by a sudden change in the ionic environment of the cell. However, vesicle production in *Enteromorpha* is reduced under extreme hyposaline and hypersaline shock (see Table II). Response to a slight fall or rise in salinity has not been fully investigated. All samples showed markedly increased granulosity of vacuoles in 0.25× seawater but *E. prolifera* and the estuarine isolate of *E. intestinalis* differed from the other two samples in showing reduced granulosity in 3× seawater (see Table II). Sensitivity to hypersaline treatment is also evident in vesicle production.

The reported differences in branch proliferation and regeneration (Reed and Russell, 1978, 1979) between ecotypes of *Enteromorpha* may be linked to the initial production of vesicles mentioned above. Differences in macroscopic features, such as branch production, may be connected with ultrastructural features, i.e. this vesiculation may be the precursor of branch production in *Enteromorpha*.

CONCLUSIONS

Ecotypes of *E. intestinalis* cannot be separated using macromorphological characters, but micromorphological responses to treatments differ between ecotypes of *E. intestinalis*. Further experimental investigation of micromorphological characters may enable existing criteria to be assessed in terms of plasticity and may lead to the identification of new characters that may be more effective in distinguishing taxa.

ACKNOWLEDGEMENTS

The first author is the recipient of a N.E.R.C. research studentship (GT4/81/ALS/21). The authors acknowledge the help of Mr. J. Smith concerning the preparation of specimens for TEM.

REFERENCES

Bliding, C. (1963). A critical survey of European taxa in Ulvales. Part I, *Capsosiphon, Percursaria, Blidingia, Enteromorpha*. *Op. bot. Soc. bot. Lund.* **8**, 1–160.
Brawley, S. H. and Wetherbee, R. (1981). Cytology and ultrastructure. In "The Biology of Seaweeds" (C. S. Lobban and M. J. Wynne, eds.), pp. 248–299. Blackwell, Oxford.
Burr, F. A., and West, J. A. (1971). Protein bodies in *Bryopsis hypnoides*. Their relationship to wound-healing and Branch septum development. *J. Ultrastruct. Res.* **35**, 476–498.
Burrows, E. M. (1959). Growth, form and environment in *Enteromorpha. J. Linn. Soc. London Bot.* **56**, 204–206.
Dangeard, P. (1957). Faculté de regénération et multiplication végétative chez les Entéromorphes. *C. r. hebd. Seanc. Acad. Sci., Ser. D* **244**, 2454–2457.
de Silva, M. W. R. N. (1970). An experimental approach to the taxonomy of the genus *Enteromorpha* (L.) Link. Ph.D. Thesis, University of Liverpool.
de Silva, M. W. R. N. and Burrows, E. M. (1973). An experimental assessment of the species *Enteromorpha intestinalis* (L.) Link and *Enteromorpha compressa* (L.) Grev. *J. mar. biol. Assoc. U.K.* **53**, 895–904.
Dreher, T. W., Grant, B. R. and Wetherbee, R. (1978). The wound response in the siphonous alga *Caulerpa simpliciuscula* C.Ag.: Fine structure and cytology. *Protoplasma* **96**, 189–203.
Goodman, C., Newall, M. and Russell, G. (1976). Rapid screening for copper tolerance in ship-fouling algae. *Int. Biodeterior. Bull.* **12**, 81–83.
Hori, T. (1973). Comparative studies of pyrenoid ultrastructure in algae of the *Monostroma* complex. *J. Phycol.* **9**, 190–199.
Koeman, R. P. T., and van den Hoek, C. (1982). The taxonomy of *Enteromorpha* Link, 1820 (Chlorophyceae) in The Netherlands. I. The section *Enteromorpha*. *Arch. Hydrobiol., Suppl.*, **63**, 279–330.
Larsen, J. (1981). Crossing experiments with *Enteromorpha intestinalis* and *E. compressa* from different European localities. *Nord. J. Bot.* **1**, 128–136.
Mathieson, A. C., Norton, T. A. and Neushul, M. (1981). The taxonomic implication of genetic and environmentally induced variation in seaweed morphology. *Bot. Rev.* **47**, 313–347.
Moss, B. and Marsland, A. (1976). Regeneration of *Enteromorpha*. *Br. Phycol. J.* **11**, 309–313.
Reed, R. H. and Russell, G. (1978). Salinity fluctuations and their influence on "Bottle Brush" morphogenesis in *Enteromorpha intestinalis* (L.) Link. *Br. phycol. J.* **13**, 149–153.

Reed, R. H. and Russell, G. (1979). Adaptation to salinity stress in populations of *Enteromorpha intestinalis* (L.) Link. *Estuar. Coastal mar. Sci.* **8,** 251–258.

Silverberg, B. A. (1975). An ultrastructural and cytochemical characterisation of microbodies in the green algae. *Protoplasma* **83,** 269–295.

Spurr, A. R. (1969). A low-viscosity resin embedding medium for electron microscopy. *J. Ultrastruct. Res.* **26,** 31–43.

15 | The Validity of Morphological and Anatomical Characters in Distinguishing Species of *Ulva* in Southern Australia

J. A. PHILLIPS

Department of Botany, Monash University, Clayton, Victoria, Australia

Abstract: Thirty-seven populations of the three common species of *Ulva* (*U. rigida* Agardh, *U. laetevirens* Areschoug, *U. stenophylla* Setchell and Gardner) in southern Australia were sampled at localities from Adelaide, South Australia, along the coastline to Walkerville, Victoria, and at St. Helens, Tasmania. The value of morphological and anatomical characters in defining these three species is examined critically. Morphology, thallus thickness, cell size, pyrenoid number, presence/absence of marginal teeth, and both cell shape and chloroplast position (in surface view and in transverse section of the upper marginal and mid regions of the thallus) are all extremely variable, showing considerable overlap among the three species. Therefore, these characters are considered to have limited taxonomic use. Cell shape and chloroplast position in transverse section of the basal regions of the thallus are constant characters and adequately separate the three species. In culture, the developmental pattern of each species is constant and characteristic and confirms the validity of the species as defined by anatomical characters. The taxonomic implications of this study and their relevance to other *Ulva* species are discussed.

INTRODUCTION

Ulva is a genus well known for its taxonomic difficulties. The morphologically simple thallus of *Ulva* presents few morphological or anatomical characters for taxonomic use and the variability of these characters is poorly understood.

Five species of *Ulva* have been identified in southern Australia: *Ulva rigida* Agardh, *U. laetevirens* Areschoug, *U. stenophylla* Setchell and Gardner, *U.*

Systematics Association Special Volume No. 27, "Systematics of the Green Algae", edited by D. E. G. Irvine and D. M. John, 1984, pp. 353–361. Academic Press, London and Orlando.
ISBN 0 12 374040 1

Copyright © by the Systematics Association
All rights of reproduction in any form reserved

lactuca Linnaeus and *U. fasciata* Delile. *U. rigida, U. laetevirens* and *U. stenophylla* are sufficiently common to enable study of geographical, seasonal and local habitat variation of the morphological and anatomical characters which are frequently used in the taxonomy of the genus.

The taxonomy of *U. rigida* requires clarification and will be the subject of a further publication. *U. rigida* Agardh and *U. rigida sensu* Bliding are considered as two distinct species, both of which are common on the southern Australian coast. It is necessary to propose a new name for *U. rigida sensu* Bliding. *U. laetevirens* Areschoug, the type locality of which is Port Phillip, Victoria, is the earliest record I have found for this species and is regarded as the valid name.

METHODS

Thirty-seven populations of *U. rigida, U. laetevirens* and *U. stenophylla* were sampled from Adelaide, South Australia to Walkerville, Victoria (about 1100 km), and at St. Helens, Tasmania. To assess seasonal variation, one population for each species was sampled at three monthly intervals for one year. Local habitat variation was assessed by sampling populations growing in rough water, moderately rough, sheltered and subtidal habitats.

For each population sample, 25 attached individuals from the centre of distribution of the population were selected. The following characters were recorded: thallus shape, thallus length and breadth, thallus thickness and the presence of marginal teeth. Thallus thickness was recorded in the upper marginal, mid and basal regions of the thallus. A subsample of 10 plants was examined for cell size, cell shape, cell arrangement, chloroplast position and pyrenoid number. These characters (except cell arrangement) were recorded from both surface view and in transverse section in the upper marginal, mid and basal regions of the thallus. In this study, the basal region of the thallus is defined as the non-rhizoid-containing region immediately adjacent to where there are rhizoids between the two cell layers.

Cultures were established from wild plants collected from different localities and habitats and in different seasons. Each species was cultured a minimum of 10 times. Zygotes, biflagellate zooids (some of which were demonstrated to be gametes) and zoospores were inoculated into separate dishes containing a modification of Provasoli's ES medium (Tanner, 1980) and placed into four temperature/daylength regimes: 15°C + 10.5:$\overline{13.5}$ h; 13°C + 9.5:$\overline{14.5}$ h; 13°C + 12.5:$\overline{11.5}$ h; and 11°C + 10.5:$\overline{13.5}$ h. Overhead lighting supplied a quantum irradiance of 60–70 μE cm^{-2} s^{-1}. Cultures were maintained up to four generations.

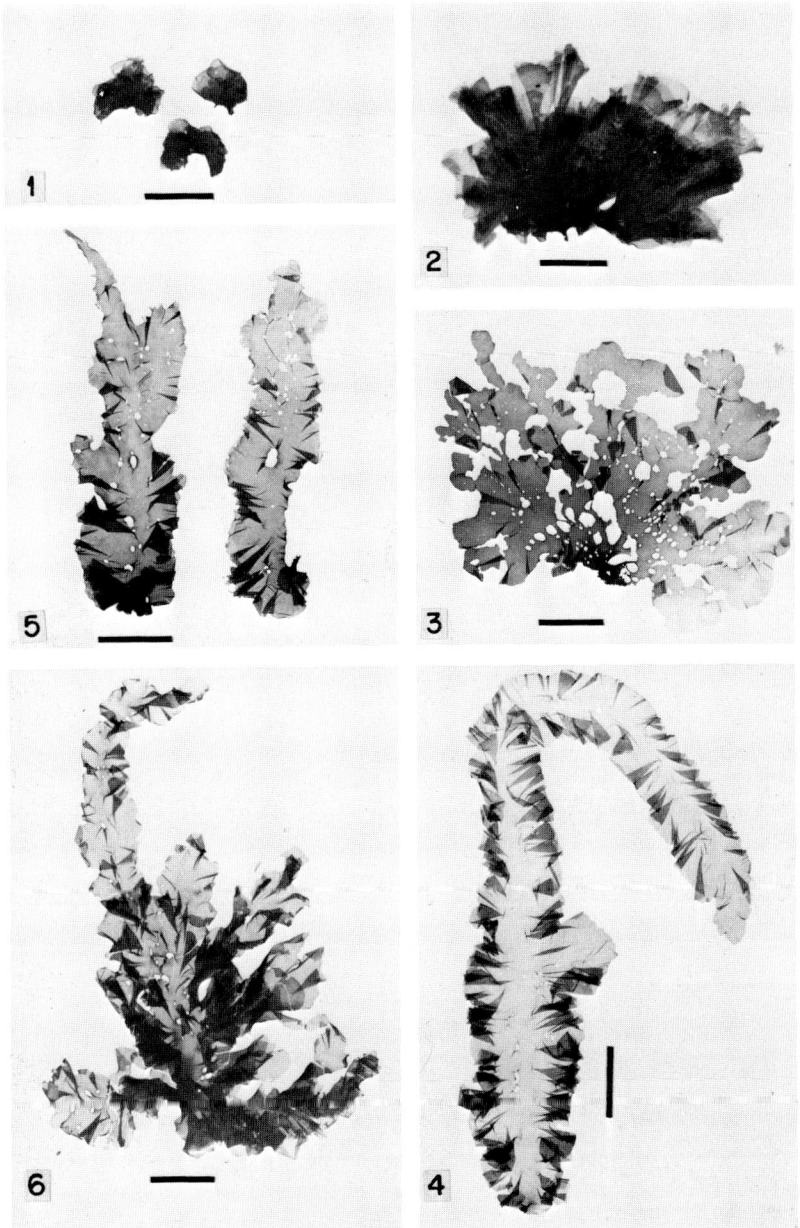

Fig. 1. Small tufted form of U. rigida (scale 4 cm).
Fig. 2. Lobed thallus of U. rigida (scale 4.2 cm).
Fig. 3. Highly reticulate simple thallus of U. rigida (scale 4.3 cm).
Fig. 4. U. stenophylla (scale 4.2 cm).
Fig. 5. Simple linear thallus of U. laetevirens (scale 4.2 cm).
Fig. 6. U. rigida thallus with elongated linear segments (scale 4.3 cm).

RESULTS

The morphology of *U. rigida* and *U. laetevirens* is extremely variable. Both species vary from small tufted forms (Fig. 1) on exposed coasts to larger, variously shaped, simple (Fig. 3) or lobed thalli (Fig. 2). It is impossible to distinguish between these two species on the basis of morphology alone. In contrast, the morphology of *U. stenophylla* is constant (Fig. 4). The thallus is simple, linear, or, less commonly, divided to the base into linear segments. The simple thallus or each segment of the branched thallus is composed of a flattened mid-region and ruffled margin. This character can only be used with caution, as simple linear forms of *U. laetevirens* (Fig. 5) and lobed forms of *U. rigida* and *U. laetevirens* in which the lobes have elongated into linear segments (Fig. 6) can superficially resemble the simple or branched forms of *U. stenophylla*, respectively.

Mean cell sizes are extremely variable and show considerable overlap among the three species (Table I). Similarly, mean thallus thickness is an extremely variable character and shows overlap among the three species (Table II). In all three species, the cells as seen in surface view are arranged

Table I. Variation in mean cell sizes in southern Australian populations of *U. rigida*, *U. laetevirens* and *U. stenophylla*.

	U. rigida	*U. laetevirens*	*U. stenophylla*
Number of populations sampled	13	10	14
Cell size—surface view (number of cells = 100)			
$\bar{x}$ cell length (μm)			
Marginal region	15–24	15–22	13–24
Mid-region	18–23	15–22	15–20
Basal region	19–25	17–22	18–24
$\bar{x}$ cell breadth (μm)			
Marginal region	11–18	12–16	10–18
Mid-region	14–17	11–16	11–14
Basal region	14–19	13–16	14–18
Cell size—transverse section (number of cells = 100)			
$\bar{x}$ cell height (μm)			
Marginal region	21–31	19–34	17–27
Mid-region	30–37	24–44	19–36
Basal region	33–45	37–65	26–46
$\bar{x}$ cell width (μm)			
Marginal region	14–19	13–19	12–19
Mid-region	14–18	12–15	13–16
Basal region	15–23	14–16	15–19

Table II. Variation in mean thallus thickness in southern Australian populations of *U. rigida*, *U. laetevirens* and *U. stenophylla* (number of plants per population sample = 25).

	U. rigida	*U. laetevirens*	*U. stenophylla*
Number of populations sampled	13	10	14
Marginal region (μm)	50–80	45–87	38–62
Mid-region (μm)	77–95	61–120	52–90
Basal region (μm)	91–142	103–166	79–136

in no obvious pattern (Fig. 7). However, short rows up to nine cells long occur in some areas of the upper marginal region of the thallus in all species. In surface view, cells of the three species are either polygonal or quadrangular in shape and the chloroplast either lies as a thin plate close to two or three anticlinal walls or lies adjacent to the outer cell wall. In transverse section of the upper marginal and mid-regions of the thallus, cell shape and the position of the chloroplast are also variable. However, in transverse section of the basal regions, these two characters are constant and distinguish each species. In *U. rigida*, cells in the basal region (Fig. 8) are rectangular in shape and the chloroplast lies adjacent to the lateral cell walls. In *U. laetevirens*, the basal cells (Fig. 9) are conical in shape, tapering towards the thallus surface. The well developed parietal chloroplast occupies the outer half to three-quarters of the cell. In *U. stenophylla*, these cells (Fig. 10) are bullet-shaped and the chloroplast lies close to the outer cell wall.

Generally, pyrenoid number varies from one to three in the three species except in the basal regions of *U. laetevirens* and *U. stenophylla*, where there are three to several pyrenoids per cell.

U. rigida does not possess microscopic marginal teeth. In *U. laetevirens* and *U. stenophylla*, they may be present. *U. laetevirens* possesses macroscopic marginal teeth when it grows in very sheltered or subtidal localities.

Cultured plants of *U. rigida* attain a maximum size of 2 cm long by 1–2 cm wide and are always cuneate (Fig. 11) or orbicular (Fig. 12) in shape. Mature cultured plants of *U. laetevirens* (Fig. 13) reach a maximum size of 16 cm long and 1–2 cm wide and are linear to oblanceolate in shape. Margins of young cultured plants less than 1–2 cm long (Fig. 14) are smooth. The development of marginal teeth occurs as growth continues. Cultured plants of *U. stenophylla* (Fig. 15) are always linear to lanceolate in shape, tapering towards the apex. They attain a maximum size of 16 cm long by 2 cm wide.

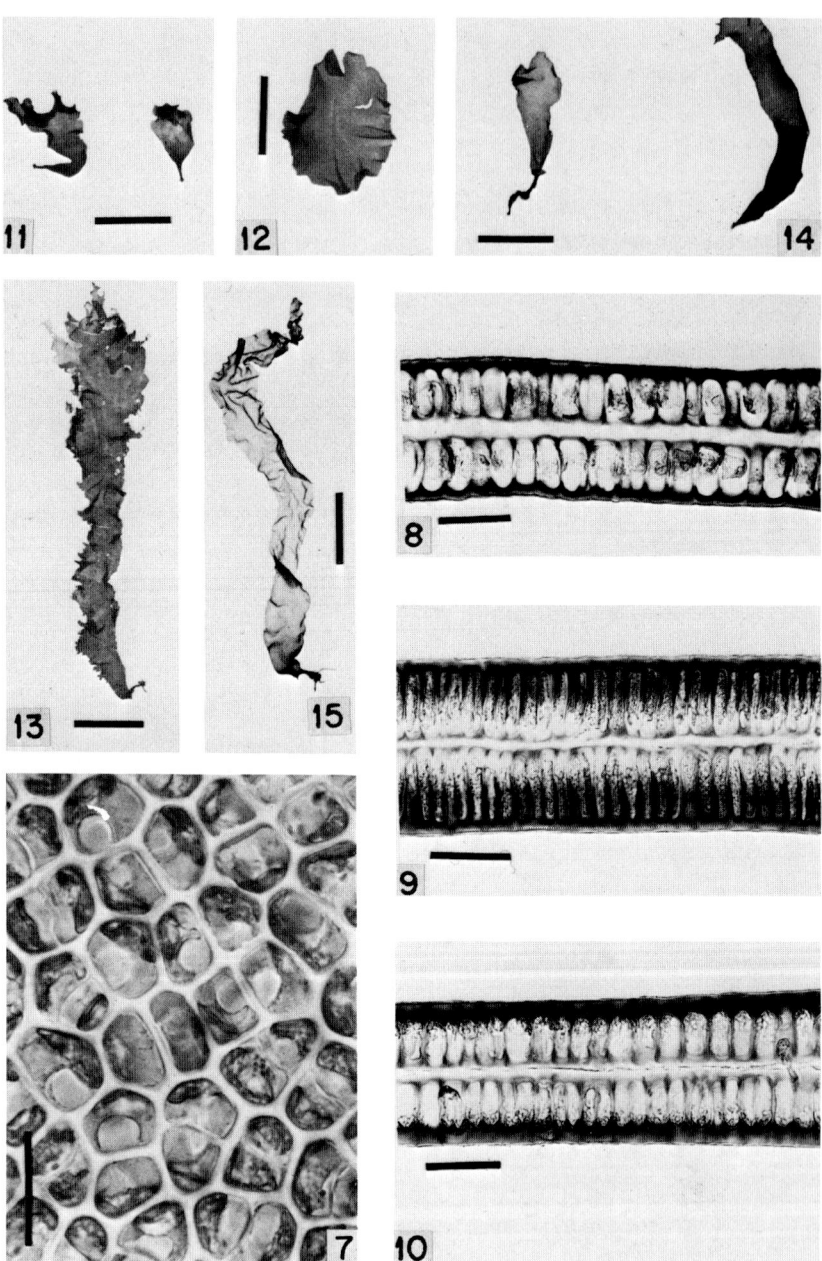

Fig. 7. Surface view of cells of *U. rigida* (scale 20 μm).
Fig. 8. Transverse section of the basal region of *U. rigida* (scale 60 μm).
Fig. 9. Transverse section of the basal region of *U. laetevirens* (scale 60 μm).
Fig. 10. Transverse section of the basal region of *U. stenophylla* (scale 50 μm).
Fig. 11. Mature cultured plants of *U. rigida* (scale 1.5 cm).

Table III. Non-variable and variable characters of U. rigida, U. laetevirens and U. stenophylla in southern Australia.

Non-variable characters

1. Cell shape in transverse section of the basal region of the thallus
2. Chloroplast position in transverse section of the basal region of the thallus

Variable characters

1. Morphology
2. Cell size from surface view and in transverse section
3. Thallus thickness
4. Cell shape and chloroplast position from surface view and in transverse section of the marginal and mid-regions of the thallus
5. Pyrenoid number
6. Presence of marginal teeth

DISCUSSION

Table III lists the non-variable and variable characters of U. rigida, U. laetevirens and U. stenophylla populations in southern Australia. Cell shape and chloroplast position in transverse section of the basal region of the thallus are reliable taxonomic characters which clearly separate these three species from each other. All other characters are exceedingly variable and/or show considerable overlap in the three species.

The findings of the present study agree with those of Tanner (1979), Steffensen (1976), and Titlyanov et al. (1975) but contrast with those of Bliding (1968) and Koeman and van den Hoek (1981). Tanner found that morphology, cell size, thallus thickness and pyrenoid number were extremely variable and could not be used to separate the seven species of Ulva on the Pacific coast of North America. Steffensen in a study of Ulva species in New Zealand and Titlyanov et al. studying U. fenestrata Postels and Ruprecht in the Sea of Japan arrived at similar conclusions. Bliding, however, in his study of the eight Ulva species in Europe considered that pyrenoid number, arrangement and sizes of cells, and thallus thickness were non-variable characters. Koeman and van den Hoek have largely followed the taxonomic concepts proposed by Bliding.

Cell shape in transverse section of the frond has been used previously to characterise U. rigida (Agardh, 1883; Papenfuss, 1960), U. laetevirens (= U.

Fig. 12. Mature cultured plant of U. rigida (scale 1 cm).
Fig. 13. Mature cultured plant of U. laetevirens (scale 2.6 cm).
Fig. 14. Young cultured plants of U. laetevirens (scale 1 cm).
Fig. 15. Mature cultured plant of U. stenophylla (scale 1.9 cm).

rigida sensu Bliding, Bliding, 1968; Vinogradova, 1974; Tanner, 1979; Koeman and van den Hoek, 1981) and *U. stenophylla* (Tanner, 1979). The conical-shaped cells of *U. laetevirens* differentiate this species from all other described species of *Ulva* in the widely separated geographical areas of Europe, the Black Sea, the Pacific coast of North America and southern Australia. Similarly, the bullet-shaped cells of *U. stenophylla* separate this species from the other species of *Ulva* on the Pacific coast of North America and southern Australia. In contrast, the rectangular cells of *U. rigida* do not differentiate it from some other species of *Ulva* such as *U. fasciata* Delile and *U. fenestrata,* which also possess rectangular cells. Therefore, the specific limits of *U. rigida* are more difficult to ascertain in localities where two or more of these species occur.

Recent studies have reported variations in chloroplast position in response to light intensity changes (Titlyanov *et al.*, 1975), in dividing cells (Løvlie, 1964) and in response to an endogenous circadian rhythm (Britz and Briggs, 1976). Accordingly, an assessment of chloroplast position in the transverse section of the basal region of the thallus of other *Ulva* species must be undertaken before this character can be applied in other geographical areas.

In culture, the developmental pattern of *U. rigida, U. laetevirens* and *U. stenophylla* is constant and characteristic for each species. There are no published accounts of the development of *U. rigida* to which the present observations can be compared. Published accounts of the development in culture of *U. laetevirens* (= *U. rigida,* Bliding, 1968) and *U. stenophylla* (Tanner, 1979) agree with what has been observed in southern Australian populations of these two species. These observations are interpreted as supporting evidence which confirms the validity of the species as defined by anatomical characters.

ACKNOWLEDGEMENTS

I wish to express my gratitude to Dr. Margaret Clayton for her support and advice during this study and in preparation of this manuscript. The research was supported by a Monash University Graduate Scholarship which is gratefully acknowledged.

REFERENCES

Agardh, J. G. (1883). Till algernes systematik. Nya bidrag. VI. Ulvaceae. *Acta Univ. lund.* **19,** 1–181.

Bliding, C. (1968). A critical survey of European taxa in Ulvales. Part II. *Ulva, Ulvaria, Monostroma, Kornmannia. Bot. Not.* **121,** 535–629.

Britz, S. J. and Briggs, W. R. (1976). Circadian rhythms of chloroplast orientation and photosynthetic capacity in *Ulva*. *Plant Physiol.* **58**, 22–27.

Koeman, R. P. T. and van den Hoek, C. (1981). The taxonomy of *Ulva* (Chlorophyceae) in the Netherlands. *Br. phycol. J.* **16**, 9–53.

Løvlie, A. (1964). Genetic control of division rate and morphogenesis in *Ulva mutabilis* Føyn. *C. r. Trav. Lab. Carlsberg* **34**, 77–168.

Papenfuss, G. F. (1960). On the genera of the Ulvales and the status of the order. *J. Linn. Soc. London, Bot.* **56**, 303–318.

Steffensen, D. A. (1976). Morphological variation in *Ulva* in the Avon-Heathcote Estuary, Christchurch. *N.Z. J. mar. freshwater Res.* **10**, 329–341.

Tanner C. (1979). The taxonomy and morphological variation of distromatic ulvaceous algae (Chlorophyta) from the Northeast Pacific. Ph.D. Thesis, University of British Columbia, Canada.

Tanner, C. (1980). *Chloropelta gen. nov.*, an ulvaceous green alga with a different type of development. *J. Phycol.* **16**, 128–37.

Titlyanov, E. A., Glebova, N. T. and Kotlyarova, L. S. (1975). Seasonal changes in the structure of the thalli of *Ulva fenestrata* P. et R. *Ekologiya* (*Sverdlovsk*) **6**, 320–324.

Vinogradova, K. L. (1974). "Ulvales (Chlorophyta) of USSR Seas." Nauka, Leningrad.

16 | Morphology and Systematics of *Stigeoclonium* Kütz. (Chaetophorales)

J. A. FRANCKE and J. SIMONS

Biological Laboratory, Free University, Amsterdam, The Netherlands

Abstract: Polymorphism in the genus *Stigeoclonium* renders the delimitation of species difficult. This polymorphism may be due to plastic responses as well as to genetic variability. Approximately 150 isolates were studied under various culture conditions and the chromosome numbers of 24 isolates were determined. The morphological variability appeared to be larger than observed by Cox and Bold (1966). Cox and Bold discerned seven morphological groups, which they considered as species; the present study discerns four morphological species, viz. *S. helveticum, S. aestivale, S. tenue* and *S. farctum*. A key for these four species is presented. The main differences between the system of Cox and Bold and this key are: the use of the germination type of the zoospore as a taxonomic criterion; and *S. subsecundum, S. pascheri* and *S. variabile sensu* Cox and Bold are rejected as distinct species. The chromosome numbers recorded were 6, 8, 10 and 12. The chromosome number of 8 was recorded throughout the entire genus, and different numbers were recorded within the same morphological species (e.g. 6, 8 and 12 in *S. tenue*). Polyploidy with a basic number of 2 may have played a significant role in the evolution of species.

INTRODUCTION

The genus *Stigeoclonium*, established by Kützing (1843), includes all attached, branched, uniseriate, filamentous green algae of which the cells of the main axis and the branches are similar in size. The thallus is heterotrichous and made up of a prostrate and an erect system of filaments. The terminal cells of the erect filaments may produce multicellular, hyaline hairs. Both prostrate and erect cells may produce rhizoids. As far as is known, the life history is haplontic and reproduction is predominantly asexual by means of quadriflagellate zoospores.

The distribution of *Stigeoclonium* species is world-wide and they grow in running or stagnant fresh water under a broad range of environmental conditions, varying from meso- to polytrophic and from unpolluted to severely polluted situations (McLean and Benson-Evans, 1974; Harding and Whitton, 1976; Francke and ten Cate, 1980; Francke and Rhebergen, 1982). Optimum development of *Stigeoclonium* in the Netherlands is during spring and autumn, when plants of several centimetres can be found as epiphytic or epilithic plants, just underneath the water surface.

A high degree of polymorphism is characteristic of *Stigeoclonium* and this renders delimitation of species difficult. This polymorphism may be due to plastic responses (McLean and Benson-Evans, 1977; Whitton and Harding, 1978; Francke, 1982) as well as to genetic variability (Francke and Rhebergen, 1982).

In major taxonomic works (Hazen, 1902; Heering, 1914; Islam, 1963; Printz, 1964) the main criteria for delimiting species are characteristics of the erect thallus, such as dimension and shape of the cells of the main axis, branching habit and hair formation. Several investigators have presented evidence as to the variability of these characters under different environmental conditions (Cox and Bold, 1966, Harding and Whitton, 1976; McLean and Benson-Evans, 1977; Francke, 1982). Cox and Bold (1966), studying *Stigeoclonium* from North America, were the first to recognize that characters of the prostrate thallus are more consistently reliable attributes than characters of the erect system. Yet the erect system is currently used for specific identification, as one usually follows the key in Printz (1964). The main criteria employed by Cox and Bold are shape, size and branching habit of the prostrate system, dimension and shape of the prostrate cells and presence and form of rhizoids. On the basis of these criteria they discerned seven morphological taxa which they considered as species, designated as *S. helveticum* Vischer, *S. aestivale* (Hazen) Collins (emend.), *S. subsecundum* (Kütz.) Kütz., *S. tenue* (Ag.) Kütz. (emend.), *S. pascheri* Cox & Bold, *S. variabile* (Nägeli) Islam, and *S. farctum* Berthold.

During a study on morphological and ecological differentiation and plasticity in the genus *Stigeoclonium,* the prostrate systems of many isolates were studied under various culture conditions (Francke, 1982) and in the natural habitats. It appeared that the classification of Cox and Bold ought to be modified in order to fit better with the observed variability.

MATERIALS AND METHODS

Approximately 150 unialgal isolates, virtually free of bacteria, were examined from a number of localities in the central western part of the Netherlands and the Friesian island of Schiermonnikoog. Information on methods of collection and isolation and on localities and field levels of some nutrients were given by Francke and ten Cate (1980), Francke (1982) and Francke and Rhebergen (1982).

The cultured prostrate systems were studied as follows. From a previously prepared clonal stock culture, a piece of algal material was transferred to a larger vessel, the bottom of which was covered by cover slips. The algae then produced zoospores, usually the night after transfer into fresh culture medium, and the zoospores settled on the cover slips and germinated. The cover slips with the germlings were then transferred to 30-ml culture jars for further development. The prostrate systems of field material were studied on glass slides which had been placed for 3–4 weeks in the outside habitats.

The nomenclature used is based upon the system of Cox and Bold (1966), who unravelled the complicated nomenclatural history.

The standard culture conditions were as follows: Wood's Hole medium (Stein, 1973) with slight modifications (Francke and ten Cate, 1980), a temperature of 16°C, a 12:12 light–dark regime and a light intensity of about 90 $\mu E\ m^{-2}\ s^{-1}$, from a cool white fluorescent tube. Initial pH was adjusted to 7.5 ± 0.25 and maintained during the culture period, which was mostly three weeks.

Chromosome numbers of 24 isolates were determined. Cytological and morphological observations were made from the same cultures. The material was fixed with Carnoy's fluid (3 parts of 96% ethyl alcohol: 1 part of glacial acetic acid), and stained applying the iron-alum aceto-carmine squash technique (Godward, 1948).

RESULTS

The isolates could be arranged in four morphological groupings, designated as *S. helveticum* (*sensu* Cox and Bold), *S. aestivale* (comprising *S. aestivale sensu* Cox and Bold and *S. subsecundum sensu* Cox and Bold), *S. tenue* (comprising *S. tenue sensu* Cox and Bold and *S. pascheri sensu* Cox and Bold) and *S. farctum* (comprising *S. farctum sensu* Cox and Bold and *S. variabile sensu* Cox and Bold).

As stated before, species delimitation is difficult due to morphological variability and plasticity, especially in the main group of Cox and Bold, comprising *S. aestivale*, *S. subsecundum*, *S. tenue*, *S. pascheri* and *S. variabile*. Despite the differences, there is a high conformity between the classification of Cox and Bold, developed on material from North America, and the classification presented here in the form of a key (Table I). An important discriminating character in this key is the germination type of the zoospore,

Table I. Key to the species of *Stigeoclonium*.

1. a. Germination of the zoospore of the erect type. Prostrate system inconspicuous—2
   b. Germination of the zoospore of the prostrate type. Prostrate system conspicuous—3
2. a. Prostrate system consisting of one holdfast cell or one or a few short rhizoids; mostly growing attached on other algae—*S. helveticum* (Fig. 1A–E)
   b. Prostrate system consisting of branched filaments, mostly with one main axis; long, slightly undulant rhizoids may be present—*S. aestivale* (Fig. 2)
3. a. Prostrate system consisting of extensive, irregular branched filaments, from which lateralia arise in an irregular manner; sometimes with slender, more or less corkscrew-like rhizoids—*S. tenue* (Fig. 3)
   b. Prostrate system consisting of a closed, pseudoparenchymatous, disc-like thallus, or of a regular branched, more open, star-like thallus in which main filaments and lateralia can be discerned (variable form)—*S. farctum* (Figs 4, 5)

which was not employed by Cox and Bold. The zoospore may germinate in two ways. Firstly, the zoospore germinates to form directly an upright filament. The basal cell of this erect filament (the original zoospore) forms a holdfast cell, or forms one to several rhizoids for attachment (*S. helveticum*), or the basal cell of this erect filament produces a prostrate filament from which the other erect filaments arise (*S. aestivale*). Secondly, the germinating zoospore grows out uni- or bilaterally to form a prostrate filament from which the erect filaments develop (*S. tenue, S. farctum*).

The *S. helveticum* isolates (Fig. 1A–D) conformed to the observations and the description of Cox and Bold. This species is characterized by its *Ulothrix*-like holdfast cells or rhizoids and the erect germination of its zoospores.

The *S. aestivale* isolates either conformed to the description of *S. aestivale* (Fig. 2A–D) or *S. subsecundum* (Fig. 2E, F), which Cox and Bold separated as two species on the basis of a more abundant production of rhizoids by the latter. However, the rhizoid production appears to be merely dependent on the stage of development of the prostrate thallus, which is determined by the culture period and conditions (Fig. 2C–F). Most isolates, cultured for three weeks under standard conditions, obtained the *S. aestivale* (*sensu* Cox and Bold) morphology (Fig. 2D). The prostrate system of algae growing outside on glass slides for a shorter or longer period of time mostly showed no rhizoid production and thus the *S. aestivale* (*sensu* Cox and Bold) morphology (Fig. 2A, B). So it seems that *S. subsecundum* is an optimally developed *S. aestivale* and therefore retention of both species is not justified.

The *S. tenue* isolates either conformed to the description of *S. tenue* (Fig. 3A, B) or *S. pascheri* (Fig. 3D), which Cox and Bold discriminated as two species on the basis of a less frequent production of rhizoids and on the inconspicuous erect system of the latter. However, these characteristics

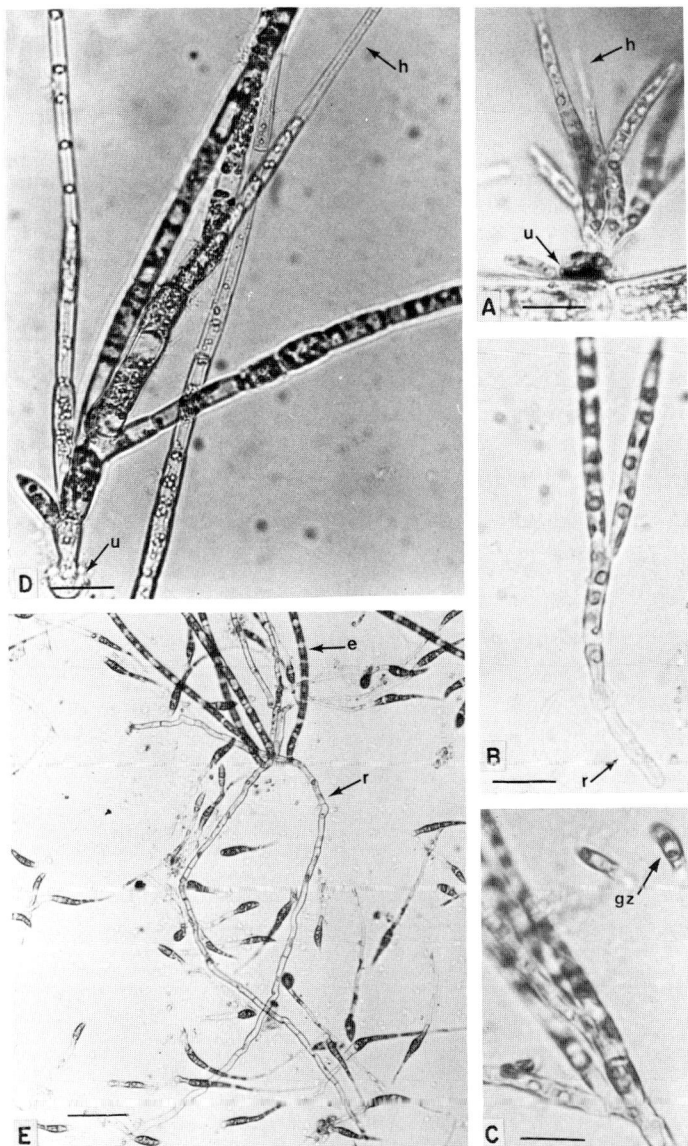

Fig. 1A–E. *Stigeoclonium helveticum* (scale bar: A–D = 20 μm; E = 80 μm). (A) Field material with *Ulothrix*-like holdfast (u) and hairs (h) growing on *Oedogonium* sp. (B) Isolate 7: two weeks culturing under standard conditions; young plant with rhizoid (r). (C) Isolate 7: three weeks culturing under standard conditions; germination of the zoospore (gz) is erect. (D) Isolate 7: three weeks culturing under low phosphorus levels; plant with *Ulothrix*-like holdfast (u), hairs (h) and a granular cell content. Three weeks culturing under standard conditions; plant with long, sparsely branched rhizoids (r), from which erect filaments (E) arise and numerous germinated zoospores.

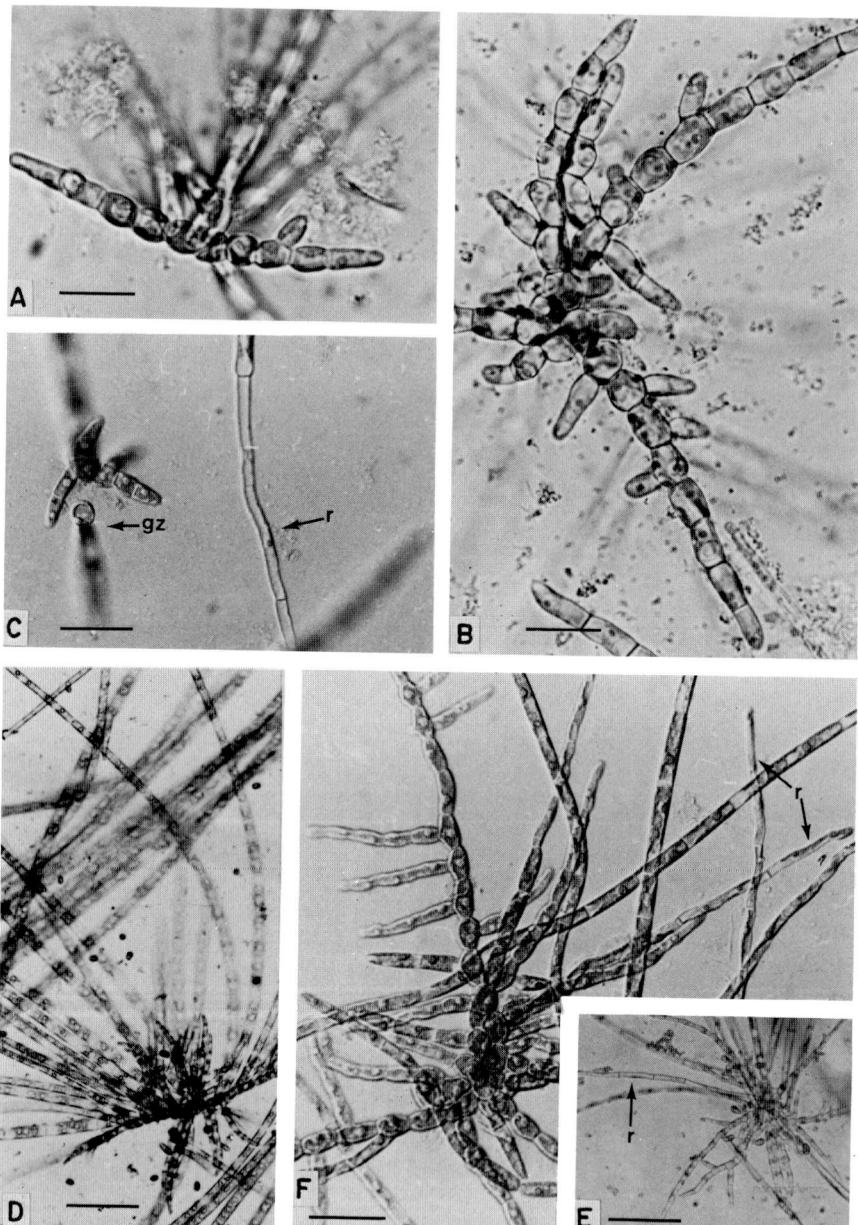

Fig. 2A–F. *Stigeoclonium aestivale* (scale bar: A, B, F = 20 μm; C = 40 μm; D, E = 80 μm). (A, B) Field material growing on a glass slide placed for four weeks in a natural habitat; a young prostrate (A) and a more developed prostrate (B) system. (C) Isolate 64: one week culturing under standard conditions; young prostrate thallus; early development of a rhizoid (r); germination of the zoospores (gz) is erect. (D) Isolate 64: three weeks culturing under standard conditions; a further developed prostrate thallus (agrees with the "aestivale" morphology of Cox and Bold). (E) Isolate 64: four weeks culturing under standard conditions; beginning of rhizoid (r) development. (F) Isolate 64: four to five weeks culturing under standard conditions; prostrate thallus with subsequent development of rhizoids (r); agrees with the "subsecundum" morphology of Cox and Bold.

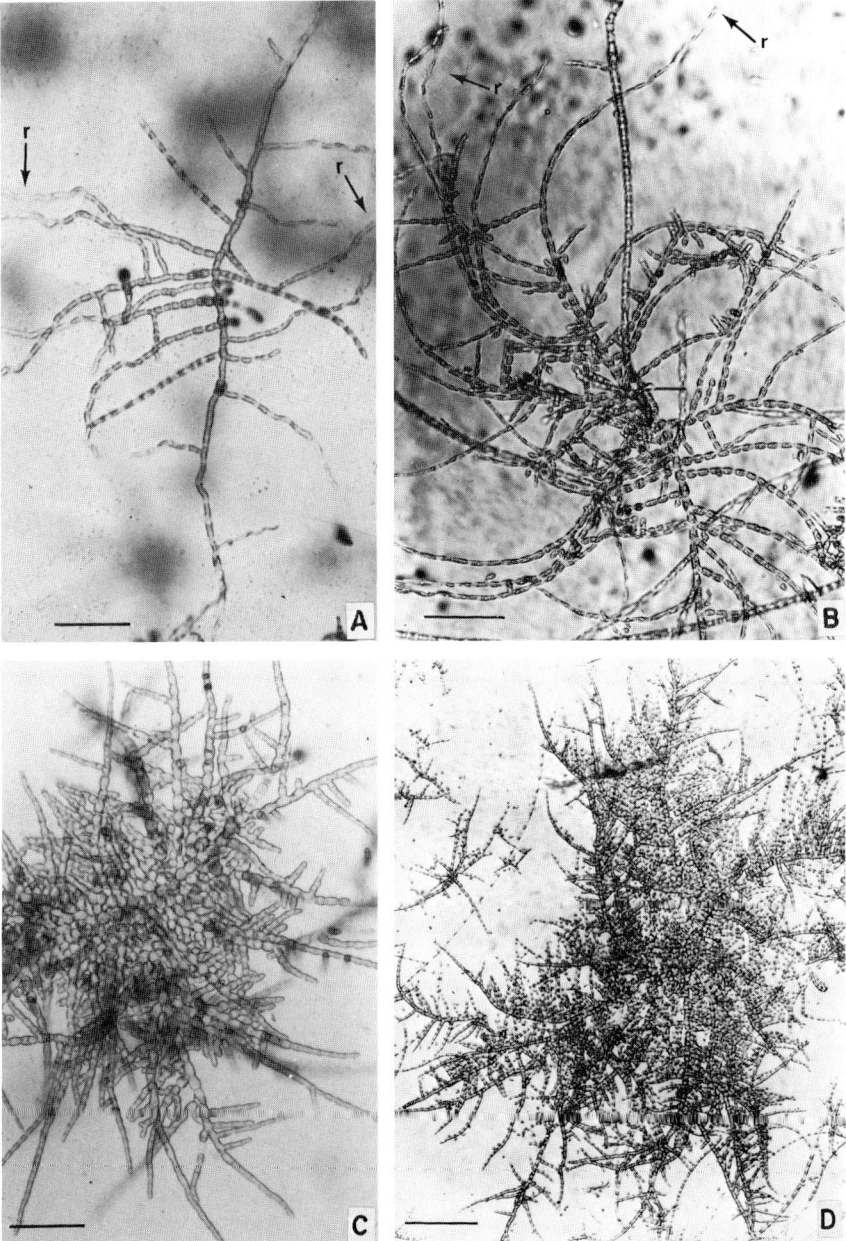

Fig. 3A–D. *Stigeoclonium tenue* (scale bar: A–C = 80 μm; D = 200 μm) (A) Isolate 19S: three weeks culturing under standard conditions; prostrate thallus of restricted growth with numerous corkscrew-like rhizoids (r). (B) Isolate 13: three weeks of culturing under standard conditions; thallus of unlimited growth with rhizoids (r). (C) Isolate 13: three weeks of culturing under high levels of nitrate-N (compare with Fig. 3B); prostrate thallus with spherical cells and no rhizoids ("pascheri" morphology of Cox and Bold). (D) Isolate 57: three weeks culturing under standard conditions; extensive spreading almost pseudoparenchymatous prostrate thallus with globular cells; rhizoids seldom produced and the erect system is extremely variable in expression ("pascheri" morphology of Cox and Bold).

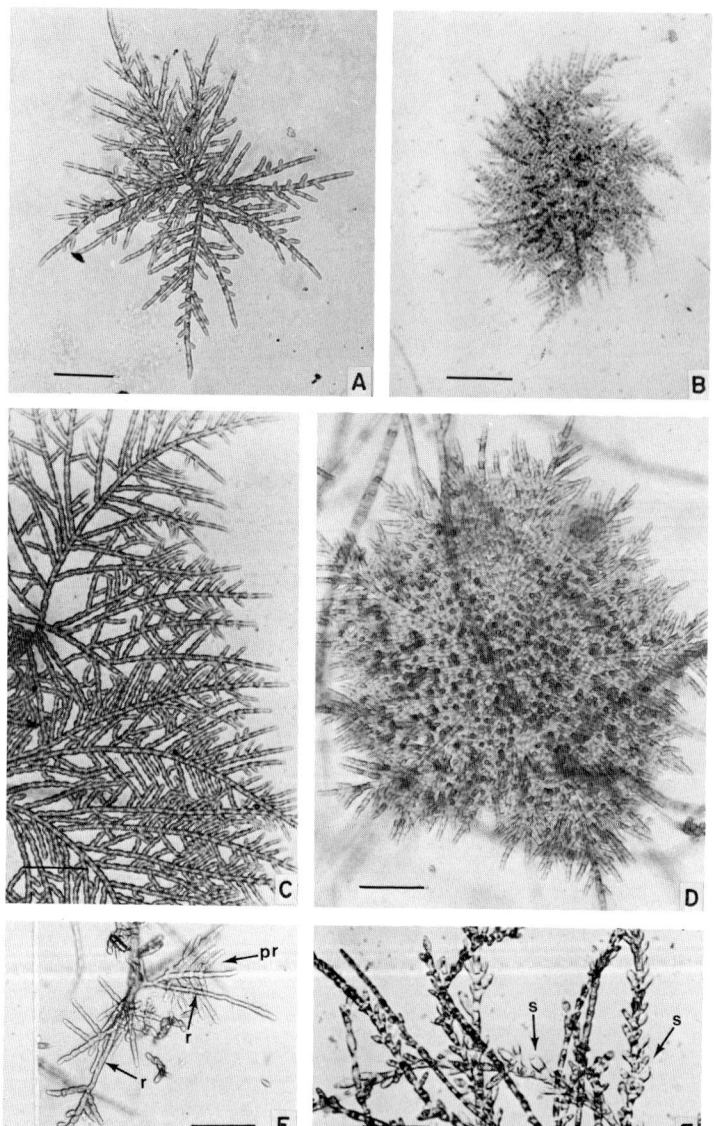

Fig. 4A–D. *Stigeoclonium farctum* (scale bar: A, C, D = 80 μm; B = 200 μm). (A) Isolate 22: two weeks culturing under standard conditions; young prostrate thallus. (B, C) Isolate 22: three weeks culturing under standard conditions; mature loosely branching prostrate thallus ("variabile" morphology of Cox and Bold). (D) Isolate 22: three weeks of culturing under high nitrate-N level; pseudoparenchymatous prostrate thallus ("farctum" morphology of Cox and Bold).

Fig. 4E, F. *Stigeoclonium farctum* var. *rivulare* (scale bars = 80 μm). (E) Field material growing on a glass slide placed for three weeks in a natural habitat; the lowermost cells of the erect filaments produce rhizoids (r), and these produce prostrate systems (pr) of filaments after adhering to the substratum. (F) Isolate 12: three weeks culturing under standard conditions; erect filaments bearing sporangia-like structures (s).

Fig. 5A–D. *Stigeoclonium farctum* (scale bars = 80 μm). (A) Isolate 68: one week culturing under standard conditions; several young prostrate thalli. (B) Isolate 68: three weeks culturing under ammonium-N levels of 14 mg l^{-1}; loosely branching prostrate thallus ("variabile" morphology of Cox and Bold). (C) Isolate 68: three weeks culturing under low phosphorus levels; loosely branching prostrate thallus. (D) Isolate 68: three weeks culturing under standard conditions; typical pseudoparenchymatous prostrate thallus (note pyrenoids of adjacent cells lying on the same level).

appeared to be extremely variable due to a high degree of genetic variability and morphological plasticity (Francke, 1982). The genetic variability leads to the distinction of a number of morphological types (Fig. 3A–D), with relatively stable characteristics under standard culture conditions, some of which might be designated as *S. pascheri* (*sensu* Cox and Bold). The genetic variability may be correlated with ecotypic differentiation in response to habitat factors, such as was demonstrated for several euryhaline *S. tenue* isolates (Francke and Rhebergen, 1982). *S. pascheri* is not retained as a separate species because both *S. pascheri* (*sensu* Cox and Bold) isolates and *S. tenue* (*sensu* Cox and Bold) isolates showed a high degree of plasticity with regard to the discriminating characters.

The *S. farctum* isolates either conformed to the description of *S. farctum* (Fig. 5D) or *S. variabile* (Fig. 4C), which Cox and Bold discerned on the basis of the disc-like pseudoparenchymatous prostrate thallus of the former. This separation is not carried through, on account of the high degree of morphological plasticity of both the *S. farctum* and *S. variabile* isolates. Depending on the culture conditions, the isolates showed a gradual variation from closed, pseudoparenchymatous prostrate systems to more open types (Figs 4A–D, 5). *S. variabile* appeared to be a morphological and ecological type of *S. farctum* (Francke, 1982). Several isolates showed deviations from the standard *S. farctum* (*sensu* Cox and Bold) morphology (Fig. 5D), viz. a form designated as *S. farctum* var. *rivulare* Butcher (Fig. 4E), with one or several pseudoparenchymatous prostrate thalli, and a form with the standard morphology except for the erect filaments that produce clumps of bulbous cells (Fig. 4F) which look like sporangia. This erect system closely resembled *S. carolinianum* of Islam (1963), but the prostrate system was insufficiently described. According to him it consisted of rhizoids.

According to different authors, *Stigeoclonium* species may form two types of asexual zoospores: a large quadriflagellate "macrozoospore" and a smaller quadri- or biflagellate "microzoospore". The macrozoospores are formed universally by every species and produce new plants directly (Islam, 1963). This was confirmed by this study: each isolate produced macrozoospores, mostly in great abundance, and only three out of 150 isolates were observed to produce both types of swarmer. These three isolates belong to *S. farctum* var. *rivulare*. The macrozoospores germinated according to the prostrate type, forming the typical *S. farctum* discs (Fig. 5D), while the microzoospore germinated according to the erect type, resulting in a prostrate thallus consisting of several pseudoparenchymatous basal systems (Fig. 4E). The *S. farctum* var. *rivulare* isolates originated from several exposed sites of a large lake (Braassemermeer). The growth form in nature was the form with several prostrate interconnected systems with rhizoids. Therefore, *S. farctum* var. *rivulare* may be an ecotype of *S. farctum* adapted to exposed situations. Butcher (1932) recorded *S. farctum* var. *rivulare* in highly calcareous waters of big streams, corresponding to exposed sites.

The results of the chromosome counts are presented in Table II while Fig. 6A, B shows the chromosomes of *S. aestivale* and *S. farctum*. The chromosome numbers determined were 6, 8, 10 and 12. Earlier records varied from 5 to 20 (Table II). Our results can at best be compared with the records of Sarma and Jayaraman (1980), who employed the classification of Cox and Bold. With the exception of *S. helveticum*, which is easily discriminated, it is impractical to compare previously published records because species identification has nearly always been based on the erect thallus.

The number of 12 for *S. helveticum* differs from that of Sarma and Jayaraman (1980) for an Indian isolate ($n = 5$) and from that of Abbas and Godward (1964) for a British isolate ($n = 20$), but agrees with the findings of Abbas and Godward (1964) for *S. helveticum* var. *minus* ($n = 12$). One of the four *S. aestivale* isolates examined was a chlorine-tolerant strain (Francke and Rhebergen, 1982) with a deviating number of 10. The chromosome number of 8 of the other *S. aestivale* isolates does not agree with the number of 10 of an Indian *S. aestivale* isolate (Sarma and Jayaraman, 1980). Regarding *S. tenue*, the number of 11 of Sarma and Jayaraman differs from our

Table II. Chromosome numbers recorded in the present study and by various authors.

	No. of isolates investigated	Chromosome numbers (n); Francke, Simons	Chromosome numbers (n); various authors
S. helveticum	3	12	20 (Abbas and Godward, 1964)
			5 (Sarma and Jayaraman, 1980)
S. helveticum var. *minus*			12 (Abbas and Godward, 1964)
S. aestivale	4	8, 10	9 (Chowdary, 1967)
			10 (Sarma and Jayaraman, 1980)
S. subsecundum	2	8	14 (Sarma and Jayaraman, 1980)
S. tenue	5	6, 8, 12	13 (Abbas and Godward, 1964)
			11 (Sarma and Jayaraman, 1980)
S. pascheri	2	6, 8	6 (R. Shayam, unpublished)
			12 (Sarma and Jayaraman, 1980)
S. variabile	2	8	16 (Abbas and Godward, 1964)
			6 (Chowdary, 1967)
			6 (Sarma and Jayaraman, 1980)
S. farctum	3	8	5 (Singh, 1954)
			8 (Abbas and Godward, 1964)
S. farctum var. *rivulare*	2	12	

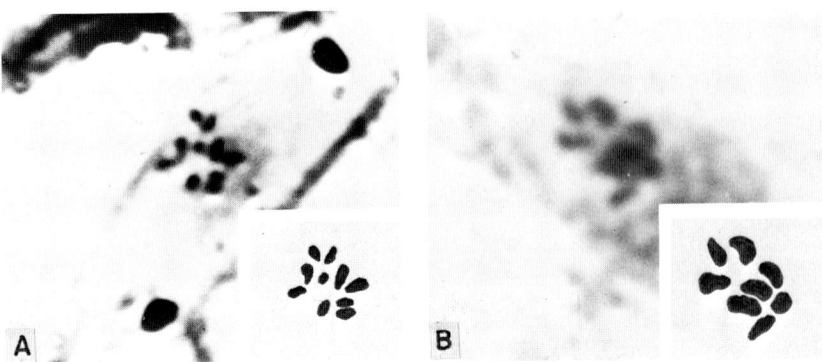

Fig. 6A, B. Chromosome numbers of *Stigeoclonium* (ca. 1500×). (A) *S. aestivale*; metaphase squash and its respective drawing ($n = 10$). (B) *S. farctum*; metaphase squash and its respective drawing ($n = 8$).

findings (6, 8, and 12). The numbers of 6 and 8 were found both in *S. tenue* and *S. pascheri*. The number of 8 recorded for our *S. variabile* isolates does not agree with the number of 6 for the Indian *S. variabile* isolate of Sarma and Jayaraman.

DISCUSSION

For practical reasons, the priority rule of nomenclature was not applied strictly, because the authors have reservations concerning the names of Cox and Bold (1966) and doubt whether the species names of Cox and Bold really correspond with previously described homonyms. In lumping two species the name of the best resembling species of Cox and Bold was chosen for the new entity. This resulted in our choosing the name *S. aestivale* instead of the formally required name *S. subsecundum*, and *S. farctum* instead of *S. variabile*.

In general, the species characteristics employed by Cox and Bold proved to be valid except for cell shape and dimensions, which are very variable, as was demonstrated by Francke (1982). Cox and Bold discerned seven species and the present study four (see Table I), yet the morphological variability appeared to be larger than observed by Cox and Bold. The reason may be that Cox and Bold examined those 20 isolates out of 60 which possessed the most diverse characteristics and rejected intermediate forms. Moreover, study of prostrate systems by Cox and Bold under culture conditions other than the standard conditions was limited.

The germination type of the zoospores appears to be a useful characteristic, easy in operation and usable in the early development of the plants. This characteristic was not used by Cox and Bold in their concept of species groups. They distinguished three groups based on the nature of the basal system consisting of (I) short, sparsely produced rhizoids, (II) a branching filament, and (III) a pseudoparenchymatous disc. Group I contains only *S. helveticum,* group III only *S. farctum,* and group II comprises the other five species. In our system only two groups are discerned (see Table I) based on the types of germination of zoospores: (1) with the erect type containing *S. helveticum,* and *S. aestivale,* and (2) with the prostrate type containing *S. tenue* and *S. farctum.*

As shown, the morphology of the prostrate system facilitates identification of species. However, the morphology of the prostrate thallus often also shows considerable variation, especially in the species *S. tenue* where a number of constant morphological types can be discerned, reflecting genetic variability. Moreover, within each species a certain amount of morphological plasticity exists, depending on the stage of development and conditions in the growth-medium. Under extreme conditions, for example a high nutrient load, the prostrate system may be disorganised into a compact system with rounded akinete-like cells as observed in *S. tenue* and *S. farctum.* Under low nutrient conditions the development of hairs on the erect filaments is often promoted. This appears to be a general phenomenon in hair-bearing algae.

The chromosome numbers recorded cannot be correlated with morphological groups, mainly because the number of 8 was recorded in all species except *S. helveticum.* Moreover, different chromosome numbers were observed in the same morphological group, for example 8 and 10 in *S. aestivale* and 6, 8 and 12 in *S. tenue.* The occurrence of three different numbers in the *S. tenue* group is supporting evidence for the relatively large amount of morphological and ecological differentiation observed in *S. tenue* (Francke, 1982). The presence of the deviating number of 10 in a chlorine-resistant strain of *S. aestivale* suggests the idea of ecotypic differentiation towards salinity tolerance in this species (Francke and Rhebergen, 1982). The records of 6 and 8 in *S. tenue* as well as in *S. pascheri* support our view that these forms are conspecific. This is also the case for *S. variabile* and *S. farctum,* where only the number of 8 was recorded.

The records of deviating numbers in the literature, especially the recent records of the Indian authors Sarma and Jayaraman (1980), point to the occurrence of geographically distinct races within the same morphological taxon.

The chromosome numbers reported till now for the various *Stigeoclonium* species present an almost uninterrupted series ranging from 5 to 20, suggesting that aneuploidy may have played a significant role in the evolution of species. On the other hand, the present study indicates that polyploidy may also have played a role in the evolution of this genus.

The presence of cytological races within morphologically defined species is an indication that for the delimitation of biological species genetic or biochemical approaches should be employed. In this connection more knowledge is also needed about the occurrence of sexual reproduction.

ACKNOWLEDGEMENTS

The investigations were supported by the Foundation for Fundamental Biological Research (BION), which is subsidized by the Netherlands Organization for the Advancement of Pure Research (ZWO). The authors would like to thank Professor Dr. M. Vroman, for critically reviewing the manuscript and A. P. van Beem for analytical help during the investigations.

REFERENCES

Abbas, A. and Godward, M. B. E. (1964). Cytology in relation to taxonomy of Chaetophorales. *J. Linn. Soc. London, Bot.* **58**, 499–507.

Butcher, R. W. (1932). Notes on the Algae from beds of rivers. *New Phytol.* **31**, 289–309.

Chowdary, Y. B. K. (1967). The chromosome numbers of some species of the genus *Stigeoclonium* Kütz. *Cytologia* **32**, 174–179.

Cox, E. R. and Bold, H. C. (1966). Phycological studies. VII. Taxonomic investigations of the genus *Stigeoclonium*. *Univ. Tex. Publ.* **6618**, 1–167.

Francke, J. A. (1982). Morphological plasticity and ecological range in three *Stigeoclonium* species (Chlorophyceae, Chaetophorales). *Br. phycol. J.* **17**, 117–133.

Francke, J. A. and Rhebergen, L. J. (1982). Euryhaline ecotypes in some species of *Stigeoclonium* Kütz. *Br. phycol. J.* **17**, 135–145.

Francke, J. A. and ten Cate, H. J. (1980). Ecotypic differentiation in response to nutritional factors in the algal genus *Stigeoclonium* Kütz. (Chlorophyceae). *Br. phycol. J.* **15**, 343–355.

Godward, M. B. E. (1948). The iron acetocarmine method for algae. *Nature (London)* **161**, 203.

Harding, J. P. C. and Whitton, B. A. (1976). Resistance to zinc of *Stigeoclonium tenue* in the field and the laboratory. *Br. phycol. J.* **13**, 65–68.

Hazen, T. E. (1902). Ulothrichaceae and Chaetophoraceae of the United States. *Mem. Torrey bot. Club.* **11**, 135–250.

Heering, W. (1914). Chlorophyceae. III. Ulothrichales, Microsporales, Oedogoniales. *In* Die Süsswasser-Flora Deutschlands, Österreichs und der Schweiz" (A. Pascher, ed), No. 6, pp. 1–250. Fischer, Jena.

Islam, A. K. M. (1963). A revision of the genus *Stigeoclonium*. *Beih. Nova Hedwigia* **10**, 1–164.

Kützing, F. T. (1843). "Phycologia generalis. . . ." Brockhaus, Leipzig.

McLean, R. O. and Benson-Evans, K. (1974). The distribution of *Stigeoclonium tenue* Kütz. in South Wales in relation to its use as an indicator of organic pollution. *Br. phycol. J.* **9**, 83–89.

McLean, R. O. and Benson-Evans, K. (1977). Water chemistry and growth form variations in *Stigeoclonium tenue* Kütz. *Br. phycol. J.* **12**, 83–88.

Printz, H. (1964). Die Chaetophoralen der Binnengewässer (eine Systematische Übersicht). *Hydrobiologia* **24**, 1–376.

Sarma, Y. S. R. K. and Jayaraman, S. (1980). Karyological studies on certain taxa of *Stigeoclonium* and *Chaetophora* (Chaetophorales, Chlorophyceae). *Phycologia* **19,** 253–259.

Singh, R. N. (1954). A comparative study of the life-cycle of two species of the genus *Stigeoclonium* (*S. farctum* and *S. amoenum*). *Rev. algol.* **1,** 42.

Stein, R. J. (1973). "Handbook of Phycological Methods, Culture Methods and Growth Measurements." Cambridge Univ. Press, London and New York.

Whitton, B. A. and Harding, J. P. C. (1978). Influence of nutrient deficiency on hair formation in *Stigeoclonium*. *Br. phycol. J.* **13,** 65–68.

17 | Taxonomical and Ultrastructural Survey of the Genus *Desmatractum* West & West (Chlorococcales)

O. L. REYMOND*

Département de Biologie Végétale, Université de Genève, Laboratoire de Microbiologie Générale, Genève, Switzerland

F. A. C. KOUWETS

Vries-Laboratorium, University of Amsterdam, Amsterdam, The Netherlands

Abstract: The genus *Desmatractum* (Treubariaceae, Chlorococcales) is composed of less than ten taxa. It has not been possible to cultivate any of them and they are rarely found. By flat-embedding allowing both observation of the very rare cells by optical microscopy and their orientation for transmission electron microscopy (TEM) both preserved and newly-collected material has been studied. The external architecture as well as certain elements of the internal ultrastructure of four taxa have been clarified. It seems that those taxa which have not been studied by electron microscopy have never or hardly ever been found again since their original description. In spite of large gaps in our knowledge of this genus, the results obtained lead us to believe that it is less artificial than it was considered to be in the past. In spite of certain general similarities with the genus *Treubaria* Bernard (Treubariaceae), this ultrastructural study clearly shows the differences which separate these two genera. A comparison with *Pachycladella* (G. M. Smith) Silva (Treubariaceae) is also made.

---

*Present address: Institut d'Histologie et d'Embryologie, Université de Lausanne, CH-1011 Lausanne, Switzerland.

Systematics Association Special Volume No. 27, "Systematics of the Green Algae", edited by D. E. G. Irvine and D. M. John, 1984, pp. 379–389. Academic Press, London and Orlando.
ISBN 0 12 374040 1   *Copyright © by the Systematics Association*
All rights of reproduction in any form reserved

INTRODUCTION

The genus *Desmatractum* W. & G. S. West (1902) emend. Pascher (1930) (Treubariaceae, Chlorococcales, Chlorophyceae) includes unicellar and free-living algae with a globose form and having a regular cell wall. The whole cell is enclosed in an extra envelope that consists of two facing cones basally connected and bearing longitudinal ridges. The cell contains a cup-shaped chloroplast with one pyrenoid. Asexual reproduction is by means of two or four zoospores each of which possesses two flagella of equal length, one or two contractile vacuoles and an eye-spot, or by autospores (see Pascher, 1930).

Nine species are included in the genus *Desmatractum* by various authors: *D. plicatum* W. & G. S. West (1902), *D. nyanzae* (Woloszynska) G. S. West (1916) [*Peniococcus nyanzae* Woloszynska (1914)], *D. bipyramidatum* (Chodat) Pascher (1930) [*Bernardinella bipyramidata* Chodat (1921)], *D. indutum* (Geitler) Pascher (1930) [*Calyptobactron indutum* Geitler (1924)], *D. elongatum* Pascher (1930), *D. obtusum* Pascher (1930), *D. delicatissimum* (Korshikov (1953), *D. elongatum* Hirose & Akiyama (in Hirose and Yamagishi, 1977), and *D. spryii* Nicholls et al. (1981).

One of these, *D. nyanzae*, was originally described as *Peniococcus nyanzae* (Scenedesmaceae), and as having a cylindrical cell with irregular margins, a median constriction and a single pyrenoid. However, the cell contains many chloroplasts and reproduces by means of a transverse division (Woloszynska, 1914). On account of the remark of West (1916) that "*Peniococcus nyanzae* would be better placed as *Desmatractum nyanzae*", Printz (1927) put the species in the genus (with G. S. West acknowledged as the author), and Philipose (1967) also mentioned the species under this name. We, on the contrary, agree with Pascher (1930) who stated that *Peniococcus* has nothing to do with *Desmatractum,* and considered the name as being not validly published.

The name *D. elongatum*, proposed by Hirose & Akiyama (in Hirose and Yamagishi, 1977) must be rejected since it is a later homonym of *D. elongatum* Pascher (1930). This is independent of whether *D. elongatum* Pascher has to be considered as merely a variety of, or even identical with, *D. bipyramidatum* Pascher. A careful study of the type material of *D. elongatum* Hirose & Akiyama could elucidate its identity, but such a study was not available at the time of this writing.

In his survey of the genus *Desmatractum*, Pascher (1930) described two new species (*D. elongatum* and *D. obtusum*) that are closely related to and markedly resemble *D. bipyramidatum,* and are possibly even identical with it and are only to be regarded as ecophenotypes. Skuja (1964) made mention of a form of *D. bipyramidatum* that is intermediate between *D. elongatum* and the typical *D. bipyramidatum*. Since *D. elongatum* and *D. obtusum* were only mentioned by Pascher (1930) from a very few localities and never found again, no definite answer on their possible identity can be given for the present.

The remaining five *Desmatractum* species, however, are of more or less rare occurrence; all have been mentioned from localities other than those from which they were originally described. So far, four out of these five species, *D. indutum, D. bipyramidatum, D. delicatissimum* and *D. spryii*, have now been studied with the electron microscope.

In the past, there has been some confusion on the taxonomic status of the species presently included in the genus *Desmatractum*. West and West (1902) placed *D. plicatum* in the chlorophycean family Palmellaceae, close to *Rhaphidium*, based on its morphological resemblance to the latter. Wille (1911) considered it as an unassigned genus near the family Oocystaceae (Protococcales, Autosporinae), as did Brunnthaller (1915). West (1916) transferred it to the family Autosporaceae. Chodat (1921) attributed *D. bipyramidatum* (as *Bernardinella bipyramidata*) to the Heterokontae, but it is not clear whether he did so on account of the cell colour ("luteo-viridis") or of the bipartite envelope ("lorica dimidiata"). Besides, he assumed that the organism concerned was certainly a resting stage of a flagellate form. Pascher (1925) also presumed it to be a heterokont, but wondered if the cell contents are possibly contracted and doubted whether it is a true vegetative cell or just a resting stage. He further noticed the resemblance to *D. plicatum*. Deflandre (1927) doubed whether *D. bipyramidatum* is a flagellate and assumed that it is a real vegetative cell, noticing the sometimes remarkably thick cell wall. Geitler (1924) attributed *D. indutum* (as *Calyptobactron indutum*) without hesitation to the Protococcaceae. Printz (1927) attributed *Desmatractum* to the protococcalean family known as the Pleurococcaceae, and *Calyptobactron* to the Oocystaceae, but considered the latter to be a heterokont (despite the presence of a pyrenoid) as, in his opinion, is *Bernardinella*; he noticed the similarity between *Bernardinella* and *Desmatractum*. Korshikov (1928), having observed a pyrenoid with starch caps and autospores with contractile vacuoles (which he considered to be reduced zoospores, = hemizoospores), placed *Bernardinella* as well as *Calyptobactron* (mispelled *Calyptrobactron* by him and many subsequent authors) in the Chlorococcaceae (Protococcales) and stated that both genera would be better placed in a separate subfamily, the Bernardinellae. He did not unite the two genera, however, because of the fact that *Calyptobactron* has true zoospores. As to the relationship between *Bernardinella* and *Desmatractum*, he noted that the latter should be studied more precisely first. Pascher (1930) united all three genera into *Desmatractum* (Protococcales) because he found true zoospores in *Bernardinella* and because of the close resemblance of *Bernardinella* to *Desmatractum*.

Fritsch (1935) placed *Desmatractum* in the chlorococcalean family Oocystaceae, Smith (1950) in the Chlorococcaceae, and Korshikov (1953) in the subfamily Treubarioideae of the Chlorococcaceae. Fott (1960) raised this subfamily to the rank of family (Treubariaceae) together with the genus *Treubaria* Bernard (1908) (cf. Reymond, 1980), distinguishing it by the possession of a segmented cell envelope. Up to the present, most taxonomists have followed this classification (e.g. Ettl, 1980 and Komárek and Fott, 1983).

MATERIALS AND METHODS

Rare or sometimes abundant, old or new, bad or well-preserved field-collected material has been studied from various countries. In order to select very rare and small cells for electron microscopy, material was flat-embedded in resin between two microscope slides, which allowed an accurate selection of cells with the light microscope (Reymond and Pickett-Heaps, 1983; see also Reymond, 1981, 1983; Reymond and Kouwets, 1981; Reymond and Skogstad, 1983). Sections were observed with the Zeiss EM 10 of the Section de Biologie de l'Université de Genève and of the Institut d'Histologie et d'Embryologie de l'Université de Lausanne.

STRUCTURE AND ECOLOGY

D. indutum, *D. delicatissimum* and *D. spryii* are long and slender planktonic species, while *D. bipyramidatum* is a relatively short and wide benthic species.

1. Desmatractum indutum

D. indutum measures 36–100 × 3–12 μm, so there exists a considerable range in cell length. The cell frequently shows a clear median constriction

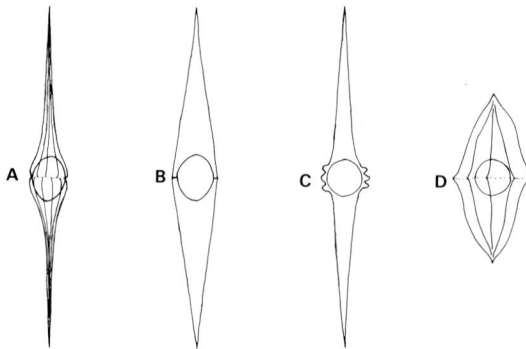

Fig. 1. (A) *Desmatractum indutum*; (B) *D. delicatissimum*; (C) *D. spryii*; (D) *D. bipyramidatum*. Observations with light microscope. For scale, see text.

17. Desmatractum West & West (Chlorococcales)

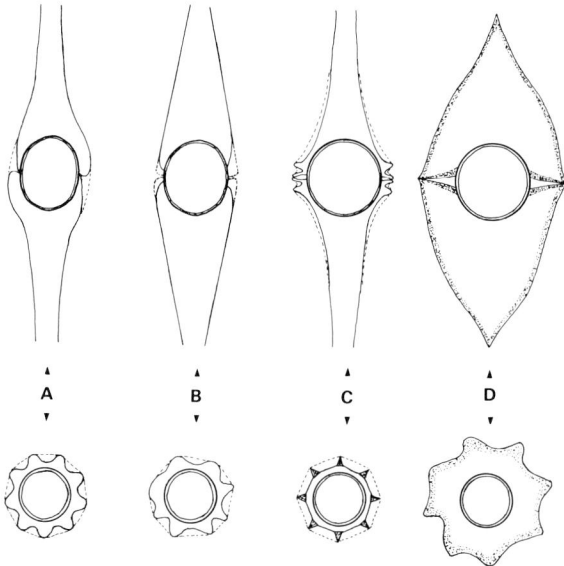

Fig. 2. Longitudinal and transversal sections: (A) *Desmatractum indutum;* (B) *D. delicatissimum;* (C) *D. spryii;* (D) *D. bipyramidatum.* From results with electron microscope. For scale, see text.

and sometimes a layer of mucilage around it (Fig. 1). Ettl (1968) made mention of 12–16 longitudinal ridges on the cell envelope, whereas Hortobagyi (1962) described a variety of this species, *D. indutum* var. *elegans,* mainly differing from the type species in having but six ridges. Bourrelly and Couté (1978) mentioned this variety with six to eight ribs. Ettl (1965), on the contrary, mentioned a form characterised by the absence of ribs and the presence of relatively short processes, so that the cells measure 30 μm at most. Reymond (1981) showed that the cells he sectioned always had nine ribs, which makes the status of var. *elegans* highly dubious (cf. Nicholls *et al.,* 1981). Furthermore, he demonstrated that one cone was rotated over about 20° with regard to the other, and that a kind of velum covers most probably the whole cell envelope (Fig. 2). Apart from the chloroplast and the nucleus, the most interesting structures that Reymond (1981) was able to find from the serial sections were a typical chlorococcalean phycoplast and a pyrenoid divided into two hemispheres by a few thylakoids. Geitler (1924) observed reproduction by means of zoospores and found that sometimes the pyrenoid divided simultaneously with the division of the chloroplast, or else one-half of the divided cell built a pyrenoid *de novo.*

Several authors noticed the resemblance between *D. indutum* and *D. plicatum.* Indeed, Geitler (1924) already stated that older cells of *D. indutum* do look like *D. plicatum.* Nicholls *et al.* (1981) even called into question the validity of *D. indutum*, because the two species seem to differ only in cell size (*D. plicatum* measures 16–22 × 6.5–9 μm). We believe that there is an obvious difference of morphology in transverse sections of these two species taken close to the level of the equator of the cell. In *D. indutum* the nine-armed star is rounded (see Reymond, 1981) whereas in *D.plicatum* the eight-armed star is spiny (see West and West, 1902). However, the real precision of the drawing of West and West is difficult to estimate. *D. plicatum* has hitherto been mentioned from paddy fields in Ceylon (West and West, 1902), from Java (Pascher, 1930) and from Mozambique (Rino, 1972), and may therefore represent a shorter, tropical, (benthic?) form of *D. indutum* (compare also the arrangements of ridges on Rino's cells). Although Geitler (1924) described *D. indutum* from a water basin in a greenhouse, this species is not restricted to tropical regions as Pascher (1930) assumed, but is probably cosmopolitan and is mentioned already from several places in Europe, the USSR, the USA and Argentina. According to Korshikov (1953) the species is rare, but if found in the plankton it occurs in large numbers. It prefers meso- to eutrophic ponds and lakes, and often occurs together with many other chlorococcalean species (e.g. Fott and Ettl, 1959). Lackey (1942) found this species (misidentified as *D. bipyramidatum*) in an American river while Nauwerck (1962), however, reported it from an oligotrophic lake in Portugal. In the Netherlands, as well as in Switzerland, it is found in meso-trophic lakes where the pH ranges from 7 to 8.

2. *Desmatractum delicatissimum*

D. delicatissimum measures 58–134 × 8–12 μm (Fig. 1). With the aid of basic fuchsine, Korshikov (1953) was able to demonstrate six or seven oblong ridges on the cell envelope which itself was covered with a thin layer of mucilage. Since none of the very few authors who reported the species subsequently could demonstrate these ridges, Reymond and Druart (1980) doubted their existence. Only Ettl (1960) mentioned a slightly different form, marked by a square optical cross-section. This possibly has something to do with the ridges on the envelope. However, recently Reymond (1983) was able to cut ultra-thin sections of two cells of *D. delicatissimum*, and confirmed the existence of about 11 ridges covered by a velum. The ridges of one cone are probably not in front of the furrows of the opposite cone as in *D. indutum*. Cross sections of these ridges are rounded near the equator of the cell (Reymond, 1983) and new observations (O. L. Reymond, unpublished) show that they are acute in narrower parts of the cone, as in *D. indutum* (Reymond, 1981, Fig. 2, right). The pyrenoid is identical to that of *D. indutum* (Fig. 2). Korshikov (1953) noticed the formation of four daughter-cells with a stigma, but not the actual sporulation. *D. deli-*

catissimum occurs as a very rare species amongst other Chlorococcales in the plankton of meso- and eutrophic ponds and lakes. According to Hortobagyi (1973) it is a stenothermal species which he found in the autumn in a eutrophic waterworks basin where the pH ranged from 7.7 to 8.7. Ettl (1960) found this forma at pH 7.2 while Reymond and Druart (1980) collected it in a mesotrophic lake rich in chlorococcalean species at a pH 8.5. *D. delicatissimum* is known from Europe, the USSR and also Jamaica (Hegewald, 1976).

3. *Desmatractum spryii*

D. spryii measures 30–85 × 7–14 μm (Fig. 1). Jensen-stained specimens show an envelope with four to eight longitudinal ridges. The most striking feature that distinguishes this species light microscopically from the others is the three transverse ribs encircling the cell in the equatorial region (Nicholls et al., 1981). However, by sectioning cells of this species, Reymond and Skogstad (1983) were able to demonstrate that in fact there are four ribs (Fig. 2), whereof the two small ribs lie rather close together. In cross section a cell showed one cone with seven ridges and an opposite cone with eight ridges. In another sectioned cell, one cone showed nine ridges and the opposite ten ridges. Each cone is covered by a velum. The ridges are present only between the apices and the outer transverse ribs. In the equatorial plane the cross sections are circular. The pyrenoid of this species also consists of two hemispheres separated by a few thylakoids. Since up to now *D. spryii* has been mentioned from only four localities, three lakes in Canada (Nicholls et al., 1981) and one in Norway (Reymond and Skogstad, 1983), little can be said on its ecology. There are, however, some similarities: all four lakes are rather shallow (3–6 m), poor in nutrients (EGV 172–270 μS/cm), and with a rather low total phosphorus content (10–37.5 ug/1). Furthermore, Reymond and Skogstad (1983) gave a pH of 7.5

4 *Desmatractum bipyramidatum*

D. bipyramidatum is a true benthic species, with much shorter and wider cells as compared with the planktonic species. The dimensions range from 8.5–18 × 17–44 μm. The cell envelope has five to 12 longitudinal ridges (Fig. 1). The cells, as sectioned by Reymond and Kouwets (1981) showed eight irregular ridges, but in contrast to the former three species it has no velum (Fig. 2). The ridges on the two cones face each other. *D. bipyramidatum* shows a phycoplast (as in *D. indutum*), and its pyrenoid is built up from two hemispheres separated by a few thylakoids. It is interesting that Korshikov (1928) had already noticed this feature in young autospores. Symbiotic-like bacteria have been found in the cytoplasm of two specimens. *D. bipyramidatum* is the most common species of the genus. It is cosmopolitan, having been mentioned from tropical (Bourrelly, 1961) as well as arctic regions (Yamagishi, 1969). It prefers an oligo- to mesotrophic marshy or boggy environment, and occurs also in the shallow parts of ponds and pools. Here it is not infrequent on bottom deposits or in amongst higher aquatics, preferring rather the mucilage produced by other algae such as desmids which are common on the same sites (see Deflandre, 1927; Thunmark, 1942; Skuja, 1964; Kouwets, 1980). In the literature, the species has been mentioned as occurring at a pH of 3.5–6.9, and exceptionally at a pH above 7.0. Lund (1942), however, remarked that the species grows very well in an enriched culture at a pH of 7–8.5.

As has been stated above, in our opinion *D. elongatum* and *D. obtusum* fall under the range of variation of *D. bipyramidatum* (cf. Korshikov, 1928; Pascher, 1930, Fig. 14b). According to the remarks of Pascher (1930), the two species mentioned occur in the same environment as *D. bipyramidatum*. Reproduction by autospores has been mentioned by Korshikov (1953) and Lund (1942), whilst Pascher (1930) noticed zoospores.

DISCUSSION

When comparing the results of the ultrastructural investigations made into four species of *Desmatractum* (Reymond, 1981, 1983; Reymond and Kouwets, 1981; Reymond and Skogstad, 1983), this genus appears to be not as artificial as we thought it to be based on the reports of light microscopical studies published by several authors. The emended description of the genus by Pascher (1930) is still correct. (cf. Reymond and Kouwets, 1981), and we can only make it a little more precise. It furthermore became evident that the supposed similarity between *Desmatractum* and *Treubaria* (Reymond and Druart, 1980) does not exist, and that these two genera can be separated with ease on the basis of their ultrastructure (Reymond, 1983). We must also mention that an ultrastructural conformity between *Desmatractum* and *Pachycladella umbrina* (G. M. Smith) Silva (1970) is highly doubtful (O. Reymond and E. Hegewald, in preparation). In two species, *D. indutum* and *D. bipyramidatum*, a phycoplast composed of centrioles and microtubules can be demonstrated, affirming the chlorococcalean status. All four species were shown to have an intraplastidial pyrenoid enclosed by a starch cap. The pyrenoid, together with the cap, is composed of two hemispheres separated by one or more thylakoids. This feature is, however, not uncommon, being known from several algal species or genera (Reymond, 1981). Already Korshikov (1928) has made mention of this structure of the

pyrenoid in *Desmatractum*. However, several authors have depicted pyrenoids surrounded by a starch cap divided into more than two parts: Ettl (1968) and Skuja (1964) for *D. bipyramidatum,* Ettl (1960) for *D. delicatissimum* and Ettl (1965, 1968) and Bourrelly and Couté (1978) for *D. indutum*. These observations could have come from some fissures in the starch hemisphere (sometimes observed with the TEM).

The structure enclosing the protoplast consists of two or three layers: a rather thick wall sometimes surrounds the proper protoplast, a wide, ribbed envelope encloses the whole cell in the equatorial plane, and this envelope is connected with the cell wall by bands of fibrils. In addition, three out of the four species examined had a thin velum covering most of the envelope. The exact shape of the envelope is very different between these four species, and is characteristic for each of them. For the moment, unfortunately, nothing can be said about the development of the envelope since none of the species examined could be cultivated, despite many trials.

REFERENCES

Bernard, C. (1908). "Protococcacées et Desmidiées d'eau douce, récoltées à Java." Landsdrukkerij, Batavia.

Bourrelly, P. (1961). Algues d'eau douce de la République de Côte de'Ivoire. *Bull. Inst. Fr. Afr. Noire, Ser. A* **23,** 283–374.

Bourrelly, P., and Couté, A. (1978). Algues d'eau douce rares ou nouvelles pour la flore française. *Rev. Algol.* [N.S.] **13,** 295–307.

Brunnthaler, J. (1915). Protococcales. *In* "Die Süsswasser-Flora Deutschland, Österreich und der Schweiz" (A. Pascher, ed) No. 5, pp. 52–205. Fischer, Jena.

Chodat, R. (1921). Algues de la région du Grand St-Bernard. I. Algues rares ou nouvelles du Plan Jupiter. *Bull. Soc. Bot. Geneve* [2] **12,** 293–305.

Deflandre, G. (1927). Sur une algue rare *Bernardinella bipyramidata* Chodat. nouvelle pour la flore française. *Arch. Bot. Mem., Caen* **1,** 220–223.

Ettl, H. (1960). Die Algenflora des Schönhengstes und seiner Umgebung. I. *Nova Hedwigia* **2,** 508–546.

Ettl, H. (1965). Die algenflora des Schönhengstes und seiner Umgebung. II. *Nova Hedwigia* **10,** 121–159.

Ettl, H. (1968). Ein Beitrag zur Kenntniss des Algenflora Tirols. *Ber. naturwiss.-Med. Ver. Innsbruck* **56,** 177–354.

Ettl, H. (1980). "Grundriss der Allgemeinen Algologie." Fischer, Stuttgart.

Fott, B. (1960). Zur Kenntniss der Gattung *Saturnella* (Chlorococcales). *Nova Hedwigia* **2,** 273–278.

Fott, B., and Ettl, H. (1959). Das Phytoplankton der Talsperre bei Sedlice. *Preslia* **31,** 213–246.

Fritsch, F. E. (1935). "The Structure and Reproduction of the Algae," Vol. I. Cambridge Univ. Press, London and New York.

Geitler, L. (1924). Ueber *Acantosphaera zachariasi* und *Calyptobactron indutum* nov. gen. et n. sp., zwei planktonische protococcaceen. *Öst. bot. Z.* **73,** 247–261.

Hegewald, E. (1976). A contribution to the algal flora of Jamaïca. *Nova Hedwigia* **28,** 45–69.

Hirose, H., and Yamagishi, T. (1977). "Illustrations of the Japanese Fresh-water Algae" (in Japanese). Uchidarokakuho Publ. Co., Tokyo.

Hortobagyi, T. (1962). Algen aus den Fischteichen von Buzsák. IV. *Nova Hedwigia* **4**, 21–53.

Hortobagyi, T. (1973). "The Microflora in the Settling and Subsoil Water Enriching Basins of the Budapest Waterworks." Akadémiai Kiadó, Budapest.

Komárek, J., and Fott, B. (1983). Das Phytoplankton des Süsswassers; von Huber-Pestalozzi. 7 Teil, 1.Hälfte. *In* "Die Binnengewässer" (A. Thienemann Schweizerbart, ed.), 1044 pp. Stuttgart.

Korshikov, O. A. (1928). Notes on some new or little known Protococcales. *Arch. Protistenkd.* **62**, 416–426.

Korshikov, O. A. (1953). Pidklas Protokokovi (Protococcineae). Vakuol'ni (Vacuolales) t² Protokokovi (Protococcales). *In* "Vznachnk Prisnovodnkh Vodorestej Ukrains'koi RSR," Vol. V, pp. 1–439. Acad. Nauk. USSR, Kiev.

Kouwets, F. A. C. (1980). Floristic and ecological notes on some little known unicellular and colony-forming algae from Dutch moorland pool complex. *Cryptogam.: Algol.* **1**, 293–309.

Lackey, J. B. (1942). The plankton algae and protozoa of two Tennessee rivers. *Am. Midl. Nat.* **27**, 191–202.

Lund, J. W. G. (1942). Contribution to our knowledge of British algae. VIII. *J. Bot., Br. Foreign* **80**, 57–73.

Nauwerck, A. (1962). Zur Systematik und Oekologie Portugiesicher Planktonalgen. *Mem. Soc. Broteriana* **15**, 5–54.

Nicholls, K. H., Nakamoto, L., and Heintsch, L. (1981). *Desmatractum spryii* sp. nov., a new member of the Chlorococcales and comments on related species. *Phycologia* **20**, 138–141.

Pascher, A. (1925). Heterokontae. *In* "Die Süsswasser-Flora Deutschlands, Österreichs und der Schweiz" (A. Pascher, ed), No. 11, pp. 1–118. Fischer, Jena.

Pascher, A. (1930). Ein grüner Sphagnum-Epiphyt und seine Beziehung zu freilebenden Verwandten (*Desmatractum, Calyptrobactron, Bernardinella*). *Arch. Protistenkd.* **69**, 637–658.

Philipose, M. T. (1967). "Chlorococcales." I.C.A.R. Monographs on Algae, New Delhi.

Printz, H. (1927). Chlorophyceae. *In* "Die natürlichen Pflanzenfamilien". (A. Engler and K. Prantl, eds.), Vol. 3, pp. 1–463. Engelmann, Leipzig.

Reymond, O. (1980). Contribution à l'étude de *Treubaria* Bernard (Chlorococcales, Chlorophyceae). *Candollea* **35**, 37–70.

Reymond, O. (1981). Contribution à l'étude de *Desmatractum* West & West (Chlorophyceae, Chlorococcales) au microscope électronique à transmission. *Arch. Sci.* **34**, 259–263.

Reymond, O. (1983). Introduction à l'ultrastructure de l'ornementation cellulaire et du pyrénoïde chez *Desmatractum delicatissimum* Koršikov (Chlorophyceae, Chlorococcales). *Arch. Sci.* **36**, 369–373.

Reymond, O., and Druart, J. C. (1980). *Desmatractum delicatissimum* Koršikov (Chlorococcales), première observation pour la France, et analogie avec *Treubaria* Bernard. *Cryptogam.: Algol.* **1**, 61–66.

Reymond, O., and Kouwets, F. A. C. (1981). Note sur l'écologie, l'ultrastructure et la taxonomie de l'algue unicellulaire *Desmatractum bipyramidatum* (Chodat) Pascher (= *Bernardinella bipyramidata*, Chodat), Chlorophyceae, Chlorococcales. *Arch. Sci.* **34**, 409–416.

Reymond, O., and Pickett-Heaps, J. D. (1983). A routine flat embedding method for electron microscopy of microorganisms allowing selection and precisely orientated sectioning of single cells by light microscopy. *J. Microsc. (Oxford)* **130**, 79–84.

Reymond, O., and Skogstad, A. (1983). Etude de quelques caractéristiques ultrastructurales et écologiques chez *Desmatractum spryii* Nicholls, Nakamoto & Heintsch (Chlorophyceae, Chlorococcales). *Arch. Sci.* **36**, 361–367.

Rino, J. A. (1972). Contribuiçao para o conhecimento das algas de agua doce de Moçambique. *Revta. Cienc. Biol., Ser. A* **2**, 51–102.
Silva, P. C. (1970). Remarks on algal nomenclature II. *Taxon* **19**, 941–945.
Skuja, H. (1964). Grundzüge der Algenflora und Algenvegetation der Fjeldgegenden um Abisko in Schwedisch-Lappland. *Nova Acta Reg. Soc. Sci. Ups.* [4] **18**(3), 1–465.
Smith, G. M. (1950). "The Freshwater Algae of the United States," (2nd ed.) McGraw-Hill, New York.
Thunmark, S. (1942). Ueber rezente Eisenocker und ihre Microorganismen-gemeinschaften. *Bull. Geol. Inst. Univ. Uppsala* **29**, 1–285.
West, G. S. (1916). "Algae," Vol. I. Cambridge Univ. Press, London and New York.
West, W., and West, G. S. (1902). A contribution to the freshwater algae of Ceylon. *Trans. Linn. Soc. London* [2] **6**(3), 123–215.
Wille, N. (1911). Conjugatae und Chlorophyceae. *In* "Die natürlichen Pflanzenfamilien" (A. Engler and K. Prantl, eds.), Part 1, Sect. 2, pp. 1–136, Engelmann, Leipzig.
Woloszynska, J. (1914). Studien über das Phytoplankton des Viktoriasees. *Hedwigia* **55**, 184–233.
Yamagishi, T. (1969). Unicellular and colonial Chlorophyceae in the Alaskan Arctic. *Gen. Educ. Rev. (Coll. Agric. Vet. Med., Nihon Univ.)* **5**, 18–29.

18 | A General Review on the Contribution of Chemotaxonomy to the Systematics of Green Algae

E. KESSLER

Institut für Botanik und Pharmazeutische Biologie, Universität Erlangen–Nürnberg, Erlangen, Federal Republic of Germany

Abstract: Chemotaxonomy can contribute to the systematics of green algae at three levels: separation of higher systematic categories; delimitation of species; and assessment of natural relationships. Chemical characters, i.e. pigments, storage products, and cell wall polysaccharides, have traditionally played a role in the separation of the green algae from other algal divisions. Thus, the Chlorophyta are characterized by the presence of chlorophyll *b*. Among the Chlorophyceae, some orders seem to have specific characters, the presence of siphonaxanthin in the Caulerpales being a typical example. Chemotaxonomy can make significant contributions to the taxonomy of morphologically ill-defined or highly variable microalgae. Thus, the taxonomy of the genus *Chlorella* has been clarified by means of biochemical and physiological characters. The natural relationships among species, genera, and possibly higher systematic categories can be assessed through studies of DNA hybridization. Work of this type has indicated that the genus *Chlorella* is heterogeneous and comprises at least three groups of species.

INTRODUCTION

Chemotaxonomy has a long tradition in bacteriology. For lack of sufficient morphological characters the utilization in bacterial taxonomy of various biochemical and physiological characters is an obvious necessity. Phycologists, on the other hand, were mainly concerned with macroalgae or with those microalgae which show visible microscopic or submicroscopic characters useful for a traditional taxonomic treatment. This has led to the

---

Systematics Association Special Volume No. 27, "Systematics of the Green Algae", edited by D. E. G. Irvine and D. M. John, 1984, pp. 391–407. Academic Press, London and Orlando.
ISBN 0 12 374040 1

Copyright © by the Systematics Association
All rights of reproduction in any form reserved

neglect by classical taxonomists of some simple, asexual microalgae, with the genus *Chlorella* being a very typical example (cf. Bold, 1968).

In the field of phycology, however, chemotaxonomy has its traditional place in the separation of the highest systematic categories of the algae. Thus, blue-green, red, brown and green algae are primarily characterized by the chemistry of their pigments, with reserve carbohydrates and cell wall polysaccharides playing an additional role (Klein and Cronquist, 1967; Lewin, 1968, 1974).

In recent times, modern methods of DNA, RNA, and protein biochemistry have provided the means for a quantitative assessment of the natural relationships among different organisms. In this area, again, bacteriology has assumed a leading position.

Thus, chemotaxonomy can contribute to the systematics of the green algae at three levels: it can help in the separation and characterization of higher systematic categories; it can provide valuable characters for the delimitation of morphologically ill-defined species of microalgae; and it can produce basic evidence for the evaluation of phylogenetic relationships among various taxa of the algae.

DELIMITATION OF HIGHER SYSTEMATIC CATEGORIES

The green algae or Chlorophyta (including the Chlorophyceae *sensu lato*, Prasinophyceae and Charophyceae) are characterized by a number of chemical properties which they have in common with mosses, ferns, and higher plants. These include the presence of chlorophyll *a* and *b* (Hager and Stransky, 1970; Meeks, 1974) and of the carotenoids β-carotin, lutein, neoxanthin, violaxanthin, antheraxanthin, and zeaxanthin (Hager and Stransky, 1970; Goodwin, 1974). Usually starch is the main storage product (Craigie, 1974; Manners and Sturgeon, 1982), and cellulose is often, but not always, the predominant cell wall polysaccharide (Mackie and Preston, 1974; Percival and McDowell, 1981).

The presence or absence of chlorophyll *b* has served as an especially important character for the assessment of the systematic position of controversial organisms. Thus, *Botryococcus* was found to contain chlorophyll *b* and starch and therefore to belong to the Chlorophyceae rather than to the Xanthophyceae (Belcher and Fogg, 1955). On the other hand, the lack of chlorophyll *b* supported the transfer of *Vaucheria* from the Chlorophyceae to the Xanthophyceae (Soma, 1960). The finding of chlorophyll *b* in *Prochloron* led to the establishment of the Prochlorophyta (Lewin, 1976; Withers *et al.*, 1978) and to the assumption that it is a prokaryotic precursor of the green algae. Recently, *Nanochlorum eucaryotum*, a newly-described and extremely small alga, was found to be eukaryotic and to contain chlorophyll *a* and *b* together with typical green algal carotenoids. As its DNA content is ten times higher than that of prokaryotic organisms, but much lower than that of eukaryotes (including *Chlorella*), *Nanochlorum* might be a primitive green alga assuming an intermediate position between prokaryotic and eukaryotic algae (Wilhelm *et al.*, 1982).

Recent systematic considerations, based on cytological and ultrastructural evidence, have led to the assumption that within the green algae two lines of evolution can be distinguished (Pickett-Heaps and Marchant, 1972; Pickett-Heaps, 1975; Stewart and Mattox, 1975; Moestrup, 1978). One line is supposed to lead from a *Chlamydomonas*-like organism to the Volvocales, Tetrasporales, Chlorococcales, some Ulotrichales, Ulvales, Chaetophorales, and Oedogoniales. The other line, which eventually evolved to the mosses and higher plants, would rise from primitive flagellates to the Conjugales, Klebsormidiales, Coleochaetales and Charales (with a side line possibly leading to the Caulerpales, Dasycladales and Siphonocladales). This scheme gains support from a variety of biochemical data. Thus, the algae of the second line seem to have in common with higher plants the presence of the peroxisomal enzyme glycolate oxidase (Frederick *et al.*, 1973) and of significant amounts of polyunsaturated C_{20} fatty acids (Attavian *et al.*, 1977); their superoxide dismutase, too, appears to be more closely related to that of the higher plants (Henry and Hall, 1977). The "lower" green algae, on the other hand, contain glycolate dehydrogenase instead of glycolate oxidase (Frederick *et al.*, 1973; Floyd and Salisbury, 1977; Bullock *et al.*, 1979). It should be stressed, however, that this biochemical evidence is rather fragmentary and based on the study of only a very limited range of species. More comprehensive and systematic studies are certainly necessary before the general validity of these results can be accepted.

According to their pigment composition, the Prasinophyceae appear to be heterogeneous. At least three different groups can be distinguished. Many of these flagellates exhibit a typical chlorophycean pigment pattern, some accumulate 2,4-divinyl-phaeoporphyrin a_5 monomethylester, and one group produces siphonaxanthin and siphonein (Ricketts, 1967, 1970, 1971), i.e. carotenoids that are otherwise typical of the Caulerpales (=Siphonales) among the Chlorophyceae. This might indicate that several lines of the green algae originated from prasinophycean flagellates (cf. Pickett-Heaps, 1975). On the other hand, the Prasinophyceae are characterized by the accumulation of mannitol rather than sucrose as a product of photosynthesis (Craigie *et al.*, 1967; Craigie, 1974; Kremer and Kirst, 1982). According to Leftley and Syrett (1973) and Bekheet and Syrett (1977), Prasinophyceae and Chlorophyceae exhibit different types of urea metabolism. Several Prasinophyceae, *Klebsormidium,* some Zygnematales and *Ulva* contain urease, like the higher plants. ATP:urea amidolyase, on the other hand, was found only in the Volvocales, Chlorococcales, Ulotrichales and Chaetophorales.

Some well-established orders of the Chlorophyceae also have special chemical properties. Thus, the Caulerpales are characterized by the presence of siphonaxanthin and siphonein, in addition to the normal chlorophycean carotenoids (Kleinig, 1969; Goodwin, 1974; Yokohama, 1981). Their cell wall contains mainly xylan and mannan (Frei and Preston, 1964, 1968; Mackie and Preston, 1974; Percival and McDowell, 1981). Some, but not all, Siphonocladales contain siphonaxanthin, but they lack siphonein (Kleinig, 1969; Goodwin, 1974). Their cell wall consists of crystalline cellulose similar to that of the higher plants (Mackie and Preston, 1974; Percival and McDowell, 1981). The Cladophorales produce as reserve carbohydrate an inulin-type fructan (Percival and Young, 1971; Craigie, 1974). Their cell wall contains crystalline cellulose; in addition, sulphated heteropolysaccharides have been found (cf. Percival and McDowell, 1981). The Dasycladales, too, have a fructan reserve carbohydrate (Percival and McDowell, 1967; Craigie, 1974). A mannan seems to be the predominant component of their cell wall (Frei and Preston, 1968; Percival and McDowell, 1981). Most of these results, however, are based on the study of only a few representatives of the respective orders, and for definitive chemotaxonomic conclusions to be reached much more comprehensive work will be required.

DELIMITATION OF GENERA AND SPECIES

Many common microalgae lack morphological characters suitable for a traditional classification. In addition, there are genera with rather conspicuous morphological features which, however, are so variable in response to environmental factors that they also present serious taxonomic problems. It is in this area, within the green algal orders represented mainly by the Chlorococcales and the Volvocales, that biochemical and physiological characters can play a decisive role in the delimitation of species.

1. *Chlorella*

The genus *Chlorella* comprises some of the commonest and most widely used green microalgae which have attained great importance not only in basic research but also, more recently, in various fields of technology. Physiologists found this genus quite frustrating due to the inability of classical taxonomy to cope with it, and the arbitrary labelling of the laboratory strains as "*C. vulgaris*", "*C. pyrenoidosa*", or "*C. ellipsoidea*". It was not appreciated until about 20 years ago that species could be defined on physiological and biochemical properties. Observations in our laboratory that two biochemical properties, namely hydrogenase activity and the production of secondary carotenoids under nitrogen-deficient conditions, were found in some, but not all strains of *Chlorella,* led to the development of a biochemical taxonomy for this genus (Kessler and Soeder, 1962; Kessler, 1967). In the course of time, the application of ten easily determined biochemical and physiological characters permitted the characterization and delimitation of 14 *Chlorella* taxa (Kessler, 1976, 1982a; Table I). The hydrogen-activating enzyme hydrogenase, which is active only in strictly anaerobic conditions, might be a biochemical relic from the early, anaerobic phase of life (cf. Kessler, 1974a) and is, therefore, of special taxonomic significance. The secondary carotenoids, which are produced by many algae under nitrogen-deficient conditions, are remarkable for two reasons. Firstly, they are synthesized only in a situation of extreme physiological stress and secondly, these carotenoids, which have been identified as astaxanthin, canthaxanthin, and echinenone (Czygan, 1968, 1982), are keto-carotenoids different from the normal "primary" carotenoids of the algae. It is interesting to note that the presence of sporopollenin, a polymeric carotenoid that was found to be a constituent of the cell wall of some strains of *Chlorella* and other algae (Atkinson *et al.,* 1972), seems to be restricted to those species which are able to synthesize secondary carotenoids (Burczyk and Hesse, 1981). Liquefaction of gelatin and hydrolysis of starch, characters commonly used in bacterial taxonomy, are due to the production of extracellular enzymes hydrolysing the respective polymeric substrates. Some *Chlorella* species ferment glucose to lactate under anaerobic conditions. Only one species, *C. protothecoides,* is unable to use nitrate as a source of nitrogen and requires thiamine for growth. These seven biochemical characters are of a qualitative nature, i.e. are either present or absent. The remaining three characters are, by contrast, quantitative and physiological. Acid tolerance and salt tolerance were found to be species-specific, the limits of growth ranging from pH 2.0 (*C. saccharophila*) to pH 6.0 (*C. homosphaera*) and from less than 1% ("0%") (*C. homosphaera*) to 5% NaCl (*C. luteoviridis*), respectively (see Table I). Whereas most *Chlorella* strains reach their upper limit of growth at 30–32°C, one species (*C. sorokiniana*) is thermophilic and capable of growth at temperatures up to 38–42°C. These data, based on the study of 88 strains, mainly from the Culture Collections at Göttingen, Cambridge, and Austin (Texas), were also used to devise a key for the identification of the *Chlorella* species (Kessler, 1978; Table II).

It should be stressed that most of these *Chlorella* species were first characterized and delimited according to biochemical and physiological criteria.

Table I. Biochemical and physiological characters of 14 *Chlorella* species (88 strains).*

Species	Hydr	Sec carot	Gelat liquef	Starch hydrol	Lact ferm	NO₃ red	Thiam	pH	NaCl (%)	Therm
C. sorokiniana Shih. & Krauss	+	−	−	−+	+	+	−	4.0	2	+
C. vulgaris Beijerinck	−	−	−	−	+	+	−	4.0	3	−
C. saccharophila (Krüger) Migula	−	−	−	+(−)	−(+)	+	−	2.0	4	−
C. 211-30 (=C. loboplora Andreyeva?)	−	−	−	+	−	+	−	4.5	0	−
C. fusca var. vacuolata Shih. & Krauss	+	+	+	+	+	+	−	3.5	3	−
C. fusca var. fusca Shih. & Krauss	+	+	+	−	+	+	−	4.0	2	−
C. fusca var rubescens (Dangeard) Kessler et al.	+	+	−	−	+	+	−	4.5	3	−
C. zofingiensis Dönz	−	+	−	−	+	+	−	5.0	1	−
C. minutissima Fott & Nováková	−	−	−	−	−	+	−	5.5	1	−
C. 211-11r (=C. mirabilis Andreyeva?)	−	−	−	+	+	+	−	4.5	1	−
C. homosphaera Skuja	+	+	−	−	−	+	−	6.0	0	−
C. kessleri Fott & Nováková	+	−	−	−	−	+	−	3.0	2	−
C. luteoviridis Chodat	−	−	−	−	−	+	−	3.0	5	−
C. protothecoides Krüger	−	−	−	−	+	−	+	4.0	4	−

*Abbreviations: Hydr, hydrogenase; Sec carot, secondary carotenoids; Gelat liquef, liquefaction of gelatin; Starch hydrol, hydrolysis of starch; Lact ferm, lactate fermentation; NO₃ red, nitrate reduction; Thiam, thiamine requirement; pH, acid tolerance (lower pH limit); NaCl %, salt tolerance; and Therm, thermophily (growth at 38°C).

Table II. A key for the identification of 14 *Chlorella* taxa (from Kessler, 1978).

1. Hydrogenase activity in anaerobiosis
    2. Synthesis of secondary carotenoids under nitrogen deficiency
        3. Liquefaction of gelatin
            4. Starch hydrolysis: *C. fusca* var. *vacuolata*
            4. No starch hydrolysis: *C. fusca* var. *fusca*
        3. No liquefaction of gelatin
            4. Limits of growth at pH 4.5 and at 3% NaCl: *C. fusca* var. *rubescens*
            4. Limits of growth at pH 6.0 and at <1% NaCl: *C. homosphaera*
    2. No secondary carotenoids synthesized under nitrogen deficiency
        3. Limit of growth at pH 3.0: *C. kessleri*
        3. Limit of growth at pH 4.0–4.5, thermophilic (growth at 38°C): *C. sorokiniana*
1. No hydrogenase activity in anaerobiosis
    2. Synthesis of secondary carotenoids under nitrogen deficiency: *C. zofingiensis*
    2. No secondary carotenoids under nitrogen deficiency
        3. No reduction of nitrate, thiamine requirement: *C. protothecoides*
        3. Reduction of nitrate, no thiamine requirement
            4. Limits of growth at pH 2.0–2.5 and at 3–4% NaCl: *C. saccharophila*
            4. Limits of growth at pH 3.0 and at 5% NaCl: *C. luteoviridis*
            4. Limit of growth at pH 4.0–4.5
                5. No starch hydrolysis, limit of growth at 3–4% NaCl: *C. vulgaris*
                5. Starch hydrolysis, limit of growth at 0–1% NaCl: *C.* spec. (211-11r; 211-30)*
            4. Limits of growth at pH 5.5 and at 1% NaCl: *C. minutissima*

---

* These two strains belong to different species which can be distinguished only according to lactate fermentation and DNA hybridization (see Tables I, III).

Fott and Nováková (1969), however, were able to show that these taxa have also species-specific morphological characters (cf. also Andreyeva, 1975) which are in complete agreement with the biochemical data. This is especially important in connection with the problem of mutations. It has been known for a long time that it is rather easy to obtain, through physical or chemical mutagenesis, pigment (and other) mutants of *Chlorella*. Thus, many mutant strains of *C. fusca* var. *vacuolata* have lost their abilities to synthesize secondary carotenoids and to liquefy gelatin (Kessler and Czygan, 1966). This has led to occasional doubts concerning the validity of biochemical characters that can easily mutate (e.g. Weber and Wettern, 1980; Czygan, 1982). It is obvious, however, that any character, no matter whether biochemical or morphological, is in principle susceptible to mutation, and it is only a combination of several characters that decides the assignment of an organism to a specific taxon.

Some other physiological characters, which are very popular in bacterial taxonomy, have been found to be unsuitable for taxonomic purposes in the genus *Chlorella*. The utilization of organic compounds as sources of carbon or nitrogen for growth is very variable and more or less strain-specific (Kessler and Czygan, 1970; Kessler, 1972a). Attempts to use such properties for the taxonomy of *Chlorella* have led to excessive "splitting" (30 taxa for 41 strains studied by Shihira and Krauss, 1965).

Of special interest is the assignment of some ecologically remarkable *Chlorella* strains. Thus, the "high-temperature" strains were found to be *C. sorokiniana* (Kessler, 1972b), whereas some strains isolated from marine environments belong to various salt-tolerant species, i.e. *C. vulgaris*, *C. saccharophila* and *C. fusca* var. *rubescens* (Kessler, 1974b). Two "zoochlorella" strains from the freshwater sponge *Spongilla* were identified as *C. sorokiniana* (= *C. vulgaris* f. *tertia*) (Kessler, 1972b) and a strain from *Heterostegina depressa* (foraminifera) as *C. saccharophila* (Lee et al., 1982). The symbiont from *Paramecium* seems to be related to *C. vulgaris* and *C. sorokiniana* (Reisser, 1975). The *Chlorella* phycobionts from several lichens belong to *C. saccharophila* (Tschermak-Woess, 1949, 1978) which includes also the species formerly known as *C. ellipsoidea* (Kessler, 1967; Fott and Nováková, 1969).

2. Prototheca

The genus *Prototheca* is generally assumed to be the colourless equivalent of *Chlorella*, and its taxonomy is faced with similar problems. Traditionally, minor morphological differences, combined with the pattern of utilization of organic carbon sources for growth, were used for the delimitation of about 15 species. Many of them, however, were later found to be synonymous (cf. Arnold and Ahearn, 1972), and recently most authors tend to accept only three species as valid (DeCamargo and Fischman, 1979; Padhye et al., 1979, cf. Kessler, 1982a). A critical examination of the utilization of organic carbon sources and of its value as a taxonomic character revealed many uncertainties and contradictions in the literature (Kessler, 1982b). The only reliable criterion seems to be the ability of *P. wickerhamii*, in contrast to *P. zopfii* and *P. stagnora*, to grow with trehalose as a source of carbon (DeCamargo and Fischman, 1979; Padhye et al., 1979; Kessler, 1982b). In addition, *P. wickerhamii* is less salt-tolerant (Kessler, 1977).

3. Scenedesmus

The coenobia of the genus *Scenedesmus* exhibit a variety of microscopic and submicroscopic characters. Some of them, however, are extremely variable, and this may in part be responsible for the fact that over 1000 taxa have been described (cf. Hegewald and Schnepf, 1979). Physiologically and biochemically, however, the 28 *Scenedesmus* strains examined so far appear rather homogeneous. All of them contain hydrogenase, produce secondary carotenoids under nitrogen-deficient conditions and are able to liquefy gelatin (Kessler and Czygan, 1967). Only the absence of starch hydrolysis in some species and certain differences in acid tolerance can serve as taxonomic characters (Kessler, 1980; cf. Hegewald, 1982; Kessler, 1982a). The polyamine pattern of a more comprehensive selection of 103 strains, however, proved again to be very uniform (Kneifel and Hegewald, 1980).

4. *Ankistrodesmus, Monoraphidium, and Keratococcus*

The genera *Ankistrodesmus* (colony-forming) and *Monoraphidium* (solitary-celled) (Komárková-Legnerová, 1969; Hindák, 1970) combine again rather evident morphological characters with a high degree of variability. A sample of 17 strains was found to be uniform in the possession of hydrogenase, the ability to synthesize secondary carotenoids, the ability to liquefy gelatin (Kessler and Czygan, 1967) and the inability to hydrolyse starch (Kessler, 1980). Only some differences in acid tolerance could be used as a physiological character (Kessler, 1980, 1982a). *Keratococcus bicaudatus* (the strain was originally labelled *Ankistrodesmus stipitatus*), on the other hand, could be clearly separated from *Ankistrodesmus* and *Monoraphidium*, as it has no hydrogenase and is unable to liquefy gelatin, but is able to hydrolyse starch (Kessler and Czygan, 1967; Kessler, 1980).

5. *Other Genera*

Certain biochemical and physiological characters have been used to support the taxonomy of a number of other genera of the Chlorococcales (cf. Kessler, 1982a): *Chlorococcum* (Bold and Parker, 1962; Czygan, 1968; McLean, 1968; Archibald and Bold, 1970; Thomas and Brown, 1970a), *Spongiococcum* and *Neospongiococcum* (Deason and Cox, 1971; Deason, 1976; Deason et al., 1977), *Characium, Pseudochlorococcum* and related algae (Archibald, 1970; Lee and Bold, 1974), *Crucigeniella, Crucigenia* and *Tetrastrum* (Hegewald and Kneifel, 1981), *Coelastrum* (Sodomková, 1969) and *Protosiphon* (Thomas and Brown, 1970b).

Among the Volvocales, the genus *Chlamydomonas* comprises a very large number of taxa. Here, too, the presence or absence of hydrogenase (cf. Kessler, 1974a) and of the ability to synthesize secondary carotenoids (Czygan, 1968), so far studied in only a very limited number of strains, might be useful chemotaxonomic characters. In addition, isoenzyme patterns (Thomas and Delcarpio, 1971), cell wall–degrading autolysines (Schlösser, 1976) and acid tolerance (Cassin, 1974) can be significant for the characterization of *Chlamydomonas* species.

A number of physiological properties have been used as supplementary characters in the taxonomy of the Chlorosarcinales (Groover and Bold, 1969). The finding of siphonaxanthin in *Microthamnion kuetzingianum* (Weber and Czygan, 1972) shows that this unusual carotenoid is not restricted to the Siphonales and to some members of the Siphonocladales and Prasinophyceae.

DETERMINATION OF NATURAL RELATIONSHIPS

Modern molecular biology has provided a number of methods which can be used for the assessment of phylogenetic relationships among different organisms. These include DNA hybridization, studies of ribosomal RNA homology, serological studies of homologous proteins and sequencing of certain proteins (e.g. cytochrome c). Their application in the field of algal systematics, however, is only in its very beginning.

Usually, the first step in the study of DNA is the determination of its base composition, i.e. its guanine + cytosine (GC) content. In bacterial taxonomy this has become a prominent species-specific character. Such data for the genus *Chlorella* (Hellmann and Kessler, 1974a) are shown in Table III. In most species the GC values are very uniform, but in some instances

Table III. Base composition of the DNA (% GC) of 14 *Chlorella* species (88 strains) and hybridization of their DNA with labelled DNA from *C. vulgaris* 211-8m and *C. fusca* var. *vacuolata* 211-8b.

Species (strains)	DNA % GC*	% Hybridization with *C. vulgaris**	% Hybridization with *C. fusca* var. *vacuolata**
C. vulgaris (17)	62	94, (64, 35)	34, (20, 13)
C. sorokiniana (16)	65, (55, 68, 78)	32, (28)	28, (16)
C. saccharophila (8)	50, (57)	(32), 17	(24), 7
C. 211-30 (=*C. lobophora*?) (1)	57	27	18
C. fusca var. *vacuolata* (10)	50	28, (14, 11)	101, (63, 30)
C. zofingiensis (4)	49, (64)	27, (13)	52, (44)
C. fusca var. *rubescens* (1)	58	27	23
C. fusca var. *fusca* (1)	55	15	17
C. luteoviridis (6)	44	16	14
C. 211-11r (=*C. mirabilis*?) (1)	58	13	11
C. minutissima (1)	43	12	10
C. kessleri (7)	56	9	8
C. homosphaera (1)	69	7	6
C. protothecoides (14)	62	8	6

* Mean values given; figures in parentheses are for minor groups of strains.

there are several distinct groups within one taxon (e.g. *C. sorokiniana*). It is striking to note that the values found within the genus *Chlorella* range from 43 to 78% GC (corresponding data for *Scenedesmus* 52–62% GC, and for *Ankistrodesmus/Monoraphidium* 63–70% GC, Hellmann and Kessler, 1974b; for *Prototheca* 62–78% GC, Kerfin and Kessler, 1978b). This indicates that the genus *Chlorella* in its present sense in part combines unrelated species.

Studies of the hybridization of the DNA from 88 *Chlorella* strains with ^{3}H-labelled DNA from *C. vulgaris* and *C. fusca* var. *vacuolata* (Kerfin and Kessler, 1978a) revealed the existence of three groups of species (Table III). The "*C. vulgaris* group" shows the highest degree of hybridization with the DNA from *C. vulgaris,* the "*C. fusca* group" exhibits a close relationship with *C. fusca* var. *vacuolata,* and the third group, with less than 20% hybridization with the DNA from either species, appears completely unrelated. Here, again, we find some groups of strains with different degrees of DNA hybridization within otherwise homogeneous taxa. This indicates the appearance of infraspecific differences at the level of the DNA. According to these results, the genus *Chlorella* seems to consist of at least three groups of species, and it cannot at present be excluded that the "third group" of species is in itself heterogeneous (cf. also Kessler, 1982a). In this connection it is interesting to note that, according to DNA hybridization, *C. fusca* var. *fusca* does not seem to be related to *C. fusca* var. *vacuolata* (see Table III). This is in agreement with the results of Fott *et al.* (1975), who showed that *C. fusca* var. *fusca* belongs, according to submicroscopic features of its cell wall, to the genus *Scenedesmus* (cf. also Hegewald and Schnepf, 1979).

Immunological studies on the soluble cytochromes *c*-553 from a number of *Chlorella* strains confirmed the close relationship of *C. fusca* var. *vacuolata, C. fusca* var. *rubescens, C. zofingiensis,* and—in contrast to the data from DNA hybridization—also *C. fusca* var. *fusca* (Kümmel and Kessler, 1980). Interestingly, the cytochromes *c*-553 of these *Chlorella* species were found to be immunologically identical with that from *Scenedesmus obliquus* (Kümmel and Kessler, 1980). This seems to indicate that not only *C. fusca* var. *fusca,* but rather the whole "*C. fusca* group" might be closely related to the genus *Scenedesmus*. Further studies on DNA hybridization, however, are necessary for a solution of these problems, and also for the clarification of the much discussed but still not proven relationship of the genera *Chlorella* (especially *C. protothecoides,* cf. Kessler 1982a,b) and *Prototheca*.

In addition, some results have been obtained with other green algae. *Chlamydomonas geitleri* and *C. reinhardtii* have a distinctly different base composition of their DNA (Tetík and Zadražil, 1982). The same is true for *Pediastrum boryanum* and *P. duplex* (Chang, 1981). Hybridization studies of DNA and ribosomal RNA indicate, quite in agreement with taxonomic expectations, that *Dunaliella* is more closely related to *Chlamydomonas* than is *Volvox* (Soh and Kochert, 1979). The similarity of ribosomal proteins from various Volvocales confirms their close relationship (Götz and Arnold, 1980b) and supports the generally assumed kinship of the colourless genus *Polytoma* with *Chlamydomonas* (Spiess and Arnold, 1976; Götz and Arnold, 1980a). Brown and Walne (1967) reported on the serological relationships among different strains and species of *Chlamydomonas*.

With regard to the removal of *Klebsormidium* from the Ulotrichales (Pickett-Heaps and Marchant, 1972; Stewart and Mattox, 1975), it is interesting to note that studies of DNA-rRNA hybridization indicate that *Ulothrix* and *Klebsormidium* are indeed not closely related (Soh and Kochert, 1979).

CONCLUSIONS

The results reviewed here show that at the level of higher systematic categories chemotaxonomy has mainly contributed supporting evidence confirming taxonomic decisions previously based on morphological and ultrastructural data. In the characterization and delimitation of the species of critical genera of morphologically ill-defined microalgae, however, chemotaxonomic work can produce a sound basis for taxonomy. The application of modern methods of DNA, RNA, and protein biochemistry, finally, brings systematics to the molecular level where ultimately all differences between the taxa have their foundation, and provides evidence that cannot otherwise be obtained. With the further development and improvement of such methods, and with their increasing availability to taxonomists, we can expect to obtain in the future most interesting and fundamental insights as to the evolutionary relationships among species, genera, and higher systematic categories of green algae.

REFERENCES

Andreyeva, V. M. (1975). "Rod *Chlorella*." Nauka, Leningrad.
Archibald, P. A. (1970). *Pseudochlorococcum*, a new chlorococcalean genus. *J. Phycol.* **6**, 127–132.
Archibald, P. A. and Bold, H. C. (1970). Phycological studies. XI. The genus *Chlorococcum* Meneghini. *Univ. Tex. Publ.* **7015**, 1–115.
Arnold, P. and Ahearn, D. G.(1972). The systematics of the genus *Prototheca* with a description of a new species *P. filamenta*. *Mycologia* **64**, 265–275.
Atkinson, A. W., Gunning, B. E. S. and John, P. C. L. (1972). Sporopollenin in the cell wall of *Chlorella* and other algae: ultrastructure, chemistry, and incorporation of ^{14}C-acetate, studied in synchronous cultures. *Planta* **107**, 1–32.
Attavian, B. N., Floyd, G. L. and Fairbrothers, D. E. (1977). Fatty acids of filamentous green algae. *Biochem. Syst. Ecol.* **5**, 65–69.

Bekheet, I. A. and Syrett, P. J. (1977). Urea-degrading enzymes in algae. *Br. phycol. J.* **12**, 137–143.
Belcher, J. H. and Fogg, G. E. (1955). Biochemical evidence of the affinities of *Botryococcus*. *New Phytol.* **54**, 81–83.
Bold, H. C. (1968). Some reflections on four decades of phycology, 1927–1967. *In* "Algae, Man, and the Environment" (D. F. Jackson, ed), pp. 1–13. Syracuse Univ. Press, Syracuse, New York.
Bold, H. C. and Parker, B. C. (1962). Some supplementary attributes in the classification of *Chlorococcum* species. *Arch. Mikrobiol.* **42**, 267–288.
Brown, R. M. and Walne, P. L. (1967). Comparative immunology of selected wild types, varieties and mutants of *Chlamydomonas*. *J. Protozool.* **14**, 365–373.
Bullock, K. W., Deason, T. R. and O'Kelley, J. C. (1979). Occurrence of glycolate dehydrogenase and glycolate oxidase in some coccoid, zoospore-producing green algae. *J. Phycol.* **15**, 142–146.
Burczyk, J. and Hesse, M. (1981). The ultrastructure of the outer cell wall-layer of *Chlorella* mutants with and without sporopollenin. *Plant Syst. Evol.* **138**, 121–137.
Cassin, P. E. (1974). Isolation, growth, and physiology of acidophilic chlamydomonads. *J. Phycol.* **10**, 439–447.
Chang, T.-P. (1981). A comparative study of *Pediastrum boryanum* (Turp.) Menegh. and *Pediastrum duplex* Meyen. *Arch. Protistenkd.* **124**, 232–243.
Craigie, J. S.(1974). Storage products. *In* "Algal Physiology and Biochemistry" (W. D. P. Stewart, ed), pp. 206–235. Blackwell, Oxford.
Craigie, J. S., McLachlan, J., Ackman, R. G. and Tocher, C. S. (1967). Photosynthesis in algae. III. Distribution of soluble carbohydrates and dimethyl-β-propiothetin in marine unicellular Chlorophyceae and Prasinophyceae. *Can. J. Bot.* **45**, 1327–1334.
Czygan, F.-C. (1968). Sekundär-Carotinoide in Grünalgen. I. Chemie, Vorkommen und Faktoren, welche die Bildung dieser Polyene beeinflussen. *Arch. Mikrobiol.* **61**, 81–102.
Czygan, F.-C. (1982). Primäre und sekundäre Carotinoide in chlorokokkalen Algen. *Algol. Stud.* **29**, 470–488.
Deason, T. R. (1976). The genera *Spongiococcum* and *Neospongiococcum* (Chlorophyceae, Chlorococcales). III. New species, biochemical characteristics and a summary key. *Phycologia* **15**, 197–213.
Deason, T. R. and Cox, E. R. (1971). The genera *Spongiococcum* and *Neospongiococcum*. II. Species of *Neospongiococcum* with labile walls. *Phycologia* **10**, 255–262.
Deason, T. R., Czygan, F.-C. and Soeder, C. J. (1977). Taxonomic significance of secondary carotenoid formation in *Neospongiococcum* (Chlorococcales, Chlorophyta). *J. Phycol.* **13**, 176–180.
DeCamargo, Z. P. and Fischman, O. (1979). Use of morpho-physiological characteristics for differentiation of the species of *Prototheca*. *Sabouraudia* **17**, 275–278.
Floyd, G. L. and Salisbury, J. L. (1977). Glycolate dehydrogenase in primitive green algae. *Am. J. Bot.* **64**, 1294–1296.
Fott, B. and Nováková, M. (1969). A monograph of the genus *Chlorella*. The fresh water species. *In* "Studies in Phycology" (B. Fott, ed), pp. 10–74. Academia, Prague.
Fott, B., Lochhead, R. and Clémençon, H. (1975). Taxonomie der Arten *Chlorella ultrasquamata* Clém. et Fott und *Chlorella fusca* Shih. et Krauss. *Arch. Protistenkd.* **117**, 288–296.
Frederick, S. E., Gruber, P. J. and Tolbert, N. E. (1973). The occurrence of glycolate dehydrogenase and glycolate oxidase in green plants. An evolutionary survey. *Plant Physiol.* **52**, 318–323.
Frei, E. and Preston, R. D.(1964). Non-cellulosic structural polysaccharides in algal cell walls. I. Xylan in siphoneous green algae. *Proc. R. Soc. London, Ser. B* **160**, 293–313.

Frei, E. and Preston, R. D. (1968). Non-cellulosic structural polysaccharides in algal cell walls. III. Mannan in siphoneous green algae. *Proc. R. Soc. London, B* **169**, 127–145.
Goodwin, T. W. (1974). Carotenoids and biliproteins. *In* "Algal Physiology and Biochemistry" (W. D. P. Stewart, ed), pp. 176–205. Blackwell, Oxford.
Götz, H. and Arnold, C.-G. (1980a). Comparative electrophoretic study on ribosomal proteins from algae. *Planta* **149**, 19–26.
Götz, H. and Arnold, C.-G. (1980b). Analysis of ribosomal proteins from various species of algae. Comparative electrophoretic study on proteins from chloroplast ribosomes. *Biochem. Physiol. Pflanz.* **175**, 1–8.
Groover, R. D. and Bold, H. C.(1969). Phycological studies. VIII. The taxonomy and comparative physiology of the Chlorosarcinales and certain other edaphic algae. *Univ. Tex. Publ.* **6907**, 1–165.
Hager, A. and Stransky, H. (1970). Das Carotinoidmuster und die Verbreitung des lichtinduzierten Xanthophyllcyclus in verschiedenen Algenklassen. III. Grünalgen. *Arch. Mikrobiol.* **72**, 68–83.
Hegewald, E. (1982). Taxonomisch-morphologische Untersuchungen von *Scenedesmus*-Isolaten aus Stammsammlungen *Algol. Stud.* **29**, 375–406.
Hegewald, E. and Kneifel, H. (1981). Amine in Algen. V. Das Vorkommen von Norspermidin und anderen Polyaminen in einigen Grünalgen. *Algol. Stud.* **28**, 313–323.
Hegewald, E. and Schnepf, E. (1979). Geschichte und Stand der Systematik der Grünalgengattung *Scenedesmus*. *Schweiz. Z. Hydrobiol.* **40**, 320–343.
Hellmann, V. and Kessler, E. (1974a). Physiologische und biochemische Beiträge zur Taxonomie der Gattung *Chlorella*. VIII. Die Basenzusammensetzung der DNS. *Arch. Microbiol.* **95**, 311–318.
Hellmann, V. and Kessler, E. (1974b). Physiologische und biochemische Beiträge zur Taxonomie der Gattungen *Ankistrodesmus* und *Scenedesmus*. III.Die Basenzusammensetzung der DNS. *Arch. Microbiol.* **100**, 239–242.
Henry, L. E. A. and Hall, D. O. (1977). Superoxide dismutases in green algae: An evolutionary survey. *In* "Photosynthetic Organelles" (S. Miyachi, S. Katoh, Y. Fujita, and K. Shibata, eds), pp. 377–382. Jpn. Soc. Plant Physiol., Tokyo.
Hindák, F. (1970). A contribution to the systematics of the family Ankistrodesmaceae (Chlorophyceae). *Algol. Stud.* **1**, 7–32.
Kerfin, W., and Kessler, E. (1978a). Physiological and biochemical contributions to the taxonomy of the genus *Chlorella*. XI. DNA hybridization. *Arch. Microbiol.* **116**, 97–103.
Kerfin, W., and Kessler, E. (1978b). Physiological and biochemical contributions to the taxonomy of the genus *Prototheca*. II. Starch hydrolysis and base composition of DNA. *Arch. Microbiol.* **116**, 105–107.
Kessler, E. (1967). Physiologische und biochemische Beiträge zur Taxonomie der Gattung *Chlorella*. III. Merkmale von 8 autotrophen Arten. *Arch. Mikrobiol.* **55**, 346–357.
Kessler, E. (1972a). Physiologische und biochemische Beiträge zur Taxonomie der Gattung *Chlorella*. VI. Verwertung organischer Kohlenstoff-Verbindungen. *Arch. Mikrobiol.* **85**, 153–158.
Kessler, E. (1972b). Physiologische und biochemische Beiträge zur Taxonomie der Gattung *Chlorella*. VII. Die Thermophilie von *Chlorella vulgaris* f. *tertia* Fott et Nováková. *Arch. Mikrobiol.* **87**, 243–248.
Kessler, E. (1974a). Hydrogenase, photoreduction and anaerobic growth. *In* "Algal Physiology and Biochemistry" (W. D. P. Stewart, ed), pp. 456–473. Blackwell, Oxford.
Kessler, E. (1974b). Physiologische und biochemische Beiträge zur Taxonomie der Gattung *Chlorella*. IX. Salzresistenz als taxonomisches Merkmal. *Arch. Microbiol.* **100**, 51–56.

Kessler, E. (1976). Comparative physiology, biochemistry, and the taxonomy of *Chlorella* (Chlorophyceae). *Plant Syst. Evol.* **125**, 129–138.

Kessler, E. (1977). Physiological and biochemical contributions to the taxonomy of the genus *Prototheca*. I. Hydrogenase, acid tolerance, salt tolerance, thermophily, and liquefaction of gelatin. *Arch. Microbiol.* **113**, 139–141.

Kessler, E. (1978). Physiological and biochemical contributions to the taxonomy of the genus *Chlorella*. XII. Starch hydrolysis and a key for the identification of 13 species. *Arch. Microbiol.* **119**, 13–16.

Kessler, E. (1980). Physiological and biochemical contributions to the taxonomy of the genera *Ankistrodesmus* and *Scenedesmus*. V. Starch hydrolysis and new assignment of strains. *Arch. Microbiol.* **126**, 11–14.

Kessler, E. (1982a). Chemotaxonomy in the Chlorococcales. *Prog. Phycol. Res.* **1**, 111–135.

Kessler, E. (1982b). Physiological and biochemical contributions to the taxonomy of the genus *Prototheca*. III. Utilization of organic carbon and nitrogen compounds. *Arch. Microbiol.* **132**, 103–106.

Kessler, E. and Czygan, F.-C. (1966). Physiologische und biochemische Beiträge zur Taxonomie der Gattung *Chlorella*. II. Untersuchungen an Mutanten. *Arch. Mikrobiol.* **54**, 37–45.

Kessler, E. and Czygan, F.-C. (1967). Physiologische und biochemische Beiträge zur Taxonomie der Gattungen *Ankistrodesmus* und *Scenedesmus*. I. Hydrogenase, Sekundär-Carotinoide und Gelatine-Verflüssigung. *Arch. Mikrobiol.* **55**, 320–326.

Kessler, E. and Czygan, F.-C. (1970). Physiologische und biochemische Beiträge zur Taxonomie der Gattung *Chlorella*. IV. Verwertung organischer Stickstoffverbindungen. *Arch. Mikrobiol.* **70**, 211–216.

Kessler, E. and Soeder, C. J. (1962). Biochemical contributions to the taxonomy of the genus *Chlorella*. *Nature (London)* **194**, 1096–1097.

Klein, R. M. and Cronquist, A. (1967). A consideration of the evolutionary and taxonomic significance of some biochemical, micromorphological, and physiological characters in the thallophytes. *Q. Rev. Biol.* **42**, 105–296.

Kleinig, H. (1969). Carotenoids of siphonous green algae: A chemotaxonomical study. *J. Phycol.* **5**, 281–284.

Kneifel, H. and Hegewald, E. (1980). Amines in algae. IV. Norspermidine and other polyamines in the genus *Scenedesmus*. *Algol. Stud.* **26**, 87–96.

Komárková-Legnerová, J. (1969). The systematics and ontogenesis of the genera *Ankistrodesmus* Corda and *Monoraphidium* gen. nov. *In* "Studies in Phycology" (B. Fott, ed), pp. 75–144. Academia, Prague.

Kremer, B. P. and Kirst, G. O. (1982). Biosynthesis of photosynthates and taxonomy of algae. *Z. Naturforsch., C: Biosci.* **37C**, 761–771.

Kümmel, H. and Kessler, E. (1980). Physiological and biochemical contributions to the taxonomy of the genus *Chlorella*. XIII. Serological studies. *Arch. Microbiol.* **126**, 15–19.

Lee, J. J., Reidy, J. and Kessler, E. (1982). Symbiotic *Chlorella* species from larger foraminifera. *Bot. Mar.* **25**, 171–176.

Lee, K. W. and Bold. H. C. (1974). Phycological studies. XII. *Characium* and some *Characium*-like algae. *Univ. Tex. Publ.* **7403**, 1–127.

Leftley, J. W. and Syrett, P. J. (1973). Urease and ATP:urea amidolyase activity in unicellular algae. *J. Gen. Microbiol.* **77**, 109–115.

Lewin, R. A. (1968). Biochemistry and physiology of algae: Taxonomic and phylogenetic considerations. *In* "Algae, Man, and the Environment" (D. F. Jackson, ed), pp. 15–26. Syracuse Univ. Press, Syracuse, New York.

Lewin, R. A. (1974). Biochemical taxonomy. In "Algal Physiology and Biochemistry" (W. D. P. Stewart, ed), pp. 1–39. Blackwell, Oxford.

Lewin, R. A. (1976). Prochlorophyta as a proposed new division of algae. *Nature (London)* **261**, 697–698.

Mackie, W. and Preston, R. D. (1974). Cell wall and intercellular region polysaccharides. In "Algal Physiology and Biochemistry" (W. D. P. Stewart, ed), pp. 40–85. Blackwell, Oxford.

McLean, R. J. (1968). New taxonomic criteria in the classification of *Chlorococcum* species. I. Pigmentation. *J. Phycol.* **4**, 328–332.

Manners, D. J. and Sturgeon, R. J. (1982). Reserve carbohydrates of algae, fungi, and lichens. *Encycl. Plant Physiol., New Ser.* **13A**, 472–514.

Meeks, J. C. (1974). Chlorophylls. In "Algal Physiology and Biochemistry" (W. D. P. Stewart, ed), pp. 161–175. Blackwell, Oxford.

Moestrup, Ø. (1978). On the phylogenetic validity of the flagellar apparatus in green algae and other chlorophyll a and b containing plants. *BioSystems* **10**, 117–144.

Padhye, A. A., Baker, J. G. and D'Amato, R. F. (1979). Rapid identification of *Prototheca* species by the API 20C system. *J. Clin. Microbiol.* **10**, 579–582.

Percival, E. and McDowell, R. H. (1967). "Chemistry and Enzymology of Marine Algal Polysaccharides." Academic Press, London.

Percival, E. and McDowell, R. H. (1981). Algal walls—composition and biosynthesis. *Encycl. Plant Physiol., New Ser.* **13B**, 277–316.

Percival, E. and Young, M. (1971). Low molecular weight carbohydrates and water-soluble polysaccharide metabolized by the Cladophorales. *Phytochemistry* **10**, 807–812.

Pickett-Heaps, J. D. (1975). "Green Algae: Structure, Reproduction and Evolution in Selected Genera." Sinauer Assoc., Sunderland, Massachusetts.

Pickett-Heaps, J. D. and Marchant, H. J. (1972). The phylogeny of the green algae: A new proposal. *Cytobios* **6**, 255–264.

Reisser, W. (1975). Zur Taxonomie einer auxotrophen *Chlorella* aus *Paramecium bursaria* Ehrbg. *Arch. Microbiol.* **104**, 293–295.

Ricketts, T. R. (1967). Further investigations into the pigment composition of green flagellates possessing scaly flagella. *Phytochemistry* **6**, 1375–1386.

Ricketts, T. R. (1970). The pigments of the Prasinophyceae and related organisms. *Phytochemistry* **9**, 1835–1842.

Ricketts, T. R. (1971). Identification of xanthophylls KI and KIS of the Prasinophyceae as siphonein and siphonaxanthin. *Phytochemistry* **10**, 161–164.

Schlösser, U. G. (1976). Entwicklungsstadien- und sippenspezifische Zellwand-Autolysine bei der Freisetzung von Fortpflanzungszellen in der Gattung *Chlamydomonas*. *Ber. dt. bot. Ges.* **89**, 1–56.

Shihira, I. and Krauss, R. W. (1965). "*Chlorella*. Physiology and Taxonomy of Forty-one Isolates." University of Maryland, College Park.

Sodomková, M. (1969). Physiologische und biochemische Charakterisierung einiger *Coelastrum*-Arten. *Arch. Protistenkd.* **111**, 223–227.

Soh, G. L. and Kochert, G. (1979). Nucleotide sequence homology of some algal ribosomal RNAs. *J. Phycol.* **15**, 276–279.

Soma, S. (1960). Chlorophyll in *Vaucheria* as a clue for the determination of its phylogenetic position. *J. Fac. Sci., Univ. Tokyo Sect. 3* **7**, 535–542.

Spiess, H. and Arnold. C. G. (1976). Comparative investigation of the cytoplasmic ribosomal proteins of *Chlamydomonas reinhardii* and *Polytoma mirum*. *Plant Sci. Lett.* **6**, 267–271.

Stewart, K. D. and Mattox, K. R. (1975). Comparative cytology, evolution and classification of the green algae with some consideration of the origin of other organisms with chlorophylls a and b. *Bot. Rev.* **41**, 104–135.

Tetík, K. and Zadražil, S. (1982). Characterization of DNA of the alga *Chlamydomonas geitleri* Ettl. *Biol. Plant.* **24**, 202–210.

Thomas, D. L. and Brown, R. M. (1970a). New taxonomic criteria in the classification of *Chlorococcum* species. III. Isozyme analysis. *J. Phycol.* **6**, 293–299.

Thomas, D. L. and Brown, R. M. (1970b). Isoenzyme analysis and morphological variation of thirty-two isolates of *Protosiphon*. *Phycologia* **9**, 285–292.

Thomas, D. L. and Delcarpio, J. B. (1971). Electrophoretic analysis of enzymes from three species of *Chlamydomonas*. *Am. J. Bot.* **58**, 716–720.

Tschermak-Woess, E. (1949). Über wenig bekannte und neue Flechtengonidien. I. *Chlorella ellipsoidea* Gerneck, als neue Flechtenalge. *Oesterr. Bot. Z.* **95**, 341–343.

Tschermak-Woess, E. (1978). Über den *Chlorella*-Phycobionten von *Trapelia coarctata*. *Plant Syst. Evol.* **130**, 253–263.

Weber, A. and Czygan, F.-C. (1972). Chlorophylle und Carotinoide der Chaetophorineae (Chlorophyceae, Ulotrichales). 1. Siphonaxanthin in *Microthamnion kuetzingianum* Naegeli. *Arch. Mikrobiol.* **84**, 243–253.

Weber, A. and Wettern, M. (1980). Some remarks on the usefulness of algal carotenoids as chemotaxonomic markers. *In* "Pigments in Plants" (F.-C. Czygan, ed), pp. 104–116. Fischer, Stuttgart.

Wilhelm, C., Eisenbeis, G., Wild, A. and Zahn, R. (1982). *Nanochlorum eucaryotum*: A very reduced coccoid species of marine Chlorophyceae. *Z. Naturforsch., C: Biosci.* **37C**, 107–114.

Withers, N., Vidaver, W. and Lewin, R. A. (1978). Pigment composition, photosynthesis and fine structure of a non-blue-green prokaryotic algal symbiont (*Prochloron* sp.) in a didemnid ascidian from Hawaiian waters. *Phycologia* **17**, 167–171.

Yokohama, Y. (1981). Distribution of the green light-absorbing pigments siphonaxanthin and siphonein in marine green algae. *Bot. mar.* **24**, 637–640.

19 | Species-Specific Sporangium Autolysins (Cell-Wall-Dissolving Enzymes) in the Genus *Chlamydomonas**

U. G. SCHLÖSSER

Sammlung von Algenkulturen, Pflanzenphysiologisches Institut, Universität Göttingen, Göttingen, Federal Republic of Germany

Abstract: The release of reproduction units from sporangia and gametangia is usually associated with a particularly striking and short-term lysis of the mother cell wall. The release of zoospores from sporangia is mediated in the genus *Chlamydomonas* by lytic enzymes ("sporangium autolysins") of extreme specificity: they are secreted only during spore liberation; their action is confined to the wall of the sporangium; and they are species-specific, i.e. the sporangium autolysins act only between related species, respectively. From the 65 active *Chlamydomonas* strains 15 autolysin groups have been distinguished. Strains of each autolysin group are also similar with respect to other characteristics, e.g. form and size of the cells, generation time, period of flagellar motility in the asexual cell cycle, zoospore release mechanism and slime production. Sporangium autolysins can therefore be used as chemotaxonomic markers characterising groups of related species, enabling us to devise a more natural classification of this, one of the largest genera of the green algae, of which more than 500 species have been described.

INTRODUCTION

More than 500 species have been described in the genus *Chlamydomonas*. This is one of the largest genera of the green algae and is divided into subgenera, sections (Gerloff, 1940) and species (Ettl, 1976) according to characteristics of the vegetative cell, i.e. morphological features of the plastid and of the cell wall, number and position of contractile vacuoles, size and form of the cell, and length of the flagella. Characteristics of reproduction

---

*Dedicated to Professor Hans Adolf von Stosch on the occasion of his 75th birthday.

Systematics Association Special Volume No. 27, "Systematics of the Green Algae", edited by D. E. G. Irvine and D. M. John, 1984, pp. 409–418. Academic Press, London and Orlando.
ISBN 0 12 374040 1   *Copyright © by the Systematics Association*
All rights of reproduction in any form reserved

are not taken into consideration apart from the mode of cell division and the morphology of gametes and zygotes in a few species.

I want to introduce a physiological characteristic of asexual reproduction into the discussion on intrageneric division which might permit us to distinguish between groups of related species.

SPORANGIUM WALL AUTOLYSIS

During asexual reproduction a cell of a *Chlamydomonas* species such as *C. reinhardtii* divides within the mother cell wall into two, four, eight or 16 zoospores, each provided with a new cell wall and functioning flagella, thus forming a sporangium. The zoospores liberate themselves by total breakdown of the sporangium wall which triggers the swelling of intrasporangial slime (Fig. 1). This wall lysis is mediated by an enzyme called "sporangium wall autolysin", the chemical action of which is being investigated at present.

Autolysin activity can be demonstrated in the following way. Cells are

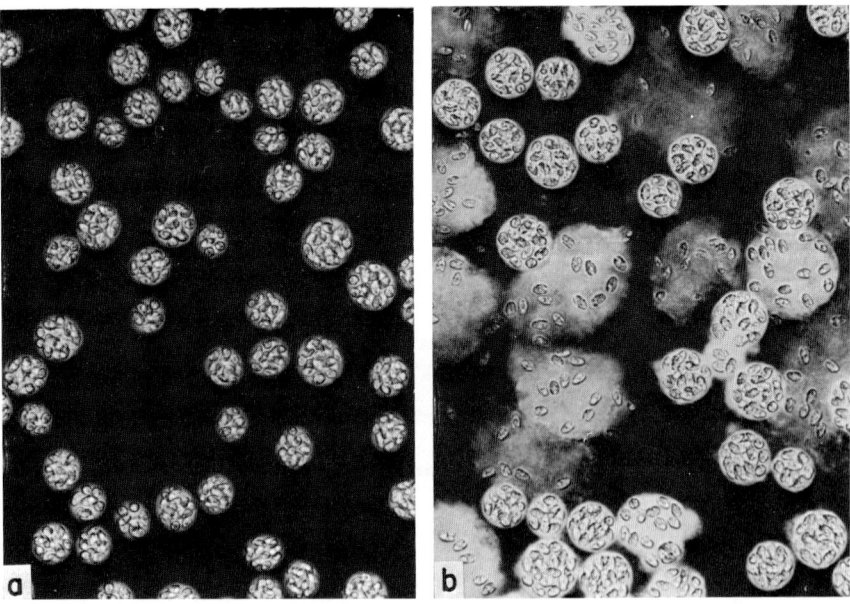

Fig. 1. Release of spores in *Chlamydomonas reinhardtii* from heat-fixed sporangia negatively stained with India ink. (a) Sporangia. (b) Swelling of intrasporangial slime during autolysin-induced dissolution of the sporangium walls (Schlösser, 1976).

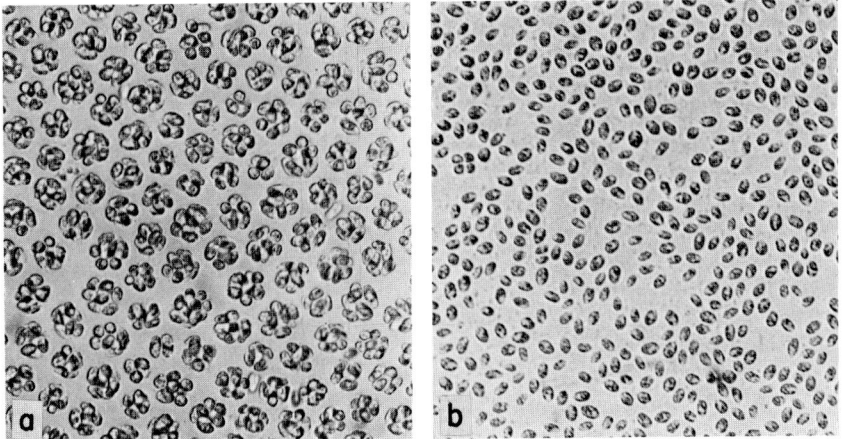

Fig. 2. Synchronized culture of *Chlamydomonas reinhardtii*. (a) Sporangia. (b) Zoospores (Schlösser, 1981).

synchronized in aerated mineral medium by light–dark changes every day and by dilution to a constant cell number at the end of each dark period. When cell divisions are completed, cells in the sporangium stage (Fig. 2a) accumulate during the dark phase for several hours. This is because the release of zoospores is slightly inhibited by aerated culture conditions. If concentrated 10-fold by centrifugation a sporangium suspension can be induced to release zoospores in quantity within 30 min by incubation in illuminated dishes without aeration (Fig. 2b). This is achieved by the total dissolution of the sporangium walls and the secretion of autolysin into the medium.

Autolysin action is demonstrable in this "sporulation medium" by a simple bioassay after separation from the zoospores by centrifugation: heat-fixed sporangia (5 min at 55°C) from 2 ml of a standard synchronized culture are incubated at 30°C with 2 ml of the sporulation medium. Activity of the autolysin is estimated quantitatively as the time required for quantitative release of the killed zoospores. Normally this is 5 to 20 min.

The wall-dissolving ability of the autolysin can also be demonstrated with an isolated and purified fraction of the sporangium wall as substrate.

SPECIES SPECIFICITY OF AUTOLYSIN ACTION

More than 100 species of *Chlamydomonas* have been tested for autolysin activity (Schlösser, 1976). A sporangium autolysin was demonstrable in all species in which both production of sporangia and release of zoospores

Table I. Autolysin groups distinguished among 65 active strains of *Chlamydomonas*.

Group no.	SAG* no.	Species	SAG* no.	Species
1	81.72	C. globosa Kroes	73.72	C. reinhardtii P. A. Dangeard
	7.73	C. incerta Pascher	18.79	C. reinhardtii P. A. Dangeard
	11-32a	C. reinhardtii P. A. Dangeard	77.81	C. reinhardtii P. A. Dangeard
	11-32b	C. reinhardtii P. A. Dangeard	54.72	C. smithii Hoshaw and Ettl
	11-32c	C. reinhardtii P. A. Dangeard	11-31	C. species
	11-32aM	C. reinhardtii P. A. Dangeard		
2	4.72	C. angulosa Dill	11.73	C. inepta Ettl
	14.72	C. debaryana Goroschankin	26.72	C. komma Gerloff
	6.79	C. debaryana Goroschankin		
	7.79	C. debaryana Goroschankin		
3	11-41	C. asymmetrica Korshikov	70.72	C. peterfii Gerloff
	11-7	C. gloeopara Rodhe and Skuja	12.83	C. species
4	11-60a	C. mexicana Lewin	37.72	C. oblonga Pringsheim
	11-60b	C. mexicana Lewin		
5	25.72	C. iyengarii Mitra		
6	9.72	C. callosa Gerloff	4.83	C. sphaeroides Gerloff

7	2.72	C. aggregata Deason & Bold	11-9	C. humicola Lucksch
	6.72	C. applanata Pringsheim	16.79	C. species
	11-36a	C. dysosmos Moewus		
8	66.72	C. species		
9	17.73	C. calleus Ettl	18.72	C. frankii Pascher
	64.72	C. elliptica Korshikov	19.72	C. frankii Pascher
10	2.75	C. gymnogama Deason	52.72	C. segnis Ettl
	9.83	C. pallidostigmatica King	1.79	C. segnis Ettl
11	69.72	C. gelatinosa Korshikov		
12	11-5/9	C. eugametos Moewus	74.72	C. species
	11-11	C. indica Mitra	75.72	C. species
	3.73	C. starrii Ettl		
13	2.79	C. aculeata Korshikov	14.73	C. pitschmannii Ettl
14	6.73	C. geitleri Ettl	40.72	C. pinicola Ettl
	22.72	C. hindakii Ettl	59.72	C. terricola Gerloff
	33.72	C. monoica Strehlow	8.73	C. terricola Gerloff
	35.72	C. noctigama Korshikov	16.73	C. species
	36.72	C. noctigama Korshikov	19.73	C. species
15	8.72	C. brannonii Pringsheim	1.73	C. species
	10.83	C. texensis King	3.79	C. species
			4.79	C. species

* Sammlung von Algenkulturen, Pflanzenphysiologisches Institut, Universität Göttingen (Schlösser, 1982).

could be synchronized. A comparative investigation in which the sporulation media of all these active strains were cross-tested against their fixed sporangia did not indicate a common sporangium autolysin in the genus, but a species specificity, i.e. the sporangium autolysins act only between morphologically similar species. From the 65 active strains 15 autolysin groups could be distinguished (Table I).

A lytic action in the bioassay among these groups has been proven in three cases: autolysins of all strains of group 2 act on all strains of group 1; those of all strains of group 3 act on all strains of group 4; and that of 19.73 C. spec. (group 14) acts on 66.72 C. spec. (group 8). This action is non-reciprocal.

The secretion of all autolysins is strictly developmental-stage-specific. They are secreted only by zoospores and act only on the sporangium wall, thus indicating a qualitative change of the wall during the cell cycle.

Protein-like characteristics of the autolysins have been demonstrated (Schlösser, 1976). Autolysins are presumed to be proteolytic enzymes since the wall of *Chlamydomonas* does not contain cellulose but rather hydroxyproline-rich glycoproteins.

AUTOLYSIN GROUP-SPECIFIC CHARACTERISTICS

Species with a common autolysin correspond in other physiological and morphological respects:

1. Form and size of the cells are largely in agreement.
2. The generation time under identical laboratory conditions is 16 to 20 h in most of the autolysin groups; but it is 12 to 14 h in all strains of the groups 1 and 15, and 20 to 24 h in all strains of group 14.
3. The period of flagellum motility ends normally before the divisions

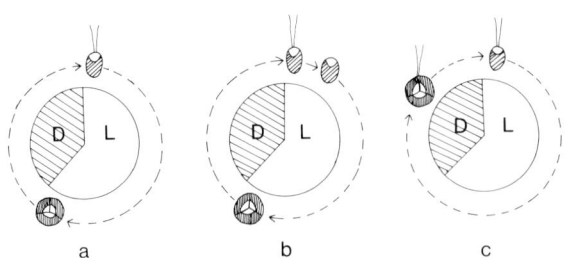

Fig. 3. Different periods of flagellum motility during the asexual cell cycle of *Chlamydomonas* species in synchronised culture. (a) Zoospores and vegetative cells flagellated. (b) Only zoospores flagellated. (c) All stages flagellated.

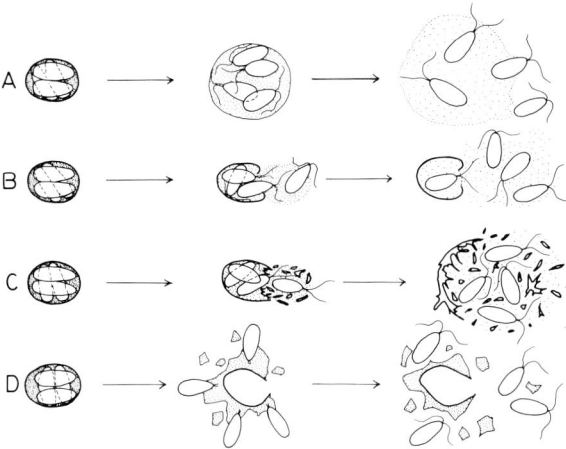

Fig. 4. Different modes of autolysin-mediated zoospore release in the genus *Chlamydomonas*. (A) Total dissolution of the sporangium wall and unlimited swelling of the sporangium slime. (B, C) Partial dissolution of the sporangium wall resulting in a hole (B) or in fragmentation into pieces (C) and unlimited swelling of the sporangium slime. (D) Partial dissolution of the sporangium wall associated with an explosive inversion; no swelling of the sporangium slime (Schlösser, 1976).

begin (Fig. 3a). Strains of the groups 2, 3, 4, 6, 9 and 10 tend to become immobile within the first hours of the cell cycle (Fig. 3b). All isolates of group 14 on the other hand remain motile during the whole asexual cell cycle (Fig. 3c), except for 36.72 (*C. noctigama*). Flagellated sporangia have been described in *C. mirabilis* Korshikov (Korshikov, 1938) and *C. cingulata* Pascher (Vlk, 1940).

4. The mode of spore release is generally by means of total dissolution of the sporangium wall (Fig. 4A) in the groups 12, 13 and 14. There is a partial lysis of the sporangium wall resulting in a hole (Fig. 4B); only in 59.72 (*C. terricola*) does the sporangium wall fragment into pieces (Fig. 4C). In all strains of group 15 a partial dissolution of the sporangium wall is associated with an explosive inversion (Fig. 4D).

5. Slime production of vegetative cells and sporangia (Fig. 5) is a feature of groups 9 and 10.

6. Stickiness of the sporangium wall characterises groups 2, 9 and 12; only in this developmental stage do cells stick together and to the culture vessel, thus indicating a change in the nature of the surface.

Until now the genus *Chlamydomonas* has been divided artificially into subgenera, sections and species (Pascher, 1927; Gerloff, 1940; Ettl, 1976). Future work on sporangium autolysin specificity might contribute a truer and more reliable assessment of natural relationships.

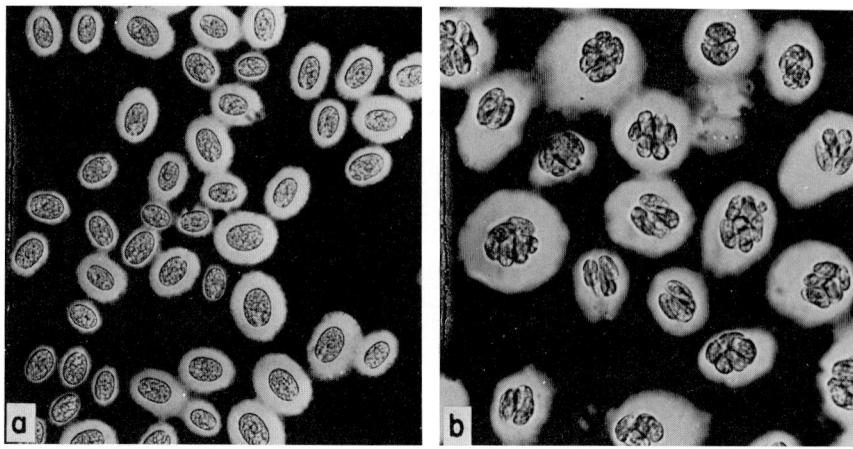

Fig. 5. Slime-producing *Chlamydomonas* species; synchronised cultures negatively stained with India ink. (a) Vegetative cells of 52.72 *C. segnis*. (b) Sporangia of 19.72 *C. frankii*.

CELL WALL AUTOLYSIS DURING REPRODUCTION: A GENERAL PHENOMENON IN ALGAE

Specific wall changes during reproduction have been very often observed microscopically in nearly all algal divisions (Fig. 6). It should be possible to correlate them with autolysin activity by the methods used in *Chlamydomonas*. This requires the following steps:

Selection of suitable cell systems with an autolytic release of reproduction units.

Methods of synchronization of both formation and release in high cell densities for the purpose of concentrating the autolysins and their special substrates in the walls.

Bioassays in order to demonstrate and control the enzyme activities.

These methods have been used successfully (for review, see Schlösser, 1981), e.g. for the demonstration of autolytic gamete release from their own walls in *Chlamydomonas reinhardtii*) Claes, 1971; Schlösser *et al.*, 1976) and the release of daughter spheroids in *Volvox carteri*. Both have turned out to be proteases (Jaenicke and Waffenschmidt, 1981). The fragmentation of a filamentous green alga, *Geminella* species, is also mediated by an autolysin (Schlösser, 1981).

I hope that these methods will make it possible to detect and isolate further autolysins of a high specificity to both developmental stages and species in other groups of algae, thereby leading to the creation of more natural groups of related species.

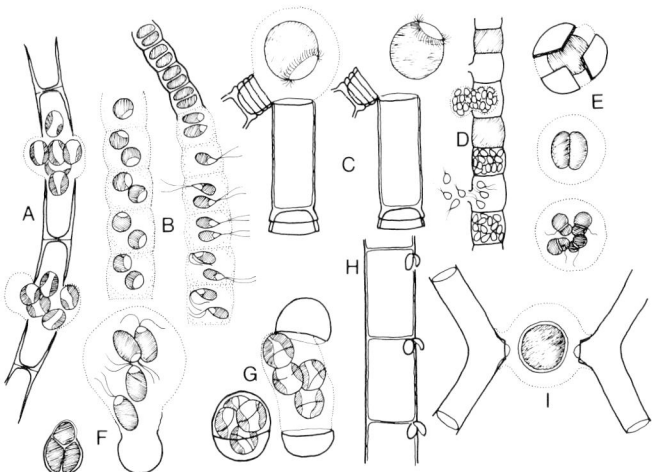

Fig. 6. Release of reproduction units from sporangia and gametangia connected with changes of the walls. (A) Aplanospores in *Tribonema* species, swelling jelly. (B) Aplanospores and zoospores in *Microspora quadrata;* unlimited swelling of the sporangium walls. (C–F) Release in a transitory vesicle of the zoospore in *Oedogonium concatenatum* (C), gametes in *Ulothrix zonata* (D), germination of zygotes in *Glenodinium lubiniensiforme* (E) and *Hydrodictyon reticulatum* (F). (G) Aplanospores in *Chlorothecium inaequale,* swelling jelly which lifts an operculum. (H) Gametangia of *Acrosiphonia* species; formation of operculum. (I) Zygote of *Mougeotia oedogonioides,* separated from the empty gamete walls by swelling slime. (A–I) redrawn after: (A) Pascher (1939), (B) Skuja (1956), (C) Hirn (1900), (D) Dodel (1876), (E) Diwald (1938), (F) Schlösser (original), (G) Pascher (1932), (H) Kornmann (1965) and (I) Czurda (1931).

ACKNOWLEDGEMENT

The support of the Deutsche Forschungsgemeinschaft is gratefully acknowledged.

REFERENCES

Claes, H. (1971). Autolyse der Zellwand bei den Gameten von *Chlamydomonas reinhardii. Arch. Mikrobiol.* **78,** 180–188.
Ettl, H. (1976). "Die Gattung *Chlamydomonas* Ehrenberg," Cramer, Vaduz.
Gerloff, J. (1940). Beiträge zur Kenntnis der Variabilität und Systematik der Gattung *Chlamydomonas. Arch. Protistenkd.* **94,** 311–502.
Jaenicke, L. and Waffenschmidt, S. (1981). Liberation of reproduction units in *Volvox* and *Chlamydomonas:* proteolytic processes. *Ber. Dtsch. bot. Ges.* **94,** 375–386.
Korshikov, O. A. (1938). Volvocineae. *In* "Vyznachnyk Prisnovodnykh Vodorostej Ukrains'koi RSR," Vol. IV, pp. 1–184. Akad. Nauk URSR, Kiev.
Pascher, A. (1927). Volvocales-Phytomonadinae. *In* "Süsswasser-Flora Deutschlands, Österreichs und der Schweiz" (A. Pascher, ed), No. 4, pp. 20–506. Fischer, Jena.

Schlösser, U. G.(1976). Entwicklungsstadien- und sippenspezifische Zellwand-Autolysine bei der Freisetzung von Fortpflanzungszellen in der Gattung Chlamydomonas. Ber. Dtsch. bot. Ges. **89**, 1–56.

Schlösser, U. G. (1981). Algal wall-degrading enzymes—autolysines. Encycl. Plant Physiol., New Ser., **13B**, 333–351.

Schlösser, U. G. (1982). Sammlung von Algenkulturen, Pflanzenphysiologisches Institut der Universität Göttingen. Ber. Dtsch. bot. Ges. **95**, 181–276.

Schlösser, U. G., Sachs, H., and Robinson, D. G. (1976). Isolation of protoplasts by means of a "species-specific" autolysine in Chlamydomonas. Protoplasma **88**, 51–64.

Vlk, W. (1940). Über die Morphologie und Entwicklung von Chlamydomonas cingulata Pascher im Vergleich mit anderen durch ihr Pyrenoid auffallenden Arten. Lotos **87**, 1–12.

20 | The Role of Extrinsic Factors in the Past and Future of Green Algal Systematics

P. C. SILVA

Department of Botany, University of California, Berkeley, California, USA

Abstract: Taxonomy is an art that involves subjective interpretation of objective data. The subjectivity of interpretation is influenced by factors related to the context (time, place, taxonomic group, historical precedents, historical sequence) or inherent in the interpreter (ability, training, experience, personality, interests, goals). Although appraisal of specific interpretations is itself subjective, it seems likely that an appreciation of factors influencing past and present taxonomic judgment would improve future interpretation. The innate ability of investigators to make durable taxonomic judgments is handicapped by increased specialization, unbalanced emphasis on new technology, and diminished ties with classical taxonomy. Moreover, some judgments are more appropriate to phylogenetics than to taxonomy. Aesthetics—especially the sense of form and balance—play a role in formulating and assessing taxonomic judgments. Didactics play an even greater role, with special favor being accorded those taxonomic conclusions that support alleged phylogenetic trends and generalizations useful in teaching. Current taxonomic practice is often strongly influenced by tradition, which in some instances may be traced to a single investigator. Historical precedents and historical sequence profoundly affect taxonomic decisions. Paradigms can be useful, but may also impede correct interpretation. The concept of any taxon is biased by the element that was first made known, despite conscious efforts to distinguish between biological and nomenclatural types.

RELATIONSHIP BETWEEN SCIENTIFIC METHOD AND SYSTEMATICS

The teaching of the scientific method to youngsters is considered in most quarters to be a good thing. Teachers feel virtuous in having opened new

---

Systematics Association Special Volume No. 27, "Systematics of the Green Algae", edited by D. E. G. Irvine and D. M. John, 1984, pp. 419–433. Academic Press, London and Orlando.
ISBN 0 12 374040 1

Copyright © by the Systematics Association
All rights of reproduction in any form reserved

doors to incipient Priestleys, Darwins, and Curies, while pupils are made to feel that they have been enlightened, indeed spared the embarrassment of going through life without appreciating the importance of careful observation, controlled experimentation, systematization of data, perception of patterns, and the proposal and testing of theories. Let me allay fears or hopes that I am going to be iconoclastic. I do not intend to speak against the scientific method, but rather to put it into the perspective of an algal systematist.

The relationship between the scientific method and systematics is not as obvious as might be expected and is not always appreciated by the practicing taxonomist. Appreciation implies sophistication, which in turn is based on experience coupled with common sense. When we are introduced to an idea so new that it lies beyond our experience, our initial reaction is necessarily unsophisticated. We generally accept it or reject it in equally unqualified terms. A basis for qualifying our opinion of the idea and for perceiving various ramifications of the idea develops only with experience. Eventually, the fervent appeal felt by some is blunted by the distaste felt by others, resulting in a rational middle ground. Consider the voluminous literature, pro-Darwin and anti-Darwin, evoked by that great thinker in the nineteenth century. Consider also the voluminous neo-Darwinian literature that is being produced in the twentieth century, with writers neither swallowing Darwin's ideas hook, line, and sinker nor rejecting them summarily.

At the present time, society is programmed to accept the scientific method as a great idea, and even members of fundamentalist religious sects safely embrace it, knowing better than many self-styled scientists that the method bears only a loose relation to ensuing interpretation. The charge of subversion or distortion is brought against fundamentalists, but I believe that the alleged mistreatment of data is often a matter of interpretation, the same interpretation that is indulged in to a greater or lesser degree by even rigorous scientists. When a teacher exposes a student to the scientific method, he or she brings cultural enrichment into someone's life but at the same time may be sowing the seeds of discontent. If the expectation arises, as it often does, that the scientific method can solve problems in life as well as in specific fields of enquiry, and if this reasonable expectation is accompanied by the unreasonable assumption that the scientific method embodies an orderly accumulation of data necessarily followed by their logical interpretation, the stage is set for disappointment. This disappointment is expressed in various ways and at various levels. In most instances, disillusionment sooner or later leads to the sophistication required for a rational attitude. In the most dramatic and tragic cases, however, there may be a loss of ability to cope with the realities of life. In a much less dramatic fashion, many students are puzzled if not disillusioned by the course of events in a particular scientific field, even if they avoid the mistake of extending the scientific method to life in general. After all, the innate abilities and the training of a scientist in no way decrease his humanity. Scientists, like all other humans, behave in a predictably unscientific manner. The real progress of science thus differs significantly from that which would seem to be in accordance with the scientific method.

With reference to the field of the systematics of green algae, or of any other group of organisms, puzzlement or disillusionment with the scientific method could and should be avoided among budding systematists by teaching them that taxonomy is not a science, but rather an art based upon results obtained through use of the scientific method. No matter how much information becomes available, no matter how discriminating the data, the ultimate goal in taxonomy is the recognition of categories and their arrangement in a system of classification, processes that are intuitive and subjective. Nor, in systematics, is subjectivity confined to these final acts of taxonomic synthesis. Rather, subjectivity is inherent in every interpretation at every step of every data-gathering investigation.

Subjectivity in systematics is influenced by factors that fall into two groups—those most closely related to the context and those most closely related to the interpreter. A taxonomic opinion or interpretation is not offered in a vacuum. Rather, it is introduced into a vast body of previously promulgated taxonomic opinions and the fate of a new opinion depends upon the context: time, place, taxonomic group, historical precedents, and historical sequence. All of these contextual factors operate together. While it seems unlikely that in the past any promulgator of a taxonomic opinion stopped to consider the context, I suggest that an appreciation of factors influencing past and present taxonomic judgment would improve future interpretation.

EXTRINSIC FACTORS RELATED TO CONTEXT

1. Time and Place

A taxonomic opinion can have no effect on taxonomic history unless it is circulated so that it can be assessed and either accepted or rejected. Taxonomic literature is replete with works of great potential taxonomic importance that were completely or almost completely ignored because of an unfavorable context. Since they had no effect, either at the time they were published or at any time before their usefulness was superseded, they cannot meaningfully be considered part of taxonomic history. An interesting part of the history of taxonomic literature, to be sure, but not a part of taxonomic history. I could cite numerous examples, but most have been so thoroughly ignored that their names are not familiar to anyone. Two British workers who suffered this fate are well known, however. John Stackhouse's imaginative division of the Linnaean genus *Fucus* into 37 genera was lost in the pages of the Memoirs of the Imperial Society of Naturalists of Moscow, published in 1809. Soon afterward, Moscow was occupied by the French under Napoleon and nearly completely destroyed by fire. The stock of the Stackhouse memoir was a casualty. For less obvious reasons, Samuel Frederick Gray's "A Natural Arrangement of British Plants," published in 1821 and including a thorough and novel treatment of the algae, was ignored. Surely every systematist has had the experience of discovering statements and articles that have been generally overlooked or ignored. On the other hand, those contributions that are introduced, by design or by luck, at the right time and in the right place have a disproportionately large impact.

2. Historical Sequence

Historical sequence is of great importance in taxonomy. The type locality of an alga biases our ideas on its morphology, biology, and geological history. We refer to a species as Japanese because it was first described from Japan and thereby imply that Japan is its center of distribution or place of origin even though it may be known from a wide area in the North Pacific. Conversely, in analyzing the Japanese flora there is a category "California species," implying that these were not only first described from California, but that they originated there. At the level of species, the part of the morphological spectrum that was first collected or studied or serves as the nomenclatural type biases our views, even though we try to remain impartial and say that we understand that the biological type and nomenclatural type do not necessarily coincide.

This problem is greatly amplified in algae with life histories that involve an alternation or sequence of heteromorphic somatic phases. What *is* the biological type of *Derbesia marina,* for example, the *Derbesia* phase or the *Halicystis* phase? In the Bryopsidaceae we have competitive taxonomic schemes, one for the gametophyte and one for the sporophyte. In formulating phylogenetic schemes it must be kept in mind that the two phases have evolved together, even though the rates of change may have differed. This coevolution must also be kept in mind when deciding on generic boundaries, since within a given group of species one of the two phases may exhibit a continuous spectrum or remain relatively unchanged while the other exhibits a marked morphological discontinuity. Lack of synchrony in the evolution of the two phases can result in rather severe taxonomic problems, as in the *Derbesia–Bryopsis* series (Rietema, 1975). At one point in that series, a *Derbesia*-like sporophyte alternates with a *Bryopsis*-like gametophyte. This morphological intermediacy is defined as the genus *Bryopsidella*.

In the red algae *Trailliella intricata* Batters is the tetrasporophyte in the life history in which *Bonnemaisonia hamifera* is the gametophyte. This statement has recently been published: "The tetrasporangial phase of the Pacific *Bonnemaisonia nootkana* (Esper) Silva is also recognized as *Trailliella intricata* and is indistinguishable from the British material. It would be incorrect to cite *Trailliella intricata* in the synonymy of *Bonnemaisonia hamifera,* in view of its association with a second species of this same genus" (Dixon and Irvine, 1977). Here we see confusion between morphological concepts and taxonomy. *Trailliella intricata* is a name that must be applied, as determined by its type material, to a single taxon. The tetrasporangial phase of *Bonnemaisonia hamifera* has coevolved with the gametophyte of that same species, not with the tetrasporangial phase of *B. nootkana*. The fact that the two tetrasporangial phases are indistinguishable in the present state of the art is perhaps confusing or even distressing, but immaterial to taxonomy. The tetraspores somehow know their correct identity, in one case producing *B. hamifera,* in the other case producing *B. nootkana*. The point of this example is that historical sequence—that is, which taxonomic scheme was introduced first, the one for the gametophyte *Bonnemaisonia* or the one for the sporophyte *Trailliella*—biases the outcome, but for taxonomic purposes both phases must be considered as a totality. Crustose tetrasporophytes in red algae offer fascinating examples in which crusts assigned to a single genus in accordance with tetrasporophyte taxonomy alternate with gametophytes assigned to two different families as defined by gametophyte taxonomy. We must strive for whole organismal taxonomy.

Generic concepts are biased by the species that was first put in the genus or serves as nomenclatural type. Generic concepts are also biased by one's field experience. When someone from the Pacific coast of North America and someone from the Mediterranean get together to discuss *Cystoseira,* there will be a problem in communication. The Californian representatives of the genus have markedly flattened secondary axes in contrast to the usual Mediterranean plant with its terete branches and branchlets. And those for whom dulse exemplified *Rhodymenia* must make a special effort to reassociate that alga with a new name (*Palmaria*) in a new family (Palmariaceae) in a new order (Palmariales) while reformulating a generic concept of *Rhodymenia* based on *R. pseudopalmata*. Our generic concepts are usually more strongly influenced by material within our experience than by a cerebral consideration of nomenclatural types.

EXTRINSIC FACTORS RELATED TO THE TAXONOMIST

1. Competency

Subjectivity inherent in the taxonomic interpreter is related to ability, training, experience, personality, interests, and goals. Partly because we live in an egalitarian age and partly because we live in glass houses, even to suggest that talent plays a part in the durability of taxonomic interpretation is unconscionably elitist. Yet anyone who carefully traces the taxonomic history of a particular group cannot escape being impressed by the unevenness of impact that individual workers have had. Not all have taken steps in the forward direction. The disservice done by the German naturalist Esper is familiar to many phycologists. Around 1800 he published inaccurate illustrations of seaweeds, often based on snippets obtained from other workers, such as Turner or Mertens. At the other end of the spectrum, those phycologists who have had reason to use the work of Marshall Avery Howe quickly gain confidence that here is someone whose opinion can be valued, someone who combines thorough scholarship with common sense. Every specialist can cite workers who have been taxonomically disastrous and others who have done their best to relieve the chaos. Diatoms and desmids have been plagued with investigators who did not know what taxonomy is all about, who have abused nomenclature, and who have lacked common sense. Most likely the fault lay more in lack of adequate training than in lack of innate ability. Systematic phycologists are currently being turned out in progressively greater numbers at institutions other than the classical centers. As new technology and new subdisciplines take more and more time away from traditional studies, young taxonomists may find themselves embarked on a career launched in the excitement of a single technique, such as electron microscopy, or microspectrophotometry, or culturing, without having had the benefit of years of field experience or a background in spermatophyte systematics. Moreover, the practice of taxonomy entails many esoteric operations, devices, and symbols that are steeped in tradition and must be understood if the taxonomy itself is to be fully appreciated. At the risk of sounding arrogant and condescending, I should like to suggest that there is often a need for student systematists to broaden their training, preferably by means of a postdoctoral fellowship at an institution oriented towards traditional taxonomy. There is a need for workshops, not in special taxonomic groups, but in general taxonomic philosophy and methodology. New techniques and viewpoints are essential to progress in systematics, but the historical context must be carefully preserved.

I should like to offer an example of the role of talent in taxonomy that is dear to my heart, since it concerns the genus upon which I based my doctoral research. The Abbot Giuseppe Olivi was Professor of Zoology at the University of Padova in the latter part of the eighteenth century. In a kind of appendix to his magnum opus, "Zoologia Adriatica," Olivi took two superficially diverse organisms, one of which had been considered a tunicate, the other a sponge, examined them anatomically, decided that they were algae, and came up with a unifying generic character that set the new genus apart from all other genera of marine algae known at that time, namely, *Fucus, Ulva,* and *Conferva.* The date was 1792. The organisms were those now known as *Codium bursa,* with the form of a tennis ball, and *C. vermilara,* with a terete, dichotomously branched thallus. I marvel at the sharpness of this observer, who concluded, so far ahead of his time, that despite great superficial diversity, these two organisms were congeneric; moreover, that they were plants. Olivi was the first to use the term "utricle" in connection with *Codium* and the first to observe gametangia (which he called "globuli") as well as the plexus of tubular filaments. In short, he grasped the fundamental structural plan of this genus. *Codium* thus became the first genus of marine algae to be narrowly and correctly defined. Unfortunately, Olivi's work was not immediately noticed by botanists, and the name that he chose for the genus—*Lamarckia*—lay out of sight while that same name was being employed elsewhere in the plant kingdom—for a genus of grasses. Eventually *Lamarckia* the grass was conserved against *Lamarckia* the alga, but in my estimation Olivi remains a minor hero because of his discerning and potentially important work.

I have already mentioned John Stackhouse as an example of someone whose work in part, at least, suffered the fate of remaining unknown. Before publishing his ill fated memoir in Moscow, however, he wrote a charming three-part treatise on British marine algae, "Nereis Britannica," published privately at Bath between 1795 and 1801. In this treatise, Stackhouse established several new genera, including *Codium.* The generic character, tersely expressed, compares unfavorably with that given by Olivi: "Fructification: invisible; frond roundish; soft and spungy, when wet; velvety, when dry." The single species was *Codium tomentosum.* Stackhouse dissected the thallus and saw the tubular construction but thought that the plastids that aggregated at the apices of the utricles were "seeds."

In 1867, long after phycology had been established as a subdiscipline of botany, another worker—W. F. R. Suringar—looked at the anatomy of *Codium.* Dealing with a collection from Japan, he noticed that the utricles had sharply pointed apices. That character alone was used as the basis of a new genus, which Suringar called *Acanthocodium.* It turns out that this character is not correlated with primary anatomical characters and that it is useful only at the level of species. Even at this level, it may be highly variable. Pointed utricles are found in three different sections of the genus.

I feel that Olivi was the ablest of the three investigators, yet if their work is put into historical context the appraisal may be more difficult. The development of knowledge is subject to opposing movements: expansion resulting from the elucidation of diversity, and contraction resulting from the perception of unity. The two factors often operate alternately, like a pendulum. Olivi was faced with diversity resulting from large numbers of species of marine algae being forced to fit the mold of the three Linnaean genera. He sought and found a unifying concept. Suringar was faced with unity resulting from Olivi's shrewd interpretation. He sought, or at least was receptive to, a diversifying concept. Had he not been so superficial, had he looked at the bases of the utricles to see how they originated from the medullary filaments or from one another, he would have found a diversifying principle that is valid today at the level of subgenus or section.

Figure 1 illustrates variations on a theme: diversity within a fundamental framework of medullary filaments, utricles, and gametangia. Nine species are drawn to the same scale and represent nine sections of the genus. Four species from the British flora are here: *Codium bursa, C. adhaerens, C. fragile* subsp. *atlanticum,* and *C. tomentosum. Codium cylindricum,* the species with the largest thallus (up to 10 m long) is shown, and at the base of its utricles are those of the species with the smallest thallus, *C. petaloideum,* a thin blade about 5 × 12 mm. The largest utricles are those of *C. megalophysum,* whose gametangia are large enough to be seen easily with the naked eye. The entire utricle system of *C. petaloideum* could fit into the medullary filament of *C. megalophysum.* The length ratio of the utricles of the two species is about 50. Perhaps no other genus of plants has as wide a dimensional spread among basic anatomical structures.

Figure 2 illustrates independent experimentation of a minor character in three phylogenetic lines. At "d" is *C. fragile* subsp. *tomentosoides,* very similar to what Suringar called *Acanthocodium fragile;* "c" is *C. spinescens,* a narrow endemic in Western Australia, in the same subgenus as *C. fragile* but readily distinguishable; "b" is *C. acuminatum* from Madagascar and Mozambique, closely related to *C. adhaerens.* Most remarkable is "a," *C. elisabethiae,* a very narrow endemic in the Azores and Madeira. In all respects it is indistinguishable from *C. bursa* except that the apices of the utricles are pointed. In those two island groups *C. elisabethiae* replaces *C. bursa.*

Closely related to the role of ability of investigators to draw durable taxonomic conclusions is their ability to obtain adherents to their point of view. We thus enter the realm of personal politics, sometimes involving conscious effort, but usually passively correlated with personality, ability, time, and position. Certain persons—almost invariably great teachers—have a profound effect on the direction of systematic studies and on the acceptance or rejection of the countless number of conclusions that form the

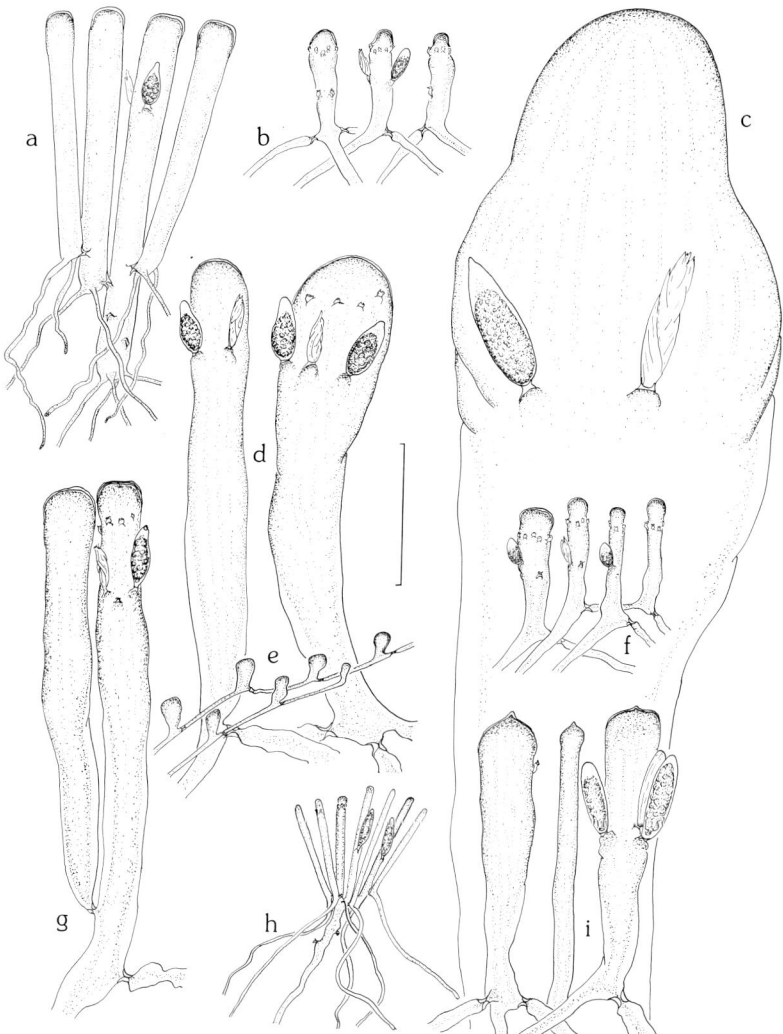

Fig. 1. Utricles from various species of *Codium* drawn to same scale (1 mm). (a) *C. ritteri*. (b) *C. laminarioides*. (c) *C. megalophysum*. (d) *C. cylindricum*. (e) *C. petaloideum*. (f) *C. tomentosum*. (g) *C. bursa*. (h) *C. adhaerens*. (i) *C. fragile* subsp. *atlanticum*.

superstructure of our taxonomic system. Students may be influenced by their professors to the point of allowing their originality to be suppressed. More important, even the ablest teachers and investigators are not consistently correct or always favored by the best ideas, and therefore there are times when prestige does science a disservice.

Students, and even some professional biologists, sometimes confuse

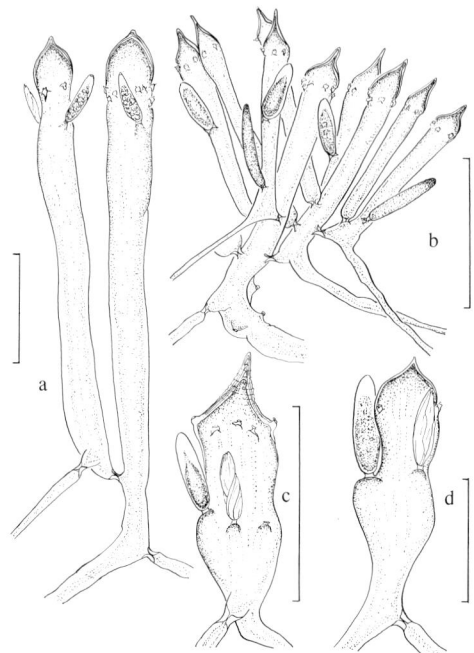

Fig. 2. Pointed utricular apices in various species of *Codium*. (a) *C. elisabethiae*. (b) *C. acuminatum*. (c) *C. spinescens*. (d) *C. fragile* subsp. *tomentosoides*. Scale = 1 mm.

competency with reputation or prestige. As someone with many years of editorial experience, I assure you that some of the worst papers are submitted by highly reputable workers. I have had visitors from abroad, who were visiting various herbaria, naively say to me: "But this determination must be correct. I checked it against a specimen in the New York Botanical Garden" (or the British Museum). The investigator did not consider whether a type specimen was involved or question the qualifications of the person who had made the determination, the overriding point being that the specimen was housed in a prestigious institution.

2. Aesthetics

As stated above, taxonomy is an art and taxonomists are thus artists—good, bad, and indifferent. They are frequently aesthetes, giving due consideration to the various components of beauty, such as form, balance, color, and texture. One undesirable manifestation of aesthetics is the repugnance of many taxonomists to leave untidy loose ends. There is a tendency to put everything in a slot, regardless of fit, and many publications fail to give a hint of the magnitude and number of unsolved problems. Although I am an ardent campaigner for good housekeeping in taxonomy, I must emphasize that the cause is not served by sweeping problems under the rug. Rather, they should be isolated on the bare floor with a sign identifying them as problems and, under the best of circumstances, a suggestion as to what might be done about them. A desire for balance often intrudes into the practice of taxonomy. Many taxonomists seem to prefer that all coordinate taxa be approximately the same size. There is also a desire that coordinate taxa in one phylogenetic line be approximately equivalent in degree of mutual distinctness to those in another line.

3. Tradition

Certain taxonomic groups have their own traditions. Some traditions have obscure roots, while others are clearly traceable to a single influential individual. In his "British Desmidieae", Ralfs (1848) recognized 162 species and 34 varieties. A quarter of a century later, Nordstedt employed subspecies, varieties, and forms. Other authors added the category subform. The tradition of recognizing a superabundance of infraspecific taxa in desmids was thus firmly established more than a century ago. The tradition was shared by *Ankistrodesmus, Pediastrum,* and *Scenedesmus,* three genera of Chlorococcales that were mistaken for desmids by Ralfs. It has logical origins. Desmids and the three chlorococcalean genera exhibit numerous measurable features. Desmids offer special problems in their variable symmetry. Lastly, desmids have attracted amateurs, who, more than professional biologists, are likely not to have a sound taxonomic philosophy. Individuals rather than populations have been described. Taxonomic tradition among diatoms is nearly identical to that for desmids.

In the very large genus *Chlamydomonas,* on the other hand, the tradition has been to describe species rather than infraspecific taxa. Undoubtedly some, but not all, of the alleged species are worth no more biologically than are infraspecific taxa in desmids. It may be asked whether populations of *Chlamydomonas* are morphologically more homogeneous than are populations of desmids.

There is a different kind of tradition in the Oedogoniales—a formalization of the presentation of data slavishly followed for many decades. The tradition had a definite beginning in a paper published in 1870 by Wittrock. A separate line was used for each dimensional descriptor. The formula was brought to perfection by Hirn in his monograph (1900), which serves as the starting point for nomenclature of the group. The formula is still employed by certain workers today. So strong is the tradition that even when it is not a matter of a new taxon and therefore there is no need for Latin, the

Wittrockian formula may appear in that language. Is this standardization of the presentation of data bad? Not completely, but can one countenance closing the door on taxonomic characters as of 1870?

A third type of tradition is what I call the tyranny of names. We were taught, we naively believed, and we in turn have taught that nomenclature is the handmaiden of taxonomy. Yet as teachers we would think twice about adopting a better taxonomic scheme if it meant the abandonment of certain personally favored names. Resistance to change is a powerful force in taxonomy.

4. Didactics

Didactical factors are important in influencing interpretation and the acceptance of an interpretation. Perhaps the best example is to be found in the red algae. Strasburger in 1894 published a very important paper summarizing the exciting developments of the preceding decade, including the discovery of the mechanism and biological significance of meiosis and the relationship between syngamy and meiosis in sexual life histories. He also pointed out the total lack of karyological information in the algae. Wolfe stepped forward and in 1904 published the first chromosome counts in algae. Working with *Nemalion*, he counted eight chromosomes in the gametophyte and 16 in the carposporophyte (the filaments that issued from the zygote). Influenced by the fact that tetrasporophytes were unknown in *Nemalion*, Wolfe convinced himself that meiosis took place during carposporogenesis and that the carpospores were haploid. He was criticized almost immediately—not so much for his cytological evidence as for his conclusion, which did not agree with the recent demonstrations of meiosis in the germination of zygospores of freshwater algae and in tetrasporogenesis of the brown alga *Dictyota*. Kylin (1916) and Cleland (1919) rectified the situation by showing conclusively that in *Nemalion* meiosis occurred in the first division of the zygote. Then in 1927, Børgesen described *Liagora tetrasporifera*, an unusual species in the Nemaliales in which tetrads of carpospores were formed, strongly suggesting sporic meiosis. This discovery greatly increased the attractiveness of believing in zygotic meiosis in *Nemalion* because the perceived phylogenetic postponement of meiosis from the zygote (*Nemalion*) to carposporogenesis (*Liagora tetrasporifera*), and thence to tetrasporogenesis (the bulk of the red algae), held great didactic appeal. Quotable phylogenetic trends are useful in teaching, it must be admitted. Through the elegant work of von Stosch in 1965 and others shortly thereafter, however, we know that the "conclusive" cytological work of Kylin and Cleland was incorrect. Carpospores of *Nemalion* germinate to produce a microscopic filamentous tetrasporophyte and thus one more quotable phylogenetic trend must be abandoned.

The exciting and scientifically laudable ideas put forth during the present symposium are certain to generate excellent examples of the role of didactics in the acceptance of an interpretation. I predict that general acceptance of the ultrastructurally based classifications proposed here will be many years in materializing. After seeing the list of characters for the various classes and orders and taking into account the amount of ultrastructural information in the head of the average undergraduate student, I foresee strong reluctance to adopt any new scheme of classification that so seriously affects traditional teachings. In a schizophrenic manner, the Ulotrichales and Ulvales will probably remain in phycological curricula in their traditional form along with an account of the major ultrastructurally perceived phylogenetic lines.

5. Paradigms

A paradigm is a didactic tool, but it also may influence our interpretation of experimental or observational data. The model of bipolar sexuality—male/female—inherited by botanists from zoologists, could be made to apply to *Volvox,* the Zygnematales, the Oedogoniales, and *Coleochaete* early on, and to other green algae later, but it only served to confuse the picture in red algae, most of which have three kinds of thalli—male, female, and tetrasporangial. In most groups of red algae, males are inconspicuous, and as long as they were overlooked it was incumbent upon investigators to fit the other two kinds of thalli into the male/female pattern. Regardless of the decision, the answer was always wrong. It was not until the paradigm of alternation of generations was proposed by Steenstrup for coelenterates in 1842, extended to vascular cryptogams and conifers by Hofmeister in 1851, and further extended to algae by Sachs in 1874 that the life history of Floridean red algae could be understood.

6. Perception of Taxonomic History

Some taxonomists have seriously considered the context of their work, but in a misdirected way. In many papers there is a section entitled "Taxonomic History." Here is a typical account: "The genus *Mastocarpus* was established by Kützing in 1843. His proposal was ignored. In 1933, however, Setchell and Gardner treated it as a subgenus of *Gigartina*. In the period 1980–1983 *Mastocarpus* was resurrected by various workers, including Perestenko and Guiry." This is not taxonomic history. Rather, it is the history of the name *Mastocarpus,* hence nomenclatural history. We can cover the same ground from a purely taxonomic point of view. "In 1843 Kützing established a genus to receive those foliose red algae that had conspicuous mammiform cystocarps and a filamentous medulla. In 1933, Setchell and Gardner selected from this group of species those that lacked tetrasporophytes and treated them as a subgenus of *Gigartina*. In the period 1980–1983, closely following the discovery that these species had a life history in which the upright gametophyte alternated with a crustose tetrasporophyte previously described as an independent genus, Kützing's genus was resurrected but with a highly modified circumscription." Obviously, a blending of taxonomic and nomenclatural history would be maximally informative.

CONCLUSION

I have mentioned several aspects of the historical context of a taxonomic interpretation and discussed how they affected both the interpretation and its reception. I strongly recommend that taxonomists consider the context in which they are working. Such appreciation of extrinsic factors might raise the quality of our taxonomic output. At the least, it would make the practice of taxonomy more enjoyable.

Our taxonomic system is a synthesis of millions of individual taxonomic decisions of varying magnitude. No taxonomist can hope personally to make more than an extremely small percentage of the decisions necessary to support the taxonomic structure that he is teaching or altering. Every taxonomist relies very heavily upon the contributions of others. Therefore, we must conclude that the perceived difference between taxonomic history and taxonomy is a matter of perspective rather than substance. Taxonomic history *is* taxonomy. What is proposed today becomes history tomorrow. As we move forward, it is essential that we bring up the rear guard.

REFERENCES

Børgesen, F. (1927). Marine algae from the Canary Islands, especially from Teneriffe and Gran Canaria. III. Rhodophyceae. Part I. Bangiales and Nemalionales. *Biol. Medd.—K. dan. Vidensk. Selsk.* **6**(6), 1–97.

Cleland, R. E. (1919). The cytology and life-history of *Nemalion multifidum*, Ag. *Ann. Bot. (London)* **33**, 323–351.

Dixon, P. S. and Irvine, L. M. (1977). "Seaweeds of the British Isles. Volume 1. Rhodophyta. Part 1. Introduction, Nemaliales, Gigartinales." British Museum (Natural History), London.

Esper, E. J. C. (1797–1808). "Icones Fucorum. . ." Raspe, Nürnberg. (For critique, see *Wassmann J. Biol.* **11**, 221–232.)

Gray, S. F. (1821). "A Natural Arrangement of British Plants . . ." Baldwin, Cradock & Joy, London.

Hirn, K. E. (1900). Monographie und Iconographie der Oedogoniaceen. *Acta Soc. sci. Fenn.* **27**, 1–394.

Hofmeister, W. (1851). "Vergleichende Untersuchungen der Keimung, Entfaltung und Fruchtbildung höherer Kryptogamen . . ." F. Hofmeister, Leipzig.

Kützing, F. T. (1843). "Phycologia generalis. . ." Brockhaus, Leipzig.

Kylin, H. (1916). Über die Befruchtung und Reduktionsteilung bei *Nemalion multifidum*. *Ber. Dtsch. bot. Ges.* **34,** 257–271.

Olivi, G. (1792). "Zoologia adriatica . . ." Bassano.

Ralfs, J. (1848). "The British Desmidieae." Reeve, Benham & Reeve, London.

Rietema, H. (1975). Comparative investigations on the life-histories and reproduction of some species in the siphoneous green algal genera *Bryopsis* and *Derbesia*. Doctoral Dissertation, Rijksuniversity of Gröningen.

Sachs, J. (1874). "Lehrbuch der Botanik . . . Vierte . . . Auflage." Engelmann, Leipzig.

Setchell, W. A. and Gardner, N. L. (1933). A preliminary survey of *Gigartina*, with special reference to its Pacific North American species. *Univ. Calif. Publ. Bot.* **17,** 255–339.

Stackhouse, J. (1795–1801). "Nereis britannica; . . ." S. Hazard, Bath.

Stackhouse, J. (1809). Tentamen marino-cryptogamicum, ordinem novum, in genera et species distributum, in Classe XXIVta Linnaei sistens. *Mém. Soc. imp. Nat. Moscou* **2,** 50–97.

Steenstrup, J. J. S. (1842). "Ueber den Generationswechsel . . ." C. A. Reitzel, Copenhagen.

Strasburger, E. (1894). The periodic reduction of the number of the chromosomes in the life-history of living organisms. *Ann. Bot. (London)* **8,** 281–316.

Suringar, W. F. R. (1867). Algarum japonicarum Musei botanici L. B. index praecursorius. *Ann. Mus. bot. Lugd.-Bat.* **3,** 256–259.

Von Stosch, H. A. (1965). The sporophyte of *Liagora farinosa* Lamour. *Br. phycol. Bull.* **2,** 486–496.

Wittrock, V. B. (1870). Dispositio Oedogoniacearum suecicarum. *Öfvers. K. Vetensk.-Akad. Förh., Stockholm* **27,** 119–144.

Wolfe, J. J. (1904). Cytological studies on *Nemalion*. *Ann. Bot. (London)* **18,** 607–630.

Taxonomic Index

A

Acanthocodium, 425
  fragile, 426
Acetabularia, 3, 17, 90, 94, 101, 139, 140, 160, 272, 332, 339
Acetabularieae, 273, 300
Acicularia, 273
Acrochaete, 131, 222
  repens, 222
Acrosiphonia, 125, 126, 127, 128, 129, 130, 145, 146, 417
Acrosiphoniaceae, 183
Acrosiphoniales, 17, 18, 158, 159, 160, 161, 171, 185
Aegagropila [section], 162, 163, 164, 165, 168, 170
Affines [section], 162, 164, 165
Agardhia [subsection], 315, 322, 325, 326, 327
Akontae, 5
Alysium, 272
Anadyomenaceae, 135
Anadyomene, 135, 137, 157, 158, 159, 164, 165, 168, 169, 170, 171, 271, 272
Anadyomeneae, 4
Angiospermeae, 4
Ankistrodesmus, 399, 401, 429
  stipitatus, 399
Ankyra, 59
Apatococcus, 216
Aphanochaetaceae, 43, 61, 207, 209, 213, 215, 224, 227
Aphanochaete, 61, 215, 218, 222, 223, 224, 226, 227
  confervicola, 215, 226
  elegans, 226
  magna, 226
  polychaete, 215
  repens, 215
  species, 226
Apjohnia, 158, 166, 167, 168, 170, 171, 172, 272
  laetevirens, 167

Archidesmidiinae, 253, 254
Arnoldiella, 171
Arnoldiellaceae, 135
Arthrodesmus, 261
Asteromonas, 31, 58, 81, 82
Astrephomene, 58
Atractomorpha, 59
Autosporaceae, 381
Autosporinae, 381
Avrainvillea, 18, 273, 275, 276, 278, 279, 280, 282, 284, 285, 286, 288, 289
  erecta, 280
  longicaulis, 284, 285
  nigricans, 280
Avrainvilleae, 286
Azolla, 214

B

Bangia, 179
Basicladia, 135, 137, 139, 171
Batophora, 136, 139, 140, 160, 273, 331, 332, 333, 334, 336, 337, 338, 339
Bernardinella, 381
  bipyramidata, 380, 381
Bernardinellae, 381
Bicosoeca (= Bicoeca), 36
Biliphyta, 8
Binuclearia, 59, 188
Blastophysa, 135, 273
Blastosporaceae, 181
Blidingia, 101, 131, 134, 191, 192
  minima, 199
Blodgettia, 272
  australis, 170
Blodgettiomyces borneti, 170
Boergesenia, 158, 170, 173, 174
Bolbocoleon, 146, 148, 180
Bonnemaisonia, 423
  hamifera, 423
  nootkana, 423
Boodlea, 158, 164, 165, 166, 170, 171, 173, 272
Boodleoides, 157, 162, 163, 164, 165

Boodleopsis, 273, 275, 276, 279, 280, 283, 286, 289
　carolinensis, 280
Bornetella, 273
Botrydiaceae, 5
Botrydium, 4, 271, 272, 273
Botryococcus, 392
Botryophora, 273
Boueina, 273
Brachiomonas, 58
Braunia [subsection], 315, 322, 325, 327
Brownia [subsection], 317
Bryobesia, 157, 164, 165, 168, 169
　johannae, 165
Bryophyta, 8, 19, 20, 23, 304
Bryopsidaceae, 141, 142, 143, 272, 286, 288, 289, 422
Bryopsidales, 99, 101, 102, 271, 286, 287, 289, 290
Bryopsidella, 275, 286, 423
Bryopsidicae, 13
Bryopsidineae, 271, 286, 287, 288, 289, 290
Bryopsidophyceae, 5, 17, 18, 21, 23, 63, 290
Bryopsis, 3, 4, 17, 18, 101, 141, 142, 160, 271, 272, 273, 274, 275, 281, 284, 286, 290, 332, 338
　hypnoides, 349
　lyngbyei, 101
　plumosa, 278
Bulbochaete, 62, 215
Byssus, 234

C

Caespitella, 182, 213
　pascheri, 213
Callipsygma, 273, 275, 276, 286, 289
Calyptobactron, 381
　indutum, 380
Camellia, 239
Capsosiphon, 125, 191
Carteria, 30, 32, 37, 38, 57, 58, 65, 104
　crucifera, 38
　obtusa, 104
Caulerpa, 3, 18, 141, 142, 143, 271, 273, 274, 275, 277, 278, 279, 281, 282, 283, 284, 285, 286, 288, 289
　bikensis, 284
　cupressoides, 281
　prolifera, 285
　racemosa var. *peltata*, 285
　serrulata, 281, 282
　simpliciuscula, 349
Caulerpaceae, 4, 141, 288
Caulerpales, 23, 121, 141, 142, 143, 145, 146, 147, 148, 160, 275, 277, 282, 289, 340, 391, 393, 394
Caulerpeae, 271, 273, 286
Caulerpieae, 4
Caulerpoideae, 286
Cedercreutziella, 213, 218
　savoniensis, 213
Cephaleuros, 90, 234, 235, 236, 238, 239, 240, 244, 245, 338
　virescens, 239, 241, 244
Chaetangiaceae, 301
Chaetobolus, 215, 218
　gibbus, 215
Chaetocladiella, 171
Chaetomnion, 215, 218, 223
　pyriferum, 215
Chaetomorpha, 135, 158, 159, 164, 165, 180, 272
　aerea, 165
Chaetonella, 171
Chaetonema, 61, 215, 218, 223, 224, 227
　irregulare, 215
　ornatum, 215
Chaetonemopsis, 215, 227
　pseudobulbochaete, 215
Chaetopeltis, 149
Chaetophora, 61, 180, 213, 218, 220, 221, 226
　elegans, 226
　globosa, 213
　incrassata, 226
　species, 226
Chaetophoraceae, 5, 43, 61, 122, 180, 207, 209, 211, 212, 213, 215, 217, 220, 227
Chaetophorales, 5, 23, 43, 49, 53, 59, 60, 61, 110, 122, 150, 180, 181, 182, 183, 185, 187, 189, 197, 200, 207, 208, 209, 210, 212, 213, 214, 216, 217, 218, 220, 222, 225, 226, 227, 228, 234, 393, 394
Chaetophoreae, 180, 215, 227
Chaetophoroideae, 180, 215
Chaetophorophyceae, 5
Chaetosiphon, 135, 136, 137, 138, 139, 273

Chaetosiphonaceae, 135, 273
Chaetosphaeridiaceae, 41, 50, 215, 227
Chaetosphaeridiales, 215
Chaetosphaeridium, 50, 81, 102, 103
  globosum, 215
Chalmasia, 273
Chamaedoris, 157, 158, 164, 165, 168, 169, 170, 171, 173, 272
  peniculum, 165
Chamaephyceae, 4
Chamaetrichon, 211, 223
Chara [subsection], 315, 316, 326, 327
Chara, 3, 17, 47, 78, 80, 303, 307, 308, 309, 310, 312, 313, 314, 315, 316, 318, 322, 324, 325, 326, 327
  brachypus, 319
  braunii, 308, 316, 319
  canescens, 304, 312, 318, 327
  connivens, 314
  contraria, 308, 316, 324
  ecklonia, 327
  evoluta, 314
  fibrosa
    f. *fibrosa* (= *preissii*), 327
    f. *hookeri*, 327
  fragifera, 319
  globularis, 308, 314, 316, 318
  hispida f. *crassicaulis*, 327
  indica, 319
  kenoyeri, 327
  myriophylla, 327
  preissii, 316
  tomentosa, 319, 327
  vulgaris, 308, 316, 324
  zeylanica, 316, 318, 324
    var. *zeylanica*, 308, 316
    f. *michauxii*, 318
    f. *zeylanica*, 308, 316
Characeae, 3, 303
Characium, 399
Charales, 21, 22, 41, 48, 50, 51, 122, 393
Chareae, 4, 303, 304, 308, 309, 312, 316, 318, 319, 320, 322, 323, 324, 325, 326
Charicae, 13
Charophyceae, 2, 14, 18, 19, 20, 21, 22, 34, 37, 39, 41, 44, 45, 47, 48, 49, 50, 51, 52, 54, 55, 57, 60, 61, 66, 80, 81, 84, 99, 102, 103, 105, 107, 109, 110, 112, 113, 123, 161, 183, 198, 200, 247, 290, 338, 392

Charophyta, 2, 13, 14, 20, 23, 297, 298, 301, 303, 304, 312, 313, 318, 319, 321, 323, 324, 326
Charophytina, 8
Charopsis [section/subsection], 315, 326, 327
Chlamydomonadales, 23, 42, 57, 58
Chlamydomonas, 3, 15, 16, 30, 31, 33, 57, 58, 64, 65, 81, 82, 87, 88, 91, 93, 200, 399, 401, 402, 409, 410, 411, 412, 414, 415, 416, 429
  aculeata, 413
  aggregata, 413
  angulosa, 412
  applanata, 413
  asymmetrica, 412
  brannonii, 413
  callosa, 412
  cingulata, 415
  culleus, 413
  debaryana, 412
  dysosmos, 413
  elliptica, 413
  eugametos, 413
  frankii, 413, 416
  geitleri, 40, 413
  gelatinosa, 413
  globosa, 412
  gloeopara, 412
  gymnogama, 413
  hindakii, 413
  humicola, 413
  incerta, 412
  indica, 413
  inepta, 412
  iyengarii, 412
  komma, 412
  mexicana, 412
  mirabilis, 415
  moewusii, 87
  monoica, 413
  noctigama, 413, 415
  oblonga, 412
  pallidostigmatica, 413
  peterfi, 412
  pinicola, 413
  pitschmannii, 413
  reinhardii (= *reinhardtii*), 80, 81, 87, 88, 90, 92, 94, 104, 149
  reinhardtii, 401, 410, 411, 412, 416
  segnis, 413, 416

Chlamydomonas (cont.)
  species, 412, 413
  sphaeroides, 412
  starrii, 413
  terricola, 413, 415
  texensis, 413
Chlamydophyceae, 12, 21, 22, 63, 64
Chlorcorona, 104
Chlorella, 3, 58, 391, 392, 393, 394, 395, 396, 397, 398, 400, 401
  ellipsoidea, 395, 398
  fusca
    var. *fusca*, 396, 397, 400, 401
    var. *rubescens*, 396, 397, 398, 400, 401
    var. *vacuolata*, 396, 397, 400, 401
  homosphaera, 395, 396, 397, 400
  kessleri, 396, 397, 400
  lobophora, 397, 400
  luteoviridis, 395, 396, 397, 400
  minutissima, 396, 397, 400
  mirabilis, 396, 400
  protothecoides, 395, 396, 397, 400, 401
  pyrenoidosa, 395
  saccharophila, 395, 396, 397, 398, 400
  sorokiniana, 395, 396, 397, 398, 400, 401
  vulgaris, 395, 396, 397, 398, 400, 401
    f. *tertia*, 398
  zofingiensis, 396, 397, 400, 401
Chlorellales, 22, 58, 64
Chlorobionta, 8
Chlorocladus, 273
Chloroclonium, 236
Chlorococcaceae, 381
Chlorococcales, 16, 17, 43, 45, 53, 57, 58, 59, 63, 110, 160, 185, 188, 189, 227, 379, 385, 393, 394, 399, 429
Chlorococcophyceae, 5
Chlorococcum, 58, 63, 64, 399
  oleofaciens, 64
Chlorocystis, 125, 126, 130
Chlorodendron, 214
Chlorodesmis, 18, 273, 275, 276, 277, 279, 280, 281, 282, 284, 285, 286, 289
  baculifera (= *bulbosa*), 280, 285
  bulbosa, 280, 281
Chlorogonium, 58
Chlorokybales, 41, 49
Chlorokybus, 49, 60, 200
  atmosphyticus, 110
Chloropelta, 131, 187, 191, 192
Chlorophyceae, 2, 4, 6, 7, 8, 14, 15, 18, 19, 20, 21, 22, 23, 34, 35, 37, 39, 40, 42, 43, 44, 45, 48, 49, 51, 52, 53, 55, 56, 58, 59, 60, 61, 63, 64, 65, 67, 81, 96, 99, 100, 103, 104, 107, 108, 110, 113, 114, 123, 144, 149, 160, 183, 200, 228, 252, 290, 332, 379, 391, 392, 393, 394
Chlorophycota, 20
Chlorophyta, 1, 2, 6, 8, 13, 20, 22, 23, 29, 30, 33, 36, 40, 123, 183, 200, 201, 252, 253, 304, 348, 391, 392
Chlorosarcina, 60, 182
Chlorosarcinaceae, 208
Chlorosarcinales, 23, 43, 57, 59, 60, 63, 110, 122, 183, 184, 187, 188, 200, 400
Chlorosarcinineae, 182
Chlorosarcinopsis, 90, 200
  dissociata, 87
  pseudominor, 78, 92, 96
Chlorospermeae, 3, 4
Chlorothecium inaequale, 417
Chlorothrix, 125
Chlorotylium, 236
Chromophyta, 8
Chromoplantae, 8
Chroolepaceae, 234
Chroolepidaceae, 5, 233, 234
Chroolepus, 234, 235
Chrysophyceae, 36
Cladocephalus, 275, 276, 280, 284, 285, 286, 288, 289
Cladophora, 14, 52, 135, 136, 137, 157, 158, 159, 161, 162, 163, 164, 165, 166, 168, 169, 170, 171, 172, 174, 180, 272, 332
  catenata, 162, 165, 170, 171
  coelothrix, 162, 165, 172
  dalmatica, 162
  fuliginosa (= *catenata*), 170
  glomerata, 136, 137, 159
  incompta, 165
  intertexta, 162, 165
  liebetruthii, 162, 165
  pachyderma, 162, 169
  pellucida, 162, 165, 169
  prolifera, 167, 168, 169, 170, 171
  retroflexa, 167, 170, 171, 172
  rupestris, 168
  sericea, 162, 168
    var. *biflagellata*, 159
Cladophoraceae, 4, 134, 135, 136, 138, 208, 272
Cladophorales, 16, 18, 22, 23, 121, 135,

157, 158, 159, 160, 161, 163, 164, 165, 168, 169, 170, 171, 174, 181, 185, 224, 272, 274, 394
Cladophorella, 171
Cladophorinae, 180
Cladophoropsis, 157, 158, 164, 165, 166, 169, 170, 171, 172, 173, 174, 272
Cladostroma, 171
Cloniophora, 213, 218, 220
  willei, 213
Closteriaceae, 253
Closterium, 254
  kuetzingii, 258
  littorale, 258
  peracerosum, 258
  setaceum, 258
  strigosum, 258
Coccobotrys, 216
Coccophyceae, 4, 5
Codiaceae, 141, 273, 286, 288, 289, 300
Codiales, 282
Codieae, 4, 273, 288
Codiolaceae, 182, 183
Codiolophyceae, 5, 13, 21, 22, 64
Codiolum, 272
Codium, 3, 4, 18, 141, 271, 272, 273, 274, 275, 278, 281, 284, 286, 288, 290, 425, 427, 428
  acuminatum, 426, 428
  adhaerens, 426, 427
  bursa, 425, 426, 427
  cylindricum, 426, 427
  elisabethiae, 426, 428
  fragile, 426
    subsp. *atlanticum*, 426, 427, 428
    subsp. *tomentosoides*, 129, 426
  laminarioides, 427
  megalophysum, 426, 427
  petaloideum, 426, 427
  ritteri, 427
  spinescens, 426, 428
  tomentosum, 425, 426, 427
  vermilara, 425
Coelastrum, 399
Coeloblaste, 4
Coelosphaeridium, 273
Coleochaetaceae, 41, 50, 180, 208, 227
Coleochaetales, 13, 41, 48, 50, 51, 61, 122, 183, 227, 393
Coleochaete, 17, 47, 50, 51, 99, 102, 106, 431
  pulvinata, 47, 50

Coleochaetophyceae, 5
Colponema, 36
Conferva, 3, 425
Confervaceae, 5, 180
Confervales, 6, 180
Conferveae, 4
Confervoideae, 3, 5
Confervophyceae, 5
Conjugales, 252, 393
Conjugataceae, 3
Conjugatae, 5, 181
Conjugatophyceae, 5, 252
Conjugatophyta, 20
Conochaete, 50
Corillionia [subsection], 325, 327
Corticatae [section], 289
Cosmarium, 255, 258, 259, 260, 261, 266
  botrytis, 265
  moniliforme, 266
  taxichondrum var. *taxichondrum*, 259
Cosmoastrum, 261
Cosmostaurastrum, 261
Crenacantha, 213, 218, 221
  orientalis, 213
Crucigenia, 399
Crucigeniella, 399
Crypticae [section], 289, 290
Cryptogamia, 3
Cryptophyta, 8
Cryptoplantae, 8
Ctenocladus, 53, 198, 233, 236, 244, 246
Ctenocladales, 22, 183, 185, 187, 207, 210, 211, 215, 216, 225, 228
Cyanophyceae, 11
Cycadophyta, 20
Cylindriastrum, 261
Cylindrocapsa, 59, 185, 188
  involuta, 194
Cylindrocapsaceae, 181, 186
Cylindrocapsales, 181, 185
Cymbomonas, 32, 33, 47
  adriatica, 32
  tetramitiformis, 32, 63
Cymopolia, 273
Cymopolieae, 273
Cystodictyon, 272
Cystoseira, 423

D

Dactylopora, 273
Dasycladaceae, 4, 273, 299, 300, 301

Dasycladales, 18, 23, 99, 101, 121, 136, 139, 140, 145, 146, 147, 148, 160, 273, 297, 298, 299, 300, 301, 302, 340, 393, 394
Dasycladeae, 4, 273
Dasycladus, 273
Decandollea [subsection], 317
Derbesia, 17, 18, 101, 141, 142, 160, 271, 272, 274, 275, 284, 286, 290, 332, 422
  marina, 422
  tenuissima, 278
Derbesiaceae, 272, 288
Derbesiales, 17, 282
Dermatophyton, 171
Desikachariya [subsection], 325, 327
Desmatractum, 379, 380, 381, 386, 387
  bipyramidatum, 380, 381, 382, 383, 384, 385, 386, 387
  delicatissimum, 380, 381, 382, 383, 384, 385
  elongatum, 380, 386
  indutum, 380, 381, 382, 383, 384, 385, 386, 387
    var. *elegans,* 383
  nyanzae, 380
  obtusum, 380, 386
  plicatum, 380, 381, 383, 384
  spryii, 380, 381, 382, 383, 385
Desmideae, 4
Desmidiaceae, 5, 251, 252, 253, 255, 256, 257
Desmidiales, 253, 254
Desmidiinae, 253, 254
Desmococcus, 216
Desvauxia [subsection], 315, 327
Diatomaceae, 4
Dichotomosiphon, 18, 52, 130, 141, 143, 271, 274, 287
  pusillus, 273, 284
  tuberosus, 273, 284, 285, 287
Dichotomosiphonaceae, 141, 273
Dicranochaetaceae, 227
Dictyosphaeria, 135, 136, 137, 157, 158, 159, 160, 161, 170, 171, 173, 174, 272
  cavernosa, 170
Dictyota, 430
Digitella, 273
Dikontae, 181
Dimorphosiphon, 273, 301
Diplopora, 273
Diplosphaera, 182

Dolichomastix, 47, 84, 111
Dorsiventrales [section], 162, 163, 164, 165, 169
Draparnaldia, 61, 179, 180, 182, 195, 199, 213, 217, 218, 220, 221, 225, 226
  glomerata, 226
  mutabilis, 213, 226
  plumosa (= *mutabilis*), 226
  species, 226
Draparnaldieae, 4
Draparnaldiella, 213, 218, 220, 221
  baicalensis, 213
Draparnaldiopsis, 52, 61, 182, 213, 217, 218, 220
  alpinus, 213
Dunaliella, 17, 31, 58, 64, 401
  primolecta, 90
  salina, 9
Dysmorphococcus, 58

E

Earthya [subsection], 317
Ectocarpales, 51
Ectochaete, 227
Elakotothrix, 182
Embryophyta, 37
Embryophytina, 8
Endoclonium, 213, 215, 218
  chroolepiforme, 213
  rivulare, 213
Endophyton, 131, 185, 189, 191, 236
Enteromorpha, 14, 101, 131, 132, 133, 134, 184, 186, 187, 191, 192, 200, 272, 343, 344, 346, 348, 349
  ahlneriana, 187
  intestinalis, 194, 199, 343, 344, 345, 346, 347, 348, 349
  linza, 199
  prolifera, 343, 344, 345, 346, 348, 349
  radiata, 187
  torta, 187
Enteromorpheae, 4
Entocladia, 131, 132, 180, 185, 189, 191, 192, 222, 225, 227, 332
  viridis, 222
Ernodesmis, 157, 164, 165, 166, 167, 168, 169, 170, 171, 172, 173, 272
  verticillata, 167
Euastrum, 255, 258, 260

Euchlorophyceae, 5
Eucophycopeltis, 235
Eudorina, 58
Euglena, 3
Euglenophyceae, 2
Euglenophyta, 2
Eugomontia, 125, 126, 128, 129, 130, 145, 147
　sacculata, 126, 129
Eusiphonales, 277

F

Flabellaria, 273, 286, 289
　petiola, 288
　petiolata, 286
Flabellarieae, 286, 288
Follicularia, 272
Fridaea (= Friedaea), 236
Friedaea, 215, 218, 221
　torrenticola, 215
Friedmannia, 53, 54, 60, 102, 103, 110, 184, 186, 187, 188, 193, 200, 247, 332, 338, 339
　israelensis, 78, 81, 195, 199
Fritschiella, 61, 182, 214, 218, 226, 227
　species, 226
　tuberosa, 89, 90, 96, 214, 226
Fucales, 51
Fucus, 4, 422, 425
Furcilla, 81

G

Gamophyta, 20
Gayralia, 125, 127, 187, 191, 192
　oxysperma, 128
Gemina [subgenus], 132
Geminella, 416
Gemmiphora, 171
Geppella, 275
Gigartina, 431, 432
Gioallenia [subsection], 317
Glaucophyceae, 5, 144
Glenodinium lubiniensiforme, 417
Gloeoplax, 225
Gloeotila, 60
Gloeotilopsis, 60
　sterilis, 60
Glomeratae [section], 162, 163

Golenkinia minutissima, 90, 96
Gomontia, 125, 236, 272
Gomontiaceae, 272
Gomontieae, 236
Gonatoblaste, 215, 221, 227
　rostrata, 215
Gonatozygon chadefaudii, 254
Gonatozygonaceae, 253
Gongrosira, 4, 183, 185, 186, 189, 191, 236
Gongrosireae, 236
Goniotrichum, 180
Grovesia [subsection], 315, 316, 325, 326, 327
Guerlesquinia [subsection], 325, 327
Gymnocodiaceae, 301
Gymnospermeae, 4
Gyroporella, 273

H

Haematococcus, 58
Hafniomonas, 37, 38, 65, 81, 104, 113, 114
Halicoryne, 273
Halicystidaceae, 272
Halicystis, 17, 272, 275, 286, 422
Halimeda, 18, 141, 160, 271, 273, 274, 275, 276, 277, 278, 279, 280, 281, 282, 283, 284, 285, 286, 288, 289, 290, 299, 301
　cryptica, 277
　incrassata, 281
　scabra, 280
　tuna, 280, 281, 282, 283, 284
Halimeda [section], 289
Halimedaceae, 286, 288
Halimedineae, 271, 286, 287, 288, 290, 291
Halimedoideae, 286
Halochlorococcum, 125
Halosphaera, 85
Hansgirgia, 235
Hartmania [subsection], 314, 327
Herposteiraceae, 227
Herposteiron naegeli, 227
Heterokontae, 6, 381
Heteromastix (= Nephroselmis), 31, 32, 47
　rotunda, 85
Heterostegina depressa, 398
Heterothallus, 235
Hormideae, 4
Hormidiella, 49
Hormidium, 180

Hydrodictyon, 4
　reticulatum, 417
Hyella [subsection], 317

I

Imahoria [subsection], 325, 326, 327
Interfilum, 60
Ireksokonia, 214, 218, 221
　formosa, 214
Isokontae, 5
Iwanoffia, 214, 219
　terrestris, 214

J

Jaoa, 211
Johnson-sea-linkia, 275, 284, 286, 289

K

Keratococcus, 399
　bicaudatus, 399
Kirchneriella, 188
Klebahniella, 214, 219, 220, 225
　elegans, 214
Klebsormidiaceae, 41
Klebsormidiales, 13, 17, 41, 49, 50, 66, 122, 183, 185, 198, 393
Klebsormidium, 49, 183, 198, 393, 402
　flaccidum, 110
Knightia [subsection], 317
Kuetzingia [subsection], 315

L

Lamarckia, 425
Laminariales, 51
Lamprothamnium, 304, 307, 308, 309, 310, 312, 313, 314, 326
　papulosum, 318
Lemna, 219
Leptosira, 189, 236
Liagora tetrasporifera, 430
Lochmium, 217, 236
Longiarticulatae [section], 157, 162, 163, 164, 165, 166, 168, 171
Loxophyceae, 82, 88
Lychnothamnus, 304, 307, 308, 309, 310, 312, 313, 314, 316, 326, 327

M

Magnolia grandiflora, 239
Mamiella, 84, 85, 105, 106, 107, 111, 112
　gilva, 105
Mamiellaceae, 105
Mantoniella, 32, 33, 46, 84, 98, 99, 101, 103, 106, 107
　squamata, 81, 82, 106
Mastocarpus, 431
Mesostigma, 15, 32, 37, 47, 85, 99, 102, 105, 106, 107, 109, 112, 113
　viride, 105
Mesotaeniaceae, 253
Micrasterias, 255, 259, 260, 261, 262, 263, 264
　americana, 263, 264
　denticulata, 264
　mahabuleshwarensis, 263
　　var. *europa*, 263
　　var. *reducta*, 263
　　var. *typica*, 263
　　var. *wallichi*, 262, 263
　papillifera, 264
　rotata, 264
　thomasiana, 264, 265, 266
　torreyi, 264
Microdictyon, 135, 137, 157, 158, 159, 164, 165, 169, 170, 171, 272
Micromonadophyceae, 14, 21, 22, 23, 40, 41, 43, 45, 46, 52, 62, 63, 65, 66
Micromonas, 31, 46, 47
　pusilla, 32, 81
Micronesicae [section], 289
Micropoa, 215, 219
　leptochaete, 215
Microspora, 59, 186
　quadrata, 417
Microsporaceae, 43, 59
Microsporales, 6, 181, 185, 186
Microsporinae, 181
Microsporopsis, 49
Microthamniaceae, 210
Microthamnion, 53, 54, 99, 102, 104, 110, 179, 210, 217, 339
　kuetzingianum, 90, 400
Migularia [subsection], 317, 326
Mizzia, 273
Monomastix, 47, 81, 98
Monoraphidium, 399, 401

Monostroma, 96, 125, 128, 129, 130, 184, 186, 187, 189, 191, 192, 348
  angicava, 186
  bullosum, 187, 188, 189, 190, 196, 200
  grevillei, 82, 128, 199
  nutidum, 186
Monostromaceae, 182
Monostromataceae, 189
Mougeotia, 17
  oedogonioides, 417
Muellaria [subsection], 317
Myxonemopsis, 214, 219, 220, 221
  crassimembranacea, 214

N

Nanochlorum, 393
  eucaryotum, 392
Nautococcus mammilatus, 58
Nemaliales, 430
Nemalion, 430
Nematophyceae, 5
Neomeris, 273
Neospongiococcum, 399
Nephroselmis (= *Heteromastix*), 15, 31, 32, 33, 47, 84, 85, 87, 90, 98, 99, 100, 102, 105, 106, 107, 112
  gilva, 84
  olivacea, 79, 84, 94, 98, 99, 106, 107, 108, 112
  pyriformis, 85, 91
  rotunda, 85, 94, 96, 98, 99
Nitella [subsection], 316, 317
Nitella, 3, 304, 307, 308, 309, 310, 314, 316, 318, 323, 324, 326
  acuminata, 308, 316, 324
  batrachospermum, 308
  confervacea, 316
  flexilis, 308, 316, 318
  gracilis, 308
  hyalina, 308, 316
  mirabilis, 318, 319
  mucronata, 308, 316
  oligospira, 308, 316
  opaca, 308, 316, 319
  pulchella, 319
  subglomerata, 319
  tenuissima, 308, 316
Nitelleae, 304, 308, 309, 318, 319, 320, 324, 325, 326

Nitellopsis, 304, 307, 308, 309, 310, 312, 313, 314, 325, 326
Nordstedtia, 215, 219, 220, 227
  globosa, 215
Nostoc, 219
  verrucosa, 214
Nylandera, 235

O

Ochlochaete, 131
  histrix, 215
Oedocladium, 62
Oedogoniaceae, 5, 181, 208
Oedogoniales, 6, 16, 17, 21, 23, 43, 49, 58, 59, 60, 61, 62, 64, 65, 100, 104, 183, 393, 429, 431
Oedogoniicae, 13
Oedogoniinae, 181
Oedogoniophyceae, 5, 21, 23, 64
Oedogonium, 62, 90, 107, 180, 367
  concatenatum, 417
Oocystaceae, 381
Opuntia [section], 289, 290
Ostreobiaceae, 141, 144
Ostreobiales, 285
Ostreobium, 141, 143, 145, 271, 273, 274, 285, 286, 288
  quekettii, 286
  reineckeri, 286
Ovulites, 273

P

Pachycladella, 379
  umbrina, 386
Pachysphaera, 84, 85
Palaeodasycladus, 273
Paleoporella, 273
Palia [subsection], 317
Palmaria, 423
Palmariaceae, 423
Palmariales, 423
Palmella, 58
Palmellaceae, 4, 381
Palmelleae, 4
Pandorina, 9, 58
Paramecium, 398
Pavlova, 36
Pavlovales, 35, 36

Pediastrum, 429
  boryanum, 401
  duplex, 401
Pedinomonas, 31, 46, 47, 77, 78, 82, 87, 88, 101, 103
  minor, 32, 101
  tuberculata, 77, 78, 83, 88, 97
Pedobesia, 275, 286
Peniaceae, 253
Penicillus, 272, 273, 275, 276, 277, 278, 279, 280, 281, 282, 283, 284, 285, 286, 289
  capitatus, 278, 280, 281, 283
  sibogae, 276, 286
Peniococcus, 380
  nyanzae, 380
Penium, 254
  margaritaceum, 254
  polymorphum, 254
  spirostriolatum, 254
Percursaria, 131, 134, 186
  dawsonii, 133, 134
  percursa, 133
Persoonia [subsection], 317, 326
Petiolata, 286
Phacotus, 58
Phaeophila, 148, 222
Phaeophyceae, 20
Phaeophyta, 12, 19
Phycopeltis, 234, 235, 236, 237, 238, 239, 240, 241, 242, 243, 244, 245
  epiphytum, 239, 242
Phyllogloea, 149
Phyllosiphon, 273
Phyllosiphonaceae, 144, 273, 274
Phyllosiphonales, 285
Phymatodocis, 254
Physolinum, 234, 235, 236
Phytophysa, 273
Piliniella, 131
Pinophyta, 20
Pithophora, 158, 171, 272
Placodermae, 5
Plantae, 9
Platydorina, 58
  caudata, 13
Platymonas, 31, 32, 37, 53, 56
Pleurastraceae, 56, 247
Pleurastrales, 22, 23, 42, 56, 60, 66, 67, 210
Pleurastrophyceae, 21, 22, 23, 34, 35, 38, 39, 42, 44, 45, 48, 49, 52, 53, 55, 56, 60, 63, 65, 66, 81, 102, 103, 104, 107, 108, 110, 113, 114, 144, 225, 228
Pleurastrum, 53, 54, 56, 110, 144, 236
  paucicellulare, 225
  terrestre, 56, 89, 90
Pleurococcaceae, 381
Pleurococcus, 182
Pleurotaenium, 265
  coronatum, 265
  ehrenbergii, 265
  mamillatum, 265
  spinulosum, 254
Polyphysa, 273
Polytoma, 58, 402
  obtusum, 94
Porphyra, 272
Prasinocladus, 53, 56, 65
Prasinophyceae, 2, 20, 21, 22, 37, 38, 47, 65, 66, 109, 392, 393, 400
Prasinophyta, 2, 13, 14, 20, 21, 23
Prasiola, 181, 185
Prasiolales, 17, 185, 186
Primicorallina, 273
Pringsheimiella, 131
Printzia, 233, 235
Prochloron, 392
Prochlorophyceae, 2
Prochlorophyta, 2, 392
Proctoria [subsection], 325, 327
Prostaurastrum, 262
Protococcaceae, 4, 381
Protococcales, 381
Protococcineae, 6
Protococcoideae, 5
Protococcophyceae, 5
Protoctista, 9
Protoderma, 223
Protomonostroma, 187, 191, 192
Protonemeae, 4
Protosiphon, 272, 399
Protosiphonaceae, 272
Prototheca, 398, 401
  stagnora, 398
  wickerhamii, 398
  zopfii, 398
Prymnesiophyceae, 36
Pseudendocloniopsis, 53, 60
Pseudendoclonium, 53, 125, 129, 131, 133, 134, 147, 183, 185, 186, 187, 189, 191, 195, 223, 236

akinetum, 186
basiliense, 194, 199
printzii, 186
prostratum, 186
Pseudobryopsis, 101, 141, 142, 160, 199, 273, 275, 286, 290, 332, 338
Pseudochaete, 214, 219
crassiseta, 214
Pseudochlorococcum, 399
Pseudochlorodesmis, 275, 276, 283, 284, 286, 289
furcellata, 283
Pseudocodiaceae, 286, 289
Pseudocodiineae, 289
Pseudocodium, 273, 275, 276, 278, 279, 280, 285, 286, 288, 289
floridanum, 278, 280, 288
Pseudodichotomosiphon, 273
Pseudodictyon, 236
Pseudopringsheimia, 125, 128
confluens, 126
Pseudoscourfieldia, 84, 85, 87, 98, 99, 113
marina, 85
Pseudostruvea, 170
Pseudotetraspora marina, 188, 189, 190
Pseudotrebouxia, 53, 56, 60, 110
Pteridophyta, 8, 20
Pteromonas, 58, 81, 82, 87
Pterosperma, 47, 85
Pyramimonas, 32, 37, 38, 47, 83, 84, 85, 87, 94, 98, 99, 101, 102, 105, 107
amylifera, 86, 101, 103, 106
parkeae, 86
tetrarhynchus, 85
Pyrobotrys, 58

R

Radiofilum, 59, 188
Rajia [subsection], 317
Raphidiastrum, 261
Raphidonema, 49, 182
Repentes [section], 157, 162, 163, 164, 165, 166
Rhabdoporella, 273
Rhaphidium, 381
Rhexinema, 225
Rhipidodesmis, 273, 275
Rhipidosiphon, 289
Rhipileae, 286

Rhipilia, 275, 276, 286, 289
Rhipiliopsis, 275, 276, 286, 289
Rhipocephalus, 272, 273, 275, 276, 277, 278, 283, 284, 286, 289
Rhipsalis [section], 289
Rhizoclonium, 135, 158, 159, 164, 165, 169, 180, 272
grande, 164, 165, 168
riparium, 165
Rhizothallus, 236
Rhodophyta, 14, 19
Rhodymenia, 423
pseudopalmata, 423
Rhyniicae, 13
Richardwoodia [subsection], 325, 327
Riddellia [subsection], 317
Rupestres [section], 162, 163, 168

S

Saccodermae, 5
Saprochaete, 214, 227
saccharophila, 214
Scenedesmaceae, 380
Scenedesmus, 398, 401, 429
obliquus, 401
Schizomeridaceae, 43, 61, 207, 209, 215, 228
Schizomeris, 53, 61, 131, 179, 215, 216, 217, 219, 224, 226, 228
leibleinii, 215, 222, 226
Scourfieldia, 32, 47, 81, 96, 98, 103
caeca, 78, 80, 82, 83, 88, 90, 91, 92, 94
Seletonellaceae, 300
Siphonaceae, 272, 274
Siphonales, 271, 272, 274, 277, 289, 400
Siphoneae, 4, 5, 271, 272, 274
Siphonineae, 6
Siphonocladaceae, 135
Siphonocladales, 16, 18, 22, 121, 134, 135, 136, 137, 138, 139, 140, 145, 146, 147, 148, 157, 170, 171, 174, 272, 274, 340, 393, 394, 400
Siphonocladeae, 272, 274
Siphonocladineae, 6
Siphonocladophyceae, 5, 174
Siphonocladus, 135, 137, 157, 158, 170, 171, 172, 173, 174, 272
tropicus, 167
Siphonoclathrus, 275

Siphonophyceae, 4, 5
Siphonophyta, 290
Siphophyceae, 5
Skvortzoviothrix, 214, 219, 220
   terrestris, 214
Smithsoniella, 53, 148, 233, 244, 246
Sorastrum, 100
Spermatozopsis, 82, 104
Sphaeroplea, 59, 157, 174, 182, 185, 272
Sphaeropleaceae, 4, 5, 43, 53, 59, 62, 272
Sphaeropleales, 43, 58, 59, 60, 62, 157, 174, 181, 185
Sphaeropleineae, 182
Sphaerosiphon, 272
Sphagnum, 225
Spirogyra, 17
Spongilla, 398
Spongiococcum, 399
Spongomorpha, 125, 129
Sporocladus, 148, 236
Staurastrum, 255, 256, 257, 258, 259, 260, 261, 262, 266
   anatinum, 257, 258
      var. *curtum*, 257
      var. *denticulatum*, 258
      var. *longibrachiatum*, 257
         f. *curtum*, 258
         f. *denticulatum*, 258
         f. *glabrum*, 258
         f. *hirsutum*, 258
         f. *longibrachiatum*, 258
         f. *paradoxum*, 258
   brasiliense, 261
   cingulum var. *affine*, 262
   denticulatum, 266
   dilatatum, 266
   dorsidentiferum, 262
   echinatum, 258
   furcigerum
      var. *reductum*, 262
      var. *simplicissimum*, 262
   gracile, 256, 257
   hirsutum, 261
   micron, 257
   muricatum, 261
   paradoxum, 256, 257, 261
   pelagicium, 262
   pentacerum, 259
   pileolatum, 261
   polymorphum, 260
   polytrichum, 261
   sebaldi var. *ornatum*, 260
   tetracerum, 259
   vestitum, 258, 259
      var. *subanatinum*, 258
Staurodesmus, 261
Stephanokontae, 5, 6
Stichococcus, 49
Stigeoclonium, 61, 90, 93, 94, 179, 180, 182, 213, 214, 215, 217, 219, 220, 221, 223, 224, 225, 226, 229, 363, 364, 366, 372, 374, 375
   aestivale, 226, 363, 364, 365, 366, 367, 373, 374, 375
   amoenum, 226
   carolinianum, 372
   chroolepiforme, 213
   farctum, 217, 226, 363, 364, 365, 366, 370, 371, 372, 373, 374, 375
      var. *rivulare*, 370, 372, 373
   helveticum, 217, 226, 363, 364, 365, 366, 367, 373, 375
      var. *minus*, 373
   huberi, 226
   nanum, 226
   pascheri, 226, 363, 364, 365, 366, 372, 373, 374, 375
   species, 226
   subsecundum, 226, 363, 364, 365, 366, 373, 374
   tenue, 214, 226, 363, 364, 365, 366, 369, 372, 373, 375
   variabile, 226, 363, 364, 365, 372, 373, 374, 375
Stomatochroon, 234, 235, 236, 237, 240, 241, 243
   lagerheimii, 235
Streptophyta, 8
Struvea, 157, 164, 165, 166, 170, 171, 172, 173, 272

T

Tellamia, 236
Tetrakontae, 181
Tetraselmidiales, 22, 23, 42, 56, 66, 67
Tetraselmidicae, 13

Tetraselmis, 31, 32, 37, 38, 53, 54, 55, 56, 63, 66, 85, 87, 91, 93, 96, 99, 106, 107, 113, 114
  cordiformis, 78, 79, 85, 86, 87, 90, 92, 95
  striata, 87
Tetraspora, 58, 149, 272
Tetrasporales, 58, 149, 189, 393
Tetrasporineae, 6
Tetrasporophyceae, 5
Tetrastrum, 399
Thallophyta, 5
Thamniochaete, 61, 213, 215, 219
  aculeata, 215
  huberi, 215
Thamniolum, 214, 227
  elegans, 214
Thyrsoporella, 273
Tieffallenia [subsection], 316, 317
Tolypella, 304, 307, 308, 309, 310, 312, 313, 314, 316, 318, 320, 324, 325, 326
  boldii, 320
  canadensis, 320
  glomerata, 308, 316, 324
  intricata, 320
  nidifica, 320
  salina, 319, 320
Tracheophyta, 20, 23
Trailliella, 423
  intricata, 423
Trebouxia, 53, 54, 56, 57, 110, 247
Trentepohlia, 19, 90, 109, 234, 235, 236, 237, 238, 240, 241, 242, 243, 331, 332, 334, 335, 336, 337, 338, 339
  aurea, 244, 246
Trentepohliaceae, 5, 52, 53, 208, 233, 234, 235, 236, 237, 238, 239, 240, 241, 242, 243, 244, 245, 246, 247
Trentepohliales, 15, 22, 23, 91, 92, 94, 102, 105, 108, 113, 114, 123, 148, 198, 247
Trentepohlieae, 236
Treubaria, 379, 381, 386
Treubariaceae, 379, 381
Treubarioideae, 381
Tribonema, 417
Trichloris, 31, 32, 33, 47, 63
Trichodiscus, 214, 219
  elegans, 214
Trichophilus, 225, 226
  polymorpha, 226

Trichosarcina, 52, 53, 125, 183, 184, 186, 187, 195, 228
  polymorpha, 194, 198, 199
Triploporella, 273
Trochiliscaceae, 301
Tydemania, 275, 276, 277, 278, 284, 286, 289
  expeditionis, 276, 277, 284

U

Udotea, 18, 141, 271, 273, 274, 276, 277, 278, 279, 282, 283, 284, 286, 289
  conglutinata, 285
  cyathiformis, 285
  flabellum, 278, 284, 285
  glaucescens, 276
  indica, 280, 281
  javensis, 276, 277, 278, 280, 281, 282
  minima (= *petiolata*), 277, 283, 288
  orientalis, 280
  papillosa, 277
  petiolata (= *minima*), 277, 280, 281, 282, 284, 286, 288
Udoteaceae, 141, 279, 288, 297, 298, 300, 301, 302
Udoteae, 288
Ulothrichaceae, 5
Ulothricheae, 179, 180
Ulothrichopsis, 60
Ulothrix, 18, 49, 53, 61, 96, 125, 127, 129, 130, 160, 179, 180, 182, 184, 185, 186, 187, 189, 191, 193, 194, 195, 198, 214, 226, 402
  belkae, 61, 214, 226
  confervicola, 226
  fimbriata, 61, 194, 214, 226
  flacca, 187, 192, 193
  gigas, 226
  implexa, 187, 191, 194, 195, 196, 200
  mucosa, 188
  pallusalsa, 187
  tenuissima, 194
  verrucosa, 194, 195, 196, 198
  zonata, 51, 110, 181, 182, 186, 187, 194, 199, 200, 332, 417
Ulotrichaceae, 180, 182
Ulotrichales, 5, 6, 17, 23, 99, 100, 101, 104, 110, 121, 122, 123, 125, 126, 130, 131,

Ulotrichales (cont.)
    133, 136, 141, 145, 146, 147, 148, 160,
    174, 179, 180, 181, 182, 183, 184, 185,
    186, 187, 188, 189, 191, 192, 194, 195,
    197, 198, 199, 200, 207, 208, 210, 214,
    228, 393, 394, 402, 431
Ulotricheae, 179, 180
Ulotrichinae, 180, 181
Ulotrichineae, 6
Ulotrichophyceae, 5, 21
Ulva, 3, 98, 100, 131, 132, 133, 147, 160,
    183, 184, 186, 187, 189, 191, 192, 193,
    198, 200, 272, 298, 332, 353, 359, 360,
    394, 425
  fasciata, 192, 354, 360
  fenestrata, 359, 360
  lactuca, 90, 92, 95, 96, 101, 192, 199, 200,
    354
  laetevirens, 353, 354, 355, 356, 357, 358,
    359, 360
  mutabilis, 191, 194
  rigida, 353, 354, 355, 356, 357, 358, 359,
    360
  stenophylla, 353, 354, 355, 356, 357, 358,
    359, 360
Ulvaceae, 4, 5, 131, 134, 180, 181, 208,
    227, 228, 271, 272
Ulvales, 23, 53, 99, 100, 101, 121, 122, 131,
    132, 133, 134, 136, 144, 145, 146, 147,
    148, 160, 181, 182, 187, 189, 191, 207,
    210, 211, 215, 216, 217, 220, 225, 227,
    393, 431
Ulvaphyceae (= Ulvophyceae), 51
Ulvaria, 131, 160, 184, 187, 191, 192, 332
  oxysperma, 191, 199, 200
Ulvella, 131
Ulvellaceae, 131, 134, 211
Ulvophyceae, 5, 14, 17, 21, 22, 23, 34, 35,
    38, 39, 41, 44, 45, 48, 49, 51, 52, 53,
    54, 55, 56, 57, 58, 60, 63, 64, 65, 66,
    81, 82, 84, 93, 103, 104, 107, 108, 110,
    113, 114, 121, 122, 123, 130, 131, 138,
    139, 141, 144, 146, 147, 148, 149, 160,
    183, 200, 228, 233, 247, 290, 291, 331,
    332, 338
Ulvophyta, 20
Uronema, 61, 183, 210, 214, 216, 219, 227
  belkae, 194
  confervicolum, 194, 214, 226

marinum, 226
schwiakofii, 226
Urospora, 18, 96, 125, 126, 127, 129, 160,
    182, 185, 194, 195, 272, 332, 339
  gregaria, 128

V

Valonia, 135, 136, 137, 157, 158, 159, 161,
    164, 165, 166, 167, 170, 171, 172, 173,
    174, 271, 272
  aegagropila, 165, 166, 170
  ocellata, 170
  ventricosa, 170
Valoniaceae, 4, 135, 272
Valoniopsis, 158, 170
Vaucheria, 4, 271, 272, 273, 392
Vaucheriaceae, 5, 273
Vaucherieae, 4
Vaucheriopsis, 273
Vermiporella, 273
Verticillatae, 273
Viridiplantae, 1, 8, 11, 13, 19, 20, 23
Vogania [subsection], 317
Volvocaceae, 4
Volvocales, 16, 17, 21, 23, 43, 58, 100, 104,
    160, 393, 394, 399
Volvocicae, 8, 13
Volvocineae, 4, 6
Volvocophyceae, 4, 5
Volvox, 3, 58, 402, 431
  carteri, 94, 416
  globator, 88, 90, 92

W

Wallmania [subsection], 315, 325, 326
Wildenowia [subsection], 315, 316, 327
Willeella [section], 168
Willeella, 168, 169
Wittrockiella, 164, 165, 168, 171
  salina, 165

X

Xanthidium, 259, 260, 261
  antilopeum, 262
  controversum var. planctonicum, 262

subhastiferum
    var. *murrayi* fac. *triquetra,* 262
  tetracentrotum, 259
Xanthophyceae, 59, 144, 392

Z

Zygnemaphyceae, 13, 22, 23, 252, 253, 254
Zygnemataceae, 5
Zygnematales, 21, 41, 49, 66, 122, 253, 254, 431
Zygnematicae, 13
Zygnematophyceae, 5, 66
Zygnema(to)phyceae, 21
Zygnematophyta, 20
Zygnemeae, 4
Zygophyceae, 5, 252
Zygophyta, 13

Editor-in-Chief, Special Volume Series
D. L. HAWKSWORTH PhD DSc FLS FIBiol

Systematics Association Publications

1. BIBLIOGRAPHY OF KEY WORKS FOR THE IDENTIFICATION OF THE BRITISH FAUNA AND FLORA 3rd edition (1967)
   Edited by G. J. Kerrich, R. D. Meikle and N. Tebble   Out of print
2. FUNCTION AND TAXONOMIC IMPORTANCE (1959)
   Edited by A. J. Cain
3. THE SPECIES CONCEPT IN PALAEONTOLOGY (1956)
   Edited by P. C. Sylvester-Bradley
4. TAXONOMY AND GEOGRAPHY (1962)
   Edited by D. Nichols
5. SPECIATION IN THE SEA (1963)
   Edited by J. P. Harding and N. Tebble
6. PHENETIC AND PHYLOGENETIC CLASSIFICATION (1964)
   Edited by V. H. Heywood and J. McNeil   Out of print
7. ASPECTS OF TETHYAN BIOGEOGRAPHY (1967)
   Edited by C. G. Adams and D. V. Ager
8. THE SOIL ECOSYSTEM (1969)
   Edited by H. Sheals
9. ORGANISMS AND CONTINENTS THROUGH TIME (1973)†
   Edited by N. F. Hughes

Published by the Association

Systematics Association Special Volumes

1. THE NEW SYSTEMATICS (1940)
   Edited by Julian Huxley (Reprinted 1971)
2. CHEMOTAXONOMY AND SEROTAXONOMY (1968)*
   Edited by J. G. Hawkes
3. DATA PROCESSING IN BIOLOGY AND GEOLOGY (1971)*
   Edited by J. L. Cutbill
4. SCANNING ELECTRON MICROSCOPY (1971)*
   Edited by V. H. Heywood
5. TAXONOMY AND ECOLOGY (1973)*
   Edited by V. H. Heywood
6. THE CHANGING FLORA AND FAUNA OF BRITAIN (1974)*
   Edited by D. L. Hawksworth
7. BIOLOGICAL IDENTIFICATION WITH COMPUTERS (1975)*
   Edited by R. J. Pankhurst

*Published by Academic Press for the Systematics Association
†Published by the Palaeontological Association in conjunction with the Systematics Association

8. LICHENOLOGY: PROGRESS AND PROBLEMS (1976)*
    Edited by D. H. Brown, D. L. Hawksworth and R. H. Bailey
9. KEY WORKS TO THE FAUNA AND FLORA OF THE BRITISH ISLES AND NORTHWESTERN EUROPE (1978)*
    Edited by G. J. Kerrich, D. L. Hawksworth and R. W. Sims
10. MODERN APPROACHES TO THE TAXONOMY OF RED AND BROWN ALGAE (1978)*
    Edited by D. E. G. Irvine and J. H. Price
11. BIOLOGY AND SYSTEMATICS OF COLONIAL ORGANISMS (1979)*
    Edited by G. Larwood and B. R. Rosen
12. THE ORIGIN OF MAJOR INVERTEBRATE GROUPS (1979)*
    Edited by M. R. House
13. ADVANCES IN BRYOZOOLOGY (1979)*
    Edited by G. P. Larwood and M. B. Abbot
14. BRYOPHYTE SYSTEMATICS (1979)*
    Edited by G. C. S. Clarke and J. G. Duckett
15. THE TERRESTRIAL ENVIRONMENT AND THE ORIGIN OF LAND VERTEBRATES (1980)*
    Edited by A. L. Panchen
16. CHEMOSYSTEMATICS: PRINCIPLES AND PRACTICE (1980)*
    Edited by F. A. Bisby, J. G. Vaughan and C. A. Wright
17. THE SHORE ENVIRONMENT: METHODS AND ECOSYSTEMS (2 Volumes) (1980)*
    Edited by J. H. Price, D. E. G. Irvine and W. F. Farnham
18. THE AMMONOIDEA (1981)*
    Edited by M. R. House and J. R. Senior
19. BIOSYSTEMATICS OF SOCIAL INSECTS (1981)*
    Edited by E. Howse and J.-L. Clément
20. GENOME EVOLUTION (1982)*
    Edited by G. A. Dover and R. B. Flavell
21. PROBLEMS OF PHYLOGENETIC RECONSTRUCTION (1982)*
    Edited by K. A. Joysey and A. E. Friday
22. CONCEPTS IN NEMATODE SYSTEMATICS (1983)*
    Edited by A. R. Stone, H. M. Platt and L. F. Khalil
23. EVOLUTION, TIME AND SPACE: THE EMERGENCE OF THE BIOSPHERE (1983)*
    Edited by R. W. Sims, J. H. Price and P. E. S. Whalley
24. PROTEIN POLYMORPHISM: ADAPTIVE AND TAXONOMIC SIGNIFICANCE (1983)*
    Edited by G. S. Oxford and D. Rollinson
25. CURRENT CONCEPTS IN PLANT TAXONOMY (1984)*
    Edited by V. H. Heywood and D. M. Moore
26. DATABASES IN SYSTEMATICS (1984)*
    Edited by R. Allkin and F. A. Bisby
27. SYSTEMATICS OF THE GREEN ALGAE (1984)*
    Edited by D. E. G. Irvine and D. M. John

*Published by Academic Press for the Systematics Association

DATE DUE